VOLUME ONE HUNDRED

CURRENT TOPICS IN DEVELOPMENTAL BIOLOGY

Heart Development

VOLUME ONE HUNDRED

CURRENT TOPICS IN DEVELOPMENTAL BIOLOGY

Heart Development

Edited by

BENOIT G. BRUNEAU
The Gladstone Institute of Cardiovascular Disease
San Francisco, CA, USA

ELSEVIER

AMSTERDAM • BOSTON • HEIDELBERG • LONDON
NEW YORK • OXFORD • PARIS • SAN DIEGO
SAN FRANCISCO • SINGAPORE • SYDNEY • TOKYO
Academic Press is an imprint of Elsevier

Academic Press is an imprint of Elsevier
525 B Street, Suite 1900, San Diego, CA 92101-4495, USA
225 Wyman Street, Waltham, MA 02451, USA
32, Jamestown Road, London NW1 7BY, UK
Linacre House, Jordan Hill, Oxford OX2 8DP, UK

First edition 2012

ISBN: 978-0-12-387786-4
ISSN: 0070-2153

For information on all Academic Press publications visit our website at elsevierdirect.com

Printed and bound in USA

12 13 14 15 9 8 7 6 5 4 3 2 1

Contents

Contributors

Phil Barnett
Department of Anatomy, Embryology and Physiology, Heart Failure Research Center, Academic Medical Center, University of Amsterdam, Amsterdam, The Netherlands

Brian L. Black
Cardiovascular Research Institute and Department of Biochemistry and Biophysics, University of California, San Francisco, California, USA

Hozana Andrade Castillo
Brazilian National Laboratory for Biosciences, Brazilian Association for Synchrotron Light Technology, Rua Giuseppe Máximo Scolfaro, Campinas, São Paulo, Brazil

Wen-Yee Choi
Department of Cell Biology, Howard Hughes Medical Institute, Duke University Medical Center, Durham, North Carolina, USA

Lionel Christiaen
Center for Developmental Genetics, Department of Biology, New York University, New York, USA

Vincent Christoffels
Department of Anatomy, Embryology and Physiology, Heart Failure Research Center, Academic Medical Center, University of Amsterdam, Amsterdam, The Netherlands

Malou van den Boogaard
Department of Anatomy, Embryology and Physiology, Heart Failure Research Center, Academic Medical Center, University of Amsterdam, Amsterdam, The Netherlands

Ramón A. Espinoza-Lewis
Cardiovascular Research Division, Department of Cardiology, Children's Hospital Boston, Harvard Medical School, Boston, Massachusetts, USA

Aibin He
Harvard Stem Cell Institute, Harvard University, Cambridge, Massachusetts, USA

Robert G. Kelly
Developmental Biology Institute of Marseilles-Luminy, Aix-Marseille Université, CNRS UMR 7288, Marseilles, France

James D. Kotick
MD Program, Leonard M. Miller School of Medicine, University of Miami, Miami, Florida, USA

Joy Lincoln
Center for Cardiovascular and Pulmonary Research, The Research Institute at Nationwide Children's Hospital, Department of Pediatrics, The Ohio State University, Columbus, Ohio, USA

David J. McCulley
Cardiovascular Research Institute and Department of Biochemistry and Biophysics, University of California, San Francisco, California, USA

Kenneth D. Poss
Department of Cell Biology, Howard Hughes Medical Institute, Duke University Medical Center, Durham, North Carolina, USA

William T. Pu
Department of Cardiology, Children's Hospital Boston, Boston, and Harvard Stem Cell Institute, Harvard University, Cambridge, Massachusetts, USA

Paul R. Riley
Department of Physiology, Anatomy and Genetics, University of Oxford, Oxford, OX1 3PT, United Kingdom

Michael Schubert
Institut de Génomique Fonctionnelle de Lyon (UCBL, CNRS UMR 5242, ENSL, INRA 1288), Ecole Normale Supérieure de Lyon, 69364 Lyon Cedex 07, France

Ian C. Scott
Program in Developmental and Stem Cell Biology, The Hospital for Sick Children, and Department of Molecular Genetics, University of Toronto, Toronto, Ontario, Canada

Tiago José Pascoal Sobreira
Brazilian National Laboratory for Biosciences, Brazilian Association for Synchrotron Light Technology, Rua Giuseppe Máximo Scolfaro, Campinas, São Paulo, Brazil

Henrique Marques Souza
Department of Histology and Embryology, Institute of Biology, University of Campinas (UNICAMP), Campinas, São Paulo, Brazil

Alberto Stolfi
Department of Histology and Embryology, Institute of Biology, University of Campinas (UNICAMP), Campinas, São Paulo, Brazil

Ge Tao
Molecular Cell and Developmental Biology Graduate Program, Leonard M. Miller School of Medicine, University of Miami, Miami, Florida, USA

Theadora Tolkin
Center for Developmental Genetics, Department of Biology, New York University, New York, USA

Sylvia Sura Trueba
Brazilian National Laboratory for Biosciences, Brazilian Association for Synchrotron Light Technology, Rua Giuseppe Máximo Scolfaro, Campinas, São Paulo, Brazil

Da-Zhi Wang
Cardiovascular Research Division, Department of Cardiology, Children's Hospital Boston, Harvard Medical School, Boston, Massachusetts, USA

José Xavier-Neto
Brazilian National Laboratory for Biosciences, Brazilian Association for Synchrotron Light Technology, Rua Giuseppe Máximo Scolfaro, Campinas, São Paulo, Brazil

Pingzhu Zhou
Department of Cardiology, Children's Hospital Boston, Boston, Massachusetts, USA

Preface

The development of the embryonic heart is a fascinating process that incorporates myriad molecular and morphogenetic events. These precisely assemble multiple different cell types into a functional beating heart. The first heart beat, early in development, is the most evident sign of life and is essential for embryonic life. The importance of the precise assembly of the embryonic heart is evident in the high incidence of congenital heart defects, which affect several thousand children each year. The past 10 years have seen a renaissance in this field of research, and in this issue of *Current Topics in Developmental Biology*, important advances in our understanding of the development of the heart are reviewed.

In Chapter 1, Scott discusses the early decisions that define the initial cardiac lineages and their differentiation, including the transcriptional cues essential for defining the cardiac cell. Kelly follows in Chapter 2 with a comprehensive look at the lineages that contribute to the developing heart and how these are regulated during cardiac morphogenesis. The concept of heart fields has been at the forefront in the past decade, but this concept has been controversial. Xavier-Neto and colleagues carefully examine the literature and offer a different perspective on the issue in Chapter 3. The vertebrate heart has been the focus of most studies on heart development, and yet important insights into fundamental aspects of cardiogenesis have been obtained from the chordate, *Ciona intestinalis*, which are reviewed in Chapter 4 by Tolkin and Christiaen.

In subsequent chapters, the regulation of cardiac morphogenesis is explored. Pu and colleagues review the very important Gata4 transcription factor, and its impact on heart development and disease in Chapter 5. Christoffels and colleagues in Chapter 6 tackle the concepts of transcriptional regulation of localized gene expression, which is the main driving force for precise cardiac morphogenesis, as well as the root of congenital heart defects. An important aspect of heart development is the formation of the valves that ensure directional blood flow from one chamber to the next. Lincoln and colleagues in Chapter 7 review recent work on understanding valve development and how this aspect of heart development is deeply relevant to human disease. Another important component of the heart is the epicardium, a layer of cells that lines the outside of the heart and provides important signals for cardiac growth. In Chapter 8, Riley reviews the current knowledge on epicardial biology, including its potential role in mammalian cardiac regeneration.

Congenital heart defects are thought to arise largely from mutations in genes encoding factors important for various aspects of heart development. McCulley and Black in Chapter 9 review the genetics of congenital heart disease, providing an extensive and thorough insight into the established and potential genetic abnormalities that result in abnormal heart formation.

Protein-coding genes have been the primary focus of studies on embryonic development, including that of the heart. In the past several years, an exciting discovery has been the involvement of small noncoding RNAs known as microRNAs, in fine-tuning important aspects of cardiac organogenesis. Espinoza-Lewis and Wang review the now-extensive body of literature on the importance and broad impact of microRNA-based regulation of cardiovascular development in Chapter 10.

Finally, the past decade has seen progress on one of the most unanticipated and exciting fields: vertebrate cardiac regeneration. Initially explored in fish and newts, studies of cardiac regeneration have extended to mammalian hearts, bringing hope that damaged hearts could be endogenously repaired. In Chapter 11, Choi and Poss review progress in understanding the biology of cardiac regeneration.

Together, these reviews synthesize the most recent and exciting concepts in heart development. The past 10 years have seen extraordinary conceptual and technical advances that have revolutionized our understanding of the morphogenesis of the heart and, most impressively, the immediate relevance to human disease has risen to the forefront from this research. I hope that the readers of this volume will be captivated and inspired by these beautifully written chapters.

Benoit G. Bruneau

CHAPTER ONE

LIFE BEFORE *NKX2.5*: CARDIOVASCULAR PROGENITOR CELLS: EMBRYONIC ORIGINS AND DEVELOPMENT

Ian C. Scott*,†

Contents

Abstract

Development of the heart, like that of other organs, requires the specification of progenitor cell populations that will ultimately form the differentiated cell types of the functional organ. A relatively recent and exciting advance in cardiac research has been the identification of cardiovascular progenitor cells (CPCs), which have the potential to form the major cell types of the heart (cardiomyocytes, smooth muscle, and endothelium/endocardium). This suggests that a common progenitor is responsible for much of heart development and has spurred great interest in use of CPC-like cells for cardiac repair. In this review, CPC development is discussed, with a focus on early events prior to the initiation of cardiac gene expression. In particular, I discuss evidence that CPC fate is established during gastrulation, well before a time when heart development has typically been studied. Pathways regulating CPC specification are

* Program in Developmental and Stem Cell Biology, The Hospital for Sick Children, Toronto, Ontario, Canada
† Department of Molecular Genetics, University of Toronto, Toronto, Ontario, Canada

Current Topics in Developmental Biology, Volume 100
ISSN 0070-2153, DOI: 10.1016/B978-0-12-387786-4.00001-4

examined. The relationship between CPC specification and migration is further discussed. Finally, how CPCs may be related to efforts to promote cardiac development by approaches including reprogramming is discussed.

1. Introduction

Development and growth of the heart has been a highly active area of research in the past few decades. Organogenesis in general requires the specification of progenitor cell(s), their migration to the organ-forming region, interactions and signaling within and between tissues, morphogenesis to form the proper organ shape, differentiation of progenitor cells to required specialized and subspecialized cell types, and later growth and functional maturation of the organ. Great progress has been made in delineating the molecular and cellular processes that drive heart tube formation and later cardiac morphogenesis and maturation. However, the earliest events of heart development—when the progenitor cells that form the heart arise—have been more difficult to examine. This has largely stemmed from an absence of markers to characterize heart progenitors.

A relatively recent and exciting advance in cardiac research has been the identification of cardiovascular progenitor cells (CPCs), which have the potential to form many of the major cell types of the heart (cardiomyocytes, endothelium/endocardium, and smooth muscle; Kattman *et al.*, 2006; Moretti *et al.*, 2006; Wu *et al.*, 2006). This suggests that a common progenitor is responsible for much of heart development and has spurred great interest in use of CPC-like cells for cardiac repair. Excellent reviews in the past year have been focused on early events of heart development (Evans *et al.*, 2010; Lopez-Sanchez and Garcia-Martinez, 2011; Meilhac and Buckingham, 2010; Vincent and Buckingham, 2010), and it is not my intention in this review to revisit events covered in detail in these works.

In this review, early events of CPC development are discussed, prior to the initiation of cardiac gene expression. It should be noted that the contribution of the cardiac neural crest, clearly a key participant in heart development (Hutson and Kirby, 2007), is not covered in this chapter. In particular, I focus on evidence from animal and *in vitro* stem cell models that CPC fate is established during gastrulation, well before a time when heart development has typically been studied. The relationship between embryonic CPCs and efforts to promote cardiac development by approaches including reprogramming will then be examined. Finally, what I see as future outstanding questions in CPC biology and therapeutic potential will be discussed.

2. Early Heart Development: Is It as Simple as Location, Location, Location?

Where does the heart come from? For decades, it has been known that cardiomyocytes arise from bilateral populations of anterior lateral plate mesoderm (ALPM; Dehaan, 1963; Stalsberg and DeHaan, 1969). These elegant chick studies showed that these bilateral populations later fuse at the midline to form the linear heart tube. Explant of tissues to this precardiac mesoderm area can induce cardiac fate (Schultheiss *et al.*, 1995; Tam *et al.*, 1997). In the more recent "molecular biology era," work from a number of groups has shown that this region of the ALPM is privileged for cardiac mesoderm development. This is achieved through a balance of signals, both stimulatory and inhibitory for cardiac differentiation. Procardiac BMPs and FGFs are secreted from the adjacent lateral endoderm and ectoderm, which promote both early myocardial fate and further differentiation of cardiomyocytes (Barron *et al.*, 2000; Reifers *et al.*, 2000; Schultheiss *et al.*, 1997; Shi *et al.*, 2000). In contrast, the neurectoderm secretes WNTs, which inhibits cardiac mesoderm formation at the ALPM stage. The secretion of WNT inhibitors in the anterior portion of the embryo further restricts cardiac fate (Marvin *et al.*, 2001; Schneider and Mercola, 2001). The integration of these signals results in cardiac mesoderm being present in specific bilateral positions in a defined anterior region of the embryo. Indeed, perturbation of these gradients of signals can result in ectopic or an absence of myocardial differentiation.

A major advance in the field of early cardiac fate arose from study of the *Drosophila melanogaster* (fruit fly) *tinman* mutant, which lacks a dorsal vessel (the fly heart equivalent, essentially a beating tube) (Bodmer, 1993). Cloning of the vertebrate *tinman* homologue, which encodes the NK-class homeodomain transcription factor *Nkx2.5* (also referred to as *Csx*) followed. Importantly, analysis of *Nkx2.5* expression revealed that it is initiated in two bilateral stripes of the ALPM, mirroring in general the precardiac mesoderm described by embryological studies (Chen and Fishman, 1996; Schultheiss *et al.*, 1995; Tonissen *et al.*, 1994). It is important to note that the *Nkx2.5*-positive ALPM population does not represent ALL of the future myocardium. At ALPM stages in multiple vertebrate species, a large portion of the future second-heart field myocardium is largely devoid of *Nkx2.5* expression (Brade *et al.*, 2007; Cai *et al.*, 2003; Hami *et al.*, 2011; Yutzey and Kirby, 2002). Conversely, the majority of, but not all *Nkx2.5*-positive cells will contribute to the heart (Goldstein and Fishman, 1998; Raffin *et al.*, 2000; Redkar *et al.*, 2001). In the fruit fly, dorsal vessel formation and *tinman* expression are dependent on FGF, BMP (*dpp* in flies), and WNT signaling (Beiman *et al.*, 1996; Frasch, 1995;

Gisselbrecht *et al.*, 1996; Wu *et al.*, 1995). This is largely conserved in vertebrates. Further, while *Nkx2.5* is not essential for initial myocardial specification (as *tinman* is in the fly), absence of *Nkx2.5* results in profound defects in heart development in mouse, with more subtle perturbations of *Nkx2.5* being causative of congenital heart defects in humans (Lyons *et al.*, 1995; Schott *et al.*, 1998). This conservation of genes in heart development has been substantiated in multiple cases and strongly supports the use of "lower" organisms with simpler heart architectures to decipher the core molecular pathways that regulate cardiac development.

If it is largely true that cardiac progenitors initiate *Nkx2.5* expression in the precardiac mesoderm, and *Nkx2.5* is the earliest marker we truly have for heart progenitors (discussed more below), does the key event in cardiogenesis coincide with the turning on of *Nkx2.5*? And if this is the case, does this mean that ANY cell found in the right position of the ALPM, at the right time, can be specified as cardiac mesoderm? Or do specialized cardiovascular progenitors exist, that once in the right location are privileged in their ability to form cardiac mesoderm? On these important questions, several decades of embryological experiments have yielded mixed conclusions. It is clear that the position of cells that will ultimately form cardiac mesoderm is restricted pregastrulation. Fate mapping studies in the chick, mouse, frog, and zebrafish have conclusively shown that cells destined to form the heart localize to specific areas of the pregastrula embryo and are further coarsely organized in a pattern that reflects the later organization of the heart (Garcia-Martinez and Schoenwolf, 1993; Keegan *et al.*, 2004; Redkar *et al.*, 2001). Placement of cells to this area pregastrulation can further bias cells to take on a cardiac fate following gastrulation (Parameswaran and Tam, 1995; Schoenwolf and Garcia-Martinez, 1995; Tam *et al.*, 1997). Viewed in this light, the key question now becomes if cardiac progenitors are specified prior to/during gastrulation, before they arrive at the ALPM. In other words, is a progenitor cell population primed during gastrulation or is cardiac fate solely the result of ALPM localization? Several explant studies, primarily in chick, have attempted to address this issue. Much of this work has concluded that ALPM localization, in apposition to the anterior endoderm, is key to cardiac differentiation (Schultheiss *et al.*, 1995). However, some work has found that specified cardiac progenitors can be isolated from the pregastrula/gastrula stage embryos (Antin *et al.*, 1994; Auda-Boucher *et al.*, 2000; Sater and Jacobson, 1989). Several confounding factors have complicated these studies. The first is the homogeneity of tissues studied: can a pure population, free of potential inductive sources of cells, be isolated and studied? How cardiac fate itself is assayed, which can heavily influence conclusions, has varied from expression of various cardiomyocyte markers to beating behavior. Further, do regulative processes result in these inductive sources being reformed in explants over the course of time? Finally, in cases where explants have been maintained in culture medium, what is a “normal” medium or signaling environment to assess fate determination? New approaches and markers to isolate and study putative cardiac progenitor populations must clearly be employed to further examine the determination of cardiac fate.

3. CPCs: The Building Blocks of the Heart

When does the ability or potential of a cell to form the lineages of the heart first arise and is this restricted? Further, how many progenitor populations are needed to make the embryonic heart? The use of novel biomarkers to isolate and study the potential of progenitor populations has started to address these questions. Work from three groups, published in 2006, demonstrated that cells with hallmarks of a CPC could be isolated from cultures of embryonic stem (ES) cells (Kattman *et al.*, 2006; Moretti *et al.*, 2006; Wu *et al.*, 2006). In the first study, an *Nkx2.5*+/c-Kit+ population was isolated from the mouse embryo that could be expanded *in vitro* (Wu *et al.*, 2006). These cells could in clonal assays form both cardiomyocytes and smooth muscle. A second group used a transgenic reporter recapitulating expression of *Isl1*, a marker of the second heart field, to isolate an *Isl1*$^{+}$/Flk1^{+}/*Nkx2.5*$^{+}$ population from differentiating ES cells (Moretti *et al.*, 2006). This population demonstrated the capacity to form all three major cardiovascular lineages (cardiomyocytes, endothelium, and smooth muscle). Importantly, in clonal assays, these “multipotent Isl1+ cardiovascular progenitors” (MICPs) differentiated via cellular intermediates that form either cardiomyocytes/smooth muscle or endothelium/smooth muscle *in vitro*, likely representing the lineage restriction of CPCs that occurs *in vivo*. MICP-like cells could be isolated from the embryonic mouse heart, further demonstrating that this population of cells does exist *in vivo*. Based on a previous demonstration that in mouse descendants of Flk1-expressing cells contribute to myocardium of the heart (Motoike *et al.*, 2003), at third group used Flk1 expression to isolate a subpopulation of mesoderm with CPC-like characteristics (Kattman *et al.*, 2006). In differentiating ES cells, mesoderm (*Brachyury*- or *T*-positive) that initially expresses high levels of Flk1 forms primitive blood and endothelial cells with a high propensity. In contrast, a later-arising Flk1-positive population readily forms cardiomyocytes, smooth muscle, and endothelium. This “CPC” population contains the vast majority of all myocardial potential in ES cell cultures. Further, using the same markers, a population of cells can be isolated from the headfold stage mouse embryo that *in vitro* demonstrates a similar cardiogenic potential. Importantly, this Flk1-positive population likely represents an earlier stage of CPC development than the *Nkx2.5*- and *Isl1*-positive ones described above. These pioneering studies established two important points. First, a cardiovascular progenitor does exist *in vivo*, which is (at the very least) multipotential with respect to cell lineages of the heart. Second, development of this cardiovascular progenitor can be (to some extent) faithfully recapitulated *in vitro* using an ES cell model.

Given the existence of CPCs, the question of when cardiac fate is first determined simplifies to one of when CPCs are first determined *in vivo*. *Nkx2.5*- or *Isl1*-positive cells found in the ALPM clearly represent heart progenitors, but are they simply a later descendent of a predetermined CPC? As discussed in the next section, a definitive marker of an earlier CPC population does not at present exist. Further, this population is likely transient in nature. While the (perhaps more mature) *Isl1*-positive population can be expanded in culture, this is not the case for "late Flk1" CPCs, which may represent an earlier stage of CPC that nevertheless cannot be maintained in an undifferentiated state in culture. An understanding of what initiates CPC fate is therefore required. This may seem to be a largely academic exercise, as multipotential cardiac progenitors have been isolated from ES cell and other sources. However, for approaches such as reprogramming and stem cell therapy (discussed further below) to move from the bench to potential bedside applications, it may be that for robust results to be obtained a proper progenitor state intermediate is required.

4. CPC Determination: The Quest for "CardioD"

The study of differentiation of pluripotent ES cells to more restricted progenitors has shown that ES cells exist in a "poised" state, with a large fraction of genes being expressed at low levels and primed for robust activation or silencing based on the proper inputs (Bernstein *et al.*, 2006; Rada-Iglesias *et al.*, 2011). Work primarily in development of the endoderm has shown that pioneer transcription factors may present a universal model for differentiation events. In ES cells, FoxD3/A1 are prebound to the promoters and enhancers of endoderm-specific genes, and with the proper differentiation signals, act as a platform to recruit the needed transcriptional coactivators to initiate robust lineage-specific gene expression (Lupien *et al.*, 2008; Xu *et al.*, 2009). Do such pioneer factors exist for cardiovascular progenitors? Work in this area was largely influenced by groundbreaking work showing that the basic helix–loop–helix (bHLH) transcription factor MyoD was alone sufficient to drive skeletal myoblast fate in fibroblast and other cell lines (Edmondson and Olson, 1989; Tapscott *et al.*, 1988). However, a survey of cardiac bHLH factors, while revealing many important regulators of heart development, such as the *Hand1/2* genes (Firulli *et al.*, 1998; Riley *et al.*, 1998; Srivastava *et al.*, 1997), did not uncover factors sufficient to promote cardiac fate. The transcriptional network regulating the differentiation of CPCs to the various cardiac lineages is covered in chapters 2 and 6 of this volume. Below, relatively recent work on the earliest transcriptional events that may promote CPC fate and behavior are discussed (summarized in Fig. 1.1).

In the absence of adequate markers of CPCs, work on CPC fate has largely been restricted to analyzing the ability to promote cardiomyocyte differentiation in stem cell lines including P19 embryocarcinoma cells and ES cells. Of course, some caution must be used when interpreting these studies, in particular with respect to P19 cells which are largely poised to differentiate as myocardium with addition of signals such as retinoic acid (McBurney *et al.*, 1982). Loss-of-function studies in these models are not temporal in nature, and therefore often cannot define the nature of a defect in cardiomyocyte development (i.e., at what level in the CPC hierarchy a gene may act). Further, studies based on overexpression of genes well beyond endogenous *in vivo* levels must be viewed with caution. Despite these caveats, tissue culture models have repeatedly shown that several transcription factors have promyocardial activity in multipotent stem cells. As discussed further below, this is especially evident for members of the serum response factor (SRF), Gata, and Mesp family of transcription factors. It is important to note that members of the Nkx2.5/2.7 class of transcription factors, while critical for later heart development and morphogenesis, do

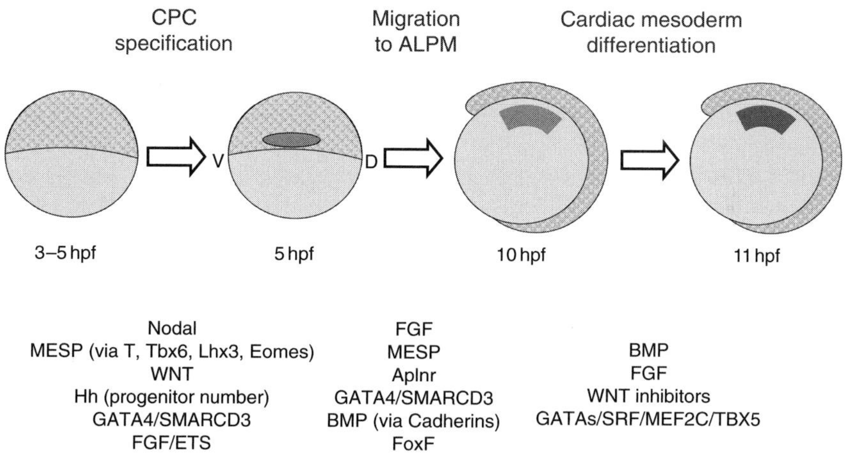

Figure 1.1 Summary of stages of CPC development *in vivo*. A schematic of CPC development using the zebrafish embryo model is shown. The area in blue, first arising at 5 h postfertilization (hpf), denotes specified CPCs. These cells migrate to the ALPM during gastrulation and initiate *Nkx2.5* expression (shown in red). Key regulators of various stages of CPC development, as discussed in the text, are shown below. (For interpretation of the references to color in this figure legend, the reader is referred to the Web version of this chapter.)

not appear to be essential for CPC specification (Lyons *et al.*, 1995; Prall *et al.*, 2007; Targoff *et al.*, 2008). This of course contrasts directly with the dramatic fly *tinman* heartless phenotype. However, in *tinman* mutants, a much broader defect in mesoderm development is evident, reflecting its broader expression in fly mesoderm than the relatively cardiac-restricted expression of vertebrate *Nkx2.5/2.7* genes (Bodmer, 1993).

SRF is a MADS box transcription factor that binds to CArG element ($CC(A/T)_6GG$) sequence motifs to activate muscle-specific gene expression in skeletal, smooth, and cardiac muscle (Chang *et al.*, 2001; Miano, 2003). SRF recruits a number of transcriptional cofactors with later roles in heart development, notably Myocardin and the homeodomain-only protein (HOP; Chen *et al.*, 2002; Shin *et al.*, 2002; Wang *et al.*, 2001). SRF further interacts with a number of cardiac-specific transcription factors during heart development, notably Nkx2.5, Gata4, MEF2c, and Tbx2/5 (He *et al.*, 2011). In particular, SRF is a key regulator of contractile proteins and assembly of the contractile apparatus. Later mutation of *Srf* in cardiomyocytes and smooth muscle results in sarcomere disassemble and death by embryonic day 11.5 in mice (Miano *et al.*, 2004; Niu *et al.*, 2008). The complete absence of SRF in *Srf*-null mice results in a lack of mesoderm formation, precluding an analysis of SRF function specifically in CPCs (Arsenian *et al.*, 1998). In contrast, *Srf−/−* ES cells retain some capacity to form mesoderm; however, their ability to form CPCs or cardiomyocytes has not been directly assayed (Weinhold *et al.*, 2000).

In both mice and the sea squirt *Ciona intestinalis*, absence of *Mesp* gene function results in an acardiac (heartless) phenotype (Kitajima *et al.*, 2000; Satou *et al.*, 2004). In mice, *Mesp1*, which encodes a bHLH transcription factor, is first expressed in the primitive streak (Saga *et al.*, 1996). A lineage-tracing approach was used to determine the fate of these early *Mesp1*-expressing cells via use of a knockin *Mesp1* allele where expression of Cre recombinase is regulated by the *Mesp1* promoter/enhancer. This experiment showed that *Mesp1*-expressing cells in the primitive streak contained future cardiac progenitors, as well as cells destined to form facial mesoderm and other anterior mesodermal derivatives (Saga *et al.*, 1999). Further, in chimeric mouse embryos containing a mixture of wild type and *Mesp1/2* double-knockout cells, *Mesp1/2* mutant cells are capable of contributing to most structures save cephalic mesoderm and heart (Kitajima *et al.*, 2000). The precise nature of the heartless phenotype in *Mesp1/2* mutant mice is difficult to interpret. In global loss of Mesp function, mesoderm does not properly leave the primitive streak during gastrulation, resulting in an absence of mesoderm in the ALPM region where cardiac mesoderm is normally induced (Kitajima *et al.*, 2000). Therefore, while Mesp function may directly promote CPC fate, it may alternatively regulate in mesoderm (including future CPCs) a prerequisite epithelial–mesenchymal transition (EMT) needed to egress from the primitive streak and migrate during gastrulation.

Mesp1 has been reported as both the earliest marker of CPCs and a key regulator of CPC fate (Bondue and Blanpain, 2010; Bondue *et al.*, 2008; Saga *et al.*, 2000). Is the evidence consistent with this assertion? At first glance, loss-of-function work in mouse and *Ciona* certainly seem to support this model. The *Ciona* model is perhaps the best available system to examine all stages of CPC development (Davidson, 2007; Davidson and Levine, 2003). The B7.5 lineage, which arises early in development, can be directly visualized using transgenic animals. The B7.5 cell undergoes an asymmetric cell division, producing both future trunk ventral cells (TVCs—muscle) and the future heart (and sphincter muscles, which may be analogous to a second heart-field population). B7.5 descendants that will form the cardiac lineage migrate anteriorly, analogous to what is seen in vertebrate species, prior to formation of the heart. Loss of *Mesp* function globally in *Ciona* (via morpholino injection) clearly prevents B7.5 cell migration to the cardiogenic region and expression of a host of cardiac markers (Satou *et al.*, 2004). In *Ciona*, a B7.5-specific *Mesp* promoter element can be used to drive gene expression only in the B7.5 cell lineage. Expression of a constitutively active form of Mesp (using a strong VP16 transactivation domain fusion protein) in B7.5 cells recruits cells that would form TVCs to a cardiac fate (Davidson *et al.*, 2005). While highly suggestive of a cell-autonomous role for Mesp in directing CPC fate, it must of course be considered that these experiments are based on overexpression analyses. *Mesp* expression itself is activated by factors including Tbx6, Lhx3, and maternal β-catenin in *Ciona* (Christiaen *et al.*, 2009; Davidson *et al.*, 2005). In turn, Mesp activates transcription of genes including *Ciona Ets*, with Ets in turn (following potentiation by FGF signaling) stimulating expression of *FoxF*, a regulator of cardiac progenitor migration (Beh *et al.*, 2007; Davidson *et al.*, 2006). Mesp activity in *Ciona* cardiac progenitor cells is therefore essential for migration and differentiation. This final point is difficult to interpret, as overexpression of a gain-of-function version of (Mesp-VP16) in B7.5 cells impairs migration, but not cardiac differentiation (Davidson *et al.*, 2005).

In ES cells, *Mesp1* expression enriches for a population with an enhanced ability to form cardiovascular lineages (Bondue *et al.*, 2011). Mesp1 has been shown to activate a number of genes in ES cells associated with EMT, including *Snai1* (Lindsley *et al.*, 2008). *Mesp1* expression itself is activated by factors including T/Brachyury (perhaps analogous to the role of Tbx6 in *Ciona*) and Eomesodermin (Costello *et al.*, 2011; David *et al.*, 2011). Transcriptome and chromatin immunoprecipitation (ChIP) approaches have both shown that Mesp1 can both bind to enhancer elements and regulate expression of a number of key cardiac development genes, including *Nkx2.5*, *Gata4*, and *Myocardin* (Bondue *et al.*, 2008, 2011).

So is Mesp1 the long-sought "CardioD"? Clearly, in the mouse and *Ciona* embryo, *Mesp1* expression is present in cells that will later give rise to the heart (Saga *et al.*, 1999; Satou *et al.*, 2004). In ES cells, Mesp1 can regulate many cardiac-specific transcription factors. However, many of these genes, in particular *Nkx2.5*, *Myocardin*, and *Mef2c*, are not expressed (at least at the level of sensitivity conferred by RNA *in situ* hybridization) until after gastrulation is completed, a time when *Mesp1* expression is no longer evident in mouse or zebrafish cardiac cells (Saga *et al.*, 1999; Sawada *et al.*, 2000). How does one account for this discrepancy? It may be that Mesp1 protein is highly stable and outlasts the depletion of *Mesp1* transcript in CPCs. Potentially, Mesp1 is a pioneer transcription factor for CPCs that potentiates cardiac gene promoters for later activation. It is interesting to note that a recent study has described *Dkk1* as a major Mesp1 target in ES cells for procardiac activity (David *et al.*, 2008). As inhibition of canonical WNT signaling is essential for cardiogenesis to occur at the ALPM (Marvin *et al.*, 2001; Schneider and Mercola, 2001), perhaps Mesp1 initiates protection from WNT signaling in CPCs coincident with their migration away from the primitive streak during gastrulation (as in the streak, proper WNT signaling is essential for mesoderm induction and patterning). A third possibility is that Mesp1 regulates migration of cells that include CPCs, and that without this migration event (and arrival at the ALPM) cardiogenesis cannot proceed. The question of the link between migration and CPC development will be further discussed in the next section. Curiously, to my knowledge, Mesp function in heart development has not been described in other model organisms, notably *Drosophila* (which has a *Mesp* homologue; Moore *et al.*, 2000) and zebrafish. Further studies in ES cells, *Ciona*, and other models will likely clarify the role of Mesp proteins in CPC development.

Assuming Mesp1 is not solely "CardioD", what other events may be required to initiate CPC development? In *Ciona* and mouse, *Gata* gene expression (*Gata4* in mouse, there is only one *Gata* gene in *Ciona*) is subsequent to that of *Mesp1* and may indeed be directly regulated by Mesp1 (Bondue *et al.*, 2008; Ragkousi *et al.*, 2011). Unlike *Nkx2.5* and other cardiac transcription factor genes, *Gata4* (and its apparent homologue *Gata5* in zebrafish) is expressed early in gastrulating mesoderm and endoderm (Heikinheimo *et al.*, 1994; Reiter *et al.*, 1999). Gata factor activity is essential for cardiogenesis in flies, mouse, zebrafish, and *Ciona*, where inhibition of *pannier*, *Gata4/6*, *Gata5/6*, and *Gata*, respectively, results in heartless embryos and an absence of *Nkx2.5* expression (Alvarez *et al.*, 2003; Holtzinger and Evans, 2007; Klinedinst and Bodmer, 2003; Peterkin *et al.*, 2007; Ragkousi *et al.*, 2011; Zhao *et al.*, 2008). Studies in P19 cells had previously shown that GATA4 promotes terminal differentiation of cardiomyocytes (Grepin *et al.*, 1995, 1997). Given the continuous expression of *Gata4* in presumptive CPCs and differentiated cardiomyocytes, could GATA factors play an earlier role in CPC fate? And if so, as *Gata4* (or *Gata5* in zebrafish) is expressed in both mesoderm and endoderm, how would cardiac specificity be conferred? An answer to this conundrum arose from analysis of *Smarcd3* (*Baf60c*), a variant subunit of the SWI/SNF chromatin-remodeling complex (Lickert *et al.*, 2004). The SWI/SNF complex, which uses a core Brg1 or Brm ATPase, alters nucleosomal arrangement on promoter/enhancer elements to facilitate DNA sequence-specific transcription factor access (Kwon *et al.*, 1994). Many of the associated subunits of the SWI/SNF complex (of which there are over 10 in vertebrates) comes in multiple "flavors," with more than one gene encoding variant subunits that can independently be found in a given SWI/SNF complex (Wang *et al.*, 1996; Wu *et al.*, 2009). Changes in expression of SWI/SNF complex members have been shown, for example, to be a potent regulator or neural progenitor differentiation in the mouse spinal chord (Lessard *et al.*, 2007; Yoo *et al.*, 2009). In the case of the SMARCD subunit of the SWI/SNF complex, expression of *Smarcd3* in mouse is highly enriched in the heart. SMARCD3 can greatly potentiate the ability of a number of cardiac transcription factors to activate target genes (Lickert *et al.*, 2004; Takeuchi and Bruneau, 2009). Further, knockdown of *Smarcd3* by shRNAi in mice results in severe defects in early heart development (Lickert *et al.*, 2004). Can SMARCD3 potentiate GATA4 procardiac activity *in vivo*? Recent work in the mouse embryo shows that this is the case (Takeuchi and Bruneau, 2009). Transfection of gastrula-stage mouse embryos with *Gata4* and *Smarcd3* expression constructs results in ectopic sites of cardiac gene expression, and in many cases (with further addition of *Tbx5*) beating extracardiac cells. Gene expression profiling of these cells showed an upregulation of many myocardial markers, with ChIP analysis further showing that GATA4/SMARCD3 co-occupy elements in the promoters/enhancers of these genes. SMARCD3 activity is essential for this GATA4 activity, as transfection of *Gata4* alone does not cause ectopic sites of beating or cardiac gene expression. This is due, in part, to SMARCD3 promoting recruitment of GATA4 to the SWI/SNF complex.

What is the nature of this GATA4/SMARCD3 activity? Is the larger complex containing GATA4 and the SWI/SNF complex (containing SMARCD3) "simply" activating a cassette of myocardial genes, leading to apparent myocardial differentiation. Development of other CPC-derived lineages (smooth muscle and endocardium) was not assayed in these experiments. However, expression of *Mesp1* was decreased in *Gata4/Smarcd3*-expressing cells, suggesting that a CPC intermediate was not achieved (see discussion of reprogramming by Gata4 below). Recent work from our group has used the zebrafish model to further examine the nature of GATA4/SMARCD3 activity (Lou *et al.*, 2011). Global overexpression of *Gata5* (the zebrafish homologue of mouse *Gata4*) and *Smarcd3* by RNA microinjection results in abnormal embryos with enlarged hearts, but no apparent extracardiac sites of myocardial differentiation. Transplantation approaches yielded far more interesting results. In zebrafish, CPCs arise at 6 hours postfertilization (hpf) from bilateral positions along the embryonic margin of the embryo, displaced 60–120° along the circumference of the embryo/yolk interface from the dorsal (shield) position (Keegan *et al.*, 2004). Transplantation of cells to the margin of a host embryo at 4 hpf can (in 10–20% of cases) result in myocardial differentiation of transplanted donor cells, as assayed by EGFP signal from a cardiomyocyte-specific *myl7:EGFP* transgene in donor cells (Fig. 1.2A; Lee *et al.*, 1994; Scott *et al.*, 2007). In comparison to wild-type donor cells, *Gata5/Smarcd3* over-expressing donor cells contribute to the heart (and differentiate as cardio-myocytes, based on *myl7:EGFP* expression) at a much higher frequency

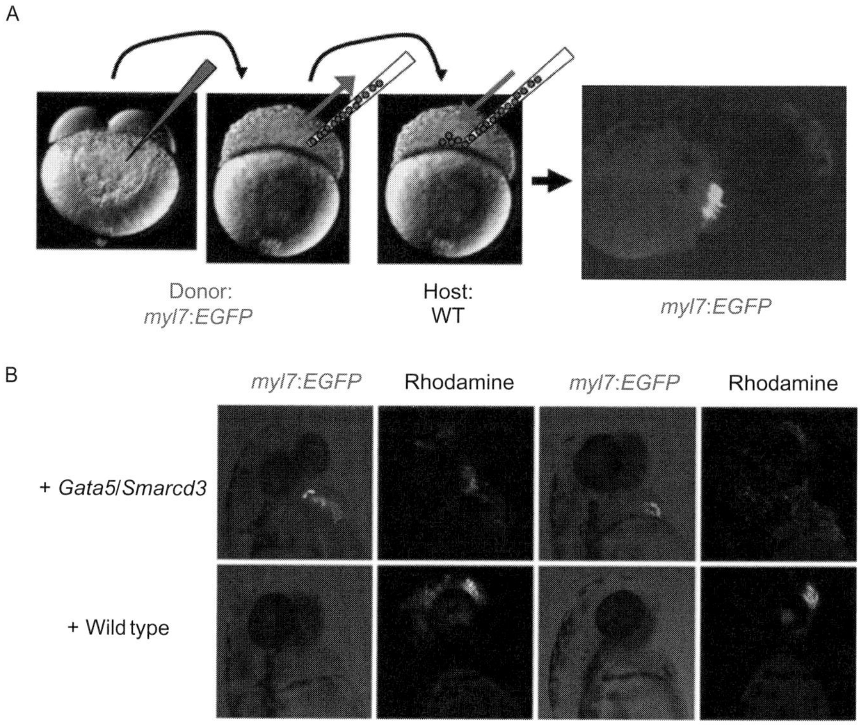

Figure 1.2 Gata5/Smarcd3 promotes a CPC-like state in zebrafish. (A) Schematics of transplantation experiments in zebrafish to assay cardiac potential. Cells from a transgenic 4-hpf *myl7:EGFP* donor embryo (injected with Rhodamine tracer at the one-cell stage) are transplanted to a host embryo at 4 hpf. EGFP signal in the heart at 24–48 hpf indicates donor cells that have differentiated as cardiomyocytes in the host heart. By manipulating host or donor cells, gene function in cardiac fate can be examined. (B) Transplants where donor cells are placed in the animal cap of host embryos. Wild-type donor cells do not take on a cardiac fate (no EGFP), with Rhodamine tracer showing cells remain in the brain/retina. Overexpression of *Gata5/Smarcd3* in donor cells promotes myocardial fate in the heart (EGFP and Rhodamine signal now evident in heart). (See Color Insert.)

(typically 50–60%), with a much greater number of donor cardiomyocytes formed per embryo (Lou *et al.*, 2011). These results suggest a cell autonomous role for GATA5/SMARCD3 in myocardial differentiation. However, as the margin of the zebrafish embryo presumably contains all the needed signals for CPC specification, these experiments alone do not allow a conclusive examination of GATA5/SMARCD3 activity.

In contrast to the margin of the zebrafish embryo (which is akin to the mouse primitive streak), the animal cap of the 4–6 hpf zebrafish embryo contains cells that will contribute to ectodermal cell lineages (forebrain, midbrain, and retina), but not cardiomyocytes (Ho and Kimmel, 1993; Kimmel *et al.*, 1990). More rigorous transplants were carried out in zebrafish in which donor cells were placed in the animal cap of host embryos (Fig. 1.2B). Remarkably, the donor cells placed in the animal cap formed cardiomyocytes in the host heart, and at a frequency (50–60% of transplants) indistinguishable from that observed in experiments where cells were placed in the margin. It is important to note that myocardial differentiation was always noted in the host heart, and not in extracardiac locations, in these experiments. Interestingly, GATA5/SMARCD3-expressing cells that had localized to the heart but had not initiated myocardial differentiation were found to express markers of endocardium and smooth muscle cell fate (again in the appropriate locations in the heart). This effect of GATA5/SMARCD3 is nonautonomous, as cotransplanted wild-type cells in these experiments are not induced to localize to the heart or differentiate to cardiac lineages (Lou *et al.*, 2011). Are GATA5/SMARCD3-expressing zebrafish cells being driven to a CPC fate in these experiments? These cells do demonstrate two hallmarks of CPCs: they can differentiate to form all three cardiac lineages (although the multipotentiality of an individual cell was not assessed) and they home to the heart-forming region of the embryo. Unlike wild-type cells placed in the animal cap, which associated with the neighboring host cells, GATA5/SMARCD3-expressing cells demonstrated behaviors associated with migration and preferentially associated with each other. The localization of these cells to the heart may simply represent the consequences of their exclusion from the animal cap region. However, as transplants to multiple locations in the embryo yield similar results, we currently favor a model where these cells are actively migrating to the heart-forming region. This would suggest that these cells undergo a CPC-like program in the absence of the endogenous signals and cues CPCs would experience in the embryo. Further, perhaps migratory behavior is a key aspect of CPC fate. The delineation of GATA5/SMARCD3 target genes in the zebrafish model will help to answer these questions.

How do these results in ES cells, mice, *Ciona,* and zebrafish translate to endogenous CPCs? Elegant clonal analysis in the mouse has suggested that CPC fate may be established very early in development, around the onset of gastrulation, with roughly 150 CPCs or CPC precursor being present at this stage (Meilhac *et al.*, 2004). How are these cells selected from a broader field? In the zebrafish, roughly 20% of cells in the "heart field", or pregastrula embryo region where CPCs are known to arise from, will contribute to the heart (Keegan *et al.*, 2004, 2005; Stainier *et al.*, 1993). Despite this, expression of genes such as *Gata5*, *Smarcd3*, and *Mesp1* is much broader at this stage, encompassing in zebrafish the majority of future mesendodermal cells that will gastrulate (our unpublished results; Ochi *et al.*, 2008; Reiter *et al.*, 1999; Sawada *et al.*, 2000). No gene to date has been found that discretely marks CPCs at gastrulation stages, with even *Mesp1* in the mouse being expressed in not only future cardiac but also cranial and other anterior mesoderm populations (Saga *et al.*, 1996, 1999). A potential explanation is that the overlay of transcription factor expression and gradients of signaling pathways may create thresholds of responses that result in distinct cell fates. Eomesodermin has been shown to directly activate *Mesp1* expression in the mouse primitive streak, but only under conditions of low Nodal signaling levels (Costello *et al.*, 2011). In contrast, at higher Nodal signal levels, Eomesodermin promotes endodermal fate. The intersection of multiple signaling gradients in the early embryo (likely Nodal, BMP, FGF, WNT, and others) may create domains of signaling environments that can be interpreted by the proper transcription factors for cell fate decisions. Considering the example of GATA4/SMARCD3, it is interesting to note that the association of MYOD/SMARCD3 with the SWI/SNF complex is regulated in skeletal muscle by a p38-mediated phosphorylation of SMARCD3 (Forcales *et al.*, 2011). In the absence of this phosphorylation, MYOD/SMARCD3, while associated with target promoter/enhancer elements, does not stably interact with the SWI/SNF complex and robustly activate gene expression. Posttranslational acetylation of GATA4 similarly boosts its transcriptional activity in cardiomyocytes (Kawamura *et al.*, 2005; Trivedi *et al.*, 2010). Perhaps a p38-mediated or other form of signal can boost GATA4/SMARCD3 activity in CPCs.

Therefore, while a *bona fide* CardioD remains undiscovered, a transcriptional network that promotes CPC fate is beginning to be uncovered. Future work will have to focus on how signaling pathways and transcription factors interact to assemble the proper transcriptional complexes on the correct target gene (which largely remain to be determined) regulatory elements.

5. Signaling Pathways and CPC Fate

An excellent recent review has described the role of multiple signaling pathways in cardiac mesoderm and heart development (Noseda *et al.*, 2011). In this section, the role of signaling pathways in the earliest events of CPC specification, and in later CPC lineage diversification, are considered.

Mesoderm induction, which is dependent on Nodal and FGF signaling (reviewed in Kimelman and Griffin, 2000), is obviously a prerequisite step for CPC development *in vivo*. As mentioned above, the level of Nodal signal a cell receives is critical for its later fate as high levels of Nodal signaling specific endoderm (Schier and Shen, 2000). Our recent work has shown that in the absence of Nodal signaling, GATA5/SMARCD3-overexpressing cells fail to leave the animal cap and undergo cardiac differentiation (Lou *et al.*, 2011). Indeed, this is the only perturbation to date we have found that blocks the ability of these cells to localize to the heart-forming region and differentiate to cardiovascular lineages. It is intriguing then that, in both mouse and zebrafish, the GATA/SMARCD3 combination does not appear to induce expression of *Mesp1*, which would be assumed to be an obligate CPC gene (Lou *et al.*, 2011; Takeuchi and Bruneau, 2009). The ability of GATA5/SMARCD3 to rescue heart development in the absence of *Mesp1* has not been formally tested, but it is possible that while Mesp1 acts early in future CPCs following mesoderm induction (as it is directly regulated by T and Eomesodermin; Costello *et al.*, 2011; David *et al.*, 2011), the critical signal for CPC fate specification is a downstream event such as forming an active GATA/SMARCD3-containing SWI/SNF complex on appropriate target gene promoters.

At the primitive streak (or margin in zebrafish), FGF signaling is essential to induce EMT and cell fate specification genes including *T* and *Tbx6* (Ciruna and Rossant, 2001). In the absence of reception of an FGF signal, cells are unable to undergo effective migration from the streak and are defective in contribution to the heart (Ciruna *et al.*, 1997). This is consistent with the work in zebrafish, which has shown that inhibition of FGF signaling early in zebrafish development at gastrulation stages results in fewer CPCs being formed (Marques *et al.*, 2008). In *Ciona*, FGF signaling is essential for heart development (Davidson *et al.*, 2006). Recent work has shown that the B7.5 daughter cell that will form the heart sends out polarized protrusions toward a relatively uniform source of FGF ligand (Cooley *et al.*, 2011). This process is regulated in part by the small GTPase Cdc42, whose activity is elevated by FoxF1 upregulation of RhoDF (Christiaen *et al.*, 2008). Following an asymmetric cell division, polarized protrusive activity in the B7.5 cell boosts the level of FGF signal in the future anterior B7.5 daughter, allowing it to take on a cardiac fate. Indeed, artificially boosting FGF/Ets activity in the posterior B7.5 daughter (which would normally form TVCs) drives it to a cardiac fate as well (Davidson *et al.*, 2006). Therefore, CPC fate involves not only the presence of a proper signaling milieu but also the appropriate competence to receive/translate these signals.

These experiments argue that signaling early in development, as future cardiac cells first under gastrulation movements, may be essential for CPC fate. An analysis of the WNT and Apelin signaling pathways further suggest that this window of development may represent a critical time period for CPC formation. A paradox in applying *Drosophila* genetic results to vertebrate heart development has been that while WNT (Wingless in fly) signaling is essential for dorsal vessel formation in the fly, WNT/β-catenin signaling prevents cardiac differentiation of mesoderm at the ALPM (Schneider and Mercola, 2001; Tzahor and Lassar, 2001; Wu *et al.*, 1995). The WNT response of CPCs is in fact biphasic: whereas early in development, WNT signaling is required for cardiac development, at later (ALPM) stages WNT inhibition is required for terminal differentiation of cardiomyocytes (Ueno *et al.*, 2007). This has been elegantly demonstrated in zebrafish, where a switch in the WNT response from being "procardiac" to "anticardiac" occurs at roughly 6–8 hpf. Similar results have been observed for the Apelin signaling pathway (Scott *et al.*, 2007; Zeng *et al.*, 2007). The G protein-coupled receptor Aplnr (also referred to as APJ, Agtrl1) regulates early heart development in zebrafish, with mutation or inhibition of *aplnrb* resulting in an absence or marked reduction of cardiomyocyte formation. This defect appears to arise early in CPC development, as *nkx2.5* expression is similarly lost. The precise nature of the *aplnrb* defect remains unclear, although work suggests that migration of CPCs to the ALPM may be deficient (our unpublished results; Scott *et al.*, 2007; Zeng *et al.*, 2007). While temporal inactivation of Aplnr is currently not feasible in zebrafish, an interesting observation is that global overexpression of the Aplnr ligand Apelin in zebrafish embryos also causes a heartless phenotype. Using this as a proxy for an absence of Aplnr signaling, a heat shock-inducible transgenic zebrafish (*hsp70:apelin*) was used to determine the developmental window over which Aplnr signaling regulates CPC development. Intriguingly, heat shock of these transgenic embryos at or prior to 5 hpf results in a heartless embryo, whereas heat shock even 1 h later at 6 hpf results in a grossly normal number of cardiomyocytes being formed (Scott *et al.*, 2007). This effect is largely confined to the cardiac mesoderm, as other mesodermal cell types appear to form normally following *apelin* overexpression or *aplnr* inhibition. While the role of Aplnr signaling in mouse heart development has not been examined in detail to date, work suggests that in ES cells this pathway may be a mediator of Nodal signals for cardiac differentiation (D'Aniello *et al.*, 2009). Taken together, these experiments suggest that in fish, a narrow window of 5–6 hpf is present where essential signals (WNT, Aplnr, and perhaps others) act to direct CPC fate. Intriguingly, this represents the time period where gastrulation movements are commencing in zebrafish.

Does a niche or signaling milieu exist in the early embryo that is required for CPC specification? As mentioned previously, CPCs appear to be distributed in a salt and pepper manner within a field of cells in the pregastrula embryo (Keegan *et al.*, 2004, 2005). Assuming that all cells in this area are being subjected to similar signals, CPC fate may be a stochastic event. Alternatively, the ability of cells to respond to these signals may be altered by cell status, as is observed in hematopoietic stem cell differentiation (Trompouki *et al.*, 2011). Signals may further have an additive, but not absolute, effect on CPC fate. Work in zebrafish has shown that Hedgehog (Hh) signal reception acts cell autonomously to promote CPC fate (Thomas *et al.*, 2008). In the absence of Hh signal, roughly 50% the normal number of CPCs are formed, as determined by fate mapping of cells in the heart field region of the pregastrula embryo. The timing and magnitude of numerous signaling pathways may therefore govern CPC fate adoption. How this information is processed, or what the gatekeeper event may be for CPCs, remains to be determined. We have, however, recently found that GATA5/SMARCD3 activity is required prior to 6hpf for CPC fate promotion, a time period which matches well with the developmental window over which WNT/Aplnr/Hh signaling acts (A. Deshwar and I. Scott, unpublished results). Future work in *Ciona* and zebrafish models, where CPCs can be traced and quantified *in vivo*, may in parallel with ES cell help to answer many of these questions.

6. A Moving Story: Early CPC Migration and Fate

Work covered in the previous section argues that migratory behavior may be a key characteristic of CPC fate, a relationship that has been established in *Ciona* (Christiaen *et al.*, 2008). In vertebrates, migration of CPCs and cardiac mesoderm has largely been studied with respect to the fusion of the bilateral *Nkx2.5*-expressing populations to form a midline heart tube. What regulates earlier migration of CPCs from the primitive streak? During gastrulation, pathways regulating anterior migration of craniocaudal mesoderm are not well understood (reviewed in Yin *et al.*, 2009). Many studies suggest that CPCs leave the primitive streak and move anteriorly with characteristics of autonomously migrating cells, only later (at ALPM stages) being organized into a collective group of cells to allow heart morphogenesis to proceed (Cui *et al.*, 2009; Inagaki *et al.*, 1993; Redkar *et al.*, 2001). WNT3a/5a signals have been shown promote migration of CPCs away from the primitive streak; however, perturbation of this pathway causes cardia bifida (two-split hearts), and not an absence of cardiac differentiation (Sweetman *et al.*, 2008; Yue *et al.*, 2008). In *Ciona*, BMP signaling is required for proper CPC migration, with low levels of BMP signal initiating migration, and higher levels leading to expression of the *Nkx2.5* homologue and an arrest in migration (Christiaen *et al.*, 2010). This is intriguing, as in zebrafish BMP signaling has been shown to regulate not only cell fate and dorsoventral patterning of the mesoderm but also cell adhesion (and presumably migratory behavior) via influences on Cadherin-mediated cell adhesion (von der Hardt *et al.*, 2007). CPCs are among the first mesodermal derivatives to exit the primitive streak during gastrulation (Parameswaran and Tam, 1995; Stainier *et al.*, 1993). Perhaps given the anterior migration required of craniocaudal mesoderm, subtle defects in the timing of their migration may result in a failure of CPCs to reach the ALPM and effectuate heart development. Alternatively, CPCs may be privileged in their ability to respond to migratory cue(s) that direct them to the ALPM. The nature of these potential cue(s) is at present unknown. In the zebrafish embryo, GATA5/SMARCD3-overexpressing cells can migrate to the heart-forming region of the embryo in the absence of either host endoderm or cardiac mesoderm (Lou *et al.*, 2011). Work in *Ciona* and zebrafish should address the question of potential migratory cues for CPCs.

Given the many similarities between CPC development *in vivo* and in ES cell models, an obvious question is one of how migration behavior fits into an ES cell environment. It has long been known that in ES cell bodies (where *in vitro* differentiation is often studied) 3D structures are formed, with primitive endoderm forming an outer layer over developing mesoderm (reviewed in Niwa, 2010). Indeed, gastrulation-like movements are observed in ES cell bodies, with this being especially apparent in cells forming cardiomyocytes (Fuchs *et al.*, 2011). Using available tools to image *Mesp1* expression live in ES cells (Bondue *et al.*, 2011), the CPC migratory process could be studied in a relatively tractable *ex vivo* system. Inhibiting candidate signaling pathways in ES and animal models may reveal pathways critical for CPC migration. Alternatively, the elucidation of transcriptional targets of putative CPC regulators, such as MESP1 and GATA5/SMARCD3, may reveal chemokine and signaling receptors that provide guidance information to CPCs. Based on the analysis of WNT, Aplnr, GATA5/SMARCD3, and MESP1 activity, it seems likely that choice of the appropriate developmental stage for these studies will be a critical issue. This may be analogous to the role of the Gata factors End1/3 in *Caenorhabditis elegans* endoderm, where these master regulators of endoderm fate control cassettes of genes responsible for both endoderm functional differentiation and cell migration (Boeck *et al.*, 2011).

7. Changing the Program: CPC Relationship to Reprogramming and Transdifferentiation

The groundbreaking discovery that a small number of transcription factors can reprogram differentiated adult cell types to a pluripotent stem cell state by Yamanaka (Takahashi and Yamanaka, 2006) has fostered an explosion of research into reprogramming cells into a number of cell lineages (reviewed in Chambers and Studer, 2011). A number of approaches, including lentiviral-mediated transcription factor and miRNA overexpression, use of transposons to avoid permanent genetic insertion, transfection of modified RNAs, and addition of small molecule inhibitors, have all been used to create induced pluripotent stem (iPS) cells (Anokye-Danso *et al.*, 2011; Lyssiotis *et al.*, 2009; Takahashi and Yamanaka, 2006; Warren *et al.*, 2010; Woltjen *et al.*, 2009). Conversion of mature cells, primarily fibroblasts, into functional neurons, hepatocytes, and hematopoietic cells (to name a few of a growing list of cell types) has been described (Huang *et al.*, 2011; Szabo *et al.*, 2010; Vierbuchen *et al.*, 2010). Can this reprogramming approach be used to further examine CPC biology, and conversely can the study of CPCs inform reprogramming work seeking to make mature cardiovascular cell types?

In the past year, the demonstration that fibroblasts can be reprogrammed to form cardiomyocytes by differing methods has been reported (Efe *et al.*, 2011; Ieda *et al.*, 2010). In the first report, a screen of 14 different transcription factors and cofactors with reported cardiac function was carried out via lentiviral-mediated transfection of cardiac fibroblasts bearing an *αMHC:EGFP* transgene (Ieda *et al.*, 2010). By removing factors with no or inhibitory effects on GFP expression, a "GMT" cocktail (GATA4, MEF2C, and TBX5) was found to be sufficient to induce αMHC expression robustly in over 20% of cells. These cells expressed a number of cardiomyocyte-specific genes, in whose promoters epigenetic marks in histone modifications and DNA methylation were altered to a state closely resembling that of endogenous cardiomyocytes. Calcium fluxes and contractile activity in a subset of these cells further indicated that these "induced cardiomyocytes" (iCMs) had adopted a cardiomyocyte-like fate. Indeed, when transplanted into a murine heart environment, cardiac fibroblasts could assume a myocardial fate in the heart following infection with GMT lentiviruses. In this study, these iCMs appear to have differentiated in the absence of a classical CPC intermediate. Using an elegant *Mesp1:Cre* lineage trace method to permanently mark cells that have initiated *Mesp1* expression at any point in their history, iCMs from cardiac fibroblasts were found to not induce *Mesp1* expression during myocardial conversion. Further, as these experiments were designed to optimize *αMHC:EGFP* expression, the ability of GMT to induce endocardial and smooth muscle cell fate was not assessed.

A parallel "direct reprogramming" approach has also recently been described that can convert mouse embryonic fibroblasts (MEFs) into cardiomyocytes (Efe *et al.*, 2011). In this case, cells were infected with the Yamanaka reprogramming factors (Oct4/Sox2/Klf4) using a system where their expression could be induced and silenced using doxycycline addition/removal. "Partial" reprogramming, in which the iPS factors are only induced for 4 days (not allowing iPS cell generation), followed by culture in procardiac (BMP4) media, was found to robustly induce formation of beating cardiomyocytes. When cells were treated with iPS factors as per the usual regime to generate pluripotent cells, cardiomyocyte differentiation was not evident. This indicated that a cellular intermediate was reached during partial reprogramming that was prone to cardiac differentiation. Indeed, gene expression analysis revealed that *Mesp1* expression was induced in MEFs undergoing direct reprogramming to cardiomyocytes, as were several other myocardial-specific genes.

The demonstration that any induction of cardiomyocyte fate from fibroblasts can be induced is of course remarkable in itself. It is early days for these approaches to cardiac reprogramming, and many questions remain to be answered. Improvements on these approaches are likely required, as in both cases specialized populations (cardiac fibroblasts or MEFs) were used that are likely more specialized in their cardiac potential or pluripotency. In contrast, when tail tip fibroblasts were used, in both cases less robust results were apparent. How can these approaches be optimized? Perhaps a combination of direct reprogramming and use of cardiac transcription factors would be beneficial. A seemingly obvious question in these studies is "when is a cardiomyocyte a cardiomyocyte?" The readout of the Ieda study was αMHC expression, a marker of myocardial fate. However, the path taken toward a cellular identity may very well restrict the future potential of these cells. It may be that looking at early markers of CPC fate as an assay may yield more robust reprogramming toward a cardiovascular fate. The cell type used for reprogramming may also have impact on the output of these experiments. One Holy Grail of reprogramming research is of course to produce mature ventricular cardiomyocytes for potential cell therapy, disease modeling, and pharmacological studies (reviewed in Yi *et al.*, 2010). Recent work has suggested, for example, that not all iPS cells may be equivalent in their ability to make mature cardiomyocytes, with those derived from ventricular cardiomyocytes forming more mature ventricular cardiomyocytes upon differentiation (Xu *et al.*, 2011). While prolonged passage of iPS cells may normalize cells to a certain extent, erasing their "memory" of their original source (Polo *et al.*, 2010), different human ES and iPS lines may vary in their response to cardiac/CPC induction culture conditions (Kattman *et al.*, 2011). What is the best cell type to use for cardiac reprogramming then? This may very well vary depending on the desired outcome (e.g., CPCs vs. mature cardiomyocytes).

From the viewpoint of CPCs, what is different about the recent studies that have induced cardiac fate in various models (Efe *et al.*, 2011; Ieda *et al.*, 2010; Lou *et al.*, 2011; Takeuchi and Bruneau, 2009) may involve the presence or absence of a progenitor state first reached in these experiments. The multipotentiality of "reprogrammed" cells was only truly tested in the zebrafish model for the cardiovascular lineage (Lou *et al.*, 2011). In contrast, a CPC-like gene expression pattern, based primarily on *Mesp1* expression, was only observed following partial reprogramming (Efe *et al.*, 2011). More detailed molecular characterization of what constitutes an early CPC will help to sort out the intermediate paths taken to cardiac fate by various reprogramming strategies. Further, a greater understanding of pathways promoting CPC fate may reveal new candidates for use in reprogramming approaches (e.g., see David *et al.*, 2009).

8. Conclusions: Future Questions and Possible Applications

As a relatively young field, the study of CPC specification and fate allocation has made tremendous strides in the past few years. Coupled with rapid advances in reprogramming and cell fate conversion, many major unanswered questions in CPC biology will hopefully be assailable in the next few years. Below, I consider some of these questions, as well as how these results may be translated to therapeutic interventions.

Several major areas of enquiry in CPC biology will likely involve (Fig. 1.3): (1) what leads to determination of CPC fate; (2) how can CPCs be maintained and expanded; (3) how can differentiation of CPCs be regulated, and more mature cardiovascular cell types made; (4) what is the

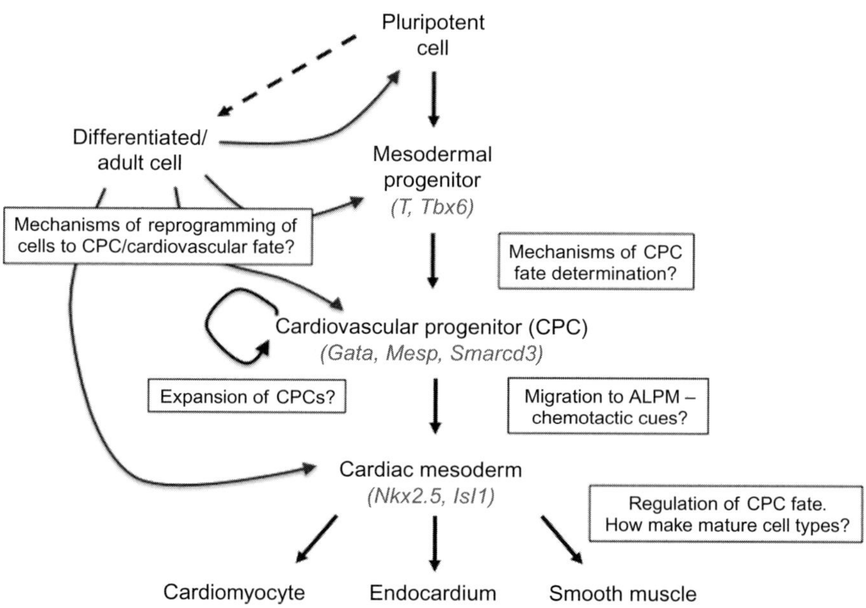

Figure 1.3 Major questions in CPC biology. A schematic of open questions in the regulation of critical steps in CPC development. (For color version of this figure, the reader is referred to the Web version of this chapter.)

relationship between embryonic CPCs and stem cell populations described in the adult heart; (5) what, if any, is an appropriate niche to allow determination and maintenance of CPCs in both the embryo and adult heart; and (6) are CPC-like cells an appropriate vehicle for cardiac repair?

One clear goal of the study of cardiac development, and in particular that of CPCs, is potential therapeutic interventions based on stem cell or differentiated cardiovascular cell addition (Sturzu and Wu, 2011; Yi *et al.*, 2010). The feasibility of this approach remains to be rigorously tested; however, an interesting question is, are CPCs an ideal platform for cardiac repair/regeneration? In the embryo, they have the ability to home to the heart-forming region and form all cardiovascular lineages. In a sense, they "know" how to carry out the job. Is this capacity maintained in the environment of an adult heart? Cardiac fibroblasts reprogrammed by GATA4/MEF2C/TBX5, if immediately placed in the mouse heart during reprogramming, appear to be competent to incorporate as functional cardiomyocytes to some extent (Ieda *et al.*, 2010). "*In vivo* reprogramming," where, for example, resident cardiac fibroblasts in the heart are directly exposed to pro-CPC factors, may also have potential. Alternatively, reprogramming endogenous circulating stem cells (e.g., hematopoietic stem cells) to a CPC fate may boost cardiac repair. Endogenous stem cell populations with CPC characteristics in the adult heart have been described in many studies (reviewed in Martin-Puig *et al.*, 2008; Wu *et al.*, 2008). These cells, if present in sufficient numbers, may be future targets for treatments that activate their ability to self-renew and robustly provide new cells for the heart. A final question is: why do our hearts not repair themselves naturally? An absence of heart regeneration in humans is not the default state in the Animal Kingdom. Indeed, the adult zebrafish heart has a remarkable regenerative capacity based on the proliferation of existing cardiomyocytes (Poss *et al.*, 2002). This is associated with a partial dedifferentiation of cardiomyocytes (Jopling *et al.*, 2010; Kikuchi *et al.*, 2010). Perhaps CPC-related pathways are involved in zebrafish heart regeneration and can be leveraged for human heart repair. In the mouse, a transient period of regenerative potential has been described in the neonatal heart (Porrello *et al.*, 2011). Perhaps this dormant potential can be reactivated in the adult.

In conclusion, the future study of CPCs and cardiac reprogramming will hopefully not only shed new light on mechanisms regulating the genesis of the building blocks of the heart but also present new opportunities to repair what is at present irreparable damage to the human heart.

ACKNOWLEDGMENTS

The author apologizes for references not cited due to space limitations. I thank Benoit Bruneau for helpful comments on this chapter. Thanks to members of the Scott lab and

Gordon Keller for providing stimulating discussions and ideas. Research in the Scott lab is funded by the Canadian Institutes of Health Research (CIHR), the Natural Sciences and Engineering Research Council of Canada (NSERC), the SickKids Foundation/Labatt Family Heart Centre, and the Richard Lewar Centre of Excellence in Cardiovascular Research.

REFERENCES

Alvarez, A. D., Shi, W., Wilson, B. A., and Skeath, J. B. (2003). pannier and pointedP2 act sequentially to regulate Drosophila heart development. *Development* **130,** 3015–3026.

Anokye-Danso, F., Trivedi, C. M., Juhr, D., Gupta, M., Cui, Z., Tian, Y., Zhang, Y., Yang, W., Gruber, P. J., Epstein, J. A., and Morrisey, E. E. (2011). Highly efficient miRNA-mediated reprogramming of mouse and human somatic cells to pluripotency. *Cell Stem Cell* **8,** 376–388.

Antin, P. B., Taylor, R. G., and Yatskievych, T. (1994). Precardiac mesoderm is specified during gastrulation in quail. *Dev. Dyn.* **200,** 144–154.

Arsenian, S., Weinhold, B., Oelgeschlager, M., Ruther, U., and Nordheim, A. (1998). Serum response factor is essential for mesoderm formation during mouse embryogenesis. *EMBO J.* **17,** 6289–6299.

Auda-Boucher, G., Bernard, B., Fontaine-Perus, J., Rouaud, T., Mericksay, M., and Gardahaut, M. F. (2000). Staging of the commitment of murine cardiac cell progenitors. *Dev. Biol.* **225,** 214–225.

Barron, M., Gao, M., and Lough, J. (2000). Requirement for BMP and FGF signaling during cardiogenic induction in non-precardiac mesoderm is specific, transient, and cooperative. *Dev. Dyn.* **218,** 383–393.

Beh, J., Shi, W., Levine, M., Davidson, B., and Christiaen, L. (2007). FoxF is essential for FGF-induced migration of heart progenitor cells in the ascidian Ciona intestinalis. *Development* **134,** 3297–3305.

Beiman, M., Shilo, B. Z., and Volk, T. (1996). Heartless, a Drosophila FGF receptor homolog, is essential for cell migration and establishment of several mesodermal lineages. *Genes Dev.* **10,** 2993–3002.

Bernstein, B. E., Mikkelsen, T. S., Xie, X., Kamal, M., Huebert, D. J., Cuff, J., Fry, B., Meissner, A., Wernig, M., Plath, K., Jaenisch, R., Wagschal, A., *et al.* (2006). A bivalent chromatin structure marks key developmental genes in embryonic stem cells. *Cell* **125,** 315–326.

Bodmer, R. (1993). The gene tinman is required for specification of the heart and visceral muscles in Drosophila. *Development* **118,** 719–729.

Boeck, M. E., Boyle, T., Bao, Z., Murray, J., Mericle, B., and Waterston, R. (2011). Specific roles for the GATA transcription factors end-1 and end-3 during *C. elegans* E-lineage development. *Dev. Biol.* **358,** 345–355.

Bondue, A., and Blanpain, C. (2010). Mesp1: A key regulator of cardiovascular lineage commitment. *Circ. Res.* **107,** 1414–1427.

Bondue, A., Lapouge, G., Paulissen, C., Semeraro, C., Iacovino, M., Kyba, M., and Blanpain, C. (2008). Mesp1 acts as a master regulator of multipotent cardiovascular progenitor specification. *Cell Stem Cell* **3,** 69–84.

Bondue, A., Tannler, S., Chiapparo, G., Chabab, S., Ramialison, M., Paulissen, C., Beck, B., Harvey, R., and Blanpain, C. (2011). Defining the earliest step of cardiovascular progenitor specification during embryonic stem cell differentiation. *J. Cell Biol.* **192,** 751–765.

Brade, T., Gessert, S., Kuhl, M., and Pandur, P. (2007). The amphibian second heart field: Xenopus islet-1 is required for cardiovascular development. *Dev. Biol.* **311,** 297–310.

Cai, C. L., Liang, X., Shi, Y., Chu, P. H., Pfaff, S. L., Chen, J., and Evans, S. (2003). Isl1 identifies a cardiac progenitor population that proliferates prior to differentiation and contributes a majority of cells to the heart. *Dev. Cell* **5,** 877–889.

Chambers, S. M., and Studer, L. (2011). Cell fate plug and play: Direct reprogramming and induced pluripotency. *Cell* **145,** 827–830.

Chang, P. S., Li, L., McAnally, J., and Olson, E. N. (2001). Muscle specificity encoded by specific serum response factor-binding sites. *J. Biol. Chem.* **276,** 17206–17212.

Chen, J. N., and Fishman, M. C. (1996). Zebrafish tinman homolog demarcates the heart field and initiates myocardial differentiation. *Development* **122,** 3809–3816.

Chen, F., Kook, H., Milewski, R., Gitler, A. D., Lu, M. M., Li, J., Nazarian, R., Schnepp, R., Jen, K., Biben, C., Runke, G., Mackay, J. P., *et al.* (2002). Hop is an unusual homeobox gene that modulates cardiac development. *Cell* **110,** 713–723.

Christiaen, L., Davidson, B., Kawashima, T., Powell, W., Nolla, H., Vranizan, K., and Levine, M. (2008). The transcription/migration interface in heart precursors of Ciona intestinalis. *Science* **320,** 1349–1352.

Christiaen, L., Stolfi, A., Davidson, B., and Levine, M. (2009). Spatio-temporal intersection of Lhx3 and Tbx6 defines the cardiac field through synergistic activation of Mesp. *Dev. Biol.* **328,** 552–560.

Christiaen, L., Stolfi, A., and Levine, M. (2010). BMP signaling coordinates gene expression and cell migration during precardiac mesoderm development. *Dev. Biol.* **340,** 179–187.

Ciruna, B., and Rossant, J. (2001). FGF signaling regulates mesoderm cell fate specification and morphogenetic movement at the primitive streak. *Dev. Cell* **1,** 37–49.

Ciruna, B. G., Schwartz, L., Harpal, K., Yamaguchi, T. P., and Rossant, J. (1997). Chimeric analysis of fibroblast growth factor receptor-1 (Fgfr1) function: A role for FGFR1 in morphogenetic movement through the primitive streak. *Development* **124,** 2829–2841.

Cooley, J., Whitaker, S., Sweeney, S., Fraser, S., and Davidson, B. (2011). Cytoskeletal polarity mediates localized induction of the heart progenitor lineage. *Nat. Cell Biol.* **13,** 952–957.

Costello, I., Pimeisl, I. M., Drager, S., Bikoff, E. K., Robertson, E. J., and Arnold, S. J. (2011). The T-box transcription factor Eomesodermin acts upstream of Mesp1 to specify cardiac mesoderm during mouse gastrulation. *Nat. Cell Biol.* **13,** 1084–1091.

Cui, C., Cheuvront, T. J., Lansford, R. D., Moreno-Rodriguez, R. A., Schultheiss, T. M., and Rongish, B. J. (2009). Dynamic positional fate map of the primary heart-forming region. *Dev. Biol.* **332,** 212–222.

D'Aniello, C., Lonardo, E., Iaconis, S., Guardiola, O., Liguoro, A. M., Liguori, G. L., Autiero, M., Carmeliet, P., and Minchiotti, G. (2009). G protein-coupled receptor APJ and its ligand apelin act downstream of Cripto to specify embryonic stem cells toward the cardiac lineage through extracellular signal-regulated kinase/p70S6 kinase signaling pathway. *Circ. Res.* **105,** 231–238.

David, R., Brenner, C., Stieber, J., Schwarz, F., Brunner, S., Vollmer, M., Mentele, E., Muller-Hocker, J., Kitajima, S., Lickert, H., Rupp, R., and Franz, W. M. (2008). MesP1 drives vertebrate cardiovascular differentiation through Dkk-1-mediated blockade of Wnt-signalling. *Nat. Cell Biol.* **10,** 338–345.

David, R., Stieber, J., Fischer, E., Brunner, S., Brenner, C., Pfeiler, S., Schwarz, F., and Franz, W. M. (2009). Forward programming of pluripotent stem cells towards distinct cardiovascular cell types. *Cardiovasc. Res.* **84,** 263–272.

David, R., Jarsch, V. B., Schwarz, F., Nathan, P., Gegg, M., Lickert, H., and Franz, W. M. (2011). Induction of MesP1 by Brachyury(T) generates the common multipotent cardiovascular stem cell. *Cardiovasc. Res.* **92,** 115–122.

Davidson, B. (2007). Ciona intestinalis as a model for cardiac development. *Semin. Cell Dev. Biol.* **18,** 16–26.

Davidson, B., and Levine, M. (2003). Evolutionary origins of the vertebrate heart: Specification of the cardiac lineage in Ciona intestinalis. *Proc. Natl. Acad. Sci. USA* **100,** 11469–11473.

Davidson, B., Shi, W., and Levine, M. (2005). Uncoupling heart cell specification and migration in the simple chordate Ciona intestinalis. *Development* **132,** 4811–4818.

Davidson, B., Shi, W., Beh, J., Christiaen, L., and Levine, M. (2006). FGF signaling delineates the cardiac progenitor field in the simple chordate, Ciona intestinalis. *Genes Dev.* **20,** 2728–2738.

Dehaan, R. L. (1963). Migration patterns of the precardiac mesoderm in the early chick embryo. *Exp. Cell Res.* **29,** 544–560.

Edmondson, D. G., and Olson, E. N. (1989). A gene with homology to the myc similarity region of MyoD1 is expressed during myogenesis and is sufficient to activate the muscle differentiation program. *Genes Dev.* **3,** 628–640.

Efe, J. A., Hilcove, S., Kim, J., Zhou, H., Ouyang, K., Wang, G., Chen, J., and Ding, S. (2011). Conversion of mouse fibroblasts into cardiomyocytes using a direct reprogramming strategy. *Nat. Cell Biol.* **13,** 215–222.

Evans, S. M., Yelon, D., Conlon, F. L., and Kirby, M. L. (2010). Myocardial lineage development. *Circ. Res.* **107,** 1428–1444.

Firulli, A. B., McFadden, D. G., Lin, Q., Srivastava, D., and Olson, E. N. (1998). Heart and extra-embryonic mesodermal defects in mouse embryos lacking the bHLH transcription factor Hand1. *Nat. Genet.* **18,** 266–270.

Forcales, S. V., Albini, S., Giordani, L., Malecova, B., Cignolo, L., Chernov, A., Coutinho, P., Saccone, V., Consalvi, S., Williams, R., Wang, K., Wu, Z., *et al.* (2011). Signal-dependent incorporation of MyoD-BAF60c into Brg1-based SWI/SNF chromatin-remodelling complex. *EMBO J.*10.1038/emboj.2011.391.

Frasch, M. (1995). Induction of visceral and cardiac mesoderm by ectodermal Dpp in the early Drosophila embryo. *Nature* **374,** 464–467.

Fuchs, C., Scheinast, M., Pasteiner, W., Lagger, S., Hofner, M., Hoellrigl, A., Schultheis, M., and Weitzer, G. (2011). Self-organization phenomena in embryonic stem cell-derived embryoid bodies: Axis formation and breaking of symmetry during cardiomyogenesis. *Cells Tissues Organs*10.1038/cr.2011.171.

Garcia-Martinez, V., and Schoenwolf, G. C. (1993). Primitive-streak origin of the cardiovascular system in avian embryos. *Dev. Biol.* **159,** 706–719.

Gisselbrecht, S., Skeath, J. B., Doe, C. Q., and Michelson, A. M. (1996). heartless encodes a fibroblast growth factor receptor (DFR1/DFGF-R2) involved in the directional migration of early mesodermal cells in the Drosophila embryo. *Genes Dev.* **10,** 3003–3017.

Goldstein, A. M., and Fishman, M. C. (1998). Notochord regulates cardiac lineage in zebrafish embryos. *Dev. Biol.* **201,** 247–252.

Grepin, C., Robitaille, L., Antakly, T., and Nemer, M. (1995). Inhibition of transcription factor GATA-4 expression blocks *in vitro* cardiac muscle differentiation. *Mol. Cell. Biol.* **15,** 4095–4102.

Grepin, C., Nemer, G., and Nemer, M. (1997). Enhanced cardiogenesis in embryonic stem cells overexpressing the GATA-4 transcription factor. *Development* **124,** 2387–2395.

Hami, D., Grimes, A. C., Tsai, H. J., and Kirby, M. L. (2011). Zebrafish cardiac development requires a conserved secondary heart field. *Development* **138,** 2389–2398.

He, A., Kong, S. W., Ma, Q., and Pu, W. T. (2011). Co-occupancy by multiple cardiac transcription factors identifies transcriptional enhancers active in heart. *Proc. Natl. Acad. Sci. USA* **108,** 5632–5637.

Heikinheimo, M., Scandrett, J. M., and Wilson, D. B. (1994). Localization of transcription factor GATA-4 to regions of the mouse embryo involved in cardiac development. *Dev. Biol.* **164,** 361–373.

Ho, R. K., and Kimmel, C. B. (1993). Commitment of cell fate in the early zebrafish embryo. *Science* **261,** 109–111.

Holtzinger, A., and Evans, T. (2007). Gata5 and Gata6 are functionally redundant in zebrafish for specification of cardiomyocytes. *Dev. Biol.* **312,** 613–622.

Huang, P., He, Z., Ji, S., Sun, H., Xiang, D., Liu, C., Hu, Y., Wang, X., and Hui, L. (2011). Induction of functional hepatocyte-like cells from mouse fibroblasts by defined factors. *Nature* **475,** 386–389.

Hutson, M. R., and Kirby, M. L. (2007). Model systems for the study of heart development and disease. Cardiac neural crest and conotruncal malformations. *Semin. Cell Dev. Biol.* **18,** 101–110.

Ieda, M., Fu, J. D., Delgado-Olguin, P., Vedantham, V., Hayashi, Y., Bruneau, B. G., and Srivastava, D. (2010). Direct reprogramming of fibroblasts into functional cardiomyocytes by defined factors. *Cell* **142,** 375–386.

Inagaki, T., Garcia-Martinez, V., and Schoenwolf, G. C. (1993). Regulative ability of the prospective cardiogenic and vasculogenic areas of the primitive streak during avian gastrulation. *Dev. Dyn.* **197,** 57–68.

Jopling, C., Sleep, E., Raya, M., Marti, M., Raya, A., and Belmonte, J. C. (2010). Zebrafish heart regeneration occurs by cardiomyocyte dedifferentiation and proliferation. *Nature* **464,** 606–609.

Kattman, S. J., Huber, T. L., and Keller, G. M. (2006). Multipotent flk-1+ cardiovascular progenitor cells give rise to the cardiomyocyte, endothelial, and vascular smooth muscle lineages. *Dev. Cell* **11,** 723–732.

Kattman, S. J., Witty, A. D., Gagliardi, M., Dubois, N. C., Niapour, M., Hotta, A., Ellis, J., and Keller, G. (2011). Stage-specific optimization of activin/nodal and BMP signaling promotes cardiac differentiation of mouse and human pluripotent stem cell lines. *Cell Stem Cell* **8,** 228–240.

Kawamura, T., Ono, K., Morimoto, T., Wada, H., Hirai, M., Hidaka, K., Morisaki, T., Heike, T., Nakahata, T., Kita, T., and Hasegawa, K. (2005). Acetylation of GATA-4 is involved in the differentiation of embryonic stem cells into cardiac myocytes. *J. Biol. Chem.* **280,** 19682–19688.

Keegan, B. R., Meyer, D., and Yelon, D. (2004). Organization of cardiac chamber progenitors in the zebrafish blastula. *Development* **131,** 3081–3091.

Keegan, B. R., Feldman, J. L., Begemann, G., Ingham, P. W., and Yelon, D. (2005). Retinoic acid signaling restricts the cardiac progenitor pool. *Science* **307,** 247–249.

Kikuchi, K., Holdway, J. E., Werdich, A. A., Anderson, R. M., Fang, Y., Egnaczyk, G. F., Evans, T., Macrae, C. A., Stainier, D. Y., and Poss, K. D. (2010). Primary contribution to zebrafish heart regeneration by gata4(+) cardiomyocytes. *Nature* **464,** 601–605.

Kimelman, D., and Griffin, K. J. (2000). Vertebrate mesendoderm induction and patterning. *Curr. Opin. Genet. Dev.* **10,** 350–356.

Kimmel, C. B., Warga, R. M., and Schilling, T. F. (1990). Origin and organization of the zebrafish fate map. *Development* **108,** 581–594.

Kitajima, S., Takagi, A., Inoue, T., and Saga, Y. (2000). MesP1 and MesP2 are essential for the development of cardiac mesoderm. *Development* **127,** 3215–3226.

Klinedinst, S. L., and Bodmer, R. (2003). Gata factor Pannier is required to establish competence for heart progenitor formation. *Development* **130,** 3027–3038.

Kwon, H., Imbalzano, A. N., Khavari, P. A., Kingston, R. E., and Green, M. R. (1994). Nucleosome disruption and enhancement of activator binding by a human SW1/SNF complex. *Nature* **370,** 477–481.

Lee, R. K., Stainier, D. Y., Weinstein, B. M., and Fishman, M. C. (1994). Cardiovascular development in the zebrafish. II. Endocardial progenitors are sequestered within the heart field. *Development* **120,** 3361–3366.

Lessard, J., Wu, J. I., Ranish, J. A., Wan, M., Winslow, M. M., Staahl, B. T., Wu, H., Aebersold, R., Graef, I. A., and Crabtree, G. R. (2007). An essential switch in subunit composition of a chromatin remodeling complex during neural development. *Neuron* **55,** 201–215.

Lickert, H., Takeuchi, J. K., Von Both, I., Walls, J. R., McAuliffe, F., Adamson, S. L., Henkelman, R. M., Wrana, J. L., Rossant, J., and Bruneau, B. G. (2004). Baf60c is essential for function of BAF chromatin remodelling complexes in heart development. *Nature* **432,** 107–112.

Lindsley, R. C., Gill, J. G., Murphy, T. L., Langer, E. M., Cai, M., Mashayekhi, M., Wang, W., Niwa, N., Nerbonne, J. M., Kyba, M., and Murphy, K. M. (2008). Mesp1 coordinately regulates cardiovascular fate restriction and epithelial-mesenchymal transition in differentiating ESCs. *Cell Stem Cell* **3,** 55–68.

Lopez-Sanchez, C., and Garcia-Martinez, V. (2011). Molecular determinants of cardiac specification. *Cardiovasc. Res.* **91,** 185–195.

Lou, X., Deshwar, A. R., Crump, J. G., and Scott, I. C. (2011). Smarcd3b and Gata5 promote a cardiac progenitor fate in the zebrafish embryo. *Development* **138,** 3113–3123.

Lupien, M., Eeckhoute, J., Meyer, C. A., Wang, Q., Zhang, Y., Li, W., Carroll, J. S., Liu, X. S., and Brown, M. (2008). FoxA1 translates epigenetic signatures into enhancer-driven lineage-specific transcription. *Cell* **132,** 958–970.

Lyons, I., Parsons, L. M., Hartley, L., Li, R., Andrews, J. E., Robb, L., and Harvey, R. P. (1995). Myogenic and morphogenetic defects in the heart tubes of murine embryos lacking the homeo box gene Nkx2-5. *Genes Dev.* **9,** 1654–1666.

Lyssiotis, C. A., Foreman, R. K., Staerk, J., Garcia, M., Mathur, D., Markoulaki, S., Hanna, J., Lairson, L. L., Charette, B. D., Bouchez, L. C., Bollong, M., Kunick, C., *et al.* (2009). Reprogramming of murine fibroblasts to induced pluripotent stem cells with chemical complementation of Klf4. *Proc. Natl. Acad. Sci. USA* **106,** 8912–8917.

Marques, S. R., Lee, Y., Poss, K. D., and Yelon, D. (2008). Reiterative roles for FGF signaling in the establishment of size and proportion of the zebrafish heart. *Dev. Biol.* **321,** 397–406.

Martin-Puig, S., Wang, Z., and Chien, K. R. (2008). Lives of a heart cell: Tracing the origins of cardiac progenitors. *Cell Stem Cell* **2,** 320–331.

Marvin, M. J., Di Rocco, G., Gardiner, A., Bush, S. M., and Lassar, A. B. (2001). Inhibition of Wnt activity induces heart formation from posterior mesoderm. *Genes Dev.* **15,** 316–327.

McBurney, M. W., Jones-Villeneuve, E. M., Edwards, M. K., and Anderson, P. J. (1982). Control of muscle and neuronal differentiation in a cultured embryonal carcinoma cell line. *Nature* **299,** 165–167.

Meilhac, S. M., and Buckingham, M. E. (2010). The behavior of cells that form the myocardial compartments of the vertebrate heart. *In* "Heart Development and Regeneration," (N. Rosenthal and R. P. Harvey, eds.), pp. 195–217. Academic Press, London.

Meilhac, S. M., Esner, M., Kelly, R. G., Nicolas, J. F., and Buckingham, M. E. (2004). The clonal origin of myocardial cells in different regions of the embryonic mouse heart. *Dev. Cell* **6,** 685–698.

Miano, J. M. (2003). Serum response factor: Toggling between disparate programs of gene expression. *J. Mol. Cell. Cardiol.* **35,** 577–593.

Miano, J. M., Ramanan, N., Georger, M. A., de Mesy Bentley, K. L., Emerson, R. L., Balza, R. O., Jr., Xiao, Q., Weiler, H., Ginty, D. D., and Misra, R. P. (2004). Restricted inactivation of serum response factor to the cardiovascular system. *Proc. Natl. Acad. Sci. USA* **101,** 17132–17137.

Moore, A. W., Barbel, S., Jan, L. Y., and Jan, Y. N. (2000). A genomewide survey of basic helix-loop-helix factors in Drosophila. *Proc. Natl. Acad. Sci. USA* **97,** 10436–10441.

Moretti, A., Caron, L., Nakano, A., Lam, J. T., Bernshausen, A., Chen, Y., Qyang, Y., Bu, L., Sasaki, M., Martin-Puig, S., Sun, Y., Evans, S. M., *et al.* (2006). Multipotent embryonic isl1+ progenitor cells lead to cardiac, smooth muscle, and endothelial cell diversification. *Cell* **127,** 1151–1165.

Motoike, T., Markham, D. W., Rossant, J., and Sato, T. N. (2003). Evidence for novel fate of Flk1+ progenitor: Contribution to muscle lineage. *Genesis* **35,** 153–159.

Niu, Z., Iyer, D., Conway, S. J., Martin, J. F., Ivey, K., Srivastava, D., Nordheim, A., and Schwartz, R. J. (2008). Serum response factor orchestrates nascent sarcomerogenesis and silences the biomineralization gene program in the heart. *Proc. Natl. Acad. Sci. USA* **105,** 17824–17829.

Niwa, H. (2010). Mouse ES cell culture system as a model of development. *Dev. Growth Differ.* **52,** 275–283.

Noseda, M., Peterkin, T., Simoes, F. C., Patient, R., and Schneider, M. D. (2011). Cardiopoietic factors: Extracellular signals for cardiac lineage commitment. *Circ. Res.* **108,** 129–152.

Ochi, H., Hans, S., and Westerfield, M. (2008). Smarcd3 regulates the timing of zebrafish myogenesis onset. *J. Biol. Chem.* **283,** 3529–3536.

Parameswaran, M., and Tam, P. P. (1995). Regionalisation of cell fate and morphogenetic movement of the mesoderm during mouse gastrulation. *Dev. Genet.* **17,** 16–28.

Peterkin, T., Gibson, A., and Patient, R. (2007). Redundancy and evolution of GATA factor requirements in development of the myocardium. *Dev. Biol.* **311,** 623–635.

Polo, J. M., Liu, S., Figueroa, M. E., Kulalert, W., Eminli, S., Tan, K. Y., Apostolou, E., Stadtfeld, M., Li, Y., Shioda, T., Natesan, S., Wagers, A. J., *et al.* (2010). Cell type of origin influences the molecular and functional properties of mouse induced pluripotent stem cells. *Nat. Biotechnol.* **28,** 848–855.

Porrello, E. R., Mahmoud, A. I., Simpson, E., Hill, J. A., Richardson, J. A., Olson, E. N., and Sadek, H. A. (2011). Transient regenerative potential of the neonatal mouse heart. *Science* **331,** 1078–1080.

Poss, K. D., Wilson, L. G., and Keating, M. T. (2002). Heart regeneration in zebrafish. *Science* **298,** 2188–2190.

Prall, O. W., Menon, M. K., Solloway, M. J., Watanabe, Y., Zaffran, S., Bajolle, F., Biben, C., McBride, J. J., Robertson, B. R., Chaulet, H., Stennard, F. A., Wise, N., *et al.* (2007). An Nkx2-5/BMP2/smad1 negative feedback loop controls heart progenitor specification and proliferation. *Cell* **128,** 947–959.

Rada-Iglesias, A., Bajpai, R., Swigut, T., Brugmann, S. A., Flynn, R. A., and Wysocka, J. (2011). A unique chromatin signature uncovers early developmental enhancers in humans. *Nature* **470,** 279–283.

Raffin, M., Leong, L. M., Rones, M. S., Sparrow, D., Mohun, T., and Mercola, M. (2000). Subdivision of the cardiac Nkx2.5 expression domain into myogenic and nonmyogenic compartments. *Dev. Biol.* **218,** 326–340.

Ragkousi, K., Beh, J., Sweeney, S., Starobinska, E., and Davidson, B. (2011). A single GATA factor plays discrete, lineage specific roles in ascidian heart development. *Dev. Biol.* **352,** 154–163.

Redkar, A., Montgomery, M., and Litvin, J. (2001). Fate map of early avian cardiac progenitor cells. *Development* **128,** 2269–2279.

Reifers, F., Walsh, E. C., Leger, S., Stainier, D. Y., and Brand, M. (2000). Induction and differentiation of the zebrafish heart requires fibroblast growth factor 8 (fgf8/acerebellar). *Development* **127,** 225–235.

Reiter, J. F., Alexander, J., Rodaway, A., Yelon, D., Patient, R., Holder, N., and Stainier, D. Y. (1999). Gata5 is required for the development of the heart and endoderm in zebrafish. *Genes Dev.* **13,** 2983–2995.

Riley, P., Anson-Cartwright, L., and Cross, J. C. (1998). The Hand1 bHLH transcription factor is essential for placentation and cardiac morphogenesis. *Nat. Genet.* **18,** 271–275.

Saga, Y., Hata, N., Kobayashi, S., Magnuson, T., Seldin, M. F., and Taketo, M. M. (1996). MesP1: A novel basic helix-loop-helix protein expressed in the nascent mesodermal cells during mouse gastrulation. *Development* **122,** 2769–2778.

Saga, Y., Miyagawa-Tomita, S., Takagi, A., Kitajima, S., Miyazaki, J., and Inoue, T. (1999). MesP1 is expressed in the heart precursor cells and required for the formation of a single heart tube. *Development* **126,** 3437–3447.

Saga, Y., Kitajima, S., and Miyagawa-Tomita, S. (2000). Mesp1 expression is the earliest sign of cardiovascular development. *Trends Cardiovasc. Med.* **10,** 345–352.

Sater, A. K., and Jacobson, A. G. (1989). The specification of heart mesoderm occurs during gastrulation in Xenopus laevis. *Development* **105,** 821–830.

Satou, Y., Imai, K. S., and Satoh, N. (2004). The ascidian Mesp gene specifies heart precursor cells. *Development* **131,** 2533–2541.

Sawada, A., Fritz, A., Jiang, Y. J., Yamamoto, A., Yamasu, K., Kuroiwa, A., Saga, Y., and Takeda, H. (2000). Zebrafish Mesp family genes, mesp-a and mesp-b are segmentally expressed in the presomitic mesoderm, and Mesp-b confers the anterior identity to the developing somites. *Development* **127,** 1691–1702.

Schier, A. F., and Shen, M. M. (2000). Nodal signalling in vertebrate development. *Nature* **403,** 385–389.

Schneider, V. A., and Mercola, M. (2001). Wnt antagonism initiates cardiogenesis in Xenopus laevis. *Genes Dev.* **15,** 304–315.

Schoenwolf, G. C., and Garcia-Martinez, V. (1995). Primitive-streak origin and state of commitment of cells of the cardiovascular system in avian and mammalian embryos. *Cell. Mol. Biol. Res.* **41,** 233–240.

Schott, J. J., Benson, D. W., Basson, C. T., Pease, W., Silberbach, G. M., Moak, J. P., Maron, B. J., Seidman, C. E., and Seidman, J. G. (1998). Congenital heart disease caused by mutations in the transcription factor NKX2-5. *Science* **281,** 108–111.

Schultheiss, T. M., Xydas, S., and Lassar, A. B. (1995). Induction of avian cardiac myogenesis by anterior endoderm. *Development* **121,** 4203–4214.

Schultheiss, T. M., Burch, J. B., and Lassar, A. B. (1997). A role for bone morphogenetic proteins in the induction of cardiac myogenesis. *Genes Dev.* **11,** 451–462.

Scott, I. C., Masri, B., D'Amico, L. A., Jin, S. W., Jungblut, B., Wehman, A., Baier, H., Audigier, Y., and Stainier, D. Y. R. (2007). The G-protein coupled receptor Agtrl1b regulates early development of myocardial progenitors. *Dev. Cell* **12,** 403–413.

Shi, Y., Katsev, S., Cai, C., and Evans, S. (2000). BMP signaling is required for heart formation in vertebrates. *Dev. Biol.* **224,** 226–237.

Shin, C. H., Liu, Z. P., Passier, R., Zhang, C. L., Wang, D. Z., Harris, T. M., Yamagishi, H., Richardson, J. A., Childs, G., and Olson, E. N. (2002). Modulation of cardiac growth and development by HOP, an unusual homeodomain protein. *Cell* **110,** 725–735.

Srivastava, D., Thomas, T., Lin, Q., Kirby, M. L., Brown, D., and Olson, E. N. (1997). Regulation of cardiac mesodermal and neural crest development by the bHLH transcription factor, dHAND. *Nat. Genet.* **16,** 154–160.

Stainier, D. Y., Lee, R. K., and Fishman, M. C. (1993). Cardiovascular development in the zebrafish. I. Myocardial fate map and heart tube formation. *Development* **119,** 31–40.

Stalsberg, H., and DeHaan, R. L. (1969). The precardiac areas and formation of the tubular heart in the chick embryo. *Dev. Biol.* **19,** 128–159.

Sturzu, A. C., and Wu, S. M. (2011). Developmental and regenerative biology of multipotent cardiovascular progenitor cells. *Circ. Res.* **108,** 353–364.

Sweetman, D., Wagstaff, L., Cooper, O., Weijer, C., and Munsterberg, A. (2008). The migration of paraxial and lateral plate mesoderm cells emerging from the late primitive streak is controlled by different Wnt signals. *BMC Dev. Biol.* **8,** 63.

Szabo, E., Rampalli, S., Risueno, R. M., Schnerch, A., Mitchell, R., Fiebig-Comyn, A., Levadoux-Martin, M., and Bhatia, M. (2010). Direct conversion of human fibroblasts to multilineage blood progenitors. *Nature* **468,** 521–526.

Takahashi, K., and Yamanaka, S. (2006). Induction of pluripotent stem cells from mouse embryonic and adult fibroblast cultures by defined factors. *Cell* **126,** 663–676.

Takeuchi, J. K., and Bruneau, B. G. (2009). Directed transdifferentiation of mouse mesoderm to heart tissue by defined factors. *Nature* **459,** 708–711.

Tam, P. P., Parameswaran, M., Kinder, S. J., and Weinberger, R. P. (1997). The allocation of epiblast cells to the embryonic heart and other mesodermal lineages: The role of ingression and tissue movement during gastrulation. *Development* **124,** 1631–1642.

Tapscott, S. J., Davis, R. L., Thayer, M. J., Cheng, P. F., Weintraub, H., and Lassar, A. B. (1988). MyoD1: A nuclear phosphoprotein requiring a Myc homology region to convert fibroblasts to myoblasts. *Science* **242,** 405–411.

Targoff, K. L., Schell, T., and Yelon, D. (2008). Nkx genes regulate heart tube extension and exert differential effects on ventricular and atrial cell number. *Dev. Biol.* **322,** 314–321.

Thomas, N. A., Koudijs, M., van Eeden, F. J., Joyner, A. L., and Yelon, D. (2008). Hedgehog signaling plays a cell-autonomous role in maximizing cardiac developmental potential. *Development* **135,** 3789–3799.

Tonissen, K. F., Drysdale, T. A., Lints, T. J., Harvey, R. P., and Krieg, P. A. (1994). XNkx-2.5, a Xenopus gene related to Nkx-2.5 and tinman: Evidence for a conserved role in cardiac development. *Dev. Biol.* **162,** 325–328.

Trivedi, C. M., Zhu, W., Wang, Q., Jia, C., Kee, H. J., Li, L., Hannenhalli, S., and Epstein, J. A. (2010). Hopx and Hdac2 interact to modulate Gata4 acetylation and embryonic cardiac myocyte proliferation. *Dev. Cell* **19,** 450–459.

Trompouki, E., Bowman, T. V., Lawton, L. N., Fan, Z. P., Wu, D. C., Dibiase, A., Martin, C. S., Cech, J. N., Sessa, A. K., Leblanc, J. L., Li, P., Durand, E. M., *et al.* (2011). Lineage regulators direct BMP and Wnt pathways to cell-specific programs during differentiation and regeneration. *Cell* **147,** 577–589.

Tzahor, E., and Lassar, A. B. (2001). Wnt signals from the neural tube block ectopic cardiogenesis. *Genes Dev.* **15,** 255–260.

Ueno, S., Weidinger, G., Osugi, T., Kohn, A. D., Golob, J. L., Pabon, L., Reinecke, H., Moon, R. T., and Murry, C. E. (2007). Biphasic role for Wnt/beta-catenin signaling in cardiac specification in zebrafish and embryonic stem cells. *Proc. Natl. Acad. Sci. USA* **104,** 9685–9690.

Vierbuchen, T., Ostermeier, A., Pang, Z. P., Kokubu, Y., Sudhof, T. C., and Wernig, M. (2010). Direct conversion of fibroblasts to functional neurons by defined factors. *Nature* **463,** 1035–1041.

Vincent, S. D., and Buckingham, M. E. (2010). How to make a heart: The origin and regulation of cardiac progenitor cells. *Curr. Top. Dev. Biol.* **90,** 1–41.

von der Hardt, S., Bakkers, J., Inbal, A., Carvalho, L., Solnica-Krezel, L., Heisenberg, C. P., and Hammerschmidt, M. (2007). The BMP gradient of the zebrafish gastrula guides migrating lateral cells by regulating cell-cell adhesion. *Curr. Biol.* **17,** 475–487.

Wang, W., Xue, Y., Zhou, S., Kuo, A., Cairns, B. R., and Crabtree, G. R. (1996). Diversity and specialization of mammalian SWI/SNF complexes. *Genes Dev.* **10,** 2117–2130.

Wang, D., Chang, P. S., Wang, Z., Sutherland, L., Richardson, J. A., Small, E., Krieg, P. A., and Olson, E. N. (2001). Activation of cardiac gene expression by myocardin, a transcriptional cofactor for serum response factor. *Cell* **105,** 851–862.

Warren, L., Manos, P. D., Ahfeldt, T., Loh, Y. H., Li, H., Lau, F., Ebina, W., Mandal, P. K., Smith, Z. D., Meissner, A., Daley, G. Q., Brack, A. S., *et al.* (2010). Highly efficient reprogramming to pluripotency and directed differentiation of human cells with synthetic modified mRNA. *Cell Stem Cell* **7,** 618–630.

Weinhold, B., Schratt, G., Arsenian, S., Berger, J., Kamino, K., Schwarz, H., Ruther, U., and Nordheim, A. (2000). Srf(-/-) ES cells display non-cell-autonomous impairment in mesodermal differentiation. *EMBO J.* **19,** 5835–5844.

Woltjen, K., Michael, I. P., Mohseni, P., Desai, R., Mileikovsky, M., Hamalainen, R., Cowling, R., Wang, W., Liu, P., Gertsenstein, M., Kaji, K., Sung, H. K., *et al.* (2009). piggyBac transposition reprograms fibroblasts to induced pluripotent stem cells. *Nature* **458,** 766–770.

Wu, X., Golden, K., and Bodmer, R. (1995). Heart development in Drosophila requires the segment polarity gene wingless. *Dev. Biol.* **169,** 619–628.

Wu, S. M., Fujiwara, Y., Cibulsky, S. M., Clapham, D. E., Lien, C. L., Schultheiss, T. M., and Orkin, S. H. (2006). Developmental origin of a bipotential myocardial and smooth muscle cell precursor in the mammalian heart. *Cell* **127,** 1137–1150.

Wu, S. M., Chien, K. R., and Mummery, C. (2008). Origins and fates of cardiovascular progenitor cells. *Cell* **132,** 537–543.

Wu, J. I., Lessard, J., and Crabtree, G. R. (2009). Understanding the words of chromatin regulation. *Cell* **136,** 200–206.

Xu, J., Watts, J. A., Pope, S. D., Gadue, P., Kamps, M., Plath, K., Zaret, K. S., and Smale, S. T. (2009). Transcriptional competence and the active marking of tissue-specific enhancers by defined transcription factors in embryonic and induced pluripotent stem cells. *Genes Dev.* **23,** 2824–2838.

Xu, H., Yi, B. A., Wu, H., Bock, C., Gu, H., Lui, K. O., Park, J. H., Shao, Y., Riley, A. K., Domian, I. J., Hu, E., Willette, R., *et al.* (2011). Highly efficient derivation of ventricular cardiomyocytes from induced pluripotent stem cells with a distinct epigenetic signature. *Cell Res* **22**(1), 142–154.

Yi, B. A., Wernet, O., and Chien, K. R. (2010). Pregenerative medicine: Developmental paradigms in the biology of cardiovascular regeneration. *J. Clin. Invest.* **120,** 20–28.

Yin, C., Ciruna, B., and Solnica-Krezel, L. (2009). Convergence and extension movements during vertebrate gastrulation. *Curr. Top. Dev. Biol.* **89,** 163–192.

Yoo, A. S., Staahl, B. T., Chen, L., and Crabtree, G. R. (2009). MicroRNA-mediated switching of chromatin-remodelling complexes in neural development. *Nature* **460,** 642–646.

Yue, Q., Wagstaff, L., Yang, X., Weijer, C., and Munsterberg, A. (2008). Wnt3a-mediated chemorepulsion controls movement patterns of cardiac progenitors and requires RhoA function. *Development* **135,** 1029–1037.

Yutzey, K. E., and Kirby, M. L. (2002). Wherefore heart thou? Embryonic origins of cardiogenic mesoderm. *Dev. Dyn.* **223,** 307–320.

Zeng, X. X., Wilm, T. P., Sepich, D. S., and Solnica-Krezel, L. (2007). Apelin and its receptor control heart field formation during zebrafish gastrulation. *Dev. Cell* **12,** 391–402.

Zhao, R., Watt, A. J., Battle, M. A., Li, J., Bondow, B. J., and Duncan, S. A. (2008). Loss of both GATA4 and GATA6 blocks cardiac myocyte differentiation and results in acardia in mice. *Dev. Biol.* **317,** 614–619.

CHAPTER TWO

The Second Heart Field

Robert G. Kelly

Contents

Abstract

Ten years ago, a population of cardiac progenitor cells was identified in pharyngeal mesoderm that gives rise to a major part of the amniote heart. These multipotent progenitor cells, termed the second heart field (SHF), contribute progressively to the poles of the elongating heart tube during looping morphogenesis, giving rise to myocardium, smooth muscle, and endothelial cells. Research into the mechanisms of SHF development has contributed significantly to our understanding of the properties of cardiac progenitor cells and the origins of congenital heart defects. Here recent data concerning the regulation, clinically relevant subpopulations, evolution and lineage relationships of the SHF are reviewed. Proliferation and differentiation of SHF cells are controlled by multiple intercellular signaling pathways and a transcriptional

Developmental Biology Institute of Marseilles-Luminy, Aix-Marseille Université, CNRS UMR 7288, Marseilles, France

Current Topics in Developmental Biology, Volume 100
ISSN 0070-2153, DOI: 10.1016/B978-0-12-387786-4.00002-6

regulatory network that is beginning to be elucidated. Perturbation of SHF development results in common forms of congenital heart defects and particular progenitor cell subpopulations are highly relevant clinically, including cells giving rise to myocardium at the base of the pulmonary trunk and the interatrial septum. A SHF has recently been identified in amphibian, fish, and agnathan embryos, highlighting the important contribution of these cells to the evolution of the vertebrate heart. Finally, SHF-derived parts of the heart share a lineage relationship with craniofacial skeletal muscles revealing that these progenitor cells belong to a broad cardiocraniofacial field of pharyngeal mesoderm. Investigation of the mechanisms underlying the dynamic process of SHF deployment is likely to yield further insights into cardiac development and pathology.

1. Introduction

The complexity of heart morphogenesis is reflected in the large spectrum of congenital heart defects that account for the majority of life-threatening congenital anomalies at birth. Our understanding of heart development was significantly modified by three papers published in 2001 documenting the existence of cardiac progenitor cells in pharyngeal mesoderm that contribute to rapid elongation of the heart tube and give rise to a major part of the definitive heart (Kelly *et al.*, 2001; Mjaatvedt *et al.*, 2001; Waldo *et al.*, 2001). Over the intervening decade, these progenitor cells, now known as the second heart field (SHF), have been the focus of intense investigation. The impetus for this research stems from the major importance of these cells in studies of cardiac progenitor cell regulation and the etiology of congenital heart defects (Epstein, 2010; Musunuru *et al.*, 2010). Here, the discovery and refinement of the SHF model are briefly reviewed, followed by discussion of the topography and regulation of these multipotent cardiovascular progenitor cells in the early embryo. In order to minimize redundancy with prior reviews (Buckingham *et al.*, 2005; Dyer and Kirby, 2009a; Evans *et al.*, 2010; Kelly and Evans, 2010; Rochais *et al.*, 2009a; Vincent and Buckingham, 2010), established data are summarized and emphasis is placed on recent findings. We then focus on two particular subpopulations of the SHF that are highly relevant clinically, as perturbation of their development results in common forms of human congenital heart defects, including conotruncal and atrial septal defects (Hoffmann *et al.*, 2009; Parisot *et al.*, 2011). The evolutionary implications of the SHF model are also reviewed, in particular with respect to recent data characterizing this population of cells in amphibians and fish. Finally, the SHF is discussed in the context of a broad cardiocraniofacial developmental field of pharyngeal mesoderm, giving rise to skeletal muscles of the head as well as the heart.

2. The Discovery of the Second Heart Field

The embryonic heart forms by the concomitant processes of heart tube elongation and rightward looping following initial myocardial differentiation in the cardiac crescent. During these steps, the length of the heart tube increases dramatically. Maria Victoria de la Cruz and colleagues used iron oxide particle labeling experiments to investigate this phenomenon in the 1970s. These experiments revealed that the distal right ventricle and outflow tract (OFT) are late added components of the avian heart and that cardiac growth occurs by addition of cells lying outside the early heart (de la Cruz *et al.*, 1977). In the mouse, Viragh and Challice observed ongoing myocardial differentiation at the arterial pole of the heart tube up to midgestation, 3 days after initial differentiation of cardiomyocytes in the cardiac crescent (Viragh and Challice, 1973). Similarly, analysis of rat and human embryological sections suggested that OFT extension occurred by incorporation of cells from a distal infundibular prolongation in the dorsal pericardial wall (deVries, 1981). Despite this prior evidence for addition of progenitor cells from outside the early heart tube, the alternate view that all the regions of the definitive heart were prefigured in the linear heart tube was widely assumed in the 1990s.

The three studies published in 2001 identified overlapping progenitor cell populations in pharyngeal mesoderm that contribute to elongation of the embryonic OFT. Two focused on the avian heart: Markwald and colleagues identified the anterior heart-forming field, predicted by de la Cruz, in mesoderm contiguous with the arterial pole of the heart (Mjaatvedt *et al.*, 2001). Kirby and colleagues identified progenitor cells in the dorsal pericardial wall that expressed *Nkx2.5* and *Gata4* as well as HNK1 during addition to the OFT (Waldo *et al.*, 2001). Fluorescent labeling confirmed movement of cells into the elongating heart and the Kirby group showed that this process required BMP driven differentiation at the distal pole of the heart tube (Waldo *et al.*, 2001). Investigation of a position effect transgene in the mouse revealed transgene transcription in pharyngeal mesoderm contiguous with the heart tube and perdurance of β-galactosidase reporter activity in OFT and right ventricular myocardium, identifying *Fgf10* at the site of transgene integration as the first molecular marker of the murine SHF (Kelly *et al.*, 2001). DiI labeling revealed movement of cells from pharyngeal mesoderm into the murine heart tube. Although contiguous with the linear heart tube, the progenitor cell populations identified in these studies lay outside the region of precardiac mesoderm shown by fate mapping in avian embryos to give rise to the early heart tube (Dyer and Kirby, 2009a; Rosenquist and DeHaan, 1966). The term anterior heart field was thus adopted, with the subpopulation in the dorsal pericardial wall giving rise to the distal OFT dubbed the secondary heart field

(Abu-Issa *et al.*, 2004; Dyer and Kirby, 2009a; Kelly and Buckingham, 2002; Waldo *et al.*, 2001). The anterior heart field is now considered to be a subpopulation of the SHF, a larger field of progenitor cells in pharyngeal mesoderm contributing not only to the arterial pole of the heart but also to the venous pole after the linear heart tube stage (Buckingham *et al.*, 2005; Dyer and Kirby, 2009a).

3. Refinement of the Second Heart Field Model

Work in the intervening decade has extensively refined the SHF model, by which a reservoir of progenitor cells in pharyngeal mesoderm progressively adds to the scaffold of the early heart tube, giving rise to the major part of the definitive heart. In this section, we will review the localization, gene expression profile, and potency of the SHF, as well as the lineage relationship between the SHF and cells giving rise to the linear heart tube.

Current understanding of SHF development can be summarized as follows. SHF cells originate in medial splanchnic mesoderm in the anterior region of the embryo, adjacent to pharyngeal endoderm (Fig. 2.1). Immediately contiguous lateral splanchnic mesoderm gives rise to the cardiac crescent and early heart tube (Schoenwolf *et al.*, 2009). As the linear heart tube forms, the dorsal mesocardium, the mesentry by which the heart tube is suspended in the ventral region of the embryo, beaks down, isolating the medially positioned progenitor cells of the SHF in the dorsal pericardial wall, underlying ventral pharyngeal endoderm. Contiguity with the heart tube is maintained at the arterial and venous poles, through which SHF cells contribute during looping morphogenesis. The position of progenitor cells along the medial lateral axis of the early embryo thus defines the localization of descendant myocytes in the definitive heart (Abu-Issa and Kirby, 2008; Kelly and Buckingham, 2002). The SHF is also patterned along the anterior–posterior embryonic axis. Anterior-SHF cells give rise to the right ventricle and OFT at the arterial pole, as has been shown by explant and fluorescent labeling experiments in avian and mouse embryos and genetic tracing experiments using Cre recombinase (Cai *et al.*, 2003; Guo *et al.*, 2011; Kelly *et al.*, 2001; Mjaatvedt *et al.*, 2001; Rana *et al.*, 2007; Robertson *et al.*, 2007; Tirosh-Finkel *et al.*, 2006; Verzi *et al.*, 2005; Waldo *et al.*, 2001; Zaffran *et al.*, 2004). SHF cells have been estimated to move at 70 μm an hour toward the poles of the avian heart tube from a proliferative centre in the dorsal pericardial wall (Van den berg *et al.*, 2009). OFT progenitor cells are juxtaposed with neural crest-derived cells in the pharyngeal region, an apposition that plays an important role in regulating arterial pole morphogenesis (Hutson and Kirby, 2007). The posterior

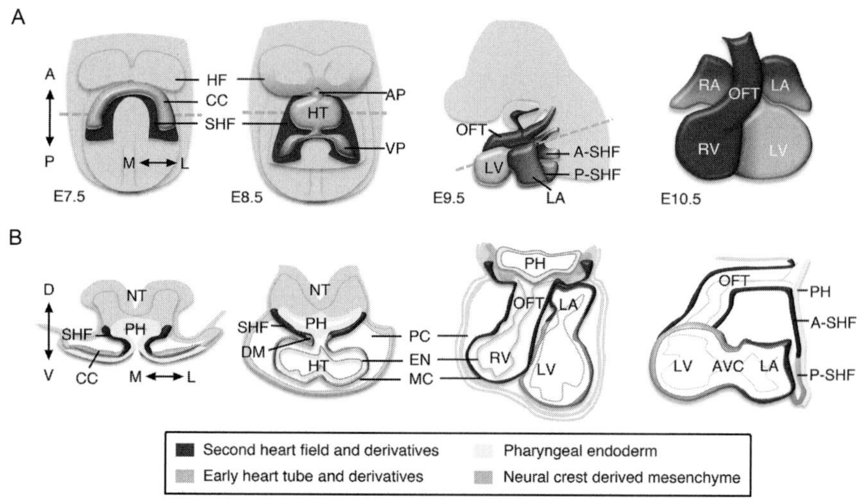

Figure 2.1 Topography of the second heart field. (A) Cartoons showing the stages of mouse heart development during which second heart field progenitors are added to the heart. From left to right: embryonic day (E) 7.5 (cardiac crescent, ventral view), E8.5 (early heart tube, ventral view), E9.5 (looped heart, left lateral view), and E10.5 (ventral view). Second heart heart field progenitor cells and their contribution to the heart are shown in dark blue and the early heart tube giving rise to the left ventricle in pink. Anterior (A) posterior (P) and medial (M) lateral (L) embryonic axes are indicated. (B) Schematized histological sections showing the location of second heart field cells and their myocardial derivatives (dark blue) at the stages shown in (A) relative to pharyngeal endoderm (yellow), myoctes derived from the early heart tube (pink), and neural crest-derived mesenchyme (orange). The first three panels show transverse sections at the levels indicated by the dotted lines in (A). The right hand panel shows a mid-saggital E9.5 section. Dorsal (D) ventral (V) and medial (M) lateral (L) embryonic axes are indicated. CC, cardiac crescent; SHF, second heart field; HF, head fold; HT, heart tube, AP, arterial pole; VP, venous pole; OFT, outflow tract; LV, left ventricle; LA, left atrium; RV, right ventricle; RA, right atrium; A-SHF, anterior component of the second heart field; P-SHF, posterior component of the second heart field; NT, neural tube; PH, pharynx; PC, pericardial cavity; DM, dorsal mesocardium; EN, enodcardium; MC, myocardium. (See Color Insert.)

component of the SHF contributes to atrial, atrial septal, and venous pole myocardium. The boundary between anterior and posterior SHF progenitor cell populations in the dorsal pericardial wall is currently poorly defined. The linear heart tube is thought to give rise predominantly to the left ventricle (Buckingham *et al.*, 2005); however, this contribution may in fact be limited to the apical part of the left ventricle and left side of the interventricular septum, since much of the left ventricular free wall has been shown to originate from *Tbx2* expressing cells in the atrioventricular region (Aanhaanen *et al.*, 2009).

The SHF differs from cells giving rise to the linear heart tube by two defining properties: continued proliferation and differentiation delay (Rochais *et al.*, 2009a). These properties allow the progressive accretion of SHF cells during heart tube elongation and are controlled by the dynamic pharyngeal signaling environment and the distinct pharyngeal mesodermal transcriptional program of the SHF. In addition to *Fgf10*, SHF cells are characterized by expression of genes encoding Fgf8 and the transcription factors Isl1, Tbx1, Prdm1, and Six1 (Cai *et al.*, 2003; Chen *et al.*, 2009; Guo *et al.*, 2011; Ilagan *et al.*, 2006; Kelly *et al.*, 2001; Robertson *et al.*, 2007). Expression of these genes is downregulated on differentiation and regulators of cardiac specification such as Nkx2.5, Gata4, and Mef2c are activated as progenitor cells approach the heart tube (Verzi *et al.*, 2005; Waldo *et al.*, 2001). Cre-labeling experiments have been highly informative in refining the SHF model and in manipulating gene function in the SHF; in particular, an *Isl1 Cre* allele expressed in right ventricular, OFT, and atrial progenitor cells and a *Mef2c* enhancer driven Cre transgene expressed in cells giving rise to the right ventricle, OFT, and interventricular septum, as well as atrial septal structures (Cai *et al.*, 2003; Goddeeris *et al.*, 2008; Verzi *et al.*, 2005). In an advance toward defining a more complete SHF transcriptome, gene expression profiling of embryonic stem (ES) cells expressing this *Mef2c* enhancer, but not a cardiac restricted *Nkx2.5* regulatory element, has identified a distinct molecular signature for the SHF (Domian *et al.*, 2009).

During ES cell differentiation, *Isl1* is activated in a subset of cells expressing the anterior mesoderm transcription factor *Mesp1*, suggesting that early cardiac progenitor lineage decisions can be studied in this system (Bondue *et al.*, 2011). ES-derived cells expressing *Isl1* and *Tbx1* are multipotent cardiovascular progenitor cells that give rise to smooth muscle and endothelial cells as well as myocardium (Bu *et al.*, 2009; Chen *et al.*, 2009; Laugwitz *et al.*, 2005). Mulitpotent progenitor cells are maintained in the SHF throughout the process of heart tube elongation and have been isolated directly from the avian SHF (Hutson *et al.*, 2010). Indeed, SHF-derived smooth muscle at the base of the aorta and pulmonary trunk appears to represent the terminal contribution of the SHF at the arterial pole of the heart (Sun *et al.*, 2007; Waldo *et al.*, 2005). Whether all cells in the SHF are multipotent or such cells are restricted to particular domains of the SHF remains to be seen. In the case of endocardial and myocardial lineages a recent study has revealed lineage heterogeneity in the endocardium and suggests that at least part of the endocardium is derived from distinct vascular endothelial progenitor cells in pharyngeal mesoderm (Milgrom-Hoffman *et al.*, 2011). Cre experiments have revealed that the epicardium and epicardially derived cells originate from *Isl1*, *Nkx2.5*, and *Six1* positive progenitor cells suggesting a potentially common origin with the SHF (Guo *et al.*, 2011; Zhou *et al.*, 2008). Among the different stem cells populations identified in the later heart, *Isl1* positive cells have been observed in the fetal and early postnatal rodent and human heart and may represent a population of resident cardiovascular stem cells derived from residual SHF cells (Bu *et al.*, 2009; Laugwitz *et al.*, 2005). Understanding how lineage choices are regulated in the SHF thus has major implications for approaches aimed at cardiac repair by manipulation of cardiac progenitor cells (Musunuru *et al.*, 2010).

The early contiguity between cells giving rise to the linear heart tube and the SHF has led to debate as to the extent to which the SHF should be considered to be a distinct progenitor cell population (Abu-Issa *et al.*, 2004; Moorman *et al.*, 2007; Xavier-Neto *et al.*, 2010). Indeed, *Isl1* is broadly expressed in early cardiac progenitors in the chick and there is evidence that murine *Isl1* is transiently expressed in cells giving rise to the linear heart tube (Ma *et al.*, 2008; Prall *et al.*, 2007; Yuan and Schoenwolf, 2000). Prior to breakdown of the dorsal mesocardium SHF cells can in fact contribute along the length of the heart tube, giving rise to the inner curvature of the embryonic heart (Kelly and Buckingham, 2002). However, in addition to differences in gene expression, arguments based on clonal analysis and evolutionary studies further suggest that late-differentiating pharyngeal mesodermal progenitor cells differ from those giving rise to the linear heart tube. Retrospective analysis of the distribution of clonally related myocytes in the developing heart has played a major role in refining the SHF model and led to the identification of two lineages of progenitor cells contributing to the embryonic heart (Meilhac *et al.*, 2004). The first lineage gives rise to the left ventricle and atrial myocardium, with a minor contribution to the right ventricle, while the second lineage gives rise to the OFT, right ventricle, and atrial myocardium, corresponding to SHF-derived parts of the heart (Meilhac *et al.*, 2004). These two lineages diverge from a common pool of cardiac progenitor cells at or prior to the time of gastrulation (Meilhac *et al.*, 2004). The onset of Isl1 expression within a subset of *Mesp1* positive differentiating ES cells and the results of labeling experiments in pregastrulation zebrafish embryos also suggest the existence of pancardiac progenitor cells in the early embryo (Bondue *et al.*, 2011; Hami *et al.*, 2011; Lazic and Scott, 2011). Recent results using clonal analysis have revealed that SHF-derived parts of the heart share a lineage relationship with craniofacial skeletal muscles and that separation of these lineages occurs after separation of the SHF from cells giving rise to the left ventricle (Lescroart *et al.*, 2010). As we shall see in Section 7, the SHF is part of a cardiocraniofacial field of pharyngeal mesoderm and appears to have been assimilated into the heart from *Islet* expressing skeletal muscle progenitor cells during vertebrate evolution (Stolfi *et al.*, 2010).

4. Regulation of Second Heart Field Deployment

The properties of continued proliferation and differentiation delay that define cardiac progenitor cells in the SHF are regulated by a complex network of intercellular signals that have been extensively studied over the past 5 years and collectively describe the niche of the SHF in the caudal pharyngeal region. These signals both regulate and are regulated by transcription factors in pharyngeal mesoderm and adjacent cell types, including pharyngeal epithelia and neural crest-derived cells, as well as autocrine signals from pharyngeal mesoderm itself (Fig. 2.2). These regulatory networks have been reviewed in detail elsewhere (Dyer and Kirby, 2009a; Rochais *et al.*, 2009a; Vincent and Buckingham, 2010) and are summarized here with emphasis on recent findings.

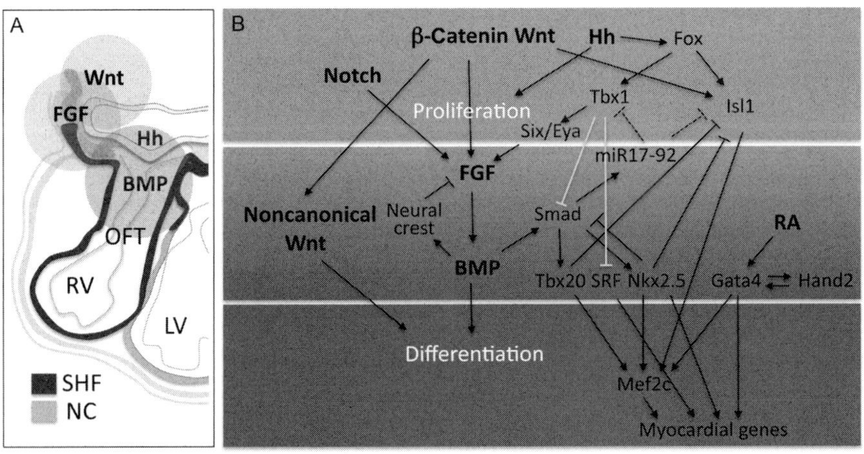

Figure 2.2 The regulation of second heart field development. (A) Cartoon showing zones of Wnt, fibroblast growth factor (FGF), Hedgehog (Hh), and bone morphogenetic protein (BMP) signaling at the arterial pole of the mouse heart. LV, left ventricle; RV, right ventricle; OFT, outflow tract; SHF, second heart field; NC, neural crest-derived cells. (B) Schema showing the network of major signaling pathways and regulatory genes impacting on progressive second heart field development during the transition from proliferating progenitor cell (top) to differentiated cardiomyocyte (bottom). Note the central position of Isl1 and Tbx1 in regulating the proliferative progenitor cell state (top), the pivotal position of FGF/BMP antagonism in controlling the balance between proliferation and differentiation (middle), and the activation of the cardiomyogenic program by a network of interacting transcription factors (bottom). Gray lines, direct protein interactions; dotted lines, microRNA silencing. See text for details. (For color version of this figure, the reader is referred to the Web version of this chapter.)

4.1. Intercellular signaling pathways regulating second heart field development

Early activation of the cardiomyogenic transcriptional program in anterior lateral splanchnic mesoderm is mediated by positive and negative signals from adjacent endoderm, overlying ectoderm, and midline structures, including bone morphogenetic protein (BMP), fibroblast growth factor (FGF), and canonical and noncanonical Wnt signaling (Evans *et al.*, 2010). These signals induce *Mesp1* positive mesodermal progenitor cells to activate cardiac transcription factors and epigenetic regulators, including *Isl1*, *Tbx5*, *Nkx2.5*, *Baf60c*, and *Gata4* that combinatorially drive cardiomyogenesis (Bruneau, 2002; van Weerd *et al.*, 2011). As the cardiac crescent and linear heart tube form, SHF cells in medial splanchnic mesoderm remain in contact with pharyngeal endoderm and their continued proliferation and delayed differentiation is regulated by canonical Wnt, FGF, and Hedgehog (Hh) signaling pathways (Fig. 2.2). The roles of these signaling pathways have been dissected using experimental manipulation in avian embryos and conditional mutagenesis in the mouse. Recent studies have addressed the relevant receptors and ligands of these pathways involved in SHF development. Wnt/β-catenin signaling plays an upstream role in promoting proliferation of SHF cells *in vivo*, in addition to promoting expansion and maintaining multipotency of Isl1-positive ES-derived cardiac progenitor cells (reviewed by Tzahor, 2007; Bu *et al.*, 2009). The role of canonical Wnt signaling in regulating *Isl1* expression and SHF proliferation and survival appears to be regulated by the Wnt coreceptor Lrp6 (Song *et al.*, 2010). While the identity and source of the Wnt ligands regulating the anterior SHF remain to be identified, Wnt2 in mesoderm adjacent to the inflow tract drives progenitor cell proliferation and atrial development in a feed-forward loop with the transcription factor Gata6 (Tian *et al.*, 2010). Sonic hedgehog, expressed in overlying pharyngeal endoderm, is required for maximal proliferation in the SHF (Dyer and Kirby, 2009b). In addition, Hh signaling impacts on neural crest development and survival as revealed by loss of function of Sonic hedgehog and conditional inactivation of the Smoothened hedgehog receptor in the mouse (Dyer and Kirby, 2009b; Goddeeris *et al.*, 2007; Hildreth *et al.*, 2009; Washington Smoak *et al.*, 2005). Fgf8 is thought to be the major FGF ligand driving SHF proliferation, with important contributions from Fgf10 and Fgf3 revealed by analysis of compound mutant embryos (Ilagan *et al.*, 2006; Park *et al.*, 2006, 2008; Urness *et al.*, 2011; Watanabe *et al.*, 2010). These FGF ligands are expressed in different cell types including pharyngeal epithelia and mesoderm and their expression is downstream of canonical Wnt signaling (Cohen *et al.*, 2007; Lin *et al.*, 2007). Conditional mutagenesis has identified mesodermal Fgf8 as an autocrine signal critical for heart tube elongation (Park *et al.*, 2008). Although SHF deployment occurs normally in *Fgf10* null embryos, loss of *Fgf10* alleles increases the severity of cardiac defects in embryos lacking *Fgf8* in the SHF (Marguerie *et al.*, 2006; Watanabe *et al.*, 2010). In *Fgf10 Fgf3* double mutant embryos, a transient decrease of *Isl1* and *Fgf10* expression is observed in the SHF at E8.5, followed by a later decrease in the maintenance of *Nkx2.5* expression, suggesting stage specific redundant roles for these FGF ligands (Urness *et al.*, 2011). Interrupted *Isl1* expression in these embryos leads to disordered SHF deployment at both poles of the heart tube, resulting in a shortened OFT and hypoplastic dorsomesenchymal protrusion (Urness *et al.*, 2011). Interestingly, in contrast to a proproliferative role in myocardial progenitor cells, FGF signaling is required for smooth muscle differentiation (Hutson *et al.*, 2010). *Fgf8* expression in the SHF is positively regulated by Notch signaling, through mesodermal expression of the Jag1 ligand (High *et al.*, 2009). The Notch target gene *Hes1*, expressed in the SHF and pharyngeal epithelia, is required for maximal proliferation in the SHF and loss of *Hes1* leads to precocious upregulation of the cyclin-dependent kinase inhibitor p27kip1 (Rochais *et al.*, 2009b).

As SHF cells approach the ventral region of the embryo and distal heart tube, they are exposed to BMP and noncanonical Wnt signals that drive differentiation (Fig. 2.2). Pro-proliferation FGF and pro-differentiation BMP signaling appear to identify core antagonistic signaling pathways regulating progressive addition of SHF cells to the heart tube (Hutson *et al.*, 2010; Tirosh-Finkel *et al.*, 2010). The balance between these pathways is influenced by neural crest cell invasion of the caudal pharyngeal region that plays a critical role in braking the proliferative effect of FGF signaling on SHF cells, in addition to a direct later role in OFT septation (Hutson *et al.*, 2006; Yelbuz *et al.*, 2002). Proliferation in the SHF is elevated after neural crest ablation, an effect that is rescued on concomitant downregulation of *Fgf8* expression (Hutson *et al.*, 2006). BMP induced targets in neural crest cells, in particular the transcription factor encoding genes *Msx1* and *Msx2*, lead through as yet unknown mechanisms to downregulation of FGF ligand expression and decreased FGF signaling (Tirosh-Finkel *et al.*, 2010); Consistent with such a model, BMP receptor and Smad signaling in neural crest cells are required for maximal OFT elongation (Jia *et al.*, 2007; Stottmann *et al.*, 2004). Msx1 and Msx2 are required for *Hand2* expression and cell survival in the SHF (Chen *et al.*, 2007). BMP signal reception is also required in *Isl1* expressing cells, promoting cardiac differentiation through transcriptional activation of microRNA 17-92, that targets SHF regulatory genes including *Isl1* and *Tbx1* (Wang *et al.*, 2010; Yang *et al.*, 2006). BMP signaling also modulates the proliferative effect of Hh signaling in the SHF (Dyer et al, 2010). Retinoic acid, noncanonical Wnt, and TGFβ signaling pathways have also been shown to be play complex roles in regulating SHF deployment. Retinoic acid, like many of the pathways involved, appears to play a series of roles, first limiting the caudal extension of the SHF along the anterior–posterior embryonic axis and later promoting *Gata4* expression and expansion of *Mef2c* enhancer expressing cells in the dorsal pericardial wall (Li *et al.*, 2010; Ryckebusch *et al.*, 2008; Sirbu *et al.*, 2008). Noncanonical Wnt signaling promotes myocardial differentiation at the arterial pole of the heart as revealed by *Wnt5a* and *Wnt11* null phenotypes (Schleiffarth *et al.*, 2007; Zhou *et al.*, 2007). Wnt11 acts downstream of Wnt/β-catenin signaling in a regulatory cascade involving ATF2 transcriptional activation of TGFβ2, required for OFT cushion development (Zhou *et al.*, 2007). Studies in the zebrafish (see below) have recently implicated TGFβ signaling in regulating proliferation and cell survival in the SHF (Zhou *et al.*, 2011); whether this role is conserved in amniotes is unknown.

4.2. Transcriptional control of second heart field development: Toward a second heart field gene regulatory network

Transcription factors such as Isl1 and Tbx1 play central roles in integrating the output of different signaling pathways during SHF development (Fig. 2.2B). Isl1 is required for development beyond the linear heart tube stage and regulates signaling pathway ligand and receptor gene expression in the pharyngeal region, and is itself a target of Wnt signaling (Cai *et al.*, 2003; Lin *et al.*, 2007). Recent data from ES cell studies implicates Isl1, together with Mesp1, in driving early cardiac specification (Bondue *et al.*, 2011; Kwon *et al.*, 2009). Tbx1 (discussed further in Section 5.1) is regulated by Hh and canonical Wnt signaling, promoting FGF ligand expression and antagonizing BMP and retinoic acid signaling (Fulcoli *et al.*, 2009; Garg *et al.*, 2001; Hu *et al.*, 2004; Huh and Ornitz, 2010; Roberts *et al.*, 2005; Vitelli *et al.*, 2002). A growing number of additional transcription factors from diverse families, including forkhead, homeobox, T-box, Gata, zinc-finger, and bHLH factors, have been found to play important roles in SHF development and single and compound mutations in these genes impact on SHF deployment. Foxc1 and Foxc2 are required for arterial pole development and have been found, together with Foxa2, to transactivate *Tbx1* regulatory elements, although the mutation of a predicted Fox binding site *in vivo* does not modulate transcription from the *Tbx1* locus (Maeda *et al.*, 2006; Seo and Kume, 2006; Yamagishi *et al.*, 2003; Zhang and Baldini, 2010). Foxc1 and Foxc2, together with Foxf1, also regulate *Isl1* through a downstream enhancer, together with Gata4 (Kang *et al.*, 2009; Kappen and Salbaum, 2009). *Mef2c* expression in the SHF is regulated by two linked enhancer elements, one activated by Isl1 and Gata4 and one by Foxh1 and Nkx2.5, revealing the importance of feedforward transcriptional circuits in the SHF (Dodou *et al.*, 2004; Von Both *et al.*, 2004). Loss of Nkx2.5, a pleiotropic regulator of cardiac development, leads to a failure of SHF deployment (Prall *et al.*, 2007). Upregulated expression of *Bmp2* and

Jarid2, encoding a histone methyl transferase, are associated with decreased proliferation in the SHF and ectopic cardiac specification in *Nkx2.5* mutant embryos (Prall *et al.*, 2007; Barth *et al.*, 2010). Tbx20 has been shown to activate the *Mef2c* anterior-SHF enhancer element synergistically with Isl1 and Gata4, involving a physical interaction with Isl1; similar interactions regulate the activity of an *Nkx2.5* SHF enhancer (Takeuchi *et al.*, 2005). *Tbx20* and *Nkx2.5* null embryos fail to downregulate *Isl1* in cardiomyocytes, indicating a shift in the role of these transcription factors on differentiation (Cai *et al.*, 2005; Prall *et al.*, 2007). The zinc-finger transcription factor Prdm1 is expressed in the anterior-SHF and pharyngeal endoderm; loss of *Prdm1* leads to cardiac defects associated with decreased *Tbx1* and *Fgf8* expression (Robertson *et al.*, 2007). Hand2 has recently been shown to be required within the SHF for right ventricular progenitor cell survival, in a reinforcing regulatory loop with Gata4 (Tsuchihashi *et al.*, 2011). Graded loss of Hand2 function using a series of SHF Cre lines leads to a spectrum of conotruncal congenital heart defects (Tsuchihashi *et al.*, 2011). Finally, the homeodomain transcription factor Six1 and its coactivator Eya1 are required for proliferation within the SHF and survival of adjacent tissues including pharyngeal epithelia and neural crest-derived mesenchyme (Guo *et al.*, 2011). Genetic interaction studies and promoter analysis position the Six/Eya transcriptional complex downstream of *Tbx1* and *Nkx2.5* and directly upstream of *Fgf8* expression (Guo *et al.*, 2011). Ongoing studies will lead to a more profound understanding of how these transcription factors integrate signaling inputs and outputs during SHF development through reconstruction of a SHF gene regulatory network.

5. Clinically Relevant Subdomains of the Second Heart Field

Direct or indirect perturbation of SHF development results in congenital heart defects. At one extreme, loss of function of *Isl1* or early ablation of key regulatory genes such as *Fgf8* results in severe heart tube elongation defects and midgestation lethality (Cai *et al.*, 2003; Park *et al.*, 2008). Importantly, however, milder defects in heart tube elongation result in alignment defects and failure to separate systemic and pulmonary circulations at birth, corresponding to a spectrum of common congenital heart anomalies seen in human patients. These include conotruncal defects resulting from OFT hypoplasia and failure of the ascending aorta to make the critical connection with the left ventricle, such as ventricular septal defects, common arterial trunk, pulmonary atresia, tetralogy of Fallot, and double outlet right ventricle, as well as venous pole anomalies including atrial and atrioventricular septal defects. In the mouse, hypomorphic alleles of genes required for normal SHF development, such as *Tbx1* or *Nkx2.5*, can result in a similar spectrum of conotruncal defects (Prall *et al.*, 2007; Zhang and Baldini, 2008). However, specific subpopulations of SHF cells contributing to the terminal stages of SHF development appear to be particularly relevant with respect to the etiology of common congenital heart defects. Here we discuss two clinically important subpopulations of the SHF giving rise to subpulmonary and atrial septal myocardium.

5.1. Future subpulmonary myocardium and conotruncal congenital heart defects

Underdevelopment of myocardium at the base of the pulmonary trunk is considered to be the primary cause of OFT alignment defects such as tetralogy of Fallot characterized by overriding aorta, ventricular septal defect, and pulmonary hypoplasia (Di Felice and Zummo, 2009; Van Praagh, 2009). Future subpulmonary myocardium is prepatterned in the SHF and its development is regulated by the transcription factor Tbx1 (Fig. 2.3A; Parisot *et al.*, 2011). *Tbx1* is the major candidate gene for DiGeorge syndrome in man, characterized by craniofacial and conotruncal congenital heart defects, and regulates differentiation delay and proliferation in the SHF (Chen *et al.*, 2009; Jerome and Papaioannou, 2001; Liao *et al.*, 2008; Xu *et al.*, 2004; Yagi *et al.*, 2003). The SHF of *Tbx1* null embryos is hypoplastic and expression of genes including *Isl1* and *Fgf8* are downregulated, while direct interaction between Tbx1 and the transcription factors SRF and Smad1 additionally contributes to the role of Tbx1 in differentiation delay (Fig. 2.2; Chen *et al.*, 2009; Fulcoli *et al.*, 2009; Kelly and Papaioannou, 2007; Liao *et al.*, 2008; Vitelli *et al.*, 2002). Tbx1 interaction targets SRF for degradation in the SHF, priming these cells for differentiation when Tbx1 is downregulated; Smad1 interaction counters pro-differentiation BMP signaling (Chen *et al.*, 2009; Fulcoli *et al.*, 2009). In a regulatory loop, BMP signaling through microRNA 17-92 ensures *Tbx1* downregulation on differentiation (Wang *et al.*, 2010). In addition, Tbx1 targets mediate the intracellular response to FGF signaling in pharyngeal mesoderm (Vitelli *et al.*, 2010). *Cre*-labeling experiments have shown that *Tbx1* is expressed in a subpopulation of SHF cells giving rise to the inferior and lateral walls of the midgestation OFT and later to myocardium at the base of the pulmonary trunk (Huynh *et al.*, 2007; Maeda *et al.*, 2006). Analysis of regionalized transgene expression domains has confirmed that precisely this subdomain of the SHF is affected in *Tbx1* null hearts (Theveniau-Ruissy *et al.*, 2008). Loss of subpulmonary myocardium in mutant hearts is associated with absence of *Sema3c* expression, encoding a neurovascular guidance molecule, and abnormal proximal coronary artery patterning (Fig. 2.3A; Theveniau-Ruissy *et al.*, 2008).

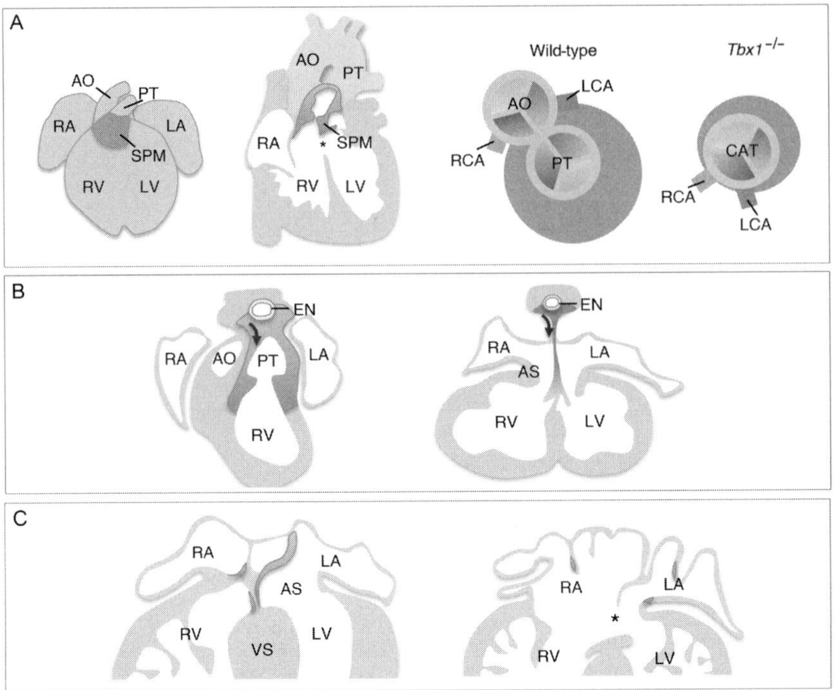

Figure 2.3 Clinically relevant subdomains of the second heart field. (A) Subpulmonary myocardium at the base of the outlet of the right ventricle (blue) in a mouse heart at E14.5 and a human heart with tetralogy of Fallot; note the hypoplastic right ventricular outlet and ventricular septal defect (asterisk). Subpulmonary myocardium is severely reduced in the common arterial trunk of *Tbx1* mutant hearts; note the abnormal ventrally positioned proximal left coronary artery (LCA) compared to the wild-type situation. (B) Cartoons showing the migration (arrows) of Hedgehog responsive cells (blue) from mesoderm adjacent to pharyngeal endoderm (yellow) into the base of the pulmonary trunk (left) and interatrial septum (right). (C) Schematized histological sections showing the contribution of progenitor cells expressing the hedgehog target gene *Gli1* to the interatrial septum (left) and the failure of atrial and atrioventricular septation (asterisk) when the hedgehog receptor smoothened is deleted from *Gli* expressing cells at early developmental stages (right). (B) and (C) adapted from Hoffmann *et al.* (2009) with permission. AO, aorta; PT, pulmonary trunk; RA, right atrium; RV, right ventricle; LA, left atrium; LV, left ventricle; SPM, subpulmonary myocardium; RCA, right coronary artery; CAT, common arterial trunk; EN, pharyngeal endoderm; AS, atrial septum; VS, ventricular septum. (See Color Insert.)

Two signaling pathways have been shown to operate specifically in this subpopulation of OFT progenitor cells. The inferior OFT wall is retinoic acid responsive and analysis of compound retinoic acid receptor mutant embryos has revealed a role in replenishment of OFT progenitor cells during SHF deployment (Li *et al.*, 2010). Retinoic acid signaling is required for Mef2c anterior-SHF enhancer maintenance and robust Gata4 expression in the SHF; mutant embryos are characterized by underdevelopment of the distal OFT and corresponding expansion of the zone of endocardial EMT, normally restricted to the proximal OFT, into the distal region of the mutant OFT (Li *et al.*, 2010). Future subpulmonary myocardial cells are also exposed to Hh signaling, revealed by Cre labeling of cells expressing the Hh target gene Gli1 (Hoffmann *et al.*, 2009). Future subaortic myocardium, in contrast, appears to be derived from the superior wall of the midgestation OFT and is characterized by elevated cell death in avian embryos, required for alignment of the ascending aorta with the left ventricle, and expression of an enhancer trap transgene integrated upstream of the gene encoding the transcription factor Hes1 (Bajolle *et al.*, 2008; Watanabe *et al.*, 1998). Analysis of the pattern of activation of this transgene at embryonic day 8, together with that of a second transgene expressed in subpulmonary myocardial progenitor cells through integration proximal to *Sema3c*, provides further evidence that different regions of the later heart are prefigured in the SHF (Bajolle *et al.*, 2008).

Nested expression of the anterior Hox genes *Hoxb1*, *Hoxa1*, and *Hoxa3*, has been documented in subpulmonary myocardial progenitor cells, suggesting that these developmental patterning genes may play a role in prepatterning progenitor cells in the SHF along the anterior–posterior embryonic axis (Bertrand *et al.*, 2011). In contrast, labeling and ablation experiments in avians have suggested that the right hand side of the secondary heart field gives rise to myocardium at the base of the pulmonary trunk, implying involvement of embryonic laterality in SHF regulation (Ward *et al.*, 2005). This is thought to contribute to rotation of SHF-derived myocardium as cells add to arterial pole (Bajolle *et al.*, 2006; Ward *et al.*, 2005). In contrast, addition of SHF-derived smooth muscle cells to the arterial pole is not accompanied by rotation (Waldo *et al.*, 2005). The question of how the SHF is prepatterned is further complicated by the finding, using clonal analysis in the mouse, that subpulmonary myocardium shares a clonal origin with muscles of facial expression on the left side of the face, whereas right-sided muscles share a clonal relationship with myocardium at the base of the aorta (Lescroart *et al.*, 2010). The transcription factor *Pitx2c* is expressed in the left but not right SHF and Cre-labeling experiments have also revealed a contribution to subpulmonary myocardium (Ai *et al.*, 2006; Liu *et al.*, 2002). Tbx1, together with Nkx2.5, controls the maintenance of asymmetric *Pitx2c* expression in cardiac progenitor cells (Nowotschin *et al.*, 2006). Pitx2 is in turn required for *Sema3c* expression in OFT myocardium (Liu *et al.*, 2002); Gata6 has been identified as another regulator of *Sema3c* expression and dicer-dependent micoRNAs have been shown to negatively regulate *Pitx2* and *Sema3c* and control regionalized cell death during OFT morphogenesis

(Kodo *et al.*, 2009; Saxena and Tabin, 2010). Further dissection of the gene regulatory network operating in future subpulmonary myocardium is required to resolve how prepatterning is encoded in the SHF.

5.2. The posterior SHF and atrial septal progenitor cells

At the venous pole of the heart, Cre labeling, DiI and explant experiments have revealed that the posterior SHF contributes to a large part of the atrial myocardium including the dorsal mesenchymal protrusion giving rise to atrial and atrioventricular septal structures (Bertrand *et al.*, 2011; Cai *et al.*, 2003; Galli *et al.*, 2008; Goddeeris *et al.*, 2008; Snarr *et al.*, 2007). Canonical Wnt signaling is required in atrial progenitor cells and the number and proliferation of posterior SHF progenitor cells is reduced in Wnt2 mutants, resulting in a failure of atrial and atrioventricular septation (Tian *et al.*, 2010). Tian and colleagues identified Gata6 as canonical Wnt target operating in a feed-forward loop to maintain proliferation of progenitor cells and regulate venous pole morphogenesis (Tian *et al.*, 2010). As at the arterial pole, distinct subpopulations of posterior SHF progenitor cells have been identified, such as those giving rise to sinus venosus and pulmonary vein myocardium or septal versus non-septal atrial myocardium (Goddeeris *et al.*, 2008; Hoffmann *et al.*, 2009; Mommersteeg *et al.*, 2010). In addition, *Pitx2* expression distinguishes and confers identity on future left atrial cells (Galli *et al.*, 2008; Liu *et al.*, 2002). Late-differentiating sinus horn myocardium is derived from *Nkx2.5* negative *Tbx18* positive progenitor cells, proliferation of which is also dependent on canonical Wnt signaling (Christoffels *et al.*, 2006; Norden *et al.*, 2011). The intersection of regulatory factors defines molecular pathways for the development of pulmonary myocardium from Pitx2 expressing cells on the left as well as the progressive restriction of pacemaker activity to right-sided venous pole myocardium, resulting in localized establishment of the sinoatrial node (Mommersteeg *et al.*, 2007a,b). In addition to *Tbx18*, *Tbx5*, *Osr1*, *Shox2*, *PDGFRA*, and the gene encoding, the membrane protein podoplanin are also required for normal development of the posterior SHF, including sinoatrial node myocardium (Blaschke *et al.*, 2007; Bleyl *et al.*, 2010; Bruneau *et al.*, 2001; Douglas *et al.*, 2009; Wang *et al.*, 2005).

Defects in posterior SHF development result in atrial and atrioventricular septal defects, major forms of congenital heart anomalies in man. Two studies have demonstrated that the primary atrial septum and dorsal mesenchymal protrusion originate from a specific subpopulation of the posterior SHF (Goddeeris *et al.*, 2008; Hoffmann *et al.*, 2009). These progenitor cells express the *Mef2c* anterior-SHF enhancer element and respond to Hh signaling from ventral pharyngeal endoderm, revealed by labeling with inducible Cre recombinase under control of the *Gli1* locus, a Hh target expressed in pharyngeal mesoderm but not in the heart itself (Goddeeris *et al.*, 2008; Hoffmann *et al.*, 2009). *Gli1-Cre* labeled cells that respond to Sonic Hh ligand as early as day 8 of mouse development migrate into the heart between E9.5 and E11.5 (Fig. 2.3B; Hoffmann *et al.*, 2009). Hh signal reception thus occurs days before septum morphogenesis, suggesting that septal and nonseptal atrial progenitor cells are prepatterned in the SHF. Activity of the Hh receptor Smoothened is required specifically in Mef2c SHF positive cells for atrial septal morphogenesis and atrioventricular septation (Goddeeris *et al.*, 2008; Hoffmann *et al.*, 2009). In the absence of Hh signal reception, these cells fail to contribute to septal structures; in addition, abnormal differentiation is observed in splanchnic mesoderm (Fig. 2.3C; Goddeeris *et al.*, 2008).

The discovery that future septal structures are not the product of positional information in the forming atria but are prepatterned in the SHF provides a new paradigm for studies of atrial septation. The extent to which mutations in genes encoding cardiac transcription factors associated with atrial septal defects in man, including *NKX2.5*, *TBX5*, and *GATA4*, affect this population of progenitor cells remains to be seen. *Gli1* expression is observed in subpulmonary myocardial as well as atrial septal progenitor cells and Hh signaling plays an important role in conotruncal development, positively regulating *Tbx1* expression (Dyer and Kirby, 2009a,b; Hoffmann *et al.*, 2009; Hu *et al.*, 2004). Moskowitz and colleagues have raised the intriguing suggestion that during vertebrate evolution those parts of the heart essential for separating the pulmonary from systemic circulation, the right ventricular outlet, and interatrial septum, may have coevolved with Hh producing ventral endoderm that gives rise to lung epithelium (Hoffmann *et al.*, 2009).

6. The Second Heart Field in Vertebrate Evolution

While SHF deployment is required to subsequently divide the amniote heart into systemic and pulmonary chambers, the SHF also exists in vertebrate species with partial or no septation, suggesting that this progenitor cell population initially evolved not to drive septation but to facilitate morphogenesis of a longer heart tube. A SHF has been identified in amphibians where it is characterized, as in amniotes, by prolonged Islet expression and delayed differentiation (Brade *et al.*, 2007; Gessert and Kuhl, 2009). Fate mapping and transplantation studies in Xenopus suggest that these cells segregate early from cells giving rise to the embryonic heart tube, subsequently residing in the pericardial roof and dorsal mesocardium, from where they contribute to the OFT but not ventricle of the amphibian heart

(Fig. 2.4A; Gessert and Kuhl, 2009; Lee and Saint-Jeannet, 2011). A corollary of these observations is that the single ventricle of Xenopus may correspond to the systemic or left ventricle of the amniote heart (Brade *et al.*, 2007; Lee and Saint-Jeannet, 2011). Myocardial differentiation of

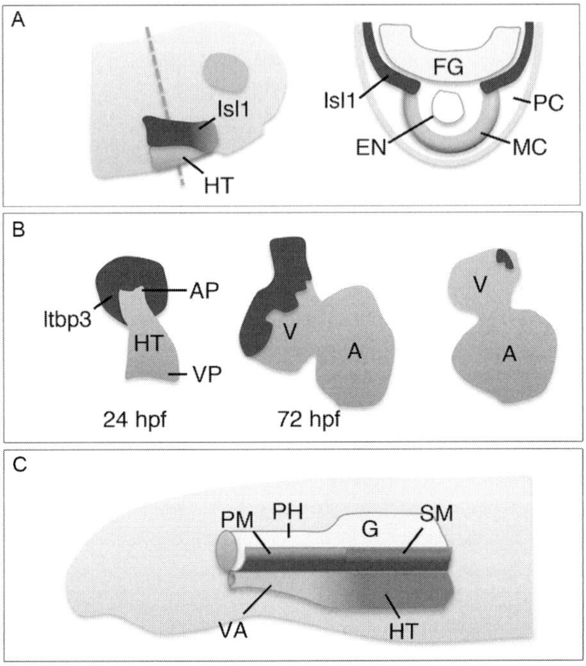

Figure 2.4 The amphibian, fish, and agnathan second heart fields. (A) Cartoons showing the location of the SHF in Xenopus. Isl1 expressing cells (blue) are contiguous with differentiated cells of the heart tube, as seen in a lateral view of a stage 31 Xenopus embryo (left). Anterior Isl1 positive cells (pale blue) are Nkx2.5 negative. A section (right) at the level of the dotted line indicates the location of the Xenopus SHF in the dorsal mesocardium and pericardial roof. Adapted from Gessert and Kuhl (2009), with permission. (B) Cartoon showing the distribution of *latent TGFβ binding protein 3* transcripts (blue) in extracardiac cells contiguous with the arterial pole of the zebrafish heart at 24 hours postfertilization (hpf) (left). *ltbp3* driven Cre recombinase allows genetic tracing of cells that previously expressed ltbp3 to the distal half of the ventricle (centre); after *ltpb3* knockdown morphant ventricles lack ltbp3 derived cardiomycytes (right). Adapted from Zhou *et al.* (2011). (C) Cartoon showing the distribution of cardiac progenitor cells in the lamprey. Isl1 expressing cells in pharyngeal and splanchnic mesoderm are shown in blue; splanchnic mesoderm also expresses the lamprey Tbx4/5 gene (green). Adapted from Kokubo *et al.* (2010), with permission. HT, heart; FG, foregut; EN, endocardium; MC, myocardium; PC, pericardial cavity; AP, arterial pole; VP, venous pole; V, ventricle; A, atrium; PH, pharynx; G, gut; VA, ventral aorta. (For interpretation of the references to color in this figure legend, the reader is referred to the Web version of this chapter.)

Nkx2.5-positive cells in the pericardial roof and dorsal mesocardium of the frog embryo is known to be inhibited by Notch signaling (Rones *et al.*, 2000). Intriguingly, transplantation experiments have revealed that the Xenopus SHF gives rise not only to the walls of the OFT but also to the conotruncal septum, a structure that contains crest-derived cells in birds and mammals, suggesting that epithelial to mesenchymal transformation of SHF-derived OFT endocardial cells is the sole source of the incomplete outlet septum in xenopus (Lee and Saint-Jeannet, 2011). In support of such a model, neural crest cells neither enter the cardiac OFT nor are required for formation of the outlet septum in Xenopus (Lee and Saint-Jeannet, 2011).

Consistent with conserved mechanisms of vertebrate heart tube extension, four recent studies have identified a population of late-differentiating cardiac cells equivalent to the SHF in zebrafish embryos. The fish heart is characterized by single atrial and ventricular chambers and a smooth muscular outlet surrounded proximally by a sleeve of myocardium (Grimes *et al.*, 2006). Previously, Reifers *et al.* had demonstrated that zebrafish *fgf8* is expressed medially to *Nkx2.5*-positive cells giving rise to the early fish heart and is later expressed in ventricular myocardium; further more, in *fgf8* mutant embryos, the ventricle is selectively truncated (Reifers *et al.*, 1998). Bakkers and colleagues subsequently investigated whether cells contributed to the arterial pole of the fish ventricle after the heart tube stage using fluorescent transgene analysis (de Pater *et al.*, 2009). They observed addition of new ventricular myocytes to the arterial pole during a late second phase of differentiation accounting for the increase in cell number during ventricular growth. Further, de Pater *et al.* showed that FGF signaling regulated addition of these cells to the fish heart, whereas, in contrast to the situation in the mouse, *Isl1* appeared to be involved only in growth of the venous pole (de Pater *et al.*, 2009).

In the past year, three papers have built on these studies and firmly established the existence of a multipotent SHF-like population contributing progenitor cells to the zebrafish heart during the second day of embryonic development (Fig. 2.4B; Hami *et al.*, 2011; Lazic and Scott, 2011; Zhou *et al.*, 2011). These studies used DiI labeling, photoconvertible fluorescence, and genetic tracing with Cre recombinase BAC and conditional reporter transgenes to demonstrate progressive addition of *Nkx2.5* positive cells outside the early heart tube to myocardium and smooth muscle cells at the arterial pole of the fish heart. As in amniotes, the zebrafish SHF is situated in the pericardial wall adjacent to the arterial pole of the heart and in addition to *Nkx2.5* expresses the *Mef2c* homologue *mef2cb* and the gene encoding latent TGFβ binding protein 3, *ltbp3* (Lazic and Scott, 2011; Zhou *et al.*, 2011). Functional studies revealed that *mef2cb* is required for late differentiation of myocardium at the arterial pole downstream of FGF signaling (Lazic and Scott, 2011), whereas *ltbp3* is required for forming the distal half of the ventricle, in addition to arterial pole endothelial and smooth muscle cells (Fig. 2.4B; Zhou *et al.*, 2011). The distal part of the zebrafish ventricle had previously been shown to be differentially responsive to retinoic acid (Stainier and Fishman, 1992). Prospective lineage studies of the origin of these late-differentiating cells suggest that they have a similar pregastrulation origin to early differentiating cells, reminiscent of the common origin of the first and second cardiac lineages in the early mammalian embryo (Hami *et al.*, 2011; Lazic and Scott, 2011). As in amniotes, BMP signaling promotes SHF proliferation and the addition of cells from the pericardial wall was significantly reduced in the absence of functional Hh signaling or *Tbx1* (Hami *et al.*, 2011). The conserved properties of the SHF opens up the possibility of using forward genetics in the zebrafish to identify and characterize novel regulators of SHF development and that may also underlie congenital heart defects in man. As a first step in this direction Zhou *et al.* demonstrated that the TGFβ pathway is required for expansion of zebrafish SHF, progenitor cells being specified but subsequently depleted on blockade of TGFβ signaling (Zhou *et al.*, 2011). Although TGFβ2 is known to be involved in OFT differentiation, an earlier role for TGFβ signaling in murine SHF development is supported by the requirement for Foxh1, a specific transducer of TGFβ-like signals, (Bartram *et al.*, 2001; Von Both *et al.*, 2004; Zhou *et al.*, 2007).

Islet expressing cells contiguous with the embryonic heart have been identified in the most primitive living vertebrate, the agnathan lamprey (Kokubo *et al.*, 2010). Lamprey *Isl1* is expressed in pharyngeal and splanchnic mesoderm contiguous with the heart tube, together with *Tbx20*, likely identifying the agnathan SHF. Lamprey *Tbx4/5* is coexpressed with *Isl1* in splanchnic mesoderm lying caudally to the pharyngeal region, suggesting that anterior and posterior domains of the SHF may be conserved features of the vertebrate SHF (Fig. 2.4C; Kokubo *et al.*, 2010).

Islet positive cells adjacent to cells giving rise to the heart have also been identified in the protochordate *Ciona intestinalis* (Stolfi *et al.*, 2010). However, in this case, the Islet expressing cells, derived from a common MesP positive lineage with cardiac progenitor cells, do not have a cardiac fate. Dynamic imaging revealed that these cells migrate dorsally to surround the opening of the peribranchial atrium of the sessile tunicate on metamorphosis where they differentiate into atrial siphon and Tbx1-positive longitudinal skeletal muscles (Fig. 2.5A; Stolfi *et al.*, 2010). While Islet itself does not appear to be required for this process, Stolfi and colleagues identified a role for the transcription factor COE in determining skeletal versus cardiac myogenic fate. The authors proposed that reallocation of multipotent cardiopharyngeal muscle precursors present in a chordate ancestor may have contributed to emergence of the vertebrate SHF. This accords evolutionary support for treating the SHF as a distinct cardiac progenitor cell population to that giving rise to the early heart tube. A model analogous to that of SHF deployment during embryonic development is thus emerging,

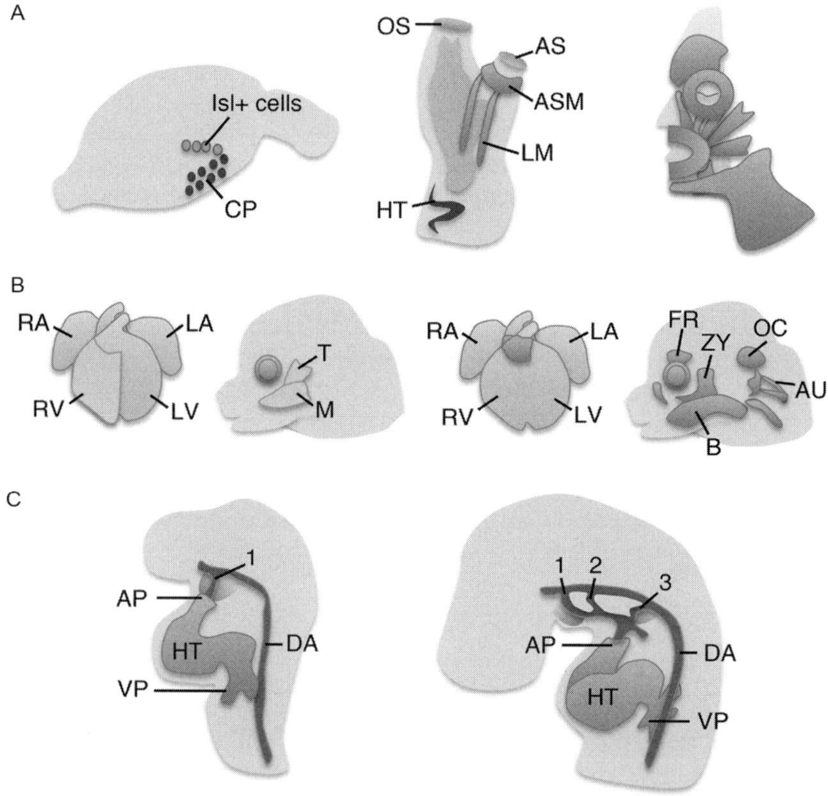

Figure 2.5 The second heart field and cardiocraniofacial development. (A) Cartoon showing separation of Islet positive cells from cardiac progenitor cells during metamorphosis of the protochordate *Ciona intestinalis* (left) and their subsequent contribution to atrial siphon and longitudinal skeletal muscles (middle). Circumferential atrial siphon muscles are comparable to human muscles of facial expression encircling the eyes and mouth (right). (B) Right ventricular myocardium shares a clonal relationship with skeletal muscles involved in mastication, derived from the first pharyngeal arch (left); myocardium at the base of the pulmonary trunk and ascending aorta shares a clonal relationship with muscles of facial expression, derived from the second pharyngeal arch (right). (C) Caudal displacement of the arterial pole of the heart during pharyngeal morphogenesis. At early stages of heart development, when right ventricular progenitor cells contribute to the heart tube, the arterial pole is positioned at the level of the first pharyngeal arch (left); later, when outflow tract progenitors are added to the heart tube, the arterial pole is positioned at the level of the second and third pharyngeal arches. (A) Adapted from Stolfi *et al.* (2010) and (B) adapted from Lescroart *et al.* (2010), with permission. CP, cardiac progenitor cells; OS, oral siphon; AS, atrial siphon; ASM, atrial siphon muscle; LM, longitunidal muscle; HT, heart; RA, right atrium; RV, right ventricle; LA, left atrium; LV, left ventricle; T, temporalis; M, master; FR, frontalis; B, buccinator; ZY, zygomaticus; OC, occipitalis; AU, auricularis; AP, arterial pole; VP, venous pole; DA, descending aorta; 1–3, first, second, and third aortic arch arteries.

by which co-option of Isl1 positive cells during vertebrate evolution has increased cardiac progenitor cell numbers to drive growth of the heart.

7. The Second Heart Field as Part of a Cardiocraniofacial Developmental Field

In amniotes, pharyngeal mesoderm gives rise to both skeletal and cardiac myogenic derivatives, containing progenitor cells for a subset of craniofacial skeletal muscles as well as SHF-derived parts of the heart (Grifone and Kelly, 2007). Skeletal muscles derived from mesoderm in the core of the pharyngeal arches are termed branchiomeric muscles. The first arch gives rise to muscles involved in opening and closing the jaw, the second arch to muscles of facial expression, and the caudal arches to muscles operating the pharynx and larynx (Schoenwolf *et al.*, 2009). These muscles, in particular facial expression muscles encircling the mouth and eyes, are highly evocative of the atrial siphon muscle and longitundinal muscles derived from Islet expressing progenitor cells in Ciona (Fig. 2.5A; Stolfi *et al.*, 2010). As in Ciona, genetic labeling has shown that these muscles are derived from *Isl1* expressing progenitor cells (Nathan *et al.*, 2008). Further, robust activation of the branchiomeric skeletal myogenic program is dependent on *Tbx1*, a key regulator of SHF development (Kelly *et al.*, 2004). Additional regulators of these muscles include Pitx2, Six1/Eya1, and the bHLH transcription factors Capsulin and MyoR; in addition, branchiomeric progenitor cells express *Nkx2.5* and *Fgf10* (Grifone and Kelly, 2007; Guo *et al.*, 2011). The transcriptional circuitry regulating branchiomeric myogenesis upstream of skeletal muscle determination factors of the MyoD family thus overlaps with that of the SHF. This differs from the genetic hierarchy upstream of myogenesis elsewhere in the embryo, trunk and appendicular muscles being derived from *Pax7* expressing progenitor cells in the somites, suggesting that embryonic mechanisms may contribute to muscle restricted myopathies (Grifone and Kelly, 2007; Tzahor, 2009; Tzahor and Evans, 2011).

Adjacent cardiac and skeletal progenitor cells in pharyngeal mesoderm do not simply share a transcriptional program. Recent clonal analysis in the mouse has demonstrated that both myogenic derivatives of pharyngeal mesoderm (SHF-derived parts of the heart and branchiomeric skeletal muscles) are derived from common progenitor cells, defining a bipotent myogenic lineage in pharyngeal mesoderm (Lescroart *et al.*, 2010). In particular, this study identified common lineage relationships between first arch derived muscles and right ventricular myocardium and between second arch derived muscles and OFT myocardium (Fig. 2.5B). These apparently disparate links reflect the sequential formation of the pharyngeal arches and initial differentiation of the heart tube at the level of the first arch where pharyngeal mesoderm gives rise to first arch derived muscles and right ventricular myocytes (Fig. 2.5C). During pharyngeal morphogenesis the arterial pole of the heart shifts caudally and OFT progenitors contribute to the heart at the level of the second arch. Importantly, divergence of branchiomeric skeletal and cardiac muscle lineages from bipotential progenitor cells occurs after the split separating the first and second cardiac lineages (Lescroart *et al.*, 2010; Meilhac *et al.*, 2004). SHF-derived myocardium therefore shares a closer lineage relationship to head muscles than to myocardium giving rise to the left ventricle. Caudal pharyngeal muscles, including neck muscles such as the trapezius, have recently been shown to be derived from occipital lateral plate mesoderm rather than somites and appear to run a Tbx1-dependent branchiomeric myogenic program (Kelly *et al.*, 2004; Theis *et al.*, 2010). Whether there is a lineage relationship between caudal pharyngeal muscles and regions of the heart derived from the posterior SHF remains to be investigated.

The above observations reveal how sequential populations of mesoderm give rise to distinct cardiac and skeletal myogenic lineages during pharyngeal morphogenesis. The SHF is thus contained within a larger field of multipotent cranial mesoderm that plays critical roles in development of the arterial pole of the heart and craniofacial morphogenesis. Alterations in the relative allocation to different myogenic derivatives within this field may have accompanied vertebrate evolution. A central property of this cardiocraniofacial field is the close interaction between mesoderm and cranial neural crest-derived cells (Hutson and Kirby, 2003). In addition to crest–mesoderm interactions during heart development, neural crest plays critical roles in craniofacial muscle morphogenesis and patterning after the specification step (Rinon *et al.*, 2007). An important clinical correlate is that perturbation of the cardiocraniofacial developmental field results in linked cardiac and craniofacial defects as seen in human genetic syndromes such as DiGeorge syndrome.

8. Perspectives

The SHF model has facilitated deconstruction of the complex events associated with early heart development and through this contributed to our current understanding of the properties of cardiovascular progenitor cells and the origins of congenital heart defects. Further excitement and surprises are to be expected from ongoing investigation of the cellular and molecular mechanisms controlling SHF development. Outstanding questions pertain to the prepatterning of future regions of the heart in the SHF. How and when are different subpopulations of the SHF established? Understanding how different congenital heart anomalies result from altered development of specific subpopulations of the SHF will be of particular importance. Further, defining gene regulatory subnetworks operative in these subdomains at a systems-wide level will contribute new candidate genes for congential heart defects. A second, related, set of questions concern how multipotency is topographically encoded within the SHF. Are all cells in the SHF multipotent or do stem cell populations reside in particular niches in the pharyngeal region? What are the mechanisms that control the divergent fates of descendents of these multipotent cells? Can common progenitor cells giving rise to skeletal and cardiac muscle be identified in pharyngeal mesoderm? Finally, addition of SHF cells to the elongating heart tube is a dynamic process and visualization of cell behavior during SHF deployment using live imaging will be required to understand how heart tube elongation takes place in the living embryo.

ACKNOWLEDGMENTS

Supported by the Fondation pour la Récherche Médicale, the Association Française contre les Myopathies, and the European Commission under the FP7 Collaborative Project CardioGeNet (Health-2007-B-223463).

REFERENCES

Aanhaanen, W. T., Brons, J. F., Dominguez, J. N., Rana, M. S., Norden, J., Airik, R., Wakker, V., de Gier-de Vries, C., Brown, N. A., Kispert, A., Moorman, A. F., and Christoffels, V. M. (2009). The Tbx2+ primary myocardium of the atrioventricular canal forms the atrioventricular node and the base of the left ventricle. *Circ. Res.* **104,** 1267–1274.

Abu-Issa, R., and Kirby, M. L. (2008). Patterning of the heart field in the chick. *Dev. Biol.* **319,** 223–233.

Abu-Issa, R., Waldo, K., and Kirby, M. L. (2004). Heart fields: One, two or more? *Dev. Biol.* **272,** 281–285.

Ai, D., Liu, W., Ma, L., Dong, F., Lu, M. F., Wang, D., Verzi, M. P., Cai, C., Gage, P. J., Evans, S., Black, B. L., Brown, N. A., *et al.* (2006). Pitx2 regulates cardiac left-right asymmetry by patterning second cardiac lineage-derived myocardium. *Dev. Biol.* **296,** 437–449.

Bajolle, F., Zaffran, S., Kelly, R. G., Hadchouel, J., Bonnet, D., Brown, N. A., and Buckingham, M. E. (2006). Rotation of the myocardial wall of the outflow tract is implicated in the normal positioning of the great arteries. *Circ. Res.* **98,** 421–428.

Bajolle, F., Zaffran, S., Meilhac, S. M., Dandonneau, M., Chang, T., Kelly, R. G., and Buckingham, M. E. (2008). Myocardium at the base of the aorta and pulmonary trunk is prefigured in the outflow tract of the heart and in subdomains of the second heart field. *Dev. Biol.* **313,** 25–34.

Barth, J. L., Clark, C. D., Fresco, V. M., Knoll, E. P., Lee, B., Argraves, W. S., and Lee, K. H. (2010). Jarid2 is among a set of genes differentially regulated by Nkx2.5 during outflow tract morphogenesis. *Dev. Dyn.* **239,** 2024–2033.

Bartram, U., Molin, D. G., Wisse, L. J., Mohamad, A., Sanford, L. P., Doetschman, T., Speer, C. P., Poelmann, R. E., and Gittenberger-de Groot, A. C. (2001). Double-outlet right ventricle and overriding tricuspid valve reflect disturbances of looping, myocardialization, endocardial cushion differentiation, and apoptosis in TGF-beta(2)-knockout mice. *Circulation* **103,** 2745–2752.

Bertrand, N., Roux, M., Ryckebusch, L., Niederreither, K., Dolle, P., Moon, A., Capecchi, M., and Zaffran, S. (2011). Hox genes define distinct progenitor sub-domains within the second heart field. *Dev. Biol.* **353,** 266–274.

Blaschke, R. J., Hahurij, N. D., Kuijper, S., Just, S., Wisse, L. J., Deissler, K., Maxelon, T., Anastassiadis, K., Spitzer, J., Hardt, S. E., Scholer, H., Feitsma, H., *et al.* (2007). Targeted mutation reveals essential functions of the homeodomain transcription factor Shox2 in sinoatrial and pacemaking development. *Circulation* **115,** 1830–1838.

Bleyl, S. B., Saijoh, Y., Bax, N. A., Gittenberger-de Groot, A. C., Wisse, L. J., Chapman, S. C., Hunter, J., Shiratori, H., Hamada, H., Yamada, S., Shiota, K., Klewer, S. E., *et al.* (2010). Dysregulation of the PDGFRA gene causes inflow tract anomalies including TAPVR: Integrating evidence from human genetics and model organisms. *Hum. Mol. Genet.* **19,** 1286–1301.

Bondue, A., Tannler, S., Chiapparo, G., Chabab, S., Ramialison, M., Paulissen, C., Beck, B., Harvey, R., and Blanpain, C. (2011). Defining the earliest step of cardiovascular progenitor specification during embryonic stem cell differentiation. *J. Cell Biol.* **192,** 751–765.

Brade, T., Gessert, S., Kuhl, M., and Pandur, P. (2007). The amphibian second heart field: Xenopus islet-1 is required for cardiovascular development. *Dev. Biol.* **311,** 297–310.

Bruneau, B. G. (2002). Transcriptional regulation of vertebrate cardiac morphogenesis. *Circ. Res.* **90,** 509–519.

Bruneau, B. G., Nemer, G., Schmitt, J. P., Charron, F., Robitaille, L., Caron, S., Conner, D. A., Gessler, M., Nemer, M., Seidman, C. E., and Seidman, J. G. (2001). A murine model of Holt-Oram syndrome defines roles of the T-box transcription factor Tbx5 in cardiogenesis and disease. *Cell* **106,** 709–721.

Bu, L., Jiang, X., Martin-Puig, S., Caron, L., Zhu, S., Shao, Y., Roberts, D. J., Huang, P. L., Domian, I. J., and Chien, K. R. (2009). Human ISL1 heart progenitors generate diverse multipotent cardiovascular cell lineages. *Nature* **460,** 113–117.

Buckingham, M., Meilhac, S., and Zaffran, S. (2005). Building the mammalian heart from two sources of myocardial cells. *Nat. Rev. Genet.* **6,** 826–835.

Cai, C. L., Liang, X., Shi, Y., Chu, P. H., Pfaff, S. L., Chen, J., and Evans, S. (2003). Isl1 identifies a cardiac progenitor population that proliferates prior to differentiation and contributes a majority of cells to the heart. *Dev. Cell* **5,** 877–889.

Cai, C. L., Zhou, W., Yang, L., Bu, L., Qyang, Y., Zhang, X., Li, X., Rosenfeld, M. G., Chen, J., and Evans, S. (2005). T-box genes coordinate regional rates of proliferation and regional specification during cardiogenesis. *Development* **132,** 2475–2487.

Chen, Y. H., Ishii, M., Sun, J., Sucov, H. M., and Maxson, R. E., Jr. (2007). Msx1 and Msx2 regulate survival of secondary heart field precursors and post-migratory proliferation of cardiac neural crest in the outflow tract. *Dev. Biol.* **308,** 421–437.

Chen, L., Fulcoli, F. G., Tang, S., and Baldini, A. (2009). Tbx1 regulates proliferation and differentiation of multipotent heart progenitors. *Circ Res.* **105,** 842–851.

Christoffels, V. M., Mommersteeg, M. T., Trowe, M. O., Prall, O. W., de Gier-de Vries, C., Soufan, A. T., Bussen, M., Schuster-Gossler, K., Harvey, R. P., Moorman, A. F., and Kispert, A. (2006). Formation of the venous pole of the heart from an Nkx2-5-negative precursor population requires Tbx18. *Circ. Res.* **98,** 1555–1563.

Cohen, E. D., Wang, Z., Lepore, J. J., Lu, M. M., Taketo, M. M., Epstein, D. J., and Morrisey, E. E. (2007). Wnt/beta-catenin signaling promotes expansion of Isl-1-positive cardiac progenitor cells through regulation of FGF signaling. *J. Clin. Invest.* **117,** 1794–1804.

de la Cruz, M. V., Sanchez Gomez, C., Arteaga, M. M., and Arguello, C. (1977). Experimental study of the development of the truncus and the conus in the chick embryo. *J. Anat.* **123,** 661–686.

de Pater, E., Clijsters, L., Marques, S. R., Lin, Y. F., Garavito-Aguilar, Z. V., Yelon, D., and Bakkers, J. (2009). Distinct phases of cardiomyocyte differentiation regulate growth of the zebrafish heart. *Development* **136,** 1633–1641.

de Vries, P. A. (1981). Evolution of precardiac and splanchnic mesdoerm in relation to the infundibulum and truncus. *In* "Perspectives in Cardiovacscular Research," (T. Pexeider, ed.), Vol. 5, pp. 31–48. Raven Press, New York.

Di Felice, V., and Zummo, G. (2009). Tetralogy of fallot as a model to study cardiac progenitor cell migration and differentiation during heart development. *Trends Cardiovasc. Med.* **19,** 130–135.

Dodou, E., Verzi, M. P., Anderson, J. P., Xu, S. M., and Black, B. L. (2004). Mef2c is a direct transcriptional target of ISL1 and GATA factors in the anterior heart field during mouse embryonic development. *Development* **131,** 3931–3942.

Domian, I. J., Chiravuri, M., van der Meer, P., Feinberg, A. W., Shi, X., Shao, Y., Wu, S. M., Parker, K. K., and Chien, K. R. (2009). Generation of functional ventricular heart muscle from mouse ventricular progenitor cells. *Science* **326,** 426–429.

Douglas, Y. L., Mahtab, E. A., Jongbloed, M. R., Uhrin, P., Zaujec, J., Binder, B. R., Schalij, M. J., Poelmann, R. E., Deruiter, M. C., and Gittenberger-de Groot, A. C. (2009). Pulmonary vein, dorsal atrial wall and atrial septum abnormalities in podoplanin knockout mice with disturbed posterior heart field contribution. *Pediatr. Res.* **65,** 27–32.

Dyer, L. A., and Kirby, M. L. (2009a). The role of secondary heart field in cardiac development. *Dev. Biol.* **336,** 137–144.

Dyer, L. A., and Kirby, M. L. (2009b). Sonic hedgehog maintains proliferation in secondary heart field progenitors and is required for normal arterial pole formation. *Dev. Biol.* **330,** 305–317.

Dyer, L. A., Makadia, F. A., Scott, A., Pegram, K., Hutson, M. R., and Kirby, M. L. (2010). BMP signaling modulates hedgehog-induced secondary heart field proliferation. *Dev. Biol.* **348,** 167–176.

Epstein, J. A. (2010). Franklin H. Epstein Lecture. Cardiac development and implications for heart disease. *N. Engl. J. Med.* **363,** 1638–1647.

Evans, S. M., Yelon, D., Conlon, F. L., and Kirby, M. L. (2010). Myocardial lineage development. *Circ. Res.* **107,** 1428–1444.

Fulcoli, F. G., Huynh, T., Scambler, P. J., and Baldini, A. (2009). Tbx1 regulates the BMP-Smad1 pathway in a transcription independent manner. *PLoS One* **4,** e6049.

Galli, D., Dominguez, J. N., Zaffran, S., Munk, A., Brown, N. A., and Buckingham, M. E. (2008). Atrial myocardium derives from the posterior region of the second heart field, which acquires left-right identity as Pitx2c is expressed. *Development* **135,** 1157–1167.

Garg, V., Yamagishi, C., Hu, T., Kathiriya, I. S., Yamagishi, H., and Srivastava, D. (2001). Tbx1, a DiGeorge syndrome candidate gene, is regulated by sonic hedgehog during pharyngeal arch development. *Dev. Biol.* **235,** 62–73.

Gessert, S., and Kuhl, M. (2009). Comparative gene expression analysis and fate mapping studies suggest an early segregation of cardiogenic lineages in Xenopus laevis. *Dev. Biol.* **334,** 395–408.

Goddeeris, M. M., Schwartz, R., Klingensmith, J., and Meyers, E. N. (2007). Independent requirements for Hedgehog signaling by both the anterior heart field and neural crest cells for outflow tract development. *Development* **134,** 1593–1604.

Goddeeris, M. M., Rho, S., Petiet, A., Davenport, C. L., Johnson, G. A., Meyers, E. N., and Klingensmith, J. (2008). Intracardiac septation requires hedgehog-dependent cellular contributions from outside the heart. *Development* **135,** 1887–1895.

Grifone, R., and Kelly, R. G. (2007). Heartening news for head muscle development. *Trends Genet.* **23,** 365–369.

Grimes, A. C., Stadt, H. A., Shepherd, I. T., and Kirby, M. L. (2006). Solving an enigma: Arterial pole development in the zebrafish heart. *Dev. Biol.* **290,** 265–276.

Guo, C., Sun, Y., Zhou, B., Adam, R. M., Li, X., Pu, W. T., Morrow, B. E., and Moon, A. (2011). A Tbx1-Six1/Eya1-Fgf8 genetic pathway controls mammalian cardiovascular and craniofacial morphogenesis. *J. Clin. Invest.* **121,** 1585–1595.

Hami, D., Grimes, A. C., Tsai, H. J., and Kirby, M. L. (2011). Zebrafish cardiac development requires a conserved secondary heart field. *Development* **138,** 2389–2398.

High, F. A., Jain, R., Stoller, J. Z., Antonucci, N. B., Lu, M. M., Loomes, K. M., Kaestner, K. H., Pear, W. S., and Epstein, J. A. (2009). Murine Jagged1/Notch signaling in the second heart field orchestrates Fgf8 expression and tissue-tissue interactions during outflow tract development. *J. Clin. Invest.* **119,** 1986–1996.

Hildreth, V., Webb, S., Chaudhry, B., Peat, J. D., Phillips, H. M., Brown, N., Anderson, R. H., and Henderson, D. J. (2009). Left cardiac isomerism in the Sonic hedgehog null mouse. *J. Anat.* **214,** 894–904.

Hoffmann, A. D., Peterson, M. A., Friedland-Little, J. M., Anderson, S. A., and Moskowitz, I. P. (2009). sonic hedgehog is required in pulmonary endoderm for atrial septation. *Development* **136,** 1761–1770.

Hu, T., Yamagishi, H., Maeda, J., McAnally, J., Yamagishi, C., and Srivastava, D. (2004). Tbx1 regulates fibroblast growth factors in the anterior heart field through a reinforcing autoregulatory loop involving forkhead transcription factors. *Development* **131,** 5491–5502.

Huh, S. H., and Ornitz, D. M. (2010). Beta-catenin deficiency causes DiGeorge syndrome-like phenotypes through regulation of Tbx1. *Development* **137,** 1137–1147.

Hutson, M. R., and Kirby, M. L. (2003). Neural crest and cardiovascular development: A 20-year perspective. *Birth Defects Res. C Embryo Today* **69,** 2–13.

Hutson, M. R., and Kirby, M. L. (2007). Model systems for the study of heart development and disease. Cardiac neural crest and conotruncal malformations. *Semin. Cell Dev. Biol.* **18,** 101–110.

Hutson, M. R., Zhang, P., Stadt, H. A., Sato, A. K., Li, Y. X., Burch, J., Creazzo, T. L., and Kirby, M. L. (2006). Cardiac arterial pole alignment is sensitive to FGF8 signaling in the pharynx. *Dev. Biol.* **295,** 486–497.

Hutson, M. R., Zeng, X. L., Kim, A. J., Antoon, E., Harward, S., and Kirby, M. L. (2010). Arterial pole progenitors interpret opposing FGF/BMP signals to proliferate or differentiate. *Development* **137,** 3001–3011.

Huynh, T., Chen, L., Terrell, P., and Baldini, A. (2007). A fate map of Tbx1 expressing cells reveals heterogeneity in the second cardiac field. *Genesis* **45,** 470–475.

Ilagan, R., Abu-Issa, R., Brown, D., Yang, Y. P., Jiao, K., Schwartz, R. J., Klingensmith, J., and Meyers, E. N. (2006). Fgf8 is required for anterior heart field development. *Development* **133,** 2435–2445.

Jerome, L. A., and Papaioannou, V. E. (2001). DiGeorge syndrome phenotype in mice mutant for the T-box gene, Tbx1. *Nat. Genet.* **27,** 286–291.

Jia, Q., McDill, B. W., Li, S. Z., Deng, C., Chang, C. P., and Chen, F. (2007). Smad signaling in the neural crest regulates cardiac outflow tract remodeling through cell autonomous and non-cell autonomous effects. *Dev. Biol.* **311,** 172–184.

Kang, J., Nathan, E., Xu, S. M., Tzahor, E., and Black, B. L. (2009). Isl1 is a direct transcriptional target of Forkhead transcription factors in second-heart-field-derived mesoderm. *Dev. Biol.* **334,** 513–522.

Kappen, C., and Salbaum, J. M. (2009). Identification of regulatory elements in the Isl1 gene locus. *Int. J. Dev. Biol.* **53,** 935–946.

Kelly, R. G., and Buckingham, M. E. (2002). The anterior heart-forming field: voyage to the arterial pole of the heart. *Trends Genet.* **18,** 210–216.

Kelly, R. G., and Evans, S. (2010). The second heart field. *In* "Cardiac Developmet and Regeneration," (R. P. Harvey and N. Rosenthal, eds.), pp. 143–169. Academic Press, Oxford.

Kelly, R. G., and Papaioannou, V. E. (2007). Visualization of outflow tract development in the absence of Tbx1 using an FgF10 enhancer trap transgene. *Dev. Dyn.* **236,** 821–828.

Kelly, R. G., Brown, N. A., and Buckingham, M. E. (2001). The arterial pole of the mouse heart forms from Fgf10-expressing cells in pharyngeal mesoderm. *Dev. Cell* **1,** 435–440.

Kelly, R. G., Jerome-Majewska, L. A., and Papaioannou, V. E. (2004). The del22q11.2 candidate gene Tbx1 regulates branchiomeric myogenesis. *Hum. Mol. Genet.* **13,** 2829–2840.

Kodo, K., Nishizawa, T., Furutani, M., Arai, S., Yamamura, E., Joo, K., Takahashi, T., Matsuoka, R., and Yamagishi, H. (2009). GATA6 mutations cause human cardiac outflow tract defects by disrupting semaphorin-plexin signaling. *Proc. Natl. Acad. Sci. USA* **106,** 13933–13938.

Kokubo, N., Matsuura, M., Onimaru, K., Tiecke, E., Kuraku, S., Kuratani, S., and Tanaka, M. (2010). Mechanisms of heart development in the Japanese lamprey, Lethenteron japonicum. *Evol Dev.* **12,** 34–44.

Kwon, C., Qian, L., Cheng, P., Nigam, V., Arnold, J., and Srivastava, D. (2009). A regulatory pathway involving Notch1/beta-catenin/Isl1 determines cardiac progenitor cell fate. *Nat. Cell Biol.* **11,** 951–957.

Laugwitz, K. L., Moretti, A., Lam, J., Gruber, P., Chen, Y., Woodard, S., Lin, L. Z., Cai, C. L., Lu, M. M., Reth, M., Platoshyn, O., Yuan, J. X., *et al.* (2005). Postnatal Isl1+ cardioblasts enter fully differentiated cardiomyocyte lineages. *Nature* **433,** 647–653.

Lazic, S., and Scott, I. C. (2011). Mef2cb regulates late myocardial cell addition from a second heart field-like population of progenitors in zebrafish. *Dev. Biol.* **354,** 123–133.

Lee, Y. H., and Saint-Jeannet, J. P. (2011). Cardiac neural crest is dispensable for outflow tract septation in Xenopus. *Development* **138,** 2025–2034.

Lescroart, F., Kelly, R. G., Le Garrec, J. F., Nicolas, J. F., Meilhac, S. M., and Buckingham, M. (2010). Clonal analysis reveals common lineage relationships between head muscles and second heart field derivatives in the mouse embryo. *Development* **137,** 3269–3279.

Li, P., Pashmforoush, M., and Sucov, H. M. (2010). Retinoic acid regulates differentiation of the secondary heart field and TGFbeta-mediated outflow tract septation. *Dev. Cell* **18,** 480–485.

Liao, J., Aggarwal, V. S., Nowotschin, S., Bondarev, A., Lipner, S., and Morrow, B. E. (2008). Identification of downstream genetic pathways of Tbx1 in the second heart field. *Dev. Biol.* **316,** 524–537.

Lin, L., Cui, L., Zhou, W., Dufort, D., Zhang, X., Cai, C. L., Bu, L., Yang, L., Martin, J., Kemler, R., Rosenfeld, M. G., Chen, J., *et al.* (2007). Beta-catenin directly regulates Islet1 expression in cardiovascular progenitors and is required for multiple aspects of cardiogenesis. *Proc. Natl. Acad. Sci. USA* **104,** 9313–9318.

Liu, C., Liu, W., Palie, J., Lu, M. F., Brown, N. A., and Martin, J. F. (2002). Pitx2c patterns anterior myocardium and aortic arch vessels and is required for local cell movement into atrioventricular cushions. *Development* **129,** 5081–5091.

Ma, Q., Zhou, B., and Pu, W. T. (2008). Reassessment of Isl1 and Nkx2-5 cardiac fate maps using a Gata4-based reporter of Cre activity. *Dev. Biol.* **323,** 98–104.

Maeda, J., Yamagishi, H., McAnally, J., Yamagishi, C., and Srivastava, D. (2006). Tbx1 is regulated by forkhead proteins in the secondary heart field. *Dev. Dyn.* **235,** 701–710.

Marguerie, A., Bajolle, F., Zaffran, S., Brown, N. A., Dickson, C., Buckingham, M. E., and Kelly, R. G. (2006). Congenital heart defects in Fgfr2-IIIb and Fgf10 mutant mice. *Cardiovasc. Res.* **71,** 50–60.

Meilhac, S. M., Esner, M., Kelly, R. G., Nicolas, J. F., and Buckingham, M. E. (2004). The clonal origin of myocardial cells in different regions of the embryonic mouse heart. *Dev. Cell* **6,** 685–698.

Milgrom-Hoffman, M., Harrelson, Z., Ferrara, N., Zelzer, E., Evans, S. M., and Tzahor, E. (2011). The heart endocardium is derived from vascular endothelial progenitors. *Development* **138,** 4777–4787.

Mjaatvedt, C. H., Nakaoka, T., Moreno-Rodriguez, R., Norris, R. A., Kern, M. J., Eisenberg, C. A., Turner, D., and Markwald, R. R. (2001). The outflow tract of the heart is recruited from a novel heart-forming field. *Dev. Biol.* **238,** 97–109.

Mommersteeg, M. T., Brown, N. A., Prall, O. W., de Gier-de Vries, C., Harvey, R. P., Moorman, A. F., and Christoffels, V. M. (2007a). Pitx2c and Nkx2-5 are required for the formation and identity of the pulmonary myocardium. *Circ. Res.* **101,** 902–909.

Mommersteeg, M. T., Hoogaars, W. M., Prall, O. W., de Gier-de Vries, C., Wiese, C., Clout, D. E., Papaioannou, V. E., Brown, N. A., Harvey, R. P., Moorman, A. F., and Christoffels, V. M. (2007b). Molecular pathway for the localized formation of the sinoatrial node. *Circ. Res.* **100,** 354–362.

Mommersteeg, M. T., Dominguez, J. N., Wiese, C., Norden, J., de Gier-de Vries, C., Burch, J. B., Kispert, A., Brown, N. A., Moorman, A. F., and Christoffels, V. M. (2010). The sinus venosus progenitors separate and diversify from the first and second heart fields early in development. *Cardiovasc. Res.* **87,** 92–101.

Moorman, A. F., Christoffels, V. M., Anderson, R. H., and van den Hoff, M. J. (2007). The heart-forming fields: One or multiple? *Philos. Trans. R. Soc. Lond. B Biol. Sci.* **362,** 1257–1265.

Musunuru, K., Domian, I. J., and Chien, K. R. (2010). Stem cell models of cardiac development and disease. *Annu. Rev. Cell Dev. Biol.* **26,** 667–687.

Nathan, E., Monovich, A., Tirosh-Finkel, L., Harrelson, Z., Rousso, T., Rinon, A., Harel, I., Evans, S. M., and Tzahor, E. (2008). The contribution of Islet1-expressing splanchnic mesoderm cells to distinct branchiomeric muscles reveals significant heterogeneity in head muscle development. *Development* **135,** 647–657.

Norden, J., Greulich, F., Rudat, C., Taketo, M. M., and Kispert, A. (2011). Wnt/{beta}-Catenin Signaling Maintains the Mesenchymal Precursor Pool for Murine Sinus Horn Formation. *Circ. Res.* **109,** e42–e50.

Nowotschin, S., Liao, J., Gage, P. J., Epstein, J. A., Campione, M., and Morrow, B. E. (2006). Tbx1 affects asymmetric cardiac morphogenesis by regulating Pitx2 in the secondary heart field. *Development* **133,** 1565–1573.

Parisot, P., Mesbah, K., Theveniau-Ruissy, M., and Kelly, R. G. (2011). Tbx1, subpulmonary myocardium and conotruncal congenital heart defects. *Birth Defects Res. A Clin. Mol. Teratol.* **91,** 477–484.

Park, E. J., Ogden, L. A., Talbot, A., Evans, S., Cai, C. L., Black, B. L., Frank, D. U., and Moon, A. M. (2006). Required, tissue-specific roles for Fgf8 in outflow tract formation and remodeling. *Development* **133,** 2419–2433.

Park, E. J., Watanabe, Y., Smyth, G., Miyagawa-Tomita, S., Meyers, E., Klingensmith, J., Camenisch, T., Buckingham, M., and Moon, A. M. (2008). An FGF autocrine loop initiated in second heart field mesoderm regulates morphogenesis at the arterial pole of the heart. *Development* **135,** 3599–3610.

Prall, O. W., Menon, M. K., Solloway, M. J., Watanabe, Y., Zaffran, S., Bajolle, F., Biben, C., McBride, J. J., Robertson, B. R., Chaulet, H., Stennard, F. A., Wise, N., *et al.* (2007). An Nkx2-5/Bmp2/Smad1 negative feedback loop controls heart progenitor specification and proliferation. *Cell* **128,** 947–959.

Rana, M. S., Horsten, N. C., Tesink-Taekema, S., Lamers, W. H., Moorman, A. F., and van den Hoff, M. J. (2007). Trabeculated right ventricular free wall in the chicken heart forms by ventricularization of the myocardium initially forming the outflow tract. *Circ. Res.* **100,** 1000–1007.

Reifers, F., Bohli, H., Walsh, E. C., Crossley, P. H., Stainier, D. Y., and Brand, M. (1998). Fgf8 is mutated in zebrafish acerebellar (ace) mutants and is required for maintenance of midbrain-hindbrain boundary development and somitogenesis. *Development* **125,** 2381–2395.

Rinon, A., Lazar, S., Marshall, H., Buchmann-Moller, S., Neufeld, A., Elhanany-Tamir, H., Taketo, M. M., Sommer, L., Krumlauf, R., and Tzahor, E. (2007). Cranial neural crest cells regulate head muscle patterning and differentiation during vertebrate embryogenesis. *Development* **134,** 3065–3075.

Roberts, C., Ivins, S. M., James, C. T., and Scambler, P. J. (2005). Retinoic acid down-regulates Tbx1 expression *in vivo* and *in vitro*. *Dev. Dyn.* **232,** 928–938.

Robertson, E. J., Charatsi, I., Joyner, C. J., Koonce, C. H., Morgan, M., Islam, A., Paterson, C., Lejsek, E., Arnold, S. J., Kallies, A., Nutt, S. L., and Bikoff, E. K. (2007). Blimp1 regulates development of the posterior forelimb, caudal pharyngeal arches, heart and sensory vibrissae in mice. *Development* **134,** 4335–4345.

Rochais, F., Mesbah, K., and Kelly, R. G. (2009a). Signaling pathways controlling second heart field development. *Circ. Res.* **104,** 933–942.

Rochais, F., Dandonneau, M., Mesbah, K., Jarry, T., Mattei, M. G., and Kelly, R. G. (2009b). Hes1 is expressed in the second heart field and is required for outflow tract development. *PLoS One* **4,** e6267.

Rones, M. S., McLaughlin, K. A., Raffin, M., and Mercola, M. (2000). Serrate and Notch specify cell fates in the heart field by suppressing cardiomyogenesis. *Development* **127,** 3865–3876.

Rosenquist, G. C., and DeHaan, R. L. (1966). Migration of precadiac cells in the chick embryo: A radioautograpohic study. *Carnegie Institute of Washington, Contributions to Embryology* **263,** 111–121.

Ryckebusch, L., Wang, Z., Bertrand, N., Lin, S. C., Chi, X., Schwartz, R., Zaffran, S., and Niederreither, K. (2008). Retinoic acid deficiency alters second heart field formation. *Proc. Natl. Acad. Sci. USA* **105,** 2913–2918.

Saxena, A., and Tabin, C. J. (2010). miRNA-processing enzyme Dicer is necessary for cardiac outflow tract alignment and chamber septation. *Proc. Natl. Acad. Sci. USA* **107,** 87–91.

Schleiffarth, J. R., Person, A. D., Martinsen, B. J., Sukovich, D. J., Neumann, A., Baker, C. V., Lohr, J. L., Cornfield, D. N., Ekker, S. C., and Petryk, A. (2007). Wnt5a is required for cardiac outflow tract septation in mice. *Pediatr. Res.* **61,** 386–391.

Schoenwolf, G. C., Bleyl, S. B., Brauer, P. R., and Francis-West, P. H. (2009). Larsen's Human Embryology. 4th edn. Elsevier Churchill Livingstone, Philadelphia.

Seo, S., and Kume, T. (2006). Forkhead transcription factors, Foxc1 and Foxc2, are required for the morphogenesis of the cardiac outflow tract. *Dev. Biol.* **296,** 421–436.

Sirbu, I. O., Zhao, X., and Duester, G. (2008). Retinoic acid controls heart anteroposterior patterning by down-regulating Isl1 through the Fgf8 pathway. *Dev. Dyn.* **237,** 1627–1635.

Snarr, B. S., O'Neal, J. L., Chintalapudi, M. R., Wirrig, E. E., Phelps, A. L., Kubalak, S. W., and Wessels, A. (2007). Isl1 expression at the venous pole identifies a novel role for the second heart field in cardiac development. *Circ. Res.* **101,** 971–974.

Song, L., Li, Y., Wang, K., and Zhou, C. J. (2010). Cardiac neural crest and outflow tract defects in Lrp6 mutant mice. *Dev. Dyn.* **239,** 200–210.

Stainier, D. Y., and Fishman, M. C. (1992). Patterning the zebrafish heart tube: Acquisition of anteroposterior polarity. *Dev. Biol.* **153,** 91–101.

Stolfi, A., Gainous, T. B., Young, J. J., Mori, A., Levine, M., and Christiaen, L. (2010). Early chordate origins of the vertebrate second heart field. *Science* **329,** 565–568.

Stottmann, R. W., Choi, M., Mishina, Y., Meyers, E. N., and Klingensmith, J. (2004). BMP receptor IA is required in mammalian neural crest cells for development of the cardiac outflow tract and ventricular myocardium. *Development* **131,** 2205–2218.

Sun, Y., Liang, X., Najafi, N., Cass, M., Lin, L., Cai, C. L., Chen, J., and Evans, S. M. (2007). Islet1 is expressed in distinct cardiovascular lineages, including pacemaker and coronary vascular cells. *Dev. Biol.* **304,** 286–296.

Takeuchi, J. K., Mileikovskaia, M., Koshiba-Takeuchi, K., Heidt, A. B., Mori, A. D., Arruda, E. P., Gertsenstein, M., Georges, R., Davidson, L., Mo, R., Hui, C. C., Henkelman, R. M., *et al.* (2005). Tbx20 dose-dependently regulates transcription factor networks required for mouse heart and motoneuron development. *Development* **132,** 2463–2474.

Theis, S., Patel, K., Valasek, P., Otto, A., Pu, Q., Harel, I., Tzahor, E., Tajbakhsh, S., Christ, B., and Huang, R. (2010). The occipital lateral plate mesoderm is a novel source for vertebrate neck musculature. *Development* **137,** 2961–2971.

Theveniau-Ruissy, M., Dandonneau, M., Mesbah, K., Ghez, O., Mattei, M. G., Miquerol, L., and Kelly, R. G. (2008). The del22q11.2 candidate gene Tbx1 controls regional outflow tract identity and coronary artery patterning. *Circ. Res.* **103,** 142–148.

Tian, Y., Yuan, L., Goss, A. M., Wang, T., Yang, J., Lepore, J. J., Zhou, D., Schwartz, R. J., Patel, V., Cohen, E. D., and Morrisey, E. E. (2010). Characterization and in vivo pharmacological rescue of a Wnt2-Gata6 pathway required for cardiac inflow tract development. *Dev. Cell* **18,** 275–287.

Tirosh-Finkel, L., Elhanany, H., Rinon, A., and Tzahor, E. (2006). Mesoderm progenitor cells of common origin contribute to the head musculature and the cardiac outflow tract. *Development* **133,** 1943–1953.

Tirosh-Finkel, L., Zeisel, A., Brodt-Ivenshitz, M., Shamai, A., Yao, Z., Seger, R., Domany, E., and Tzahor, E. (2010). BMP-mediated inhibition of FGF signaling promotes cardiomyocyte differentiation of anterior heart field progenitors. *Development* **137,** 2989–3000.

Tsuchihashi, T., Maeda, J., Shin, C. H., Ivey, K. N., Black, B. L., Olson, E. N., Yamagishi, H., and Srivastava, D. (2011). Hand2 function in second heart field progenitors is essential for cardiogenesis. *Dev. Biol.* **351,** 62–69.

Tzahor, E. (2007). Wnt/beta-catenin signaling and cardiogenesis: Timing does matter. *Dev. Cell* **13,** 10–13.

Tzahor, E. (2009). Heart and craniofacial muscle development: A new developmental theme of distinct myogenic fields. *Dev. Biol.* **327,** 273–279.

Tzahor, E., and Evans, S. M. (2011). Pharyngeal mesoderm development during embryogenesis: Implications for both heart and head myogenesis. *Cardiovasc. Res.* **91,** 196–202.

Urness, L. D., Bleyl, S. B., Wright, T. J., Moon, A. M., and Mansour, S. L. (2011). Redundant and dosage sensitive requirements for Fgf3 and Fgf10 in cardiovascular development. *Dev. Biol.* **356,** 383–397.

van den Berg, G., Abu-Issa, R., de Boer, B. A., Hutson, M. R., de Boer, P. A., Soufan, A. T., Ruijter, J. M., Kirby, M. L., van den Hoff, M. J., and Moorman, A. F. (2009). A caudal proliferating growth center contributes to both poles of the forming heart tube. *Circ. Res.* **104,** 179–188.

Van Praagh, R. (2009). The first Stella van Praagh memorial lecture: The history and anatomy of tetralogy of Fallot. *Semin. Thorac. Cardiovasc. Surg. Pediatr. Card. Surg. Annu.* 19–38.

van Weerd, J. H., Koshiba-Takeuchi, K., Kwon, C., and Takeuchi, J. K. (2011). Epigenetic factors and cardiac development. *Cardiovasc. Res.* **91,** 203–211.

Verzi, M. P., McCulley, D. J., De Val, S., Dodou, E., and Black, B. L. (2005). The right ventricle, outflow tract, and ventricular septum comprise a restricted expression domain within the secondary/anterior heart field. *Dev. Biol.* **287,** 134–145.

Vincent, S. D., and Buckingham, M. E. (2010). How to make a heart: The origin and regulation of cardiac progenitor cells. *Curr. Top. Dev. Biol.* **90,** 1–41.

Viragh, S., and Challice, C. E. (1973). Origin and differentiation of cardiac muscle cells in the mouse. *J. Ultrastruct. Res.* **42,** 1–24.

Vitelli, F., Taddei, I., Morishima, M., Meyers, E. N., Lindsay, E. A., and Baldini, A. (2002). A genetic link between Tbx1 and fibroblast growth factor signaling. *Development* **129,** 4605–4611.

Vitelli, F., Lania, G., Huynh, T., and Baldini, A. (2010). Partial rescue of the Tbx1 mutant heart phenotype by Fgf8: Genetic evidence of impaired tissue response to Fgf8. *J. Mol. Cell. Cardiol.* **49,** 836–840.

von Both, I., Silvestri, C., Erdemir, T., Lickert, H., Walls, J. R., Henkelman, R. M., Rossant, J., Harvey, R. P., Attisano, L., and Wrana, J. L. (2004). Foxh1 is essential for development of the anterior heart field. *Dev. Cell* **7,** 331–345.

Waldo, K. L., Kumiski, D. H., Wallis, K. T., Stadt, H. A., Hutson, M. R., Platt, D. H., and Kirby, M. L. (2001). Conotruncal myocardium arises from a secondary heart field. *Development* **128,** 3179–3188.

Waldo, K. L., Hutson, M. R., Ward, C. C., Zdanowicz, M., Stadt, H. A., Kumiski, D., Abu-Issa, R., and Kirby, M. L. (2005). Secondary heart field contributes myocardium and smooth muscle to the arterial pole of the developing heart. *Dev. Biol.* **281,** 78–90.

Wang, Q., Lan, Y., Cho, E. S., Maltby, K. M., and Jiang, R. (2005). Odd-skipped related 1 (Odd 1) is an essential regulator of heart and urogenital development. *Dev. Biol.* **288,** 582–594.

Wang, J., Greene, S. B., Bonilla-Claudio, M., Tao, Y., Zhang, J., Bai, Y., Huang, Z., Black, B. L., Wang, F., and Martin, J. F. (2010). Bmp signaling regulates myocardial differentiation from cardiac progenitors through a MicroRNA-mediated mechanism. *Dev. Cell* **19,** 903–912.

Ward, C., Stadt, H., Hutson, M., and Kirby, M. L. (2005). Ablation of the secondary heart field leads to tetralogy of Fallot and pulmonary atresia. *Dev. Biol.* **284,** 72–83.

Washington Smoak, I., Byrd, N. A., Abu-Issa, R., Goddeeris, M. M., Anderson, R., Morris, J., Yamamura, K., Klingensmith, J., and Meyers, E. N. (2005). Sonic hedgehog is required for cardiac outflow tract and neural crest cell development. *Dev. Biol.* **283,** 357–372.

Watanabe, M., Choudhry, A., Berlan, M., Singal, A., Siwik, E., Mohr, S., and Fisher, S. A. (1998). Developmental remodeling and shortening of the cardiac outflow tract involves myocyte programmed cell death. *Development* **125,** 3809–3820.

Watanabe, Y., Miyagawa-Tomita, S., Vincent, S. D., Kelly, R. G., Moon, A. M., and Buckingham, M. E. (2010). Role of mesodermal FGF8 and FGF10 overlaps in the development of the arterial pole of the heart and pharyngeal arch arteries. *Circ. Res.* **106,** 495–503.

Xavier-Neto, J., Davidson, B., Simoes-Costa, M. S., Castro, R. A., Castillo, H. A., Sampaio, A. C., and Azambuja, A. P. (2010). Evolutionary origins of hearts. *In* "Cardiac Developmet and Regeneration," (R. P. Harvey and N. Rosenthal, eds.), pp. 3–45. Academic Press, Oxford.

Xu, H., Morishima, M., Wylie, J. N., Schwartz, R. J., Bruneau, B. G., Lindsay, E. A., and Baldini, A. (2004). Tbx1 has a dual role in the morphogenesis of the cardiac outflow tract. *Development* **131,** 3217–3227.

Yagi, H., Furutani, Y., Hamada, H., Sasaki, T., Asakawa, S., Minoshima, S., Ichida, F., Joo, K., Kimura, M., Imamura, S., Kamatani, N., Momma, K., *et al.* (2003). Role of TBX1 in human del22q11.2 syndrome. *Lancet* **362,** 1366–1373.

Yamagishi, H., Maeda, J., Hu, T., McAnally, J., Conway, S. J., Kume, T., Meyers, E. N., Yamagishi, C., and Srivastava, D. (2003). Tbx1 is regulated by tissue-specific forkhead proteins through a common Sonic hedgehog-responsive enhancer. *Genes Dev.* **17,** 269–281.

Yang, L., Cai, C. L., Lin, L., Qyang, Y., Chung, C., Monteiro, R. M., Mummery, C. L., Fishman, G. I., Cogen, A., and Evans, S. (2006). Isl1Cre reveals a common Bmp pathway in heart and limb development. *Development* **133,** 1575–1585.

Yelbuz, T. M., Waldo, K. L., Kumiski, D. H., Stadt, H. A., Wolfe, R. R., Leatherbury, L., and Kirby, M. L. (2002). Shortened outflow tract leads to altered cardiac looping after neural crest ablation. *Circulation* **106,** 504–510.

Yuan, S., and Schoenwolf, G. C. (2000). Islet-1 marks the early heart rudiments and is asymmetrically expressed during early rotation of the foregut in the chick embryo. *Anat. Rec.* **260,** 204–207.

Zaffran, S., Kelly, R. G., Meilhac, S. M., Buckingham, M. E., and Brown, N. A. (2004). Right ventricular myocardium derives from the anterior heart field. *Circ. Res.* **95,** 261–268.

Zhang, Z., and Baldini, A. (2008). *In vivo* response to high-resolution variation of Tbx1 mRNA dosage. *Hum. Mol. Genet.* **17,** 150–157.

Zhang, Z., and Baldini, A. (2010). Manipulation of endogenous regulatory elements and transgenic analyses of the Tbx1 gene. *Mamm. Genome* **21,** 556–564.

Zhou, W., Lin, L., Majumdar, A., Li, X., Zhang, X., Liu, W., Etheridge, L., Shi, Y., Martin, J., Van de Ven, W., Kaartinen, V., Wynshaw-Boris, A., *et al.* (2007). Modulation of morphogenesis by noncanonical Wnt signaling requires ATF/CREB family-mediated transcriptional activation of TGFbeta2. *Nat. Genet.* **39,** 1225–1234.

Zhou, B., von Gise, A., Ma, Q., Rivera-Feliciano, J., and Pu, W. T. (2008). Nkx2-5- and Isl1-expressing cardiac progenitors contribute to proepicardium. *Biochem. Biophys. Res. Commun.* **375,** 450–453.

Zhou, Y., Cashman, T. J., Nevis, K. R., Obregon, P., Carney, S. A., Liu, Y., Gu, A., Mosimann, C., Sondalle, S., Peterson, R. E., Heideman, W., Burns, C. E., *et al.* (2011). Latent TGF-beta binding protein 3 identifies a second heart field in zebrafish. *Nature* **474,** 645–648.

CHAPTER THREE

An Unauthorized Biography of the Second Heart Field and a Pioneer/Scaffold Model for Cardiac Development

José Xavier-Neto,* Sylvia Sura Trueba,* Alberto Stolfi,† Henrique Marques Souza,† Tiago José Pascoal Sobreira,* Michael Schubert,‡ *and* Hozana Andrade Castillo*

Contents

* Brazilian National Laboratory for Biosciences, Brazilian Association for Synchrotron Light Technology, Rua Giuseppe Máximo Scolfaro, Campinas, São Paulo, Brazil
† Department of Histology and Embryology, Institute of Biology, University of Campinas (UNICAMP), Campinas, São Paulo, Brazil
‡ Institut de Génomique Fonctionnelle de Lyon (UCBL, CNRS UMR 5242, ENSL, INRA 1288), Ecole Normale Supérieure de Lyon, 69364 Lyon Cedex 07, France

Current Topics in Developmental Biology, Volume 100
ISSN 0070-2153, DOI: 10.1016/B978-0-12-387786-4.00003-8

Abstract

The identification of subpharyngeal cardiac precursors has had a strong influence on the way we think about early cardiac development. From this discovery was born the concept of multiple heart fields. Early support for the concept came from gene expression, genetic retrospective fate mapping, and gene targeting studies, which collectively suggested the existence of a second heart field (SHF) on the basis of specific Islet-1 (Isl-1) expression, presence of two cardiac ancestral lineages, and compatible cardiac knockout phenotypes, respectively. A decade after the original studies, support for the SHF concept is dwindling. This is because in all bilaterian models studied, Isl expression in heart progenitors is not SHF-specific, because lineage data are best explained by alternative models including an older, truly ancestral, lineage of cardiac pioneers with unrestricted contribution to all cardiac segments and, finally, because the inflow-to-outflow segmental nature of the early vertebrate peristaltic heart has been reaffirmed with novel, less invasive, methodologies. Altogether, the paradigms derived from the discovery of subpharyngeal cardiac progenitors helped us shift from relatively simple models, which rely predominantly either on patterning, gene expression patterns or lineages, to a much more sophisticated body of knowledge in which all these parameters must be accounted. Thus, it is well possible that due consideration of the key elements contained in the inflow/outflow, pioneer/scaffold, ballooning, and SHF hypotheses may provide us with a unified framework of the early stages of cardiac development. Here, we advance into this direction by suggesting an intuitive model of early heart development based on the concept of an inflow/outflow scaffold erected by cardiac pioneers, one that is required to assemble all the subsequent cell contribution that emigrates from cardiac progenitor areas.

1. Introduction

Very few observations have been as influential to our ideas about cardiac development as the discovery of a group of cardiac progenitors in the subpharyngeal mesoderm. This finding gave rise to two major interpretations about the role of these progenitors in cardiac development and, eventually, originated the concept of novel cardiac fields (Abu-Issa *et al.*, 2004; Buckingham *et al.*, 2005; Kelly *et al.*, 2001; Mjaatvedt *et al.*, 2001; Waldo *et al.*, 2001).

The two major interpretations that derived from this initial discovery reflect two fundamentally different views on the meaning attributed to the existence of subpharyngeal cardiac progenitors. In the text that follows, we discuss these ideas under the label of organized hypotheses only for ease of discussion, because, in fairness, it is not possible to frame all individual opinions and contributions into these strict categories.

In one interpretation, dubbed here as the strong interpretation, the novel cardiac field, the second heart field (SHF), is seen as a revolutionary concept, which requires a complete revision of the logic that underlies our understanding of cardiac development. It can be argued that in the strong interpretation the traditional model of regulative cardiac development is substituted by a mosaic-like form of ontogenesis in which cell fate is predominantly controlled by distinct lineages of cell progenitors. In the other interpretation, here dubbed for contrast as the light interpretation, cardiac progenitors from the subpharyngeal mesoderm are embraced in a much less ambitious, more integrative, framework of heart development, one that accommodates the fundamental findings of classical experiments and the notions of embryonic field patterning (Abu-Issa *et al.*, 2004; Buckingham *et al.*, 2005; Mommersteeg *et al.*, 2010; Moorman *et al.*, 2007; Orts-Llorca and Jimenez Collado, 1967).

Why do we need yet another review on the SHF? We believe that it is important to clarify that, no matter how dominant the SHF concept may seem to be nowadays, it has not come to be uniformly interpreted or universally accepted (Abu-Issa *et al.*, 2004; Brade *et al.*, 2007;

Buckingham *et al.*, 2005; Gessert and Kuhl, 2009; Moorman *et al.*, 2007; Xavier-Neto *et al.*, 2010). We also believe that, 10 years after the initial observations, there is valuable and sufficient material for a critical, independent, assessment of how strong and light interpretations have lived up to expectations. Moreover, because earlier, independent, analyses of the SHF concept did not have access to all the evidence that we now have at our disposal, it was not possible to conclude beyond the level of a reasonable doubt whether the SHF concept was, or was not, fully supported by experimental evidence. Here we hope that by discussing the data that originated the SHF concept under a historic and evolutionary perspective, we will provide the reader with the opportunity to critically assess the pros and cons of this revolutionary idea.

Finally, it is our opinion that the experiments that supported the initial views about the SHF do not require radical departures from the classical ideas of cardiac development. Instead, these can now be easily integrated into an intuitive model of early heart development based on the simple concept of an inflow/outflow scaffold erected by cardiac pioneers, one that is required to assemble all the subsequent cell contribution that emigrates from cardiac progenitor areas.

2. Subpharyngeal Cardiac Precursors

In 2001, three independent studies suggested a puzzling feature of cardiac development, namely, the existence of extracardiac progenitors of outflow tract and right ventricle cells in the subpharyngeal mesoderm.

Two studies in the chick embryo showed that cells from the subpharyngeal mesoderm contributed to the cardiac outflow tract (Mjaatvedt *et al.*, 2001; Waldo *et al.*, 2001). In one study, the progenitor region was defined as an area just anterior to the heart tube, while in the other study, the progenitor region was a comparatively larger territory, not only anterior to the heart tube, but also encompassing mesoderm from the pharyngeal arches. These new sources of myocardial cells in the sub pharyngeal mesoderm were dubbed secondary heart field in one study and anterior heart field in the other study (Mjaatvedt *et al.*, 2001; Waldo *et al.*, 2001). An anterior group of cardiac progenitors was also found in the pharyngeal mesoderm of transgenic mice through the analysis of expression patterns of a LacZ reporter under control of the Fgf10 (Fibroblast Growth Factor 10) locus. In these mice, LacZ-expressing cells migrated through the pharyngeal arches and ended up in the outflow tract and right ventricle. Corroborating these results, fate mapped embryos indicated that mesodermal cells dorsal to the heart contribute to the outflow tract. Based on these results, the authors proposed that the embryonic heart was composed of cells derived from two myocardial precursor populations: one lateral mesoderm progenitor area, which gives rise to the early heart tube and inflow region, as well as another population of Fgf10-expressing cells in the splanchnic and pharyngeal mesoderm, which gives rise to the outflow tract and right ventricle (Kelly *et al.*, 2001).

These findings were intriguing because they suggested the existence of an extracardiac source of heart cells, a radically different concept from the then accepted model of cardiac development in which all heart segments originated from the traditional cardiac precursor area, that is, the bilateral fields of the lateral mesoderm, which later form the cardiac crescent (Dehaan, 1963a; Rosenquist and DeHaan, 1966).

2.1. Using soft words and hard arguments

Like every effort to summarize the myriad of opinions about heart development, the following one is a just a sampling, but a sampling of the most extreme views within the spectrum, which will be useful to establish the boundaries of the various interpretations about the role of subpharyngeal cardiac progenitors in heart development (Table 3.1).

2.1.1. Strong

We perceive the strong interpretation as a revolutionary hypothesis of cardiac development. In the strong view, the discovery of extracardiac precursors is seen as the tip of the iceberg of a body of knowledge that demands extensive changes in established paradigms. What are then the distinctive elements of the strong view?

It can be argued that the strong view requires the acceptance of the idea that the heart is built from two distinct sources of progenitors cells: the well-known crescent derived from the bilateral precursors of the lateral mesoderm (Rawles, 1952) and the novel subpharyngeal area (Kelly *et al.*, 2001; Mjaatvedt *et al.*, 2001; Waldo *et al.*, 2001). In the strong view, this dichotomy is supported by marker-specific expression and by a lineage-directed mode of development based on two distinct clonal populations often loosely associated to the traditional heart forming area and to the sub pharyngeal mesoderm. The strong view can also be readily identified by a typical dismissal of the segmental nature of the heart, which is implicit in the downplay of the evidence for anterior–posterior (AP) patterning (a.k.a. craniocaudal, inflow/outflow, or venous–arterial) and of the signaling mechanisms that establish cardiac AP patterns (see Rosenthal and Xavier-Neto, 2000; Xavier-Neto *et al.*, 2001).

2.1.2. Light

We recognize the light interpretation as an inclusive view, which tries to integrate the unexpected finding of apparent extracardiac precursors in the existent knowledge structure of cardiac development. What are then the specific features of the light interpretation?

Table 3.1 Distinguishing features of strong and light interpretations of the second heart field

	Sources of myocardial cells	Requires fundamental alterations of the model	Fields defined on the basis of specific markers	Fields defined on the basis of retrospective lineages	Regional segmentation	References
Light	One	No	No	No	Yes	Abu-Issa *et al.* (2004), Moorman *et al.* (2007)
Strong	Two	Yes	Yes	Yes	No	Buckingham *et al.* (2005), Meilhac *et al.* (2004)

The light interpretation acknowledges the existence of a single, but complex, cardiac field located in the traditional heart-forming areas of the lateral mesoderm, rather than two distinct cardiac fields. In the light interpretation, cardiac fates are imposed on precursors through a combination of factors that depend on initial location and on patterning. The light interpretation embraces the concept of cardiac segmentation but does not accept the concepts of two separate fields defined on the basis of specific marker expression or of two original clonal lineages (Abu-Issa *et al.*, 2004; Moorman *et al.*, 2007; Perez-Pomares *et al.*, 2009).

Below, we discuss the pros and cons of these two interpretations. As it will be clear, we believe that the light interpretation offers a more moderate and inclusive point of view than the strong interpretation. Nonetheless, it is our opinion that the discovery of subpharyngeal cardiac progenitors can be best accommodated into a new model of cardiac development based on the idea that cardiac pioneers first set up an inflow/outflow-patterned scaffold, which is responsible for the early function of the early peristaltic vertebrate heart, as well as for the integration of all subsequent contribution that will shape the peristaltic organ into a chambered heart (see below).

3. Models of Cardiac Development

3.1. Mosaic, regulative or a patterned mosaic?

The strong interpretation depends on vigorous notions of lineage-directed development but leaves behind many features of developmental patterning, so that it is possible to interpret it as a type of ontogenesis akin to the mosaic model of development. The mosaic model is a form of development in which cell potency and fate are controlled directly as a function of descendency from a limited number of original cell lineages established at the early cleavage stages. As such, it contrasts with regulative development, in which cell fate is regulated by positional parameters via signaling mechanisms, including the well-known forms of axial signaling, which pattern all organ progenitor fields (Wolpert and Tickle, 2011).

In the strong interpretation, cardiac ontogenesis is explained on the basis of two major lineages: the first lineage, the one responsible for the formation of the whole left ventricle, fractions of right ventricle, AV canal, and atrium; and the second lineage, which includes all of the outflow tract and other cardiac segments, but the left ventricle (Buckingham *et al.*, 2005; Meilhac *et al.*, 2004). However, lineages alone are not sufficient to understand cardiac development and one can argue that any form of development that does not include patterning is an incomplete one (Lawrence and Levine, 2006). As far as the strong interpretation is concerned, a major issue left behind is segmental identity. In other words, how is it, for instance, that a right ventricular cell is specified as a ventricular cell rather than as an atrial cell or *vice versa*? The strong interpretation recognizes that both first and second lineages give rise to cells with these right ventricular and atrial phenotypes but offers no explanations for the generation of these specific cell types.

More importantly, however, no mosaic model of development can account for the outcome of classic experiments of cardiac field rotation or transplantations. In one experiment, the chicken cardiac field was removed, rotated 180° in the AP axis, and returned to matched embryos. If the procedure was performed before stage HH8, normally polarized hearts were often obtained, while none of them were found after HH8 (Orts-Llorca and Jimenez Collado, 1967; Patwardhan *et al.*, 2000). In other experiment, cells fated to give rise to posterior heart segments assumed electrophysiological characteristics of anterior heart segments when transplanted to the prospective ventricle (Inagaki *et al.*, 1993). Collectively, these studies indicate that when early cardiac progenitors are rotated about the AP axis or transplanted into different segment levels, they assume identities of their new environments, rather than those typical of their origins. This demonstrates an extraordinary regulative plasticity that is unaccounted for in the strong interpretation.

The available experience suggests that axial patterning information is a prerequisite for organ development and that signaling involves the transmission of information from specific centers in the embryo, which often deliver their signals through various soluble extracellular mediators (Wolpert and Tickle, 2011). In this sense, it is useful to consider that the Isl-1-expressing progenitor population, proposed to form a distinct SHF, is not at all insulated from extra cellular signals of the BMP-Smad1 pathway, which governs the traditional heart-forming region. By targeting the mouse Nkx2-5 locus, Prall *et al.* (2007) induced a massive activation of Bmp/Smad signaling that not only crossed the Isl-1-expressing population to reach the traditional cardiac-forming area but also projected as far as the ectoderm. This indicates that there is no evidence for clonal restriction between Isl-1-espressing cells and traditional cardiac-forming area.

In summary, there are few defensible reasons to believe that vertebrate cardiac identities are primarily established on the strict basis of progenitor lineages, as would be predicted from a mosaic model. Rather, the evidence suggests that lineage information, although necessary (Cui *et al.*, 2009), is not sufficient for cardiac morphogenesis. Any complete model of cardiac development must incorporate patterning (e.g., in the form of AP, DV, or LR patterning). It is patterning that endows cardiac development with an extensive regulatory capability, which is consistent with more modern views of development in which both lineage and regulation work together to gradually establish cell fates (Lawrence and Levine, 2006).

3.2. Retrospective fate mapping highlights the intrinsic unity of the cardiac progenitor population

One of the pillars supporting the strong interpretation is the view that retrospective fate mapping established that heart development can be sufficiently represented by two distinct ancestral cell lineages: a first lineage, which gives rise to regions that form the early cardiac primordium (i.e., the whole left ventricle plus fractions of right ventricle, AV canal, and atrium); and a second lineage, which includes all cardiac segments, but the left ventricle) (Buckingham *et al.*, 2005).

The important point to be made here is that this view of cardiac development is not complete. As depicted in Fig. 3.1, early cardiac development can only be adequately described through retrospective clonal analysis if three, rather than two ancestral lineages are considered (see Moorman *et al.*, 2007; Xavier-Neto *et al.*, 2010). These are the first and second lineages, plus the neglected, often forgotten, myocardial founder cells of Meilhac *et al.* (2004) (Fig. 3.1A).

Myocardial founders constitute the earliest and are among the bigger clones identified in the retrospective analysis of early cardiac development (the very large clones). Myocardial founders occupy all the cardiac segments from the 8.5 dpc mouse tube, including the proximal outflow tract, the embryonic right ventricle, the embryonic left ventricle, the atrioventricular canal, the body of the atrium, and the left and right atria (Meilhac *et al.*, 2004) (Fig. 3.1B).

Myocardial founders are rare. In fact, only three of such clones were detected (Meilhac *et al.*, 2004), and this may have contributed to their neglect in subsequent analyses. However, the myocardial founders represented in these very large clones are rare not because they are an irrelevant oddity of cardiac development, but because they are formed in a small cardiac primordium containing only a limited number of cells, which reduces the probability that these cells can be detected by a random event of recombination such as the one that activates LacZL expression by the mouse reporter gene (Meilhac *et al.*, 2004).

Myocardial founders provide the key to a unified view of cardiac development (see below). By its unrestricted contribution to all segments of the early 8.5 dpc mouse heart, they indicate that there is indeed a very early group of cardiac progenitors ancestral even to the two cardiac lineages emphasized by Meilhac *et al.*, 2004, which contain the pioneer cells for all the cardiac segments present at the early looping stages of the mouse heart. This is in stark contrast with the strong interpretation, which claims that the heart is a composite formed from contributions of two distinct fields: one containing essentially precursors of left ventricle, AV canal, and atria, and other containing precursors for all cardiac segments, but the left ventricle (Buckingham *et al.*, 2005; Meilhac *et al.*, 2004).

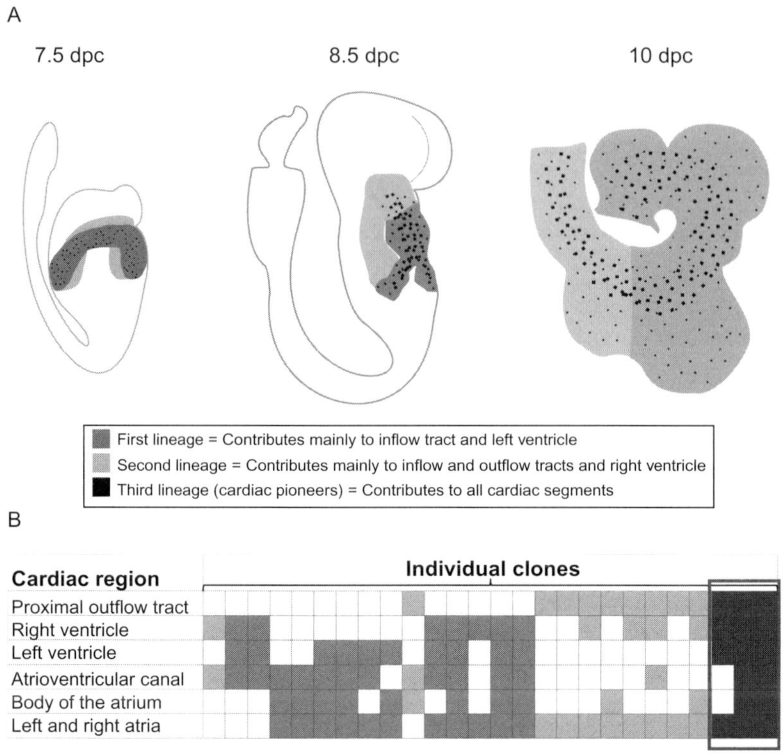

Figure 3.1 Cardiac progenitor lineages. (A) Schematic representation of embryonic mouse cardiac progenitor domains and cardiac territories highlighting the three cardiac-forming lineages defined by retrospective clonal analysis at 7.5, 8.5, and 10 dpc (days post-coitum) as described by Meilhac *et al.* (2004). The first lineage (blue) gives rise to the whole left ventricle, plus fractions of right ventricle, AV canal, and atrium. The second lineage (yellow) forms all cardiac segments, but the left ventricle. The third lineage (dark blue) is the ancestral one and corresponds to the myocardial founder cells, which are present in and contribute to outflow tract, right ventricle, left ventricle, atrioventricular canal, the body of the atrium and left and right atria (red box in B). The chambered heart scheme at 10.5 dpc was adapted from Christoffels *et al.* (2000). (B) Adaptation of a scheme illustrating individual clonal contributions to cardiac segments. Color codes represent three distinct categories of cell lineages characterized by retrospective analyses. Blue, first lineage; yellow, second lineage; dark blue, third lineage (cardiac pioneers). Note that cardiac pioneers are present in all cardiac segments (scheme modified from Meilhac *et al.*, 2004). (For interpretation of the references to color in this figure legend, the reader is referred to the Web version of this chapter.)

In summary, a closer look at the results of retrospective fate mapping indicates that cardiac morphogenesis requires one initial, pioneer, progenitor population with the ability to give rise to cells in all cardiac segments.

4. A Historical Account of the SHF Concept

4.1. Is the SHF a novel cardiac progenitor domain?

There have been multiple attempts to decide whether cardiac progenitors in the subpharyngeal mesoderm deserve, or not, the label of SHF. Some of these efforts involved arguments over the precise developmental meaning of a "field," while some others placed emphasis on the distinct gene regulatory networks that underlie morphogenesis from subpharyngeal cardiac precursors (Black, 2007; Brade *et al.*, 2007; Gessert and Kuhl, 2009; Kelly, 2005; Moorman *et al.*, 2007; Perez-Pomares *et al.*, 2009).

Here, we believe that the matter rests over far more simple and intuitive perceptions. As we see it, the question boils down to whether subpharyngeal cardiac precursors comprise, or not, an autonomous, independent, progenitor population, physically, or otherwise, distinct (see Fig. 3.2) from the traditional cardiac precursor areas previously defined in the epiblast, mesendoderm, and anterior mesoderm (Dehaan, 1963a; Garcia-Martinez and Schoenwolf, 1993; Rosenquist and DeHaan, 1966).

With the advantage of hindsight, it is now clear that an answer to this fundamental question had already been provided by one of the original studies (Kelly *et al.*, 2001), although this could only be fully appreciated in 2003, following a detailed description of mouse cardiac expression patterns for the LIM homeodomain transcription factor Islet-1 (Isl-1), then considered a genetic marker for cardiac precursors outside the traditional cardiac progenitor area (Cai *et al.*, 2003).

A careful analysis of expression patterns from endogenous Isl-1 and Fgf10 genes, as well as from a LacZ transgene trapped near the Fgf10 locus, shows that, rather than forming an independent, physically restricted, population of progenitors, subpharyngeal cardiac precursors are in fact a cranial extension of heart progenitors already present at the heart-forming regions of the mesoderm (Kelly *et al.*, 2001). From the patterns described by Cai *et al.*, 2003, it is possible to determine that Isl-1 expression spreads through the whole AP extent of the splanchnic mesoderm dorsal to the looped heart, such that there is constant physical continuity between subpharyngeal cardiac precursors and cardiac progenitors that occupy the traditional heart-forming areas of the mesoderm (Buckingham *et al.*, 2005; Cai *et al.*, 2003).

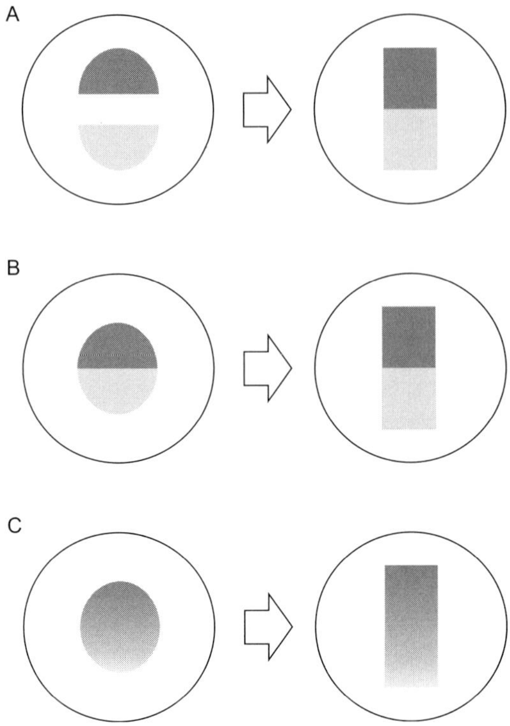

Figure 3.2 Models for generation of organ diversity. (A) Two physically distinct progenitor fields contribute to the development of different structures from the same organ. (B) Differential gene expression divides a single progenitor field, resulting in two distinct progenitor domains, which give rise to different structures from the same organ. (C) One single field is patterned, resulting in a complex field, which gives rise to different structures from the same organ. (For color version of this figure, the reader is referred to the Web version of this chapter.)

In summary, we conclude that the SHF is not a novel cardiac progenitor area as proposed in the strong interpretation, but an anterior extension of the traditional heart-forming areas of the mesoderm.

4.2. Is the SHF a cardiac progenitor region defined on the basis of specific gene expression?

After the study by Cai *et al.* (2003), it was no longer possible to claim that cardiac progenitors in the subpharyngeal mesoderm (anterior heart field or secondary heart field) formed a physically restricted cardiac progenitor area distinct from the traditional heart-forming areas of the mesoderm. To accommodate this fact, it was later suggested that the anterior and secondary heart fields would be best understood as a SHF, set apart from the traditional cardiac progenitor area, which was then rechristened as the first heart field (Buckingham *et al.*, 2005).

At this point, the SHF concept was useful because SHF progenitors did indeed seem to express a set of specific markers (Buckingham *et al.*, 2005). In this sense, one particular genetic marker, Isl-1, seemed to embody the whole dichotomy between the SHF and the traditional cardiac progenitor area. Double *in situ* hybridization (ISH) in the mouse embryo using probes chosen to represent cells from the SHF (Isl-1) and from the traditional heart-forming area (α-MLC2a) revealed two juxtaposed regions shaped into the familiar form of an inverted crescent. The inner crescent marked by Isl-1 expression was assumed to represent the SHF, while the external crescent marked by α-MLC2a expression was taken as the traditional heart-forming region at 7.0 dpc (Cai *et al.*, 2003).

The case for Isl-1 as a *bona fide* marker of the SHF was further strengthened by analysis of the cardiac Isl-1-null phenotype. The hearts of these embryos lack outflow tract, right ventricle and display a small, stunted, central atrium. This result suggests that Isl-1 is necessary for normal morphogenesis of all these structures, which are exactly the same ones suggested to depend on the SHF (Cai *et al.*, 2003). Further, Cre-LoxP fate map studies showed that cardiac precursors that expressed the LacZ gene under the ultimate control of the Isl-1 locus were preferentially found in the proximal outflow tract, right ventricle, AV canal and atria, although it was evident that labeled cells also occupied the left ventricle (Cai *et al.*, 2003).

Appropriate conclusions at the time were that the Isl-1-expressing region included the subpharyngeal cardiac progenitors (former anterior heart field or secondary heart field), plus the cells that remained in the posterior domain of the inner crescent and that, together, they could be conveniently regarded as a progenitor domain (the SHF) distinct from the traditional heart-forming region.

It turns out, however, that there were fewer reasons to infer, especially in face of a limited methodology such as standard ISH and of the heterogeneity of the markers chosen (a transcription factor, Isl-1, for the SHF and a structural protein, α-MLC2a, for the traditional heart-forming region), that there was no overlap between the SHF and the traditional heart-forming region, a conclusion that would soon be questioned by a reassessment of the older literature, by new results coming from Cre-LoxP-fate mapping and by analysis of other model species.

4.3. Is the SHF a progenitor population defined by expression of specific markers?

Retrospectively, one of the first hints that something was amiss with the strong interpretation was contained in a study of Isl-1 expression in the chick mesoderm published before the discovery of subpharyngeal cardiac precursors. Yuan and Schoenwolf (2000) showed that, rather than specifically label cardiac progenitors of the SHF, Isl-1 is first expressed in the anterior lateral mesoderm before formation of the coelom, which divides the lateral mesoderm into somatic and splanchnic layers.

Whether the chick Isl-1 pattern represented a conserved amniote feature or an avian-specific character, it was not known at the time. However, additional evidence against the idea that specific Isl-1 expression supports the strong concept came from genetic fate-mapping studies using an Isl-1-Cre driver and floxed LacZ-reporter mice (Cai *et al.*, 2003; Sun *et al.*, 2007). In these studies, most cardiac cells that activate the Isl-1-gene were found at the proximal outflow tract, right ventricle, and atria. However, it was also clear that labeled cells formed a small, but conspicuous, contingent of the left ventricle. Naturally, this represented an evident inconsistency of the strong concept, because, in this interpretation, the left ventricle is always portrayed as being exclusively derived from the traditional heart-forming areas of the mesoderm (Buckingham *et al.*, 2005; Meilhac *et al.*, 2004).

Looking back, it is easy to see that the inconsistencies above were not sufficiently compelling to force a general revision of the strong concept. However, stronger evidence in that direction came in 2007 when Prall and colleagues demonstrated that mouse Isl-1, much like chick Isl-1, is expressed throughout the lateral mesoderm, which gives rise to the somatic and splanchnic mesoderm, the latter being the origin of both the SHF and of the traditional heart-forming areas (Prall *et al.*, 2007). The findings of Prall *et al.* (2007) have recently been independently confirmed by a detailed depiction of the time course of Isl-1 and Nkx2-5 expression during early mouse cardiac development (Mommersteeg *et al.*, 2010).

Consistent with the much more general role for Isl-1 in mesoderm development as predicted by the studies above, it was shown that the Isl-1 activation domain was far larger than the one obtained in the original Cre-LoxP fate mapping. Using a Gata-4-based recombination reporter more susceptible to Cre recombination than the standard Rosa26fslz allele, Ma *et al.* (2008) showed that cells that activate Isl-1 contribute massively to all cardiac segments. Importantly, this included the left ventricle, in which Isl-1 seems to play a relevant role, as demonstrated by the severe left ventricular underdevelopment produced by Isl-1 deficiency. Therefore, it seems that the preferential labeling of cells in the proximal outflow tract, right ventricle, and atria detected in earlier studies was more the result of inefficient recombination of the Rosa26 allele utilized than any intrinsic clonal restriction of cardiac progenitors in the SHF. This particular result emphasized how difficult it is to draw definitive conclusions on the basis of a powerful, but complex and capricious technology such as genetic fate mapping. Thus, there is little evidence to support the claim that the SHF is defined on the basis of specific marker expression.

In summary, rather than being a specific marker for a distinct cardiac population (the SHF), Isl-1 expression in cardiac progenitors is being currently understood not only as a player in the ancestral bilaterian core transcription factor network necessary to assemble a pumping organ (Mann *et al.*, 2009a), but also as a component that introduced a fourth, temporal, dimension to the process of pump specification, growth, and complexity (Moorman *et al.*, 2007; Perez-Pomares *et al.*, 2009). It is possible to speculate that these two roles are controlled by at least two, perhaps more, distinct gene regulatory networks: the network controlling early developmental Isl expression in the splanchnic mesoderm of all bilaterian models studied, except in lampreys and, possibly, in amphioxus; as well as a late developmental module represented by expression in subpharyngeal pump progenitors from all bilaterian models investigated, but, perhaps, in amphioxus (see below Fig. 3.3).

5. Evolution of Isl Expression in Cardiac Development

5.1. Origins of Isl-1-expression in cardiac progenitors: Rewinding, not rewriting, the tape of cardiac evolution

Developmental and gene expression studies from key vertebrate and invertebrate species are now available and allow for a preliminary analysis of the evolutionary context in which Isl-1-expressing progenitor populations of the vertebrate heart evolved.

In an evolutionary analysis like this, multiple species are investigated to ascertain, among many other things, whether a particular character, in this case Isl-1-expressing cardiac progenitors, can be interpreted as ancestral, or derived, in a given group of species.

However, before we can address this subject, it is important to realize that, although many developmental paradigms such as the SHF stemmed from the study of amniote models such as mammals and avians, it is not always possible, or correct, to assume that these mechanisms are always present in more primitive vertebrates, invertebrate chordates, or other invertebrates (Xavier-Neto *et al.*, 2007). This is because mammalian and avian body plans are relatively new when compared to the primitive body plans of fishes and amphibians and, therefore, it is likely that mammals and avians evolved novel developmental mechanisms, or modified ancestral ones to such a point that it may not be possible to recognize typical mammalian, or avian features in the ontogenesis of more primitive body plans (Xavier-Neto *et al.*, 2010).

In sum, it is important to ask whether Isl-1 expression patterns are being analyzed in the natural developmental context of more primitive vertebrate

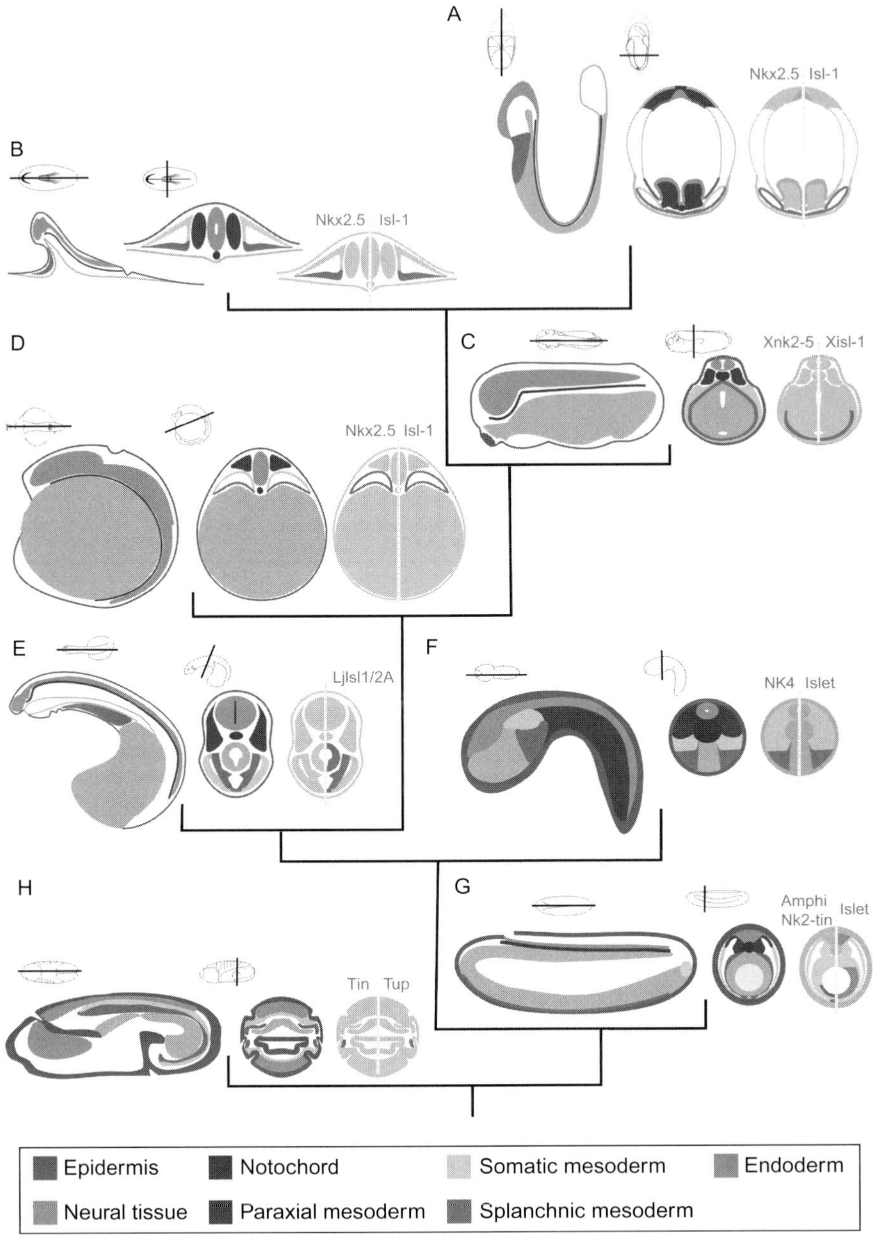

Figure 3.3 Isl and NK4 ortholog expression during circulatory pump ontogenesis in bilaterian model organisms. Schemes of longitudinal and transverse sections of embryos at post-segmentation stages. Section level is indicated above each drawing. Embryonic territories are color-coded: epidermis (dark blue), neural tissue (light blue), notochord (violet), paraxial mesoderm (red), somatic mesoderm (light pink), splanchnic mesoderm (dark pink) and endoderm (orange). Right transversal sections show the

species, or if we are automatically projecting amniote models in the biology of such basal vertebrates.

With these caveats, it is possible to integrate the knowledge obtained from the study of mice, chicken, frogs, teleosts, cyclostomes, ascidians, and flies to make informed speculations about the position and role of Isl-1-expressing cardiac precursors during the early heart development of the vertebrate ancestor.

5.2. A comparative analysis of Isl expression in vertebrate chambered heart development

5.2.1. Mice

In the mouse, it is well established that Isl-1 is expressed in both somatic and splanchnic mesoderm in 7.0–7.5 dpc embryos (Mommersteeg *et al.*, 2010; Prall *et al.*, 2007) (Fig. 3.3A). Later, Isl-1 expression is progressively lost in cardiac progenitors, but the dynamics are heterogeneous across the cardiac field. In a lateral subset, loosely reminiscent of the traditional cardiac-forming region, Isl-1 expression is lost, while Nkx2-5 expression is maintained. In contrast, in a medial subset, similar to the proposed SHF, Isl-1 expression is only gradually dissipated, while Nkx2-5 expression is simultaneously activated (Mommersteeg *et al.*, 2010). Importantly, the data leaves no doubt that cardiac progenitors coexpress Isl-1 and Nkx2-5 in early cardiac development (Fig. 3.3A).

---

endogenous expression of NK4 orthologs on the left side of the embryo and of Isl orthologs on the right side to highlight the substantial overlap between these expression patterns. (A) Schematic sections of mouse embryo at 7.5 dpc, showing Isl-1 and Nkx2.5 coexpression in somatic and splanchnic mesoderm. (B) Sections of chicken embryo at HH8 showing Isl-1 and Nkx2.5 in the splanchnic mesoderm. (C) Illustrations of a *Xenopus* embryo at NF 25 with Isl-1 and Nkx2.5 expression domains overlapping in the cardiogenic-splanchnic mesoderm. (D) Schematic sections of zebrafish embryo at 16 hpf displaying Isl-1 and Nkx2.5 coexpression in splanchnic and somatic mesoderm. (E) Scheme of a lamprey embryo at stage 25 this is the initial stage of Isl-1/2 expression restricted to the myocardium of the linear heart tube and to the subpharyngeal mesoderm. Lamprey Nkx2.5 ortholog expression is as yet unavailable. (F) Illustration of *Ciona intestinalis* mid-tailbud stage with *Ciona* Isl-1/2 (islet) weakly expressed in some pump precursors (medial trunk ventral cells, TVCs), while the *Ciona* Nkx2.5 ortholog (NK4) is expressed in all TVCs, indicating that *Islet* and *NK4* are coexpressed in pump precursors. (G) Schematic representations of an amphioxus embryo at late neurula stage (18 h) show AmphiNk2-tin expression in splanchnic mesoderm and ventral endoderm. Isl-1/2 ortholog is expressed in neural plate and endoderm, suggesting that Nk2-tin and Isl-1/2 are coexpressed in endoderm-derived tissues. (H) Schematic representation of *Drosophila* embryo (ventral side up) at late stage 11, when Tailup (Tup), the fly Isl-1/2 ortholog, is expressed with Tinman (Tin), the fly NK4 gene, which is expressed in all myocardial and pericardial cells of cardiac mesoderm. (See Color Insert.)

5.2.2. Chicken

In the chick embryo, Isl-1 is expressed throughout the lateral mesoderm before coelomic formation (HH5+) and, thereafter, Isl-1 expression concentrates in the splanchnic mesoderm. At stage HH8, Isl-1 is coexpressed with Nkx2-5 in the splanchnic and in the cardiogenic mesoderm (Fig. 3.3B), but, subsequently, disappears from cardiac mesoderm. Therefore, as discussed above, it is now clear that, as far as mouse and chick cardiac fields are concerned, Isl-1 is expressed in all early cardiac precursors (Nathan *et al.*, 2008; Yuan and Schoenwolf, 2000) and that cardiac progenitors coexpress Isl-1 and Nkx2-5.

5.2.3. Frogs

The amniote patterns of Isl-1 and Nkx2-5 expression described above are remarkably similar to the ones found in the amphibian *Xenopus laevis*. *Xenopus* Isl-1 is expressed in a crescent-shaped progenitor domain of the neurula (Nieuwkoop-Faber (NF) stages 17/18). *Xenopus* Nkx2-5 is expressed throughout this Isl-1 domain, but for its anterior-most part. This extensive overlapping expression of XIsl-1 and XNkx2.5 in the early cardiac primordium persists at early tail bud stages (NF25) (Fig. 3.3C). With further development, Isl-1 expression concentrates in the anterior dorsolateral splanchnic mesoderm and is not observed in the cardiac tube proper of stage 30 embryos. While it has not yet been determined whether Isl-1 and Nkx2-5 are expressed in the same cells, it is clear that the early amphibian cardiac progenitor domain is structured as a single anatomic field that contains heterogeneous, but overlapping groups of cardiac precursors (Brade *et al.*, 2007).

5.2.4. Zebrafish

Sufficiently detailed zebrafish Isl-1 expression patterns for all the relevant stages of cardiac development are yet to be published. In spite of this, it is possible to draw some preliminary conclusions from published reports and from data available in public data bases (Bradford *et al.*, 2011; de Pater *et al.*, 2009; Hami *et al.*, 2011). These data indicate that Isl-1 is expressed in splanchnic and somatic mesoderm, roughly corresponding to the medial region of the bilateral cardiac progenitor area at 16 hpf (Hami *et al.*, 2011) (Fig. 3.3D). At this stage of development, zebrafish Nkx2.5 is also expressed in the bilateral cardiac precursors, but in a broader lateral domain than the one of Isl-1-expressing cells (Leung *et al.*, 2008; Lin and Xu, 2009) (Fig. 3.3D). At 19 hpf, before morphogenesis of the heart tube, Isl-1 expression is found at the lateral cardiac-forming region, which corresponds, approximately, to the atrial progenitor domain (de Pater *et al.*, 2009). Further, from 24 to 48 hpf, Isl-1 is expressed in the proximal arterial pole, while at 72 hpf, it is only observed in the bulbus arteriosus and is completely absent from myocardial cells, while Nkx2.5 is expressed throughout the heart tube at these stages (Hami *et al.*, 2011; Tu *et al.*, 2009). Therefore, these data show that zebrafish Isl-1 and Nkx2.5 are coexpressed in the early bilateral cardiac field.

5.2.5. Lampreys

In lampreys, the situation contrasts with the other vertebrate models in that there seems to be no evidence for expression of the lamprey Isl-1/2 homolog in cardiac progenitors before the heart tube stage, although there is conspicuous muscle actin expression in the two bilateral progenitor areas that form the heart tube. Despite this, the Isl-1/2 homolog is expressed in the myocardium of the linear heart tube and in the subpharyngeal mesoderm (Fig. 3.3E), domains in which Tbx4/5 and Tbx20 are also active (Kokubo *et al.*, 2010). Unfortunately, there is as yet no published report of lamprey Nkx2-5 ortholog expression.

5.3. A comparative analysis of Isl expression during the ontogenesis of invertebrate circulatory pumps

More often than not, it is convenient to analyze developmental mechanisms first characterized in amniotic vertebrates such as mice and chicken in the most inclusive fashion, with the help of outgroups, such as invertebrate chordates and other invertebrates, to try and establish whether these amniote mechanisms are really ancestral in vertebrates, or evolutionary novelties of amniotes. The evolutionary context in which Isl-1 came to be involved with cardiac ontogenesis is one of these situations.

Before multiple deuterostome genomes become available, the organism of choice for such analyses was undoubtedly the fruit fly *Drosophila melanogaster*, an ecdysozoan protostome. *Drosophila* has been a crucial model for the gene regulatory networks that underpin vertebrate cardiogenesis. However, the heuristic value of cross-comparisons between vertebrate cardiac ontogenesis and *Drosophila* dorsal vessel development has to be weighed against the fact that *Drosophila* and vertebrates are only very distantly related, since their evolutionary lines diverged around 990 million years ago (Hedges and Kumar, 2003). Thus, in contrast to a popular misconception, *Drosophila* is not a primitive organism in relation to vertebrates, but, instead, a very derived species in relation to its own evolutionary branch (Schaeffer *et al.*, 2008).

As an alternative, arthropod models that branch more basally in the phylogenetic tree might be used as an attempt to identify features conserved from their last common ancestor with vertebrates that might have been lost in derived organisms, such as the fruit fly. In any case, however, great care must be exercised before it can be concluded that arthropod characters similar to amniote ones represent an ancestral trend (Xavier-Neto *et al.*, 2007).

The relatively recent decoding of invertebrate chordate (tunicates and cephalochordates) and ambulacrarian (hemichordates and echinoderms) genomes opened the exciting possibility to study the origins of the vertebrate chambered heart in its real evolutionary context, which is deuterostome evolution (Dehal *et al.*, 2002; Freeman *et al.*, 2008; Putnam *et al.*, 2008). In this sense, it has been increasingly accepted that vertebrates are more closely related to tunicates than to cephalochordates and that the solitary, pericardially encased, striated muscle-driven pump of tunicates is a closer homolog of the vertebrate chambered heart than the multiple, nonpericardial, smooth muscle-like-driven peristaltic vessels of cephalochordates (Delsuc *et al.*, 2006, 2008; Glenn Northcutt, 2005; Northcutt, 2005; Passamaneck and Di Gregorio, 2005; Simoes-Costa *et al.*, 2005).

5.3.1. Ascidian tunicates

In the ascidian tunicate, *Ciona intestinalis*, progenitor cells from the peristaltic pump share a common origin with precursors of the atrial siphon muscles (ASMs, not to be confused with the atrial cardiac chamber of vertebrates), which surround the excurrent openings. More specifically, the cells that give rise to the peristaltic pump and to the ASMs are direct descendents of the B7.5 lineage, which gives rise to an anterior-most, mixed, group of peristaltic pump and ASM precursors known as trunk ventral cells (TVCs) as well as a posterior group of tail skeletal muscle cells. Live imaging of the *C. intestinalis* larva showed that medial TVCs are peristaltic pump precursors and lateral TVCs are ASM progenitors (Stolfi *et al.*, 2010).

At the tail bud stage (~16 hpf), the *Ciona* Isl-1/2 ortholog is transiently and weakly expressed in some pump precursors (medial TVCs). In contrast, Isl-1/2 expression is strong in ASM precursors (lateral TVCs) (Fig. 3.3F), which migrate dorsally before metamorphosis to surround the atrial siphon. Important to our analysis, the *Ciona* Nkx2.5/2.3/2.6/2.7/2.8/2.9/2.10 ortholog (Holland *et al.*, 2003) herein dubbed, Ci-NK4, is expressed in all TVCs (Fig. 3.3F). Eventually, Ci-NK4 expression is lost in the late juvenile period, only to reappear, much later, at the onset of metamorphosis, when it is restricted to the definitive peristaltic pump (Davidson and Levine, 2003; Stolfi *et al.*, 2010). In summary, giving the mosaic mode of ascidian development, it is safe to conclude that Isl-1 and Ci-NK4 are coexpressed in pump precursors.

Collectively, the above-described findings led to the suggestion that the relationship between pump precursors and ASMs is homologous to the one between the traditional cardiac-forming region and the amniote subpharyngeal mesoderm, which gives rise to a subset of craniofacial muscles (Nathan *et al.*, 2008).

Homologizing characters between tunicates and vertebrates has never been easy, mostly because tunicate body plans, development, and genomes are simplified and highly divergent from the other chordate body plans (Dehal *et al.*, 2002; Hotta *et al.*, 2007; Passamaneck and Di Gregorio, 2005). In the case of the tunicate atrial siphon, it is particularly difficult to trace definitive homology relationships because in the ascidian body plan, but not in the body plan of thaliaceans (a distinct group in the tunicate subphylum), the AP axis was secondarily twisted in the adult to place the oral and aboral ends of the digestive tract at the level of the pharynx, which makes literal comparisons with the vertebrate pharynx difficult.

It is also pertinent to note that more than one vertebrate affinity is possible for the tunicate atrial siphon. The most explored homology is between the tunicate atrial siphon primordium and the vertebrate otic placode (reviewed in Kourakis *et al.*, 2010, and references therein). This homology is supported by morphological and topological characters, as well as by gene expression patterns and FGF signaling requirements for induction, but the homology is not accepted by everyone (Northcutt, 2005). Thus, the compact structure of the tunicate body makes it difficult to establish definitive homology relationships between the atrial siphon and/or its associated elements and candidate vertebrate structures. Of note, the muscles that operate the oral siphon, into which water enters toward the mouth and pharynx, do not derive from B7.5 cells. Despite these caveats, we used the similarities proposed between pump precursors plus ASMs in tunicates and traditional cardiac-forming region plus the amniote subpharyngeal mesoderm in vertebrates (Stolfi *et al.*, 2010) to orient our comparative analyses of bilaterian transcription factor expression.

5.3.2. The cephalochordate amphioxus

The circulatory system of amphioxus is driven by peristaltic vessels composed of smooth muscle-like myocardial cells (Holland *et al.*, 2003; Northcutt, 2005; Schubert *et al.*, 2006; Simoes-Costa *et al.*, 2005). Gene expression studies have indicated that the amphioxus NK4-type ortholog NK2-tin is transcribed in precursors of the myocardium of the amphioxus heart (Fig. 3.3G) (Holland *et al.*, 2003). Since amphioxus Bmp2/4, like its *Drosophila* and vertebrate orthologs, is detectable in the same or adjacent cells (Panopoulou *et al.*, 1998), it has been proposed that the overall transcriptional network governing early cardiogenesis might be conserved between flies, cephalochordates, and vertebrates (Holland *et al.*, 2003). According to this hypothesis, one would expect amphioxus NK2-tin and Isl to be coexpressed in cardiac mesoderm precursors at some point in development. However, mesodermal expression of amphioxus Isl has so far not been reported (Jackman and Kimmel, 2002; Jackman *et al.*, 2000; Schubert *et al.*, 2004, 2005), which strongly suggests that territories of NK2-tin and Isl coexpression in developing amphioxus are limited to endoderm-derived tissues (Fig. 3.3G). Therefore, more data are needed before any safe conclusions can be drawn about mesoderm Isl expression patterns in the amphioxus.

5.3.3. *Drosophila*

In *Drosophila*, the circulatory pump is a peristaltic vessel localized in the dorsal midline of the embryo, the dorsal vessel, which is formed by inner myocardial and outer pericardial cells. The fly Isl ortholog, Tailup (Tup), is initially expressed in the ectoderm, at stage 10, and is required for expression of the NK4 ortholog, Tinman (Tin), in mesoderm. At mid-stage 11, when the specification of cardiac cells initiates, Tup starts to be expressed in cardiac mesoderm, in cells that coexpress Tin and Eve and, thus, are pericardial precursor cells. At late stage 11, Tup is coexpressed with Tin all over the cardiac mesoderm, in all myocardial and pericardial cells (Fig. 3.3H). Eventually, the expression of Tup persists in all myocardial cells but is lost in some pericardial cells, while Tin expression is lost in two of the six hemisegmental myocardial cells (Mann *et al.*, 2009a). These results indicate that, at some point in development, the progenitor domains of chambered hearts of vertebrates, as well as the precursors of the peristaltic pumping organs of tunicates and arthropods coexpress Isl and NK4-type homologs (Mann *et al.*, 2009a).

5.4. Origins of Isl-1-expression in cardiac progenitors: A synthesis

To make objective inferences about the genetic programs that bilaterian animals deploy to build their circulatory pumps, we represented mesoderm and endoderm Isl-1/2, NK4, and GATA-4/5/6 ortholog expression as characters in a cladistic analysis in which the phylogenetic topology is fixed according to the known evolutionary relationships between these animals (Figs. 3.4, 3.5, 3.6). To do that, we selected specific progenitor domains of circulatory pumps from flies, ascidians, lampreys, teleosts, amphibians, avians, and mammals, so as to establish a timeline of progressive commitment to the cardiac phenotype, from the nascent mesoderm, up to the linear peristaltic tube stage shared by most bilaterians, also including other domains developmentally linked to cardiac progenitors (e.g., somatic mesoderm), or important for their differentiation (e.g., endoderm underlying cardiac progenitors). In this way, mesoderm-specific progenitor domains such as anterior mesoderm, anterior lateral mesoderm, splanchnic mesoderm, cardiac precursors, early differentiating myocytes, and straight heart muscle tube, as well as subpharyngeal mesoderm were scored for the presence or absence of Isl-1/2 (Fig. 3.4), NK4 (Fig. 3.5) and GATA-4/5/6 (Fig. 3. 6) expression and organized in an

A

Isl ortholog	Pan-mesodermal	Anterior lateral mesoderm	Anterior somatic mesoderm	Anterior splanchnic mesoderm	All cardiac precursors	Early endoderm	Early-differentiating myocytes	Straight heart tube	Subpharyngeal mesoderm	Looped heart
Mouse	YES	YES	YES	YES	YES	NO	YES	NO	YES	NO
Chick	NO	YES	NO	YES	YES	YES	YES	NO	YES	NO
Xenopus	NO	YES	NO	YES	NO	YES	NO	NO	YES	NO
Lamprey	NO	NO	NO	NO	NO	NO	NO	YES	YES	YES
Zebrafish	NO	NO	YES	YES	NO	NO	YES	NO	YES	NO
Ciona	NO	YES	YES	YES	NO	NO	NO	NO	YES	*******
Drosophila	NO	YES	NO	YES	NO	NO	YES	YES	YES	*******

******* indicates absence of that structure in the referred organism

B

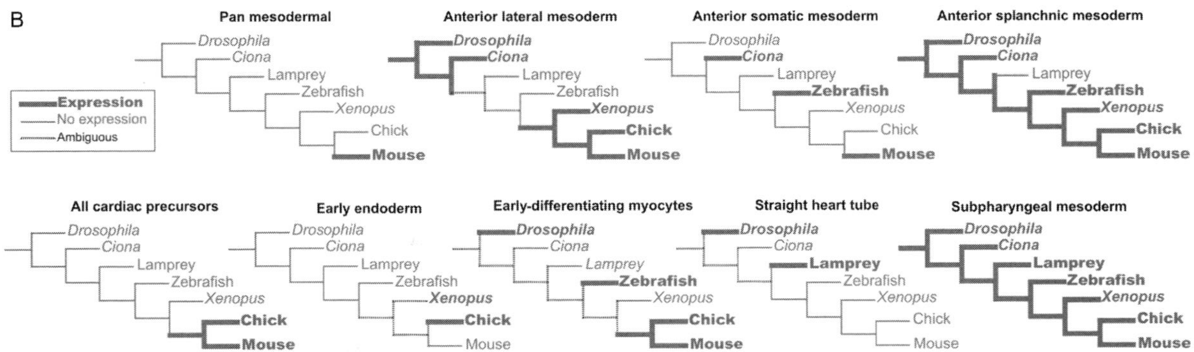

Figure 3.4 Isl expression domains in bilaterian circulatory pump ontogenesis. (A) Presence or absence of Isl ortholog expression in the embryonic progenitor domains of bilaterian circulatory pumps. (References of genes and expression patterns are available upon request.) (B) Presence or absence of Isl expression in various embryonic domains superimposed on cladograms displaying accepted phylogenetic relationships between model organisms. (For color version of this figure, the reader is referred to the Web version of this chapter.)

A

NK4 ortholog	Pan-mesodermal	Anterior lateral mesoderm	Anterior somatic mesoderm	Anterior splanchnic mesoderm	All cardiac precursors	Early endoderm	Early-differentiating myocytes	Straight heart tube	Subpharyngeal mesoderm	Looped heart
Mouse	NO	YES	YES	YES	NO	NO	YES	YES	YES	YES
Chick	NO	YES	YES	YES	NO	YES	YES	YES	YES	YES
Xenopus	NO	YES	NO	YES	NO	YES	YES	YES	YES	YES
Zebrafish	NO	YES	Unavailable	YES	NO	YES	YES	YES	NO	YES
Ciona	NO	YES	NO	YES	NO	YES	YES	YES	YES	*******
Drosophila	YES	YES	YES	YES	NO	NO	YES	YES	YES	*******

******* indicates absence of that structure in the referred organism

B

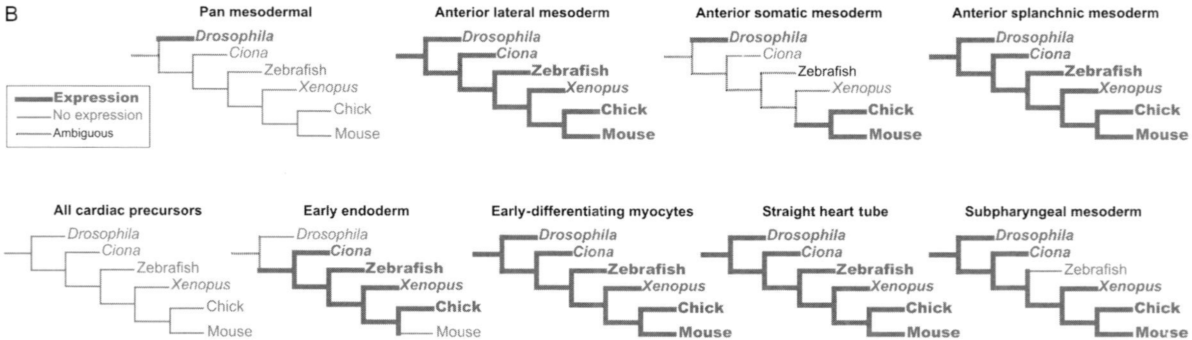

Figure 3.5 NK4 expression domains in bilaterian circulatory pump ontogenesis. (A) Presence or absence of NK4 ortholog expression in the embryonic progenitor domains of bilaterian circulatory pumps. (References of genes and expression patterns are available upon request.) (B) Presence or absence of NK4 expression in various embryonic domains superimposed on cladograms displaying accepted phylogenetic relationships between model organisms (For color version of this figure, the reader is referred to the Web version of this chapter.)

A

GATA ortholog	Pan-mesodermal	Anterior lateral mesoderm	Anterior somatic mesoderm	Anterior splanchnic mesoderm	All cardiac precursors	Early endoderm	Early-differentiating myocytes	Straight heart tube	Subpharyngeal mesoderm	Looped heart
Mouse	NO	YES	YES	YES	YES	YES	YES	YES	YES	YES
Chick	NO	YES	NO	YES	YES	YES	YES	YES	YES	YES
Xenopus	NO	YES	NO	YES	YES	YES	YES	YES	YES	YES
Zebrafish	NO	YES	YES	YES	NO	YES	YES	YES	YES	NO
Ciona	NO	YES	YES	YES	YES	YES	YES	Unavailable	NO	*******
Drosophila	NO	YES	YES	YES	YES	NO	YES	YES	NO	*******

******* indicates absence of that structure in the referred organism

B

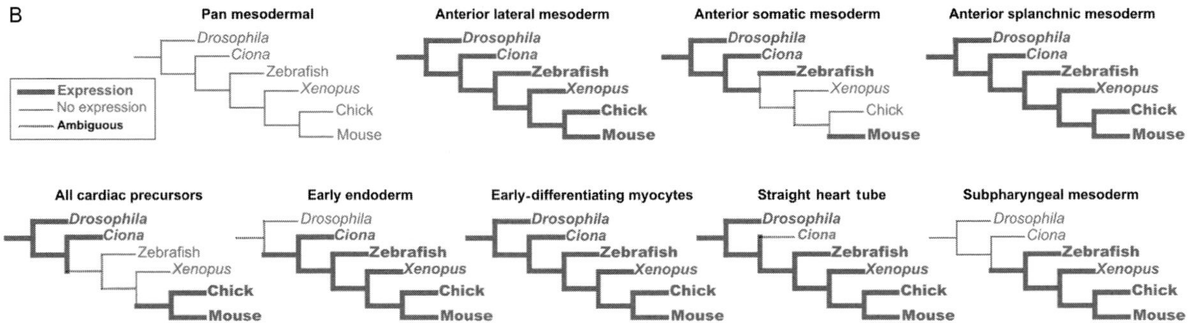

Figure 3.6 GATA expression domains in bilaterian circulatory pump ontogenesis. (A) Presence or absence of GATA expression in the embryonic progenitor domains of bilaterian circulatory pumps. (References of genes and expression patterns are available upon request.) (B) Presence or absence of GATA expression in various embryonic domains superimposed on cladograms displaying accepted phylogenetic relationships between model organisms. (For color version of this figure, the reader is referred to the Web version of this chapter.)

intuitive series, which associated gene expression with evolutionary time, developmental stage, and commitment to the cardiac phenotype.

It is easy to see that among all cardiac progenitor domains that express Isl orthologs, only the early splanchnic mesoderm and the late subpharyngeal mesoderm expression are shared by all bilaterian animals listed in Fig. 3.4. Therefore, it is likely that splanchnic and subpharyngeal Isl expression reflect ancestral patterns of mesoderm expression in the bilaterial ancestor. Contrasting with these highly conserved patterns, it is interesting to see that both panmesodermal and pancardiac Isl expression domains are recent evolutionary trends of amniotes. The first pattern seems to be a mammalian novelty or at least a mouse-specific character, while the second appears to be an attribute of avians and mammals.

In summary, data on the diversity of Isl ortholog expression across bilaterians have a direct bearing on the major questions posed about the strong interpretation. Our analysis provides direct evidence that Isl expression is not a specific marker of subpharyngeal precursors of the SHF. Rather, there is solid evidence that Isl expression is first observed throughout the anterior splanchnic mesoderm of all bilaterians.

Isl ortholog expression patterns seem to be much more variable in bilaterians than expression patterns for NK4 and GATA-4 orthologs. As indicated on Fig. 3.3 and Fig. 3.5, NK4 expression in the anterior lateral mesoderm, in the anterior splanchnic mesoderm, in the subpharyngeal mesoderm, in the early differentiating cardiomyocytes, and in the cardiomyocytes from the straight peristaltic tube are remarkably conserved across the bilaterians listed in Fig. 3.5. The pattern of conservation of GATA-4 expression across bilaterians is very similar to the one listed for NK4, except in the subpharyngeal mesoderm, where GATA-4, but not NK4, expression is conserved across vertebrates (compare Fig. 3.5 to Fig. 3.6). Also, it is interesting to note that none of the bilaterian animals listed coexpresses NK4 and GATA-4 genes throughout all cardiac precursors (Fig. 3.5 and Fig. 3.6), indicating that a complex pump field is a very old attribute of bilaterians.

In summary, today there is enough information to indicate that the concept of a specific Isl-expressing SHF population is untenable in mice and in chicken, the species in which it was first proposed. Moreover, our comparative evolutionary analysis also fails to support the idea that Isl, NK4, or GATA-4/5/6 expression marks in anyway special cardiac progenitor populations in any of the bilaterian model species investigated.

Perhaps more importantly, our comparative evolutionary analysis highlights the inherent dangers of presuming ancestral some specific characters that were first described in amniotes. Thus, as we show in Figs. 3.3, 3.4, 3.5 and 3.6, there is little, if at all, evolutionary support for the SHF concept. This is because in all bilaterian animals listed there are extensive, early waves of Isl expression in the lateral or splanchnic mesoderm.

6. Subpharyngeal Cardiac Precursors and Segmental Models of Myocardial Cell Regionalization

6.1. The axial organization of cardiac segments versus subpharyngeal cardiac precursors: Are these contradictory?

One of the outstanding contradictions of current models of cardiac development is between the classic idea of organization of cardiac segments along the AP axis and the SHF interpretation. In the past 10 years, the classic views of cardiac development, which are heavily influenced by the segmental organization of the heart, have been steadily displaced in favor of a lineage model, which is, by its turn, strongly influenced by the SHF interpretation. In the text below, we analyze the reasons behind this trend from a critical and historical standpoint and show that the existence of subpharyngeal cardiac progenitors does not conflict with the AP segmentation of cardiac progenitors, which has been recently reevaluated and shown to be an established feature of cardiac development.

6.2. The AP organization of cardiac segments is questioned

Until a few years ago, it was widely surmised that the heart grows through the gradual proliferation of its complement of early heart tube myocytes, although, in retrospect, the literature already had good indications that this was not necessarily the case (Arguello *et al.*, 1975; Stalsberg and DeHaan, 1969; Viragh and Challice, 1973). Now we know for a fact that the heart grows chiefly by addition of undifferentiated progenitors from the lateral mesoderm at its caudal pole and by incorporation of precursors from the splanchnic mesoderm at its cranial pole (van den Berg *et al.*, 2009).

When this new paradigm of cardiac growth is associated with the SHF interpretation, in which a medial group of Isl-1-expressing cardiac precursors contributes to almost all cardiac segments with apparent disregard for their AP organization, it raises at least two pertinent questions: (1) is the evidence for AP patterning of cardiac segments really solid? and (2) if cardiac segmentation indeed exists, what is its role in cardiac development in light of the major contributions of subpharyngeal Isl-1-expressing progenitors to most segments of the heart?

6.3. Do cardiac fate maps reveal an AP organization of cardiac segments?

A lingering cause for controversy and misunderstanding is represented in the fact that cardiac segment progenitors at the vertebrate lateral mesoderm rarely display an axis of orientation that is exactly aligned to the AP axis of the embryo. Also, in hindsight, it is now clear that recent doubts about the segmental AP nature of amniote cardiac precursors were fueled as much as by the SHF interpretation, as by a specific result in a fate map study. Redkar *et al.* (2001) suggested that the AP disposition of cardiac progenitors was unexpectedly labile up to stage HH8, a finding that contrasted markedly with fate maps by DeHaan and Rosenquist, which showed stable positions for cardiac progenitors as early as stage HH6. The challenging findings of Redkar *et al.* (2001) were reexamined in subsequent chicken cardiac fate maps, which, together with fate-mapping initiatives in other vertebrate models, have contributed to clarify those issues as discussed below.

6.4. Are there any roles for cardiac AP segmentation in heart development?

Reasons behind the recent, strong, opposition raised against cardiac AP segmentation and its role in cardiac development have been discussed (Buckingham *et al.*, 2005). This influential opinion provided solid arguments against an apparently unified segmental model in which each building block of cardiac progenitor tissue is an independent, clonally restricted population distributed along the AP axis of the primitive streak and lateral mesoderm, one that does not mix with others, sports a unique gene regulatory network and gives rise to a particular cardiac segment according to their rostrocaudal position (Buckingham *et al.*, 2005).

However, the very particular segmental model represented in this challenge is only an extreme view of the significance of cardiac AP organization to heart development, one that does not reflect the majority opinion of those who believe the segmental organization of heart progenitors is important for cardiac development. Thus, the proposition that is apparently refuted is only superficially similar to the ones that have been advanced and, as discussed below, the challenge disproves neither the segmental cardiac organization, nor its role in cardiac development.

6.5. Recent fate maps confirm the AP organization of cardiac progenitor segments

More recent reevaluation of cardiac fate maps in the chick embryo using traditional, invasive, techniques such as fluorescent die microinjection (Hochgreb *et al.*, 2003), as well as the novel and much less invasive combination of fluorescent construct electroporation with computational time-lapse microscopy (Cui *et al.*, 2009) emphatically reaffirm the classic organization of cardiac segments along the AP axis as determined by explant-labeling techniques (Dehaan, 1963a,b; Rosenquist and DeHaan, 1966; Stalsberg and DeHaan, 1969).

Cui *et al.* (2009) and Hochgreb *et al.* (2003) showed that progenitors of the chicken cardiac outflow are distributed in the cranial-most levels of the anterior lateral mesoderm, while sinoatrial precursors occupy the caudal-most positions, with ventricular forerunners tucked in between. Moreover, and equally important, these two studies showed that vertical intermixing in the cardiac field has practically ceased as early as stage HH7 (Cui *et al.*, 2009; Hochgreb *et al.*, 2003). Thus, it is likely that Redkar *et al.* (2001) overestimated cell mobility inside the cardiac field when they described significant overlapping of inflow and outflow cell fates at these stages (Cui *et al.*, 2009). The causes of such overestimation are not known, but conceivably, they could result from different genetic backgrounds or from experimental manipulation.

All in all, the recent maps by Cui *et al.* (2009) are refreshingly similar to the ones built by Dehaan and colleagues (Cui *et al.*, 2009; Dehaan, 1963a,b; Rosenquist and DeHaan, 1966; Stalsberg and DeHaan, 1969) almost 50 years ago and should put to rest any doubts about this feature of amniote cardiac development.

6.6. Inflow/outflow is the real axis of cardiac segmentation

In retrospect, perhaps it would have been better to speak of an inflow (sinus venosus+atria)/outflow (ventricles+outflow tract) axis of cardiac segmentation, rather than of an AP axis of cardiac axial segmentation. In the chick embryo, in which the most informative cardiac fate maps are available (Cui *et al.*, 2009; Dehaan, 1963a,b; Garcia-Martinez and Schoenwolf, 1993; Hochgreb *et al.*, 2003; Redkar *et al.*, 2001; Rosenquist and DeHaan, 1966; Stalsberg and DeHaan, 1969), it is apparent that the axis linking inflow to outflow progenitors is tilted and curved from the craniocaudal axis. This deviation is very likely a vestige left on the evolutionarily-derived flat, discoidal, avian embryo from the primitive mesoderm organization of an ancestral vertebrate in which embryonic layers were distributed over all or most of the surface of a spherical egg (judging from the general layout of fish and amphibian eggs). In effect, inspection of early embryos from frogs, teleosts, and lampreys indicates that segmental cardiac progenitor domains in these species are oriented neither in a medial to lateral nor in an anterior to posterior fashion, but in an oblique manner in which the domain of outflow precursors is cranial to the domain of inflow precursors.

6.7. The role of inflow/outflow segmentation in cardiac development

The evidence for cardiac AP segmentation has been essentially interpreted as a statistical statement about the localization of different cardiac precursors in the lateral mesoderm, rather than evidence for clonal restriction in the cardiac progenitor population (Buckingham *et al.*, 2005; Cui *et al.*, 2009; Dehaan, 1963a,b; Garcia-Martinez and Schoenwolf, 1993; Hochgreb *et al.*, 2003; Redkar *et al.*, 2001; Rosenquist and DeHaan, 1966; Stalsberg and DeHaan, 1969). Further, even in the more elaborated models, the segmental disposition of cardiac progenitors in the AP axis is interpreted as a pre-pattern, sufficient only for the broad distinction between inflow (sinus venosus + atria) and outflow (ventricles + outflow tract) domains, rather than for the precise distinction between individual cardiac chambers or segments (Hochgreb *et al.*, 2003; Moss *et al.*, 1998; Rosenthal and Xavier-Neto, 2000; Simoes-Costa *et al.*, 2005; Xavier-Neto *et al.*, 1999, 2001).

What would be the major roles of this simple division of cardiac progenitors into inflow and outflow domains? We have been arguing that establishment of the vertebrate circulation is fundamentally a double task. First, it is necessary to form blood vessels and blood cells and, second, a functional pumping organ, the straight heart tube. Thus, the primitive heart forms only after the basic circulatory plan is set down, which places very sharp limits to the morphogenetic possibilities available to the primordial heart tube in that the heart must receive blood at its posterior (inflow) pole and return it through its anterior (outflow) end. As such, the initial distinction between outflow and inflow extremities poles of the primitive, but fully functional, straight heart tube is critical for two main reasons: for adequate coupling with venous and arterial systems; and for the establishment of an effective, anti-regurgitation system based on the alternance of fast- and slow-conductive domains of cardiac muscle along the inflow/outflow axis (de Jong *et al.*, 1992).

In summary, vertebrates organize their cardiac progenitors in a variety of orientations relative to the craniocaudal axis in such a way that reflects the tridimensional shapes of their embryos. Even so, there is still a phase in their development in which progenitors of anterior cardiac segments, such as the outflow tract, find themselves in a cranial section of the lateral mesoderm relative to sinoatrial precursors. Depending on the degree of offset of the cardiac inflow/outflow axis from the craniocaudal axis, there may, or may not, be substantial overlap between inflow and outflow progenitors along the craniocaudal axis, which is the case in zebrafish embryos and chicken embryos, respectively (Simões-Costa M.S. *et al.*, unpublished; Xavier-Neto *et al.*, 2010).

7. Strong or Light?

At this time, it is appropriate to stop and analyze the evidence for and against the light and strong versions of the SHF concept in order to pass an informed judgment about them. It appears to us that, from all the data exposed above, the light interpretation represents a more realistic and defensible position than its strong counterpart. In our view, the light interpretation is more plausible because the pillars which supported the strong interpretation, namely, the existence of two distinct cardiac progenitor fields defined on the basis of specific marker expression and the presence of two specific heart-forming lineages have simply failed to be confirmed in the same original animal models on which it was originally suggested, as well as in additional phylogenetically relevant species, which encompass the whole group of bilaterian animals. Further, we feel that the light interpretation is a more parsimonious hypothesis in the sense that it does not require the suspension of belief in classic landmarks of knowledge about cardiac development, such as its segmental organization or its regulative nature.

Finally, it can be argued that the light interpretation represents a natural evolution from the strong interpretation. This is because, in the light interpretation, extracardiac progenitors from the subpharyngeal mesoderm are viewed not in the light of the initial discoveries, but through the prism of subsequent findings that limited the original interpretations.

8. A New Model of Early Cardiac Development

8.1. Reconciling the segmented nature of the heart with the complex organization of the cardiac field: The pioneer/scaffold model

The idea of cardiac inflow/outflow segmentation is not at all incompatible with the existence of subpharyngeal, Isl-1-expressing, cardiac progenitors. The model which reconciles these two fundamental notions of cardiac development will be outlined here, although its full description is not in the scope of this chapter.

To survive and thrive, vertebrate embryos, even fish ones, need to set in motion an efficient circulatory flow very early in development, long before morphogenesis can proceed to form the definitive vertebrate pump, the mature chambered heart. Thus, the early competence of the vertebrate heart is not a mere curiosity, but a real necessity of developing vertebrates.

The key to understand the scaffold/pioneer model of cardiac development is the realization that, during ontogenesis, vertebrate embryos actually form two completely different types of circulatory pumps. What is sometimes missed is the fact that the straight heart tube, which forms after bilateral cardiac precursors join at the midline, is not at all a simple miniature of the adult heart, as it was originally suggested in early models harking back to Davis (1927). Rather than a scaled-down chambered organ, the heart tube is, in fact, a completely different circulatory pump, a peristaltic organ, one that neither employs the same mechanisms as the chambered heart nor contains all of its progenitors.

8.2. A scaffold formed by pioneer cardiomyocytes

If we accept that the early embryonic cardiac tube is functional and that it does not contain all progenitors of the mature chambered heart, then we cannot escape from the conclusion that this pioneer, peristaltic, heart tube must play at least two fundamental roles for the early embryo: to provide the pumping it needs to keep development going on and to serve as a scaffold over which the other cardiac cells will join to transform it into the definitive chambered pump, the mature heart.

8.2.1. Cardiac pioneers

It is readily apparent that the pioneer/scaffold model of cardiac development offers a unique perspective to reconcile the segmental organization of the early straight heart tube with the lineage data revealed by retrospective fate-mapping analyses. As we discussed before, a complete understanding of retrospective fate mapping in the mouse embryo requires that we acknowledge the existence of myocardial founder cells (i.e., the very large clones of Meilhac *et al.*, 2004). Myocardial founder cells represent the earliest, or pioneer, clones to contribute to the mouse heart. It is important to remind that because they give rise to cardiomyocytes located from the extreme inflow pole, until the farthermost portions of the outflow tract, these pioneer clones are not restricted in their capacity to contribute to all cardiac segments (Meilhac *et al.*, 2004). Therefore, we propose that the very large clones described by Meilhac *et al.* (2004) (a.k.a. myocardial founder cells) represent the cardiac pioneers, which will initiate the unidirectional circulation and provide a platform to receive cardiac cells that ingress later into the heart.

8.2.2. Scaffold

Having characterized the pioneer side of the model, it is time to advance our notions about the scaffold. As discussed above, the inflow to outflow progenitor organization of the segmental straight heart tube has been repeatedly confirmed by fate-mapping studies performed with different techniques spanning almost half a century, and thus, it must be considered as a solid conclusion. What we suggest now is that these fate maps along the inflow–outflow axis concern the early peristaltic cardiac tube, but not necessarily the mature heart (De la Cruz and Markwald, 1998, and references therein), although we note that an organized inflow/outflow contribution to the latter has not been ruled out (Abu-Issa and Kirby, 2008; Abu-Issa *et al.*, 2004; Xavier-Neto *et al.*, 2010). Therefore, the scaffold component of the model is represented in a cardiac tube, which is organized according to inflow/outflow rules (Hochgreb *et al.*, 2003). We argue here that this inflow/outflow organization of the scaffold serves at least two roles: as the linchpin of the unidirectional function of the early peristaltic pump (de Jong *et al.*, 1992); as well as a living platform to receive and to integrate other cardiac cells entering via lateral and subpharyngeal mesoderm.

The role played by the scaffold in determining an efficient pump function is well known. The early vertebrate peristaltic tube displays a characteristic organization of myocyte automatism along the inflow to outflow axis. Again, this organization is not a curious accident of early heart development, but a critical functional feature of the tubular, peristaltic, heart. The early peristaltic heart of vertebrates is constructed so as to impart a venous (inflow) to arterial (outflow) direction to the circulation. This is primarily achieved because sinoatrial myocytes display a faster beating rate than ventricular, or outflow tract, myocytes. This gradient of automaticity is thus one of the key electrophysiological properties of the tubular scaffold of the primitive heart (de Jong *et al.*, 1992).

The part played by the scaffold as a reception platform is much less understood and explored. However, the concept is intuitive and encompasses the idea that, upon reaching the heart tube, cardiac progenitors and/or cardiomyocytes find not only an appropriate niche for the further progress of their differentiation programs, but also a source of information for the natural progression of their development toward the most appropriate fate for the specific domain in which they are found.

8.3. Acquisition of segmental identity by the late-integrating cardiac progenitor population

There are at least two general possibilities to explain how cardiac cells that do not belong to the inflow/outflow-patterned early heart tube (including Isl-1-expressing progenitors) could receive positional information and acquire segmental identity. These two possibilities diverge in the extent of commitment to inflow/outflow fates of the late-integrating cardiac progenitor population and are summarized below.

First, it is possible that Isl-1-expressing cardiac progenitors at the sub-pharyngeal and lateral mesoderm are specified by the same patterning mechanisms that communicate inflow or outflow fates to cardiac progenitors of the traditional heart-forming region, which form the tubular scaffold (see Simoes-Costa *et al.*, 2005; Xavier-Neto *et al.*, 2010) (Fig. 3.4A). In this view, the late-integrating cardiac progenitors are specified early, but delay their commitment to inflow/outflow fates to after their incorporation into the scaffold.

Alternatively, all cells entering the heart after the formation of an inflow/outflow-patterned cardiac scaffold (including Isl-1-expressing progenitors) would be naïve to inflow/outflow signals. In this view, the late-integrating cardiac progenitor populations would acquire their regional inflow/outflow segment identity according to local information obtained from surrounding cells at their specific landing sites in the tubular scaffold, in a sort of community effect for which there is some preliminary evidence (Inagaki *et al.*, 1993) (Fig. 3.7B).

In this first model, caudal Isl-1-expressing progenitors would receive posteriorizing signals, commit to inflow fates, but stay *in situ*, proliferating and delaying their differentiation to, much later in development, ingress into the nascent, inflow, compartment of the tubular scaffold. In contrast, cranial Isl-1-expressing cells would escape the initial signals for inflow patterning

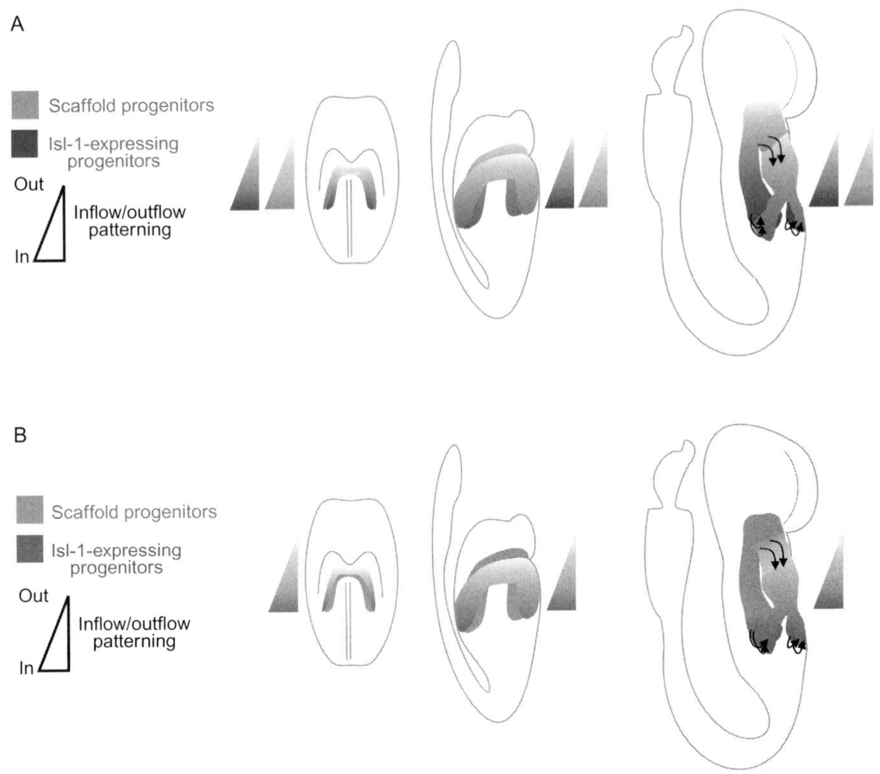

Figure 3.7 Two alternative modes of cardiac inflow/outflow patterning for the pioneer/scaffold model. (A) Global inflow/outflow patterning: All cardiac progenitors including those from the traditional heart-forming region (that contains the scaffold progenitors) and Isl-1-expressing progenitors are specified by the same patterning mechanisms early in development, but the late-integrating, Isl-1-expressing, cardiac progenitors delay their commitment to inflow/outflow fates until their incorporation into the heart tube/scaffold. (B) Scaffold-only inflow/outflow patterning: Cardiac scaffold progenitors located at the traditional heart-forming regions are patterned in the inflow/outflow axis early in development, while Isl-1-expressing progenitors are naïve to inflow/outflow signals at this stage. In this view, the descendents of Isl-1-expressing cells acquire their inflow/outflow fate according to local signals obtained from surrounding cells as they migrate to the tubular cardiac scaffold or, later, to the chambered heart. (For color version of this figure, the reader is referred to the Web version of this chapter.)

and, rather than staying put, they would be displaced in block to anterior subpharyngeal positions, from which they would proliferate and eventually migrate toward the outflow pole of the tubular scaffold. In summary, this model proposes that the late-differentiating, Isl-1-expressing, contingent of cardiac progenitors is already specified, but not determined, to their inflow or outflow fates. Upon switching off Isl-1 expression, these cardiac progenitors would migrate into the tubular scaffold and restart their differentiation programs to finally commit irreversibly to their assigned fates on site. It is evident that this model requires some sort of migration bias of anterior, subpharyngeal, cardiac progenitor cells toward outflow segments of the tubular heart and of posterior, progenitors toward the inflow segments (Fig. 3.7A).

In the alternative model, the late differentiating, contingent of cardiac progenitors is supposed to be completely naïve to the inflow/outflow patterning mechanisms that instruct the cardiac precursors that form the straight heart tube. In this view, the late-differentiating anterior (i.e., cardiac progenitors at the subpharyngeal mesoderm) and posterior precursors would migrate into the tubular scaffold and eventually receive inflow/outflow specification signals from the cardiac cells that are in the vicinity of their landing sites at the tubular scaffold (Fig. 3.7B).

The actual mechanisms that impart a segmental fate onto the late-integrating contingent of cardiac precursors are, very likely, a mix between the two general possibilities described above. Given the topology of anterior Isl-1-expressing cardiac progenitors in the subpharyngeal mesoderm, it is probable that these cells display both a bias toward outflow phenotypes and toward migration to the outflow tract and right ventricle. By the same anatomic reasons, it is well possible that posterior, Isl-1-expressing, cardiac progenitors display a propensity to assume inflow phenotypes and to migrate into the inflow tract. Further, it is well possible that the clues imparted to the late-integrating, Isl-1-expressing, group of cardiac progenitors are labile and as such, sensitive to the competing influences of local inflow/outflow segments in which they disembark, which will add another dimension to the process of commitment to cardiac segmental fates.

9. Concluding Remarks

The research programs driven by the discovery of subpharyngeal cardiac progenitors helped us to establish a transition from paradigms heavily influenced by classical views in which the outlines of cardiac development were understood on the basis of simple schemes based on embryonic morphology, patterning, gene expression patterns, or cell lineages, to a more solid body of knowledge in which all these parameters must be accounted for in an evolutionary context.

Interestingly, on one hand it is apparent that all modern hypotheses concerning early cardiac development (e.g. inflow/outflow, ballooning, and multiple cardiac field hypothesis (Abu-Issa *et al.*, 2004; Buckingham *et al.*, 2005; Christoffels *et al.*, 2000; Xavier-Neto *et al.*, 2001)) contain some solid elements that cannot be ignored in any sensible synthesis. On the other hand, however, it is increasingly clear that none of these hypotheses can explain, *per se*, all the known facts of cardiac development. Therefore, it is quite possible that a timely revision of the aforementioned models will produce more realistic, "tamed", versions of these hypotheses that will be more amenable to a synthesis. There is a lot to do in this account, but it is already possible to envision that an inflow/outflow patterning concentrated on the early differentiating cardiac progenitors that form the peristaltic tube; a ballooning hypothesis without the emphasis on the literal ballooning mode of chamber expansion and on the primitive myocardium, as opposed to cardiac progenitors from the lateral and sub pharyngeal mesoderm; and the recognition of the inherent heterogeneity within the cardiac field without the creation of novel cardiac fields will provide us with an unified framework of the early stages of cardiac development.

SOURCES OF FUNDING

Fundação de Amparo à Pesquisa do Estado de São Paulo (FAPESP) Grants 06/50843-0/08/51612-8; Conselho Nacional do Desenvolvimento Científico e Tecnológico CNPq Grants 302238/2010-7, 680.010/2009-2 and 303061/2011-1; Agence Nationale de la Recherche (ANR) grants ANR-09-BLAN-0262-02 and ANR-11-JSV2-002-01 and funding from the Centre National de la Recherche Scientifique (CNRS).

ACKNOWLEDGMENTS

The authors would like to thank Benoit Bruneau for the opportunity and Jose Maria Pérez-Pomares for critical reading and stimulating discussions.

REFERENCES

Abu-Issa, R., and Kirby, M. L. (2008). Patterning of the heart field in the chick. *Dev. Biol.* **319,** 223–233.

Abu-Issa, R., *et al.* (2004). Heart fields: One, two or more? *Dev. Biol.* **272,** 281–285.

Arguello, C., *et al.* (1975). Experimental study of the formation of the heart tube in the chick embryo. *J. Embryol. Exp. Morphol.* **33,** 1–11.

Black, B. L. (2007). Transcriptional pathways in second heart field development. *Semin. Cell Dev. Biol.* **18,** 67–76.

Brade, T., *et al.* (2007). The amphibian second heart field: Xenopus islet-1 is required for cardiovascular development. *Dev. Biol.* **311,** 297–310.

Bradford, Y., *et al.* (2011). ZFIN: Enhancements and updates to the Zebrafish Model Organism Database. *Nucleic Acids Res.* **39,** D822–D829.

Buckingham, M., *et al.* (2005). Building the mammalian heart from two sources of myocardial cells. *Nat. Rev. Genet.* **6,** 826–835.

Cai, C. L., *et al.* (2003). Isl1 identifies a cardiac progenitor population that proliferates prior to differentiation and contributes a majority of cells to the heart. *Dev. Cell* **5,** 877–889.

Christoffels, V. M., *et al.* (2000). Chamber formation and morphogenesis in the developing mammalian heart. *Dev. Biol.* **223,** 266–278.

Cui, C., *et al.* (2009). Dynamic positional fate map of the primary heart-forming region. *Dev. Biol.* **332,** 212–222.

Davidson, B., and Levine, M. (2003). Evolutionary origins of the vertebrate heart: Specification of the cardiac lineage in Ciona intestinalis. *Proc. Natl. Acad. Sci. USA* **100,** 11469–11473.

Davis, C. L. (1927). Development of the human heart from its first appearance to the stage found in embryos of twenty paired somites. *Contrib. Embryol.* **19,** 245–283.

de Jong, F., *et al.* (1992). Persisting zones of slow impulse conduction in developing chicken hearts. *Circ. Res.* **71,** 240–250.

De la Cruz, M. V., and Markwald, R. R. (1998). Living Morphogenesis of the Heart. Birkhäuser, Boston, MA.

de Pater, E., *et al.* (2009). Distinct phases of cardiomyocyte differentiation regulate growth of the zebrafish heart. *Development* **136,** 1633–1641.

Dehaan, R. L. (1963a). Migration patterns of the precardiac mesoderm in the early chick embrvo. *Exp. Cell Res.* **29,** 544–560.

Dehaan, R. L. (1963b). Regional organization of pre-pacemaker cells in the cardiac primordia of the early chick embryo. *J. Embryol. Exp. Morphol.* **11,** 65–76.

Dehal, P., *et al.* (2002). The draft genome of Ciona intestinalis: Insights into chordate and vertebrate origins. *Science* **298,** 2157–2167.

Delsuc, F., *et al.* (2006). Tunicates and not cephalochordates are the closest living relatives of vertebrates. *Nature* **439,** 965–968.

Delsuc, F., *et al.* (2008). Additional molecular support for the new chordate phylogeny. *Genesis* **46,** 592–604.

Freeman, R. M., Jr., *et al.* (2008). cDNA sequences for transcription factors and signaling proteins of the hemichordate Saccoglossus kowalevskii: Efficacy of the expressed sequence tag (EST) approach for evolutionary and developmental studies of a new organism. *Biol. Bull.* **214,** 284–302.

Garcia-Martinez, V., and Schoenwolf, G. C. (1993). Primitive-streak origin of the cardiovascular system in avian embryos. *Dev. Biol.* **159,** 706–719.

Gessert, S., and Kuhl, M. (2009). Comparative gene expression analysis and fate mapping studies suggest an early segregation of cardiogenic lineages in Xenopus laevis. *Dev. Biol.* **334,** 395–408.

Hami, D., *et al.* (2011). Zebrafish cardiac development requires a conserved secondary heart field. *Development* **138,** 2389–2398.

Hedges, S. B., and Kumar, S. (2003). Genomic clocks and evolutionary timescales. *Trends Genet.* **19,** 200–206.

Hochgreb, T., *et al.* (2003). A caudorostral wave of RALDH2 conveys anteroposterior information to the cardiac field. *Development* **130,** 5363–5374.

Holland, N. D., *et al.* (2003). AmphiNk2-tin, an amphioxus homeobox gene expressed in myocardial progenitors: Insights into evolution of the vertebrate heart. *Dev. Biol.* **255,** 128–137.

Hotta, K., *et al.* (2007). A web-based interactive developmental table for the ascidian Ciona intestinalis, including 3D real-image embryo reconstructions: I. From fertilized egg to hatching larva. *Dev. Dyn.* **236,** 1790–1805.

Inagaki, T., *et al.* (1993). Regulative ability of the prospective cardiogenic and vasculogenic areas of the primitive streak during avian gastrulation. *Dev. Dyn.* **197,** 57–68.

Jackman, W. R., and Kimmel, C. B. (2002). Coincident iterated gene expression in the amphioxus neural tube. *Evol. Dev.* **4,** 366–374.

Jackman, W. R., *et al.* (2000). Islet reveals segmentation in the Amphioxus hindbrain homolog. *Dev. Biol.* **220,** 16–26.

Kelly, R. G. (2005). Molecular inroads into the anterior heart field. *Trends Cardiovasc. Med.* **15,** 51–56.

Kelly, R. G., *et al.* (2001). The arterial pole of the mouse heart forms from Fgf10-expressing cells in pharyngeal mesoderm. *Dev. Cell* **1,** 435–440.

Kokubo, N., *et al.* (2010). Mechanisms of heart development in the Japanese lamprey, Lethenteron japonicum. *Evol. Dev.* **12,** 34–44.

Kourakis, M. J., *et al.* (2010). Key steps in the morphogenesis of a cranial placode in an invertebrate chordate, the tunicate Ciona savignyi. *Dev. Biol.* **340,** 134–144.

Lawrence, P. A., and Levine, M. (2006). Mosaic and regulative development: Two faces of one coin. *Curr. Biol.* **16,** R236–R239.

Leung, T., *et al.* (2008). The orphan G protein-coupled receptor 161 is required for left-right patterning. *Dev. Biol.* **323,** 31–40.

Lin, X., and Xu, X. (2009). Distinct functions of Wnt/beta-catenin signaling in KV development and cardiac asymmetry. *Development* **136,** 207–217.

Ma, Q., *et al.* (2008). Reassessment of Isl1 and Nkx2-5 cardiac fate maps using a Gata4-based reporter of Cre activity. *Dev. Biol.* **323,** 98–104.

Mann, T., *et al.* (2009a). The Drosophila homolog of vertebrate Islet1 is a key component in early cardiogenesis. *Development* **136,** 317–326.

Meilhac, S. M., *et al.* (2004). The clonal origin of myocardial cells in different regions of the embryonic mouse heart. *Dev. Cell* **6,** 685–698.

Mjaatvedt, C. H., *et al.* (2001). The outflow tract of the heart is recruited from a novel heart-forming field. *Dev. Biol.* **238,** 97–109.

Mommersteeg, M. T., *et al.* (2010). The sinus venosus progenitors separate and diversify from the first and second heart fields early in development. *Cardiovasc. Res.* **87,** 92–101.

Moorman, A. F., *et al.* (2007). The heart-forming fields: One or multiple? *Philos. Trans. R. Soc. Lond. B Biol. Sci.* **362,** 1257–1265.

Moss, J. B., *et al.* (1998). Dynamic patterns of retinoic acid synthesis and response in the developing mammalian heart. *Dev. Biol.* **199,** 55–71.

Nathan, E., *et al.* (2008). The contribution of Islet1-expressing splanchnic mesoderm cells to distinct branchiomeric muscles reveals significant heterogeneity in head muscle development. *Development* **135,** 647–657.

Northcutt, R. G. (2005). The new head hypothesis revisited. *J. Exp. Zool. B Mol. Dev. Evol.* **304,** 274–297.

Orts-Llorca, F., and Jimenez Collado, J. (1967). Determination of heart polarity (arterio venous axis) in the chicken embryo. *Roux. Arch. Entwicklungsmechanik* **158,** 146–163.

Panopoulou, G. D., *et al.* (1998). AmphiBMP2/4, an amphioxus bone morphogenetic protein closely related to Drosophila decapentaplegic and vertebrate BMP2 and BMP4: Insights into evolution of dorsoventral axis specification. *Dev. Dyn.* **213,** 130–139.

Passamaneck, Y. J., and Di Gregorio, A. (2005). Ciona intestinalis: Chordate development made simple. *Dev. Dyn.* **233,** 1–19.

Patwardhan, V., *et al.* (2000). The rostro-caudal position of cardiac myocytes affect their fate. *Dev. Dyn.* **218,** 123–135.

Perez-Pomares, J. M., *et al.* (2009). Building the vertebrate heart—An evolutionary approach to cardiac development. *Int. J. Dev. Biol.* **53,** 1427–1443.

Prall, O. W., *et al.* (2007). An Nkx2-5/Bmp2/Smad1 negative feedback loop controls heart progenitor specification and proliferation. *Cell* **128,** 947–959.

Putnam, N. H., *et al.* (2008). The amphioxus genome and the evolution of the chordate karyotype. *Nature* **453,** 1064–1071.

Rawles, M. E. (1952). Transplantation of normal embryonic tissues. *Ann. N. Y. Acad. Sci.* **55,** 302–312.

Redkar, A., *et al.* (2001). Fate map of early avian cardiac progenitor cells. *Development* **128,** 2269–2279.

Rosenquist, G. C., and DeHaan, R. L. (1966). Migration of precardiac cells in the chick embryo: A radiautographic study. *Contrib. Embryol.* **263,** 112–121.

Rosenthal, N., and Xavier-Neto, J. (2000). From the bottom of the heart: Anteroposterior decisions in cardiac muscle differentiation. *Curr. Opin. Cell Biol.* **12,** 742–746.

Schaeffer, S. W., *et al.* (2008). Polytene chromosomal maps of 11 Drosophila species: The order of genomic scaffolds inferred from genetic and physical maps. *Genetics* **179,** 1601–1655.

Schubert, M., *et al.* (2004). Retinoic acid influences anteroposterior positioning of epidermal sensory neurons and their gene expression in a developing chordate (amphioxus). *Proc. Natl. Acad. Sci. USA* **101,** 10320–10325.

Schubert, M., *et al.* (2005). Retinoic acid signaling acts via Hox1 to establish the posterior limit of the pharynx in the chordate amphioxus. *Development* **132,** 61–73.

Schubert, M., *et al.* (2006). Amphioxus and tunicates as evolutionary model systems. *Trends Ecol. Evol.* **21,** 269–277.

Simoes-Costa, M. S., *et al.* (2005). The evolutionary origin of cardiac chambers. *Dev. Biol.* **277,** 1–15.

Stalsberg, H., and DeHaan, R. L. (1969). The precardiac areas and formation of the tubular heart in the chick embryo. *Dev. Biol.* **19,** 128–159.

Stolfi, A., *et al.* (2010). Early chordate origins of the vertebrate second heart field. *Science* **329,** 565–568.

Sun, Y., *et al.* (2007). Islet 1 is expressed in distinct cardiovascular lineages, including pacemaker and coronary vascular cells. *Dev. Biol.* **304,** 286–296.

Tu, C. T. (2009). Nkx2.7 and Nkx2.5 function redundantly and are required for cardiac morphogenesis of zebrafish embryos. *PLoS One* **4,** e4249.

van den Berg, G., *et al.* (2009). A caudal proliferating growth center contributes to both poles of the forming heart tube. *Circ. Res.* **104,** 179–188.

Viragh, S., and Challice, C. E. (1973). Origin and differentiation of cardiac muscle cells in the mouse. *J. Ultrastruct. Res.* **42,** 1–24.

Waldo, K. L., *et al.* (2001). Conotruncal myocardium arises from a secondary heart field. *Development* **128,** 3179–3188.

Wolpert, L., and Tickle, C. (2011). Principles of Development. Oxford University Press, New York.

Xavier-Neto, J. (2010). Evolutionary origins of hearts. *In* "Heart Development and Regeneration," (N. Rosenthal and R. P. Harvey, eds.), pp. 3–45. Elsevier, London.

Xavier-Neto, J., *et al.* (1999). A retinoic acid-inducible transgenic marker of sino-atrial development in the mouse heart. *Development* **126,** 2677–2687.

Xavier-Neto, J., *et al.* (2001). Retinoid signaling and cardiac anteroposterior segmentation. *Genesis* **31,** 97–104.

Xavier-Neto, J., *et al.* (2007). Parallel avenues in the evolution of hearts and pumping organs. *Cell. Mol. Life Sci.* **64,** 719–734.

Yuan, S., and Schoenwolf, G. C. (2000). Islet-1 marks the early heart rudiments and is asymmetrically expressed during early rotation of the foregut in the chick embryo. *Anat. Rec.* **260,** 204–207.

CHAPTER FOUR

Development and Evolution of the Ascidian Cardiogenic Mesoderm

Theadora Tolkin *and* Lionel Christiaen

Contents

Abstract

The heart and other blood pumping organs are close to being universally essential in the animal kingdom. These organs present a large anatomical, morphological, and cellular diversity, which is thought to have arisen by building developmental modules on a conserved core of ancestral heart regulatory

Center for Developmental Genetics, Department of Biology, New York University, New York, USA

Current Topics in Developmental Biology, Volume 100

ISSN 0070-2153, DOI: 10.1016/B978-0-12-387786-4.00011-7

units. In this context, studies using the ascidian model system *Ciona intestinalis* offer a distinctive set of theoretical and experimental advantages, which we herein discuss in details. Development of the heart and related muscles in *Ciona* has been analyzed with a cellular to subcellular resolution unprecedented in Chordate model systems. Unique derived developmental characters of the cardiogenic mesoderm appear to be shared between *Ciona* and vertebrates. Notably, accumulating evidence point to an early Chordate origin of the cardiopharyngeal population of mesoderm cells that may have provided the foundation for the emergence of the second heart field in higher vertebrates.

1. Introduction

Young biologists can only be awestruck upon learning about the intricacy and refinement of the Amniote heart organization and physiology. This complexity arises from a modular developmental program, the alterations of which lead to congenital heart disease in about 1% of the new born humans (reviewed, e.g., in Buckingham *et al.*, 2005; Vincent and Buckingham, 2010). The four-chambered heart that permits directional blood flow within independent systemic and pulmonary circulations exists only in Amniotes; however, anatomical innovations that accompanied the diversification of the heart in higher vertebrates elaborated on a preexisting chordate developmental program (Koshiba-Takeuchi *et al.*, 2009). Indeed, the core gene regulatory program for heart development is conserved throughout the animal kingdom, and studies using the *Drosophila* model system have provided novel insights into the molecular bases of heart development in all animals including humans (Olson, 2006; Reim and Frasch, 2010).

More recently, a Tunicate, the ascidian *Ciona intestinalis* has emerged as a simple model system to study the cellular and molecular basis of early heart development in Chordates (Davidson, 2007). Molecular phylogenomic studies have shown that Tunicates, including ascidians, are the closest living relatives to vertebrates (Delsuc *et al.*, 2006; Putnam *et al.*, 2008). As such, ascidian larvae display the features of a typical chordate body plan: they possess a notochord that is restricted to the tail (hence the name Urochordates), flanked dorsally by a hollow central nervous system that forms by neurulation, laterally by skeletal muscle cells, and ventrally by the endoderm; during metamorphosis, the anterior endoderm forms pharyngeal gills slits. The ascidian heart does not function before metamorphosis, and in adults, it consists of a simple U-shaped tube that beats peristaltically in a reversible orientation to pump blood through an open circulatory system (reviewed in Davidson, 2007).

In spite of sharing chordate synapomorphies, ascidians have changed markedly since their divergence from the vertebrate lineage. One striking feature is the ascidian-specific metamorphosis that profoundly affects the body plan and transforms a swimming nonfeeding chordate larva into a sessile filter–feeding adult. Another phylum-specific feature is a drastic reduction of the cell numbers of the early embryo, which is accompanied by fixed cleavage patterns that are very stable over long evolutionary distances among ascidians. Ascidians initiate gastrulation with only about 110 cells, they develop rapidly, hatch, and form swimming larvae with only about 2600 cells. This cellular simplicity allows developmental studies to be carried out with cellular to subcellular resolution (Hotta *et al.*, 2007; Tassy *et al.*, 2006). Lineage tracing showed that a single pair of blastomeres, called B7.5 in the pregastrula 64-cell stage embryo, gives birth to the rudiment of the heart in the juvenile (Hirano and Nishida, 1997; Satou *et al.*, 2004). By the time of metamorphosis, descendents of the B7.5 lineage have given rise to anterior tail muscles (ATM), atrial siphon muscles (ASM), longitudinal body wall muscles (LoM), and the heart rudiment (Hirano and Nishida, 1997; Satou *et al.*, 2004; Stolfi *et al.*, 2010).

The recent emergence of molecular tools and resources to observe and manipulate gene function *in vivo* in ascidians has sparked a renewed interest for ascidian developmental biology (Satoh, 2003). The genome sequence of *C. intestinalis* and its sibling species *Ciona savignyi* were published in the early 2000s (Dehal *et al.*, 2002; Hill *et al.*, 2008; Small *et al.*, 2007; Vinson *et al.*, 2005); comprehensive annotations were compiled using large amounts of express sequence tags (ESTs) and state-of-the-art databases created to provide user-friendly access to this wealth of sequence data (Satou *et al.*, 2002a,b, 2005, 2006; Tassy *et al.*, 2010). Gene expression and function could also be analyzed using whole mount *in situ* hybridization, microinjection of antisense oligonucleotides, and electroporation of synthetic plasmids to drive expression of recombinant proteins in the cells of interest (Christiaen *et al.*, 2009a–d; Corbo *et al.*, 1997; Satou *et al.*, 2001a–c). Whole genome expression and tiling arrays, fluorescence-activated cell sorting and chromatin immunoprecipitation techniques are now being used to investigate cell- or lineage-specific developmental gene regulatory networks (Azumi *et al.*, 2003; Christiaen *et al.*, 2008, 2009a–d; Kubo *et al.*, 2010).

Here, we present an in-depth discussion of our current understanding of the cellular and molecular mechanisms that determine the heart versus muscle precursors fate specification and that control the behavior of heart progenitor cells within the cardiogenic mesodermal lineage during ascidian embryogenesis.

1.1. Specification of B7.5 lineage heart founder cells

Lineage studies in the solitary ascidians *Halocynthia roretzi* and *C. savignyi* showed that the heart is derived from the B7.5 blastomeres in the 64-cell embryo (Hirano and Nishida, 1997; Satou *et al.*, 2004). These cells also give rise to anterior tail muscles (ATMs), atrial siphon muscles (ASMs), and

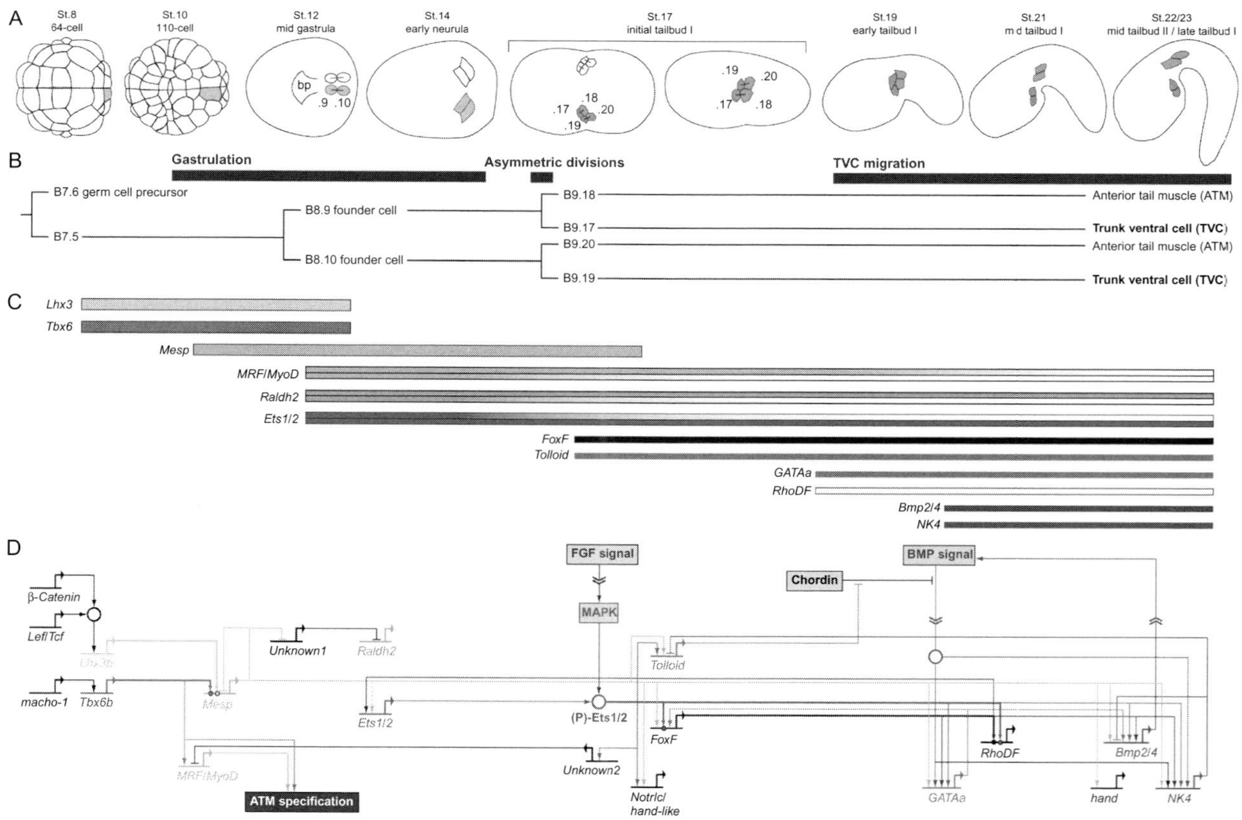
A
St.8
64-cell
St.10
110-cell
St.12
mid gastrula
bp
.9 .10
St.14
early neurula
St.17
initial tailbud I
.17 .18 .19 .20
St.19
early tailbud I
St.21
mid tailbud I
St.22/23
mid tailbud II / late tailbud I
B
Gastrulation
Asymmetric divisions
TVC migration
B7.6 germ cell precursor
B7.5
B8.9 founder cell
B8.10 founder cell
B9.18
B9.17
B9.20
B9.19
Anterior tail muscle (ATM)
Trunk ventral cell (TVC)
Anterior tail muscle (ATM)
Trunk ventral cell (TVC)
C
Lhx3
Tbx6
Mesp
MRF/MyoD
Raldh2
Ets1/2
FoxF
Tolloid
GATAa
RhoDF
Bmp2/4
NK4
D
β-Catenin
Lef/Tcf
macho-1
Tbx6b
Mesp
Unknown1
Raldh2
MRF/MyoD
ATM specification
FGF signal
MAPK
(P)-Ets1/2
Ets1/2
Unknown2
Tolloid
FoxF
Notrlc/
hand-like
Chordin
BMP signal
RhoDF
GATAa
Bmp2/4
hand
NK4

LoMs (Hirano and Nishida, 1997; Stolfi *et al.*, 2010). Large-scale *in situ* hybridization screens identified hundreds of genes showing expression in the B7.5 blastomeres but only *Mesp* is expressed exclusively in the B7.5 blastomeres of early gastrula-stage embryos (Fujiwara *et al.*, 2002; Imai *et al.*, 2004; Miwata *et al.*, 2006; Tassy *et al.*, 2010). *Mesp* encodes a bHLH transcription factor orthologous to vertebrate *Mesp1, Mesp2,* and *pMesogenin1* (Satou *et al.*, 2003, 2004). In the mouse, *Mesp1* is required in the early embryo for subsequent heart development (Kitajima *et al.*, 2000; Saga *et al.*, 1999, 2000). In *Ciona*, Mesp knockdown using antisense morpholino oligonucleotides (morpholino, or MOs) inhibited heart development and the B7.5 blastomeres gave birth exclusively to anterior tail muscles (Satou *et al.*, 2004). These data showed that *Mesp* is required for heart but not for primary muscle development in the B7.5 lineage.

1.1.1. Regulation of *Mesp* in the B7.5 blastomeres

Mesp is required for heart development but does not seem sufficient since half of the B7.5 progeny gives birth to anterior primary muscle cells (it has not been determined whether ectopic expression of *Mesp* in the ascidian embryo would be sufficient to promote mislocalized formation of heart tissue in different embryonic lineages). These data suggest that Mesp confers the competence to form heart tissue (see Section 2 for a discussion of the heart inducer(s) in ascidians). Therefore, in ascidian embryos, the competence to form heart tissue is restricted to a single pair of blastomeres as early as the 110-cell stage.

The initial study of *Mesp* function and regulation revealed that both β-catenin and the muscle determinant macho-1, the main maternal factors that specify cell fates in the vegetal hemisphere, are required for *Mesp* activation in the B7.5 blastomeres (Figs. 4.1D, 4.2; Satou *et al.*, 2004). Previous studies revealed that the effects of macho-1, which is encoded by a localized mRNA that is excluded from B7.5 cells following asymmetric divisions (Fig. 4.2), are mediated by some of the macho-1 primary targets, the

Figure 4.1 Overview of early cardiac progenitors specification in *Ciona intestinalis*. (A) Schematic representations of early embryo showing colored B7.5 lineage cells. Orange: B7.5 and daughter cells; green: trunk ventral cells (TVCs); red: anterior tail muscles (ATMs). Stage numbers are according to Hotta *et al.* (2007). (B) B7.5 lineage during the stages represented above. The main events are indicated in blue. (C) Temporal gene expression profiles in the B7.5 lineage cells. For *MRF/MyoD*, *Raldh2*, and *Ets1/2*, expression is split between early pan-B7.5 and late ATM (upper stripe) or TVC (lower stripe) specific expression. (D) Provisional Gene Regulatory Network (see the text for details). Unknown regulators are invoked to account for possible indirect regulatory inputs of Mesp on *Raldh2* and FGF-MAPK-Ets1/2 on MRF/MyoD. Other regulatory connections are based on experimental evidence. Direct connections supported by *cis*-regulatory analyses are marked by a dot on the target gene. (See Color Insert.)

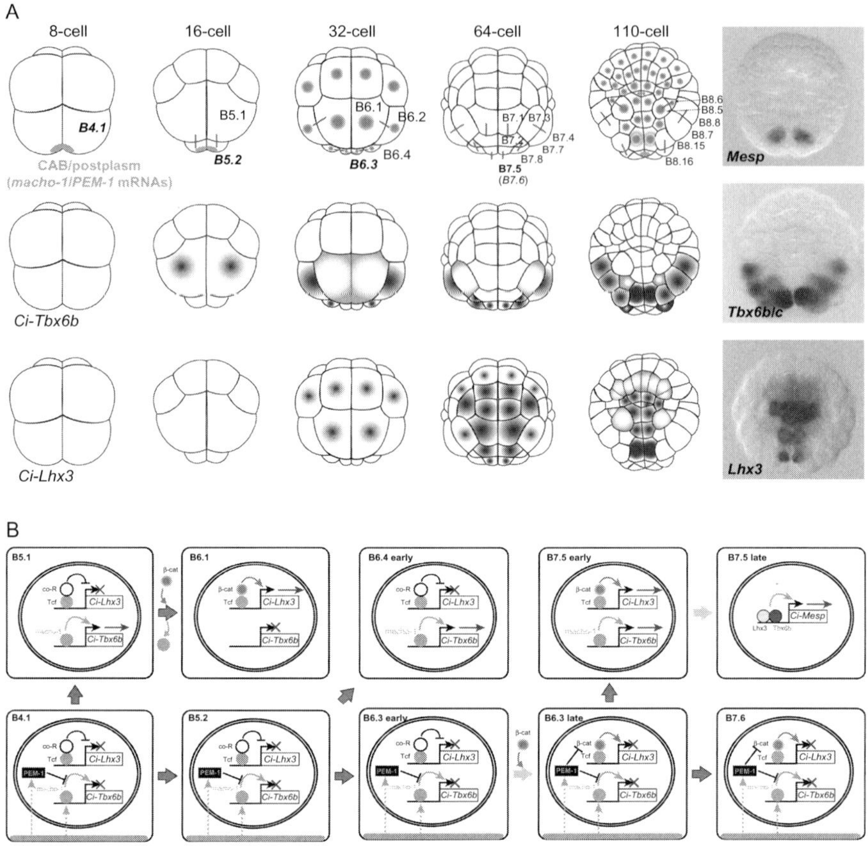

Figure 4.2 Cellular and molecular mechanisms of restricted *Mesp* activation. (A) Detailed expression profiles of *Lhx3* and *Tbx6b*, the main upstream activators of *Mesp*, shows that their spatiotemporal overlap is tightly restricted to B7.5 cells in 64-cell stage embryos. Brown dots in the upper panels indicate nuclear β-catenin according to Imai *et al.* (2000), which lacked a detailed description of the 64-cell stage. Red lines in the upper panels connect sister cells within the B-line and their names are indicated. (B) Tentative model describing the role of PEM-1-induced global transcriptional silencing, asymmetric segregation of the postplasm and late nuclearization of β-catenin for the restricted spatiotemporal coexpression of Tbx6b and Lhx3 in the B7.5 blastomeres. See the text for details. (For interpretation of the references to color in this figure legend, the reader is referred to the Web version of this chapter.)

T-box family transcription factors Tbx6b and -c (Yagi *et al.*, 2005). In keeping with a prevalent role for Tbx6 downstream of macho-1, a minimal B7.5 enhancer of *Mesp* was shown to contain three candidate Tbx6 binding sites termed B to D, which were all able to bind a Tbx6c recombinant protein in electromobility shift assays (Davidson *et al.*, 2005). Sites B and C were also required for reporter gene expression in transient transfection assays

(Davidson *et al.*, 2005). These data strongly suggest that Tbx6b and/or -c function downstream of macho-1 to activate *Mesp* in the B7.5 blastomeres.

Tbx6b transcripts are detected, at least transiently, in all of the B-line mesenchyme and primary muscle precursors starting in 16-cell embryos and until the early gastrula stage (Fig. 4.2A; Imai *et al.*, 2004; Takatori *et al.*, 2004; Yagi *et al.*, 2005). Therefore, transcriptional activation by Tbx6b/c alone is not sufficient to explain the restricted expression of *Mesp*. Morpholino-mediated knockdown of β-catenin, a key determinant in the vegetal hemisphere, inhibited *Mesp* expression (Satou *et al.*, 2004). Similarly, knockdown of the β-catenin target Fgf9/16/20 provoked a mild (less than twofold) reduction of *Mesp* transcripts' abundance in late gastrula embryos (Imai *et al.*, 2006). Therefore, it has been proposed that *Fgf9/16/20* mediates the effects of β-catenin on *Mesp* expression. Using misexpression of Tbx6b and β-catenin in the animal hemisphere, we observed that the a- and b-line neural cells, which are known to receive an FGF9/16/20 signal (Bertrand *et al.*, 2003), expressed higher levels of ectopic *Mesp* than the a- and b-epidermal precursors (Christiaen *et al.*, 2009a–d). This observation is compatible with an activating role of FGF9/16/20 for *Mesp* expression. However, in these misexpression assays, Tbx6b alone was not sufficient to induce ectopic activation of *Mesp*, even though the cells presumably perceived the FGF signal; the effects of β-catenin on *Mesp* expression appeared to be cell-autonomous and the expression pattern of *Fgf9/16/20* in early gastrula embryos is inconsistent with an instructive role restricting *Mesp* expression to the B7.5 blastomeres. For example, FGF9/16/20 induces *Twist-like-1* expression in the B7.7 blastomeres, which also express *Tbx6b* but do not turn on *Mesp* (Imai *et al.*, 2002, 2003). Thus, it seems that FGF9/16/20 upregulates *Mesp* expression in B7.5 cells but does not contribute to restricting it to the B7.5 blastomeres.

The LIM homeobox gene *Lhx3* is a primary β-catenin target that mediates its effects in the endoderm precursors (Satou *et al.*, 2001a–c). *Lhx3* is also expressed in B7.5 cells but not in mesenchyme or primary muscle precursors (Christiaen *et al.*, 2009a–d; Imai *et al.*, 2004). Misexpression of Tbx6b throughout the vegetal hemisphere caused ectopic activation of *Mesp* primarily in the endoderm precursors (Christiaen *et al.*, 2009a–d). A combination of misexpression and gene knockdown assays showed that a Tbx6b and Lhx3 synergy activated *Mesp* specifically in B7.5 blastomeres, which are the only cells that coexpress Lhx3 and Tbx6b in the 110-cell stage embryo, when *Mesp* expression starts (Fig. 4.2; Christiaen *et al.*, 2009a–d).

The minimal *Mesp* enhancer contains putative Lhx3 binding sites that overlap previously identified Tbx6 sites and were required for reporter gene expression in the B7.5 cells (Christiaen *et al.*, 2009a–d). These data indicate that cooperative DNA binding may contribute to the observed synergy between the two necessary *trans*-activators of *Mesp*, thus encoding a developmental logic—the restricted overlap of two necessary activators—in the *cis*-regulatory DNA of an essential heart specification gene.

Careful examination of the *Tbx6b* and *Lhx3* expression patterns indicates that they may be transiently coexpressed in other blastomeres of the early embryo (Fig. 4.2A). For instance, in 32-cell stage embryos, *Lhx3* starts to be expressed in B6.1 where *Tbx6b* transcripts appear to be fading away following an early expression in B5.1 (the mother cell of B6.1); similarly, in 64-cell stage embryos, *Lhx3* starts to be expressed in B7.3 when *Tbx6b* transcripts are vanishing following expression in B6.2 (the mother of B7.3; Fig. 4.2). Therefore, exquisite temporal regulation of *Tbx6b* and *Lhx3* expression may be required to achieve restricted activation of *Mesp* in the B7.5 cells. Why would *Lhx3* activation be delayed compared to *Tbx6b* in earlier stage blastomeres?

An essential feature of the molecular and cellular system that regulate temporal gene expression in the posterior vegetal blastomeres of the early ascidian embryo is a generalized repression of transcription in cells that inherit postplasmic mRNAs, including *macho-1* and *PEM-1* (Nishida and Sawada, 2001; Sardet *et al.*, 2003; Tomioka *et al.*, 2002; Yoshida *et al.*, 1996). PEM-1 has recently been shown to mediate global repression of transcription by inhibiting Serine 2 phosphorylation in the C-terminal domain of RNA polymerase II in both *C. intestinalis* and *H. roretzi* (Kumano *et al.*, 2011; Shirae-Kurabayashi *et al.*, 2011). As a result, macho-1 can activate *Tbx6b* expression following asymmetric division of B4.1 and only in the B5.1 daughter cells, since its sister B5.2 cell inherits the repressive *PEM-1*-containing postplasm (Fig. 4.2B). It is of note that the asymmetric segregation of *macho-1* mRNAs prevents further protein synthesis in B5.1, thus possibly explaining why *Tbx6b* is only transiently expressed in this blastomere (if the macho-1 protein is rapidly degraded, e.g., Fig. 4.2B).

On the other hand, nuclear localization of β-catenin, which is required for *Lhx3* activation, is barely detectable at the 16-cell stage and becomes evident by the late 32-cell stage (Imai *et al.*, 2000; Fig. 4.2). Thus, only the B7.5 blastomeres can turn on simultaneously *Tbx6b* and *Lhx3* at the 64-cell stage presumably because (1) asymmetric division of B6.3 excludes the postplasm thus permitting transient activation of *Tbx6b* and (2) high levels of β-catenin have accumulated in the nucleus by the time transcriptional repression is released, which permits synchronous activation of *Lhx3* (Fig. 4.2B). In summary, the competence to form heart tissue is restricted to the B7.5 cells at the 110-cell stage through a combination of localized maternal factors, and a timed response to maternal factor effectors dependent on progressive segregation of germline transcriptional silencers.

1.1.2. Function of Mesp in the heart founder cells

In early gastrula-stage embryos, the B7.5 blastomeres undergo a first round of symmetrical division that gives rise to a pair of cells on each side of the embryo (Fig. 4.1A). These cells have been referred to as the heart founder cells (Cooley *et al.*, 2011). *Mesp* expression is maintained in the founder cells throughout gastrulation but becomes undetectable by standard *in situ* hybridization toward the end of neurulation (Satou *et al.*, 2004). Morpholino knockdown combined with systematic qRT-PCR and *in situ* hybridization analyses suggested that *Mesp* inhibits its own expression, which could explain why it becomes downregulated toward the end of neurulation (Imai *et al.*, 2006; Satou *et al.*, 2004). *In situ* hybridization assays also indicate that *Lhx3* and *Tbx6b* expressions rapidly disappear from B7.5 blastomeres during gastrulation, which may also contribute to limiting the duration of *Mesp* expression (Imai *et al.*, 2004; Kobayashi *et al.*, 2010; Takatori *et al.*, 2004; Yagi *et al.*, 2005). Further studies would be required to determined whether *Mesp* feeds back negatively on itself and/or its upstream activators and also if *Tbx6b* and *Lhx3* expressions fail to be maintained because their upstream activators are no longer available in B7.5 lineage cells.

During neurulation, pairs of B8.10 and B8.9 founder cells on each side of the embryo undergo an asymmetric division that gives rise to prospective trunk ventral cells (TVCs) and ATMs (Figs. 4.1A and 4.3; Davidson and Levine, 2003; Davidson *et al.*, 2006; Satou *et al.*, 2004). Shortly thereafter, the TVCs migrate to the ventral side of the trunk where part of their progeny forms the heart (see Section 4 for details). When the B7.5 lineage

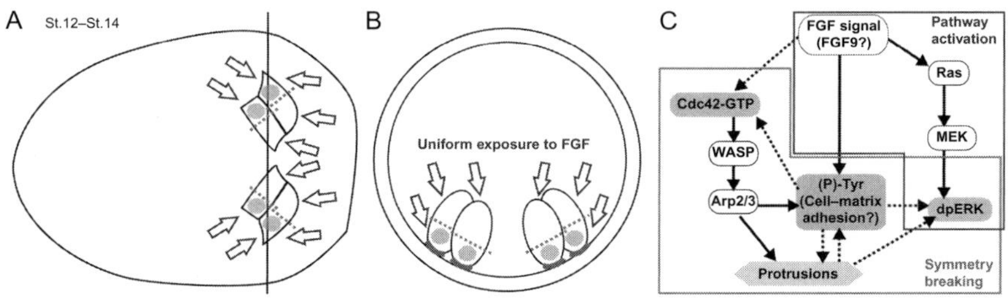

Figure 4.3 Model for asymmetric induction of the TVC fate in heart founder cells. (A, B) Dorsal (A) and lateral (B) views of stage 12–14 embryos showing the B8.9 and B8.10 blastomeres in their native positions in the embryo. Blue outlined arrows indicate the uniform FGF signal. Green dotted line indicates the prospective division plan, shifted toward the future TVC. Red arcs in (B) indicate the possible cell–matrix adhesion complexes marked by P-Tyr staining at the contact with the epidermis. The gray shaded area indicates localized MAPK activation marked by dpERK staining. (C) Schematic representation of the two main inputs that lead to asymmetric ERK activation and downstream TVC induction: blue box, activation of the Ras-MEK-ERK pathway downstream of the FGF receptor; green box, symmetry breaking correlating with an internal cytoskeletal asymmetry and localized cell–matrix adhesion of the prospective TVC membrane in contact with the epidermis. Solid arrows indicate functional interactions documented in Cooley *et al.* (2011), dotted arrows indicate possible interactions that have yet to be evaluated to clarify how cytoskeletal asymmetry restricts MAPK activation to the prospective TVCs. (For interpretation of the references to color in this figure legend, the reader is referred to the Web version of this chapter.)

is followed with Di-I in embryos that have been injected with an anti-*Mesp* MO, no TVCs are observed, and the juvenile heart does not form. Instead, the entire B7.5 lineage remains in the anterior tail, where all cells differentiate into tail muscles (Satou *et al.*, 2004). Further, expression of marker genes is lost specifically in the TVCs after *Mesp* MO knockdown. These genes include *NK4*, *GATAa*, *hand*, and *Notrlc/Hand-like*, the *Ciona* homologs of conserved "heart kernel" regulators—*Nkx2.5/tinman*, *GATA4/5/6/Pannier*, *hand* (Satou *et al.*, 2004)—as well as additional genes, such as *FoxF*, *Bmp2/4*, *Tolloid*, *ATP2A1/2/3*, and an uncharacterized EST *00152*, found to be expressed in TVCs and other tissues by large-scale *in situ* hybridization screens (Imai *et al.*, 2004, 2006; Satou *et al.*, 2001a–c, 2004). These data clearly indicate that Mesp function is required for proper TVC migration and subsequent heart development.

However, the role of Mesp in TVC specification and migration is probably, at least in part, indirect because *Mesp* is expressed throughout the B7.5 lineage while only the TVCs express heart markers such as *NK4*, *GATAa*, and *Notrlc/hand-like*. Therefore, it is thought that Mesp renders the founder cells competent to form heart tissue. For instance, an FGF signal, mediated by an ERK MAP Kinase and the transcription factor Ets1/2 is required for TVC fate induction (Davidson *et al.*, 2006; see Section 1.1). In this regard, it is possible that Mesp activity determines the TVC-specific response to the FGF-ERK-Ets1/2 pathway. It has been proposed that Mesp directly upregulates *Ets1/2* expression in the founder cells (Davidson, 2007), but a direct regulatory connection between Mesp and *Ets1/2* has yet to be established experimentally.

Nevertheless, this alleged regulatory connection would not account for all the roles of Mesp in the heart founder cells. First, upregulation and activation of *Ets1/2* is not sufficient to induce the TVC fate in the absence of Mesp. Indeed, it is not known whether *Ets1/2* is normally expressed and activated in Mesp morphant embryos, but it is clear that the FGF-ERK-Ets1/2 pathway is employed reiteratively in the early ascidian embryo to induce the a- and b-line neural precursors (Bertrand *et al.*, 2003), the notochord (Yasuo and Hudson, 2007), the endoderm (Shi and Levine, 2008), and the various mesenchyme lineages (Miya and Nishida, 2003). In this regard, Mesp probably acts as a B7.5 lineage-specific selector gene to provide TVC-specific response to an otherwise pleiotropic signal.

Second, forced expression of modified versions of Mesp did not mimic the Mesp morphant phenotype. In these studies, the DNA binding basic helix–loop–helix (bHLH) domain of Mesp was fused to the VP16 *trans*-activation domain and the resulting chimera was overexpressed in the B7.5 lineage cells using the *Mesp cis*-regulatory DNA (Davidson *et al.*, 2005). Characterization of the Mesp:VP16 expressing cells showed that TVCs failed to migrate in a majority embryos, but these cells always retained or showed ectopic expression of *Notrlc/Hand-like*. Further, Mesp:VP16 caused the formation of ectopic beating heart tissue in 13% of the analyzed juveniles, showing that heart specification could be uncoupled from TVC migration (Davidson *et al.*, 2005). These observations contrast with the phenotypes observed by targeted expression of a constitutive activator form of Ets1/2, Ets:VP16, which caused ectopic activation of *Hand-like* and the four B7.5 lineage cells to migrate into the trunk (Davidson *et al.*, 2006). Therefore, the Mesp:VP16 phenotype cannot be explained by upregulation of Ets1/2 and additional roles for Mesp must be invoked.

Raldh2 (retinaldehyde dehydrogenase 2) encodes a key enzyme for retinoic acid synthesis. In *Ciona*, *Raldh2* is expressed specifically in the B7.5 lineage cells during gastrulation and becomes restricted to the ATMs by the tailbud stage (Nagatomo and Fujiwara, 2003; Fig. 4.1C). This B7.5-specific expression pattern opens the possibility that *Raldh2* activation requires Mesp function. It is not known whether *Raldh2* expression is reduced in Mesp morphants, but it was disrupted in Mesp:VP16 expressing embryos (Davidson *et al.*, 2005). Thus, because Mesp:VP16 is thought to function exclusively as a transcription activator, it appears that wild-type Mesp can have opposite effects on B7.5 lineage gene expression and TVC migration. In keeping with this possibility, whole genome transcription profiling of B7.5 lineage cells expressing Mesp:VP16 showed a significant downregulation of TVC-specific regulators and effectors of cell migration such as GATAa or the small Rho GTPase RhoDF (see Christiaen *et al.*, 2008; Section 3.2).

The simplest hypothesis is that Mesp functions as a dual transcriptional regulator: working as a transcriptional activator upstream of *Ets1/2* and TVC markers such as *Hand-like* or a transcriptional repressor upstream of genes involved more specifically in TVC migration and ATM markers such as *Raldh2*. It is of note that, in mouse embryonic stem cells, Mesp1 directly binds to and activates conserved cardiogenic regulatory genes such as *hand2*, *Nkx2.5*, and *GATA4*, while it appears to directly repress endodermal (*Nodal*, *Goosecoid*, *FoxA2*) and primitive streak (FGF8, Brachyury/T) markers following inducible overexpression (Bondue *et al.*, 2008).

2. Induction of the Heart Progenitor Fate by an FGF-Ras-MAPK-Ets Pathway

One key feature of heart progenitor specification is the asymmetric division of founder B8.9 and B8.10 cells into smaller ventroanterior TVCs and larger dorsoposterior ATMs (Cooley *et al.*, 2011; Davidson *et al.*, 2006). Asymmetric division was first detected by observing the unequal segregation of the mitochondria-rich myoplasm into the ATMs (Davidson *et al.*, 2006). Cytochalasin-mediated inhibition of cell divisions and morphogenetic movements blocked exclusion of the myoplasm from B7.5 lineage cells that could still be induced to express the TVC-specific FoxF>RFP marker, thus showing that the presence of the myoplasm does not preclude TVC induction (Davidson *et al.*, 2006).

Multiple tissues in the early ascidian embryo are induced by the combined actions of lineage-specific "selector" genes and a shared signaling pathway involving the FGF9/16/20 ligand, the small GTPase M-Ras (Keduka *et al.*, 2009), the MAP kinase ERK1/2, and the DNA binding transcription factor Ets1/2 (e.g., Bertrand *et al.*, 2003; Miya and Nishida, 2003). *Ci-Ets1/2* is an ortholog of the vertebrate *Ets1* and *Ets2* genes and contains an Ets and a pointed domain, as well as a conserved ERK docking site and a MAPK phosphorylation site (Bertrand *et al.*, 2003; Yagi *et al.*, 2003). The observation that *Ets1/2* is upregulated in the heart founder cells indicated that a candidate FGF-Ras-MAPK-Ets1/2 signaling pathway could be involved in TVC specification (Davidson *et al.*, 2006; Miya and Nishida, 2003).

2.1. Activation of Ets1/2 specifies TVC fate

Targeted expression of a dominant negative form of the FGF receptor (dnFGFR) using the *Mesp cis*-regulatory DNA blocked cell migration and TVC-specific expression of both *Hand-like* and a FoxF>RFP reporter construct (see Section 3.1). All B7.5 descendants expressing dnFGFR formed ATMs thus showing that an FGF signal is required in addition to Mesp for TVC specification (Davidson *et al.*, 2006). Targeted expression of a dominant negative form of human H-Ras and treatment of early gastrula-stage embryos with the MEK1/2 inhibitor U0126 also blocked TVC migration, *Hand-like* expression and activation of the FoxF>RFP reporter, which indicated that FGF-mediated activation of the Ras-MEK1/2 pathway is required for TVC induction (Davidson *et al.*, 2006).

Targeted expression of constitutive activator and repressor forms of Ets1/2 indicated that it mediates the FGF-MAPK signal during TVC induction. An Ets:WRPW repressor fusion blocked TVC migration and marker gene expression (Davidson *et al.*, 2006), partially mimicking the dnFGFR and U0126 treatments. These observations are consistent with a requirement for Ets1/2 downstream of FGF-MAPK. Overexpression of Ets:WRPW also caused the B7.5 cells to scatter along the tail and elongate suggesting precocious gastrulation defects that may reflect additional roles for Ets1/2 and/or nonspecific effects of the Ets:WRPW chimera (Lionel Christiaen unpublished observations). Conversely, targeted expression of a constitutive activator Ets:VP16 fusion induced all B7.5 lineage cells to migrate into the trunk and express TVC markers. Further, this phenotype was dominant over the dnFGFR, thus strongly suggesting that Ets1/2 activation mediates the effects of the FGF-MAPK pathway during TVC induction (Davidson *et al.*, 2006).

2.2. Timing of TVC induction

Notably, dpERK immunostaining labeled specifically the small ventroanterior TVCs at the time of asymmetric cell division, thus showing that the unequal cleavage coincides with asymmetric activation of the MEK/ERK pathway that induces the TVCs (Davidson *et al.*, 2006). The timing of FGF-mediated TVC induction was further elucidated using sequential dissociation assays (Cooley *et al.*, 2011; Davidson *et al.*, 2006). When embryos electroporated with B7.5 lineage marker Mesp>GFP and the TVC-specific reporter FoxF>RFP were dissociated at stage 12 or earlier (5 hours postfertilization, hpf), the GFP positive cells failed to express RFP. RFP expression could be rescued by treatment of isolated cells with basic FGF (bFGF) and this effect was blocked by U0126, which showed that the FGF-MAPK pathway is sufficient to induce the TVC fate in the Mesp-expressing B7.5 lineage cells (Davidson *et al.*, 2006). When embryos were dissociated at stage 14 or later (6–7 hpf), half of the Mesp>GFP positive B7.5 lineage turned on the *FoxF* reporter. Treatment with U0126 at stage 14 (~6 hpf at 18 °C) or later failed to inhibit TVC induction. These data indicate that TVC induction already occurred by stage 14. Immunostaining of the diphosphorylated form of ERK1/2 (dpERK) detected MAPK activation at 6.5 hpf specifically in the newly born TVCs. Taken together, these observations showed that the TVC-inducing FGF signal is perceived and mediated by the MEK-ERK pathway in the B7.5 lineage cells between stages 12 and 14 (Cooley *et al.*, 2011; Davidson *et al.*, 2006).

2.3. TVC induction correlates with cytoskeletal polarity

Asymmetric induction of TVC fate was recently found to correlate with cytoskeletal polarity of the heart founder cells (Cooley *et al.*, 2011). Polarized protrusions invade the underlying epidermis around stage 14, just prior to asymmetric founder cell division. The ventroanterior position of these protrusions corresponds to the localization and activation of the small Rho GTPase Cdc42 (Cooley *et al.*, 2011). This polarity also correlates with ventroanterior anti-phosphotyrosine (pTyr) immunostaining, which usually labels regions of integrin-mediated cell–matrix adhesion such as focal adhesion complexes (Maher *et al.*, 1985; Zaidel-Bar and Geiger, 2010; Zamir and Geiger, 2001). Polarized pTyr staining was observed as early as stage 13 and altered by inhibition of WASP or Arp2/3 activity, suggesting that a Cdc42-activated WASP-Arp2/3 pathway contributed to asymmetric distribution of pTyr staining. Notably, targeted expression of dnFGFR, which blocked TVC induction, also abolished asymmetric distribution of pTyr staining and unequal cleavage, thus resulting in the formation of equally sized ATMs (Davidson *et al.*, 2006; L.C., unpublished observations). These data taken together with evidence for a uniform TVC-inducing FGF signal perceived at stages 12–13, suggest a model whereby the FGF and the Cdc42-WASP-Arp2/3 pathway drive polarized cell–matrix adhesion, asymmetric cell division and restrict sustained activation of the MAPK pathway to smaller ventroanterior TVCs precursors (Fig. 4.3).

In cell–matrix adhesion complexes, integrins and other signaling molecules including FGF, PDGF, and TGF-β can activate the tyrosine kinase FAK (Hunger-Glaser *et al.*, 2004; Thannickal *et al.*, 2003). It has been suggested that FAK can act in conjunction with multiple signaling inputs, including the FGFR-Ras-MEK pathway, to modulate the duration and extent of ERK signaling (Schlaepfer *et al.*, 1999). In the case of TVC induction in *Ciona* embryos, it is possible that the formation of adhesion complexes on the presumptive TVC membrane is required for sustained asymmetric activation of ERK upstream of Ets1/2. Elucidation of the detailed functional connections between the above components will require further experimentation; in particular, it is not clear whether polarized protrusive activity is directly responsible for restricted activation of ERK and/or for the asymmetric division of the founder cells or whether it is another cellular consequence of a common upstream cause such as localized cell–matrix adhesion.

3. A TVC-specific Gene Regulatory Network for Cell Migration and Heart Specification

Manipulations of Mesp and the FGF-MAPK-Ets1/2 pathway indicated that TVC induction comprises a single regulatory unit with a coherent output: the expression of heart marker genes and the collective migration of the TVC to the ventral midline. However, it has been possible to experimentally uncouple the transcriptional aspects of heart specification from the migratory behavior characteristic of the TVC (Beh *et al.*, 2007; Davidson *et al.*, 2005). Whereas targeted expression of a constitutively active form of *Ets1/2* can induce collective cell migration as well as transcription of TVC marker genes throughout the B7.5 lineage (Beh *et al.*, 2007; Davidson *et al.*, 2006), forced expression of Mesp:VP16, a constitutive activator form of Mesp, blocked TVC migration but not heart specification, which resulted in the formation of ectopic contractile heart tissue (Davidson *et al.*, 2005). Therefore, the transcriptional inputs that induce TVC fate in a subset of B7.5 lineage cells impinge on heart identity and cell migration as two separable gene network modules.

3.1. Regulation and function of FoxF during TVC specification and cell migration

The forkhead/winged helix transcription factor *FoxF* is expressed in the trunk epidermis and the TVCs, but not in the ATMs, from neurulation through the late tailbud stage (Fig. 4.1; Beh *et al.*, 2007; Imai *et al.*, 2004). Knockdown of *Mesp* by morpholino injection abolished *FoxF* expression in the TVCs but not in the epidermis (Imai *et al.*, 2006). In keeping with this observation, separate *cis*-regulatory elements were found to control tissue-specific expression of *FoxF* in the TVC and epidermal cells (Beh *et al.*, 2007). The TVC enhancer contains an E-box sequence that proved necessary for reporter gene expression and could, in principle, be bound by Mesp. Further experiments will be required to establish a direct link between Mesp and *FoxF* expression in the TVC, even more so since the E-box is not perfectly conserved with *C. savignyi* and may not coincide with an optimal Mesp binding site (Beh *et al.*, 2007). An argument against a direct activator role for Mesp is that targeted expression of the Mesp:VP16 chimera actually reduced *FoxF* expression in the B7.5 lineage cells (Beh *et al.*, 2007).

The FoxF TVC-specific enhancer also contains three putative Ets1/2 binding sites. Mutational analysis indicated that these sites are necessary for reporter gene expression (Beh *et al.*, 2007). Expression of Ets:VP16 using the *Mesp* enhancer also caused the entire B7.5 lineage to express *FoxF* and to migrate into the trunk. In the *cis–trans* complementation assays, Ets:VP16 bypasses the need for the elusive additional B7.5 lineage-specific activators of FoxF, which strongly supports the hypothesis that phosphorylated Ets1/2 binds and directly *trans*-activates the FoxF TVC enhancer (Figs. 4.1D and 4.4).

Functional studies indicated that FoxF acts primarily as a regulator of cell migration in the TVCs. When cell migration is blocked using the Mesp>Mesp:VP16 construct, those anterior B7.5 lineage cells that do not migrate into the trunk also do not express *FoxF* but do express *hand-like*, and can differentiate into beating heart tissue (Beh *et al.*, 2007; Davidson *et al.*, 2005). When the function of FoxF is manipulated directly by expressing an activator form of FoxF (Mesp>FoxF:VP16), the whole B7.5 lineage can be induced to migrate into the trunk. Conversely, expression of a dominant repressor form (FoxF:WRPW) inhibited TVC migration, but not the TVC-specific expression of *Hand-like*, *NK4*, or *GATAa*, whose expression can persist in nonmigrating anterior B7.5 cells. Expression of the FoxF:WRPW construct can also result in the formation of ectopic beating heart tissue in a small proportion of the juveniles. These results indicate that FoxF is required for migration of B7.5 cells but may be dispensable for their differentiation into cardiac tissue.

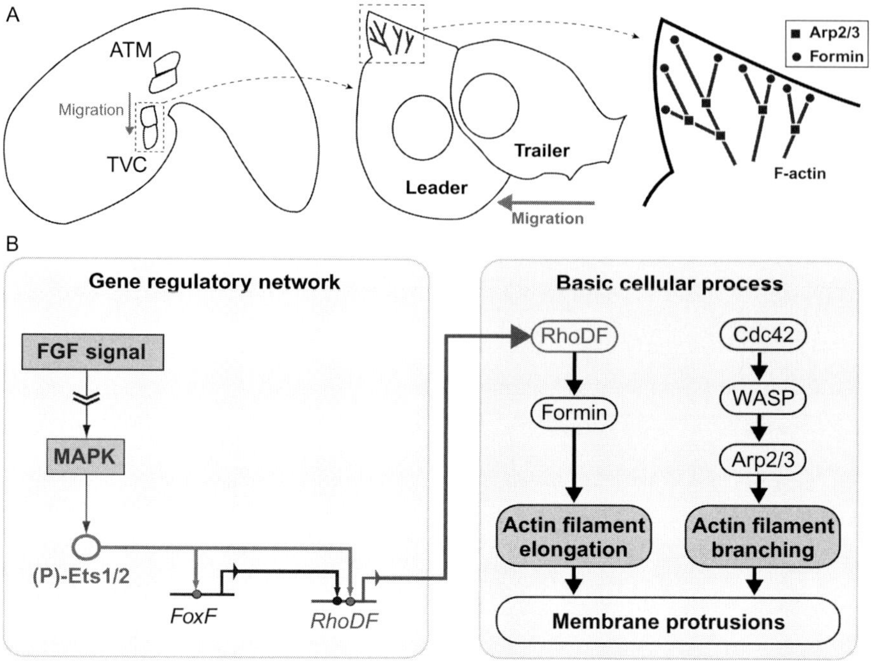

Figure 4.4 Schematic and simplified representation of the transcription–migration interface in the TVCs. (A) Mid-tailbud embryo showing the migrating TVCs and elongating ATMs, a closeup on the TVCs showing the collective cell polarity evidence for the relative positions of the leader and trailer TVC along the migratory path and a schematic representation a membrane protrusion and underlying actin-meshwork resulting from the activities of Arp2/3 and a formin. (B) Simplified model showing the feedforward loop that activates *RhoDF* expression downstream of Ets1/2 and FoxF and hypothetical protein–protein interactions leading to the actin-filament branching and elongation activities that promote membrane protrusions. (For color version of this figure, the reader is referred to the Web version of this chapter.)

Finally, in an epistasis test, the activator FoxF:VP16 was not able to rescue migration defects of an dominant negative Ets:WRPW fusion, indicating that FoxF may not be sufficient to regulate all aspects of migration and that Ets1/2 also functions in parallel to FoxF for the transcriptional control of TVC migration.

3.2. Timed deployment of the heart kernel regulators in the TVC

Despite numerous differences in heart morphology, physiology and morphogenesis, vertebrates, flies, and ascidians all express a handful of orthologous genes in precardiac tissues during heart specification. These genes comprise a functional subnetwork of transcription factors coding genes including *Nkx2.5/Tinman*, *GATA4/-5/-6/Pannier*, and *Hand* and have been referred to as the "heart regulatory kernel" (Davidson *et al.*, 2006). In *Ciona*, the orthologs of heart kernel genes, *NK4*, *GATAa*, *Hand*, and *Hand-like/NoTrlc*, are coexpressed specifically in the TVCs at the tailbud stage. This coexpression depends on Mesp function, as knockdown of *Mesp* by morpholino abolishes their expression specifically in the TVCs (Davidson *et al.*, 2006; Satou and Satoh, 2006; Satou *et al.*, 2004). Similarly, targeted inhibition of FGF signaling abolished TVC-specific expression of *NK4*, *GATAa*, and *Hand-like*, further demonstrating that both Mesp and FGF-MAPK-Ets1/2 signaling are required for all aspects of TVC fate specification (Christiaen *et al.*, 2008; Davidson *et al.*, 2006). In contrast, inhibition of FoxF function did not alter *Hand-like* expression, which is consistent with a predominant role for FoxF in regulating TVC migration but not heart specification (Beh *et al.*, 2007; Christiaen *et al.*, 2008).

The effects of dominant negative FoxF on the expression of *NK4* and *GATAa* were somewhat more variable and seemed to correlate with TVC migration: in embryos where TVC migration was strongly inhibited upon overexpression of FoxF:WRPW, *NK4* and *GATAa* were less likely to be expressed (Beh *et al.*, 2007; L.C., unpublished observations). The meaning of these observations was further elucidated in a study pointing to the role of BMP signaling in coordinating gene expression and cell migration in the TVCs. When the TVCs are born, they are located in a region of the posterior trunk at the intersection of the dorsal Chordin and ventral BMP2/4 expression domains in the neighboring epidermis (Christiaen *et al.*, 2010; Imai *et al.*, 2004; Pasini *et al.*, 2006). Shortly after the TVCs are born and start expressing FoxF and before they begin to migrate, they express *tolloid*, a metalloprotease that presumably cleaves Chordin (Hopkins *et al.*, 2007), thus possibly releasing extracellular BMP ligands and promoting signaling at the TVC membrane (Christiaen *et al.*, 2010).

Inhibition of BMP signaling using targeted expression of a truncated, dominant negative, type I receptor inhibited *NK4* expression and, to a lesser degree, *GATAa* expression (Christiaen *et al.*, 2010). These data suggested that TVC-specific expression of *GATAa*, the single ortholog of the vertebrate GATA4/5/6 family of zinc-finger transcription factors and the earliest of the heart kernel genes to be expressed in *Ciona*, requires lower levels of BMP signaling than *NK4* (Christiaen *et al.*, 2010; Ragkousi *et al.*, 2011). In this way, *GATAa* is turned on in the TVCs after the asymmetric division of the B7.5 daughter cells and downstream of *FoxF*, but before the TVCs migrate (Ragkousi *et al.*, 2011).

Targeted expression of a dominant repressor form of GATAa, GATAa:WRPW, inhibited TVC migration and blocked *FoxF*, *NK4*, *Bmp2/4* and its own endogenous expression (Ragkousi *et al.*, 2011). Like the dominant negative FoxF, GATA:WRPW did not block *Hand-like* expression, suggesting that GATAa operated in the same *Hand-like* independent subnetwork as FoxF (Fig. 4.1D).

The dominant negative form of FoxF inhibited *GATAa* expression and *vice versa* (Beh *et al.*, 2007; Ragkousi *et al.*, 2011). *FoxF* is expressed earlier than *GATAa*, thus it is possible that GATAa is involved in the maintenance of *FoxF* expression in the TVCs but morpholino-mediated gene knockdown experiments and *cis*-regulatory analyses of the FoxF TVC enhancer will be required to address this possibility. Epistasis assays will also be required to precisely dissect the individual functions of FoxF and GATAa independently of one another.

The effects of GATAa:WRPW on *NK4* and *Bmp2/4* expression are consistent with a role for GATAa in the activation of *NK4* and *Bmp2/4* since their expression starts later than that of *GATAa*, once the TVC have begun to migrate (Ragkousi *et al.*, 2011). The dominant negative BMP receptor had mild and variable inhibitory effects on TVC migration (Christiaen *et al.*, 2010). Embryos showing more severely inhibited TVC migration were also more likely to lack *NK4* expression. This effect could be reverted by targeted expression of a constitutively active form of the BMP receptor, which also inhibited TVC migration to variable extents, but maintained sustained expression of *NK4* in the TVCs. These data, together with the observation that TVC seemed to approach *Bmp2/4* expressing cells as they migrate, suggested that TVC migration contributes to an increase in BMP-SMAD1/5/8 signaling that is required for sustained *NK4* expression (Christiaen *et al.*, 2010). Preliminary observations suggested that NK4 may feedback negatively on *Tolloid* and *Bmp2/4* expression and possibly also on the TVC migration machinery (Christiaen *et al.*, 2010). Taken together, these hypotheses imply that BMP signaling increases over time as TVC migrate and that this change in signaling intensity eventually contributes to a switch of migratory mode. At present, this idea remains to be thoroughly tested using biosensors of BMP-SMAD1/5/8 signaling, transcriptional profiling and functional manipulations of BMP-SMAD1/5/8 downstream targets.

In summary, the complex regulatory interactions described above suggest a model where (1) TVC induction is rapidly followed by activation of *FoxF*, *Tolloid*, and *GATAa* expression downstream of Mesp and activated Ets1/2; (2) FoxF and GATAa maintain each other's expression and activate effector genes required for TVC migration; (3) TVC migration and *Tolloid* expression favor an increase in BMP-SMAD signaling that contributes to the upregulation of later TVC genes such as *NK4*, which could (4) feedback negatively on TVC migration and their upstream regulators (Fig. 4.1D).

3.3. Whole genome transcription profiles of migrating TVCs

The observations that inhibition of Mesp, Ets1/2, or FoxF function affects TVC migration indicate that transcriptional inputs determine the ability of heart progenitors to migrate into the trunk. Whole genome transcription profiling using fluorescence-activated cell sorting and microarrays was used to identify migration effector genes that are regulated specifically in the B7.5 lineage cells downstream of Mesp, FGF-MAPK-Ets1/2, and FoxF (Christiaen *et al.*, 2008).

Global analysis of the microarray data suggested that most of the gene expression changes occurring in tailbud stage TVCs correlate with cell migration rather than heart differentiation (i.e., most TVC candidate genes were downregulated in FoxF:WRPW and Mesp:VP16 backgrounds, which do not completely block heart differentiation). These observations were consistent with the timing of the developmental phenomenology, whereby TVC migrate between initial and mid-tailbud stages (st.19 to st.23), while the heart differentiates only after metamorphosis.

The microarray data largely confirmed regulatory interactions that were previously established using whole mount *in situ* hybridization (e.g., dominant negative FoxF inhibited *NK4* and *GATAa* but not *Hand-like* expression). Novel possible transcriptional regulators and interactions were also identified and are now being tested. For instance, a conservative analysis identified 54 transcription factors predicted to be regulated downstream of FGF-MAPK-Ets1/2 in the TVC, including 25 candidate FoxF downstream direct or indirect targets such as *GATAa*, *NK4*, and *Ets1/2*, thus indicating that FoxF may feedback on the TVC-specific expression of *Ets1/2* (Fig. 4.1D).

These data also showed that tail muscle regulators such as *MRF/MyoD* are upregulated upon targeted expression of the dominant negative FGF receptor, which indicated that the tail muscle regulatory program is actively repressed downstream of FGF-MAPK-Ets1/2 (Fig. 4.1D).

B7.5 lineage-specific whole genome transcription profiling identified a large number of candidate migration effector genes regulated downstream of FGF-MAPK-Ets1/2 and FoxF. Whole mount *in situ* hybridization verified that 50/56 (89%) candidate migration effector genes were conspicuously expressed in the TVCs at the mid-tailbud stage. These genes comprised diverse molecular and functional classes ranging from histone lysine methyltransferase to membrane receptors and extracellular proteins (Christiaen *et al.*, 2008). Orthologs of genes that had been involved in cell migration in other systems were also found to be upregulated in the TVC downstream of FGF-MAPK-Ets1/2 and FoxF. These genes included those encoding the adhesion molecules integrin-α3, integrin-β3, and talin; the PAR complex member aPKC; the receptor tyrosine kinases FGFR; Eph1 and Eph4; four RhoGEFs and three RhoGAPs; and the small RhoGTPase RhoDF, which control the dynamics of the actin-filament meshwork (Christiaen *et al.*, 2008). These data suggested that the TVC-specific transcription regulators control of the expression of genes involved in multiple aspects of cell migration such as cell polarity, membrane protrusions, and cell–matrix adhesion. However, expression of the universal small RhoGTPases RhoA, Rac1, and Cdc42 encoding genes did not change significantly upon manipulation of FoxF or the FGF receptor. The same was true for the genes encoding their downstream effectors WAVE, WASP, Arp2/3, and Diaphanous.

RhoDF was the only activator of actin polymerization that seemed to be expressed in the TVCs in direct response to FGF-MAPK-Ets1/2 and FoxF. *cis*-regulatory DNA analysis identified a TVC-specific enhancer spanning the *RhoDF* second exon and third intron. This enhancer contains putative Ets1/2 and FoxF binding sites, which were required for reporter gene expression thus indicating that Ets1/2 and FoxF provide direct transcriptional inputs for *RhoDF* expression in the TVCs (Figs. 4.1D and 4.4; Christiaen *et al.*, 2008). Functional assays using splice-blocking morpholinos and targeted expression of dominant negative and constitutively active mutants indicated that RhoDF contributes to TVC migration by promoting the formation of membrane protrusions in conjunction with Cdc42 and upstream of the formin Diaphanous (Christiaen *et al.*, 2008; L.C., unpublished). Targeted expression of the dominant negative FoxF inhibited the formation of membrane protrusions, which could be restored by coexpression of wild-type RhoDF or a constitutively active RhoDF mutant (Christiaen *et al.*, 2008). These data support the hypothesis that FoxF contributes to the formation of membrane protrusions primarily by upregulating the expression of RhoDF, which then functions together with constitutively expressed proteins to promote the actin-filament polymerization underlying lamellipodia formation. The cellular phenotypes of the above manipulations of FoxF and RhoDF suggest a model where Cdc42 promotes the branching of the actin meshwork underlying lamellipodia formation, while RhoDF promotes the nucleation and elongation of individual actin filaments (Fig. 4.4B; Christiaen *et al.*, 2008).

These data provided fundamental insights into the structure and function of the transcription–migration interface that links the TVC gene regulatory network to their migratory behavior. Hundreds of TVC-specific candidate migration effector genes remain to be analyzed to gain a comprehensive understanding of this "interface," which is in essence a system-level property of the heart progenitors.

3.4. Surrounding tissues controlling heart fate specification and TVC migration

The gene regulatory network discussed above unfolds by definition within the B7.5 lineage cells themselves and determines their intrinsic properties. For example, TVC-specific expression of signaling molecules or downstream components will confer a competence to respond to an extrinsic signal in a B7.5 or TVC-specific manner. This universal concept of developmental biology has been discussed above as it applies to TVC fate specification, which combines the inductive action of an FGF signal with the competence to form migrating heart progenitors provided by Mesp.

When they are born, the TVCs occupy a pivotal position at the tail-trunk boundary of the initial tailbud embryo, they are surrounded by multiple tissues including the epidermis, the posterior endoderm, the primary tail muscles, the notochord, and the B7.7 lineage mesenchyme (Christiaen *et al.*, 2008; Ragkousi *et al.*, 2011; L.C., unpublished observations). Therefore, it is likely that one or more of these surrounding tissues provide(s) molecular cues that help the TVCs navigate toward their final destination where the left and right pairs meet on the ventral side of the trunk, apposed between the epidermis and the endoderm (Christiaen *et al.*, 2008; Davidson *et al.*, 2005).

So far, a single article has reported on the role of surrounding tissues in heart progenitors migration and specification (Ragkousi *et al.*, 2011). In this study, an enhancer from the *Titf-1/Nkx2.1* gene was used to target expression of the dominant negative GATAa:WRPW specifically to the endoderm. This manipulation resulted in endoderm morphogenesis defects that could be observed at the early tailbud stage. In these embryos, the TVCs migrated anteriorly and expressed TVC marker genes but they failed to converge at the midline thus resulting in a phenotype reminiscent of the *cardia bifida* observed in vertebrates upon alteration of endoderm development. These results demonstrate that proper endoderm development is required for the convergence of TVCs at the midline and formation of a single heart primordium (Ragkousi *et al.*, 2011). It is not clear at present whether these TVC convergence defects are due to a failure of the endoderm to "signal" or to more strictly mechanical constraints caused by the abnormal endoderm morphogenesis.

4. Heart versus ASM Fate Specification

Lineage analyses in both *H. roretzi* and *C. intestinalis* species indicated that the TVC give birth to the heart as well as the ASMs and longitudinal muscles of the body wall (Hirano and Nishida, 1997; Stolfi *et al.*, 2010). By the late tailbud stage (st.24), TVCs have stopped their migration, converged at the midline, and started to divide (Davidson *et al.*, 2005; Ragkousi *et al.*, 2011; Stolfi *et al.*, 2010). The first divisions are relatively synchronous and occur essentially along the mediolateral axis (Fig. 4.5A). This first division is asymmetric, leading to the formation of large lateral cells and smaller median cells (Davidson *et al.*, 2005; Stolfi *et al.*, 2010). The large lateral cells undergo a second asymmetric division also oriented along the mediolateral axis. This second asymmetric division was shown to separate the small median heart precursors from the larger lateral TVCs that constitute the ASM and longitudinal muscle precursors (referred to as ASM for simplicity; Stolfi *et al.*, 2010). Following this initial separation, the small median heart precursors continue to divide without obvious asymmetry or specific

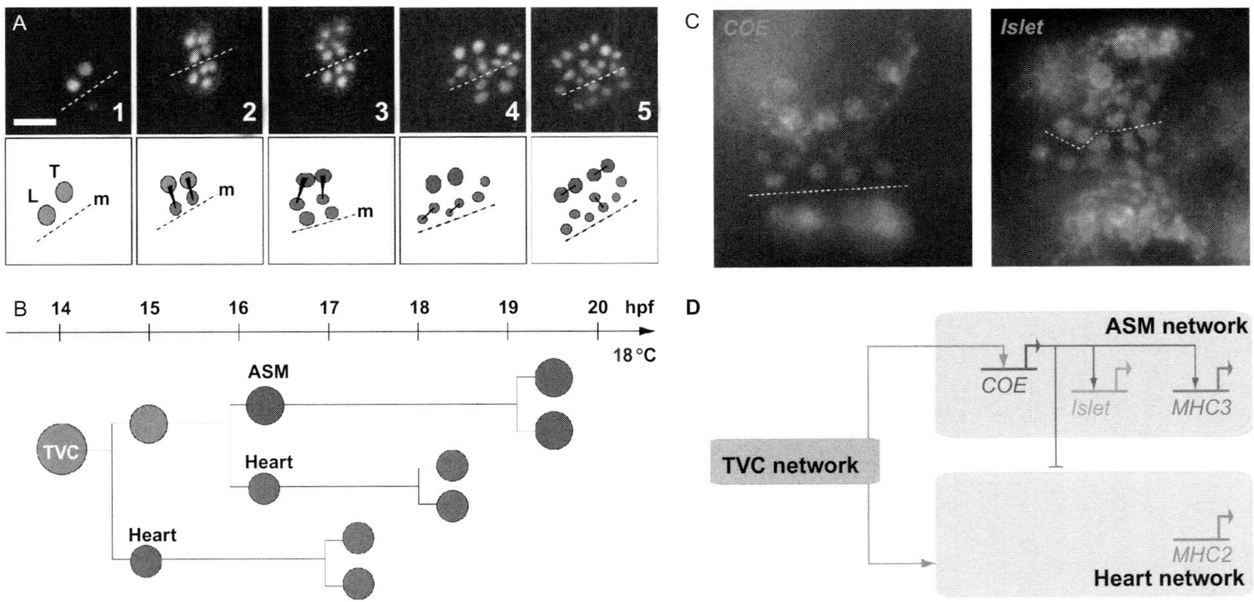

Figure 4.5 Heart versus ATM fate specification. (A) After they stop migrating, TVCs (green dots) divide asymmetrically into heart (red dots) and ASM precursors (blue dots). "1" to "5" indicate successive time points extracted from time-lapse movies in which TVCs express an Histone2B–GFP fusion (Stolfi *et al.*, 2010). Black segments in lower panels connect sister cells. m, midline (dotted line); L, Leader; T, Trailer. Scale bar: 20 μm. (B) Cell lineage showing the TVCs, ASMs, and heart precursors. hpf, hours postfertilization. (C) Double fluorescent *in situ* hybridization and immunostaining on 21 hpf larvae expressing the Mesp>lacZ transgene (red dots mark nuclear β-galactosidase in TVCs) with *COE* and *Islet* specific digoxigenin-labeled antisense RNA probes, respectively. Note the ASM-specific expression of *COE* and *Islet* (*Islet* may also be expressed at a weaker level in the median heart precursors). On the left panel, the *COE*-positive cells initiate migration toward the atrial siphon placode. White dotted line: midline. At this stage, each side displays eight heart and four ASM precursors. (D) Simplified network showing regulatory connections between the heart TVC network and the later ASM and heart-specific networks (Stolfi *et al.*, 2010). (See Color Insert.)

orientation (Fig. 4.5B). The ASM precursors undergo one more symmetric division, along the anteroposterior axis, before they start migrating toward the dorsolateral atrial siphon placode, where they form a ring of cells underlying the placodal epithelial rosette of cells (Stolfi *et al.*, 2010).

The ASM precursors were found to express the transcription factors COE and Islet (Fig. 4.5C). Targeted expression of the HLH transcription factor COE using the FoxF minimal TVC enhancer inhibited some of the TVC divisions, forced TVCs to migrate toward the dorsal side of the larva, promoted ectopic activation of an *Islet* ASM enhancer, blocked the formation of heart tissue and expression of the heart-specific marker *myosin heavy chain 2* (*mhc-2*). Conversely, targeted expression of a dominant repressor form of COE, COE:WRPW, blocked ASM migration and *Islet* enhancer activation, inhibited expression of the ASM marker *mhc-3*, and provoked expansion of the heart tissue and *mhc-2* expression. These results indicated that the transcriptional activator function of COE is necessary for ASM specification and sufficient to inhibit heart fate specification. Thus, it is possible that the default fate of the TVCs is to form heart tissue but that this program is overridden by COE and/or its downstream targets in the ASM precursors (Fig. 4.5D).

ASM divisions and migration occur for the most part after hatching, in swimming larvae that will soon attach to a substrate and undergo a dramatic metamorphosis. During metamorphosis, cells in the initial cluster of heart precursors continue to divide and presumably start to form a single-compartment heart but the morphogenetic details of this process have not been studied carefully. During the same period, cells that will form the longitudinal muscles of the body wall migrate away from the ASM ring and start expressing *Tbx1/10* (Stolfi *et al.*, 2010). This last example of differential fate specification coupled with cell migration and specific gene expression could be the third example of a reiteration in the B7.5 lineage whereby cells divide asymmetrically, only one of the daughter cell turns on specific transcription regulator(s) and starts migrating.

The cellular and molecular mechanisms that control the asymmetry and orientation of TVC division and couple them to differential activation of COE remain largely unknown. These are currently being explored and could provide insights into conserved mechanisms of heart versus head muscle fate specification in chordates.

5. Evolutionary Considerations

A comprehensive discussion of the evolution of the heart and pumping organs extends beyond the scope of this review and has been the central topic of several excellent articles (e.g., Fishman and Olson, 1997; Olson,

2006; Xavier-Neto *et al.*, 2007). Here, we briefly discuss the phylogenetic distribution and evolutionary significance of the main features of ascidian cardiogenic mesoderm development presented in the previous sections.

5.1. Conservation of the heart kernel genes with modifications

Hearts and other blood or hemolymph pumping organs are of ancient origin, as they can be found in varying forms in all extant Bilaterian phyla (Xavier-Neto *et al.*, 2007). Concordantly, key cardiogenic transcription regulators and signaling pathways have somewhat conserved roles during heart development from *Drosophila* to humans. The panbilaterian myocardial determination kernel includes conserved regulatory interactions between orthologs of *Nkx2.5/tinman*, *Mef2*, *Hand*, *GATA4/-5/-6/Pannier*, *Tbx6/Doc1/-2/-3*, and *Tbx20/H15/mid*. These genes are all expressed at one point or another in the cardiogenic mesoderm during heart specification in flies and vertebrate model systems (reviewed in Davidson and Erwin, 2006; Olson, 2006; Reim and Frasch, 2010).

As mentioned above, the *Ciona* orthologs of conserved cardiogenic regulators are expressed in the TVCs but specific differences have been observed that point to possible changes in the temporal deployment of a conserved core heart regulatory network. For example, *Ciona* possesses a *Tbx20* ortholog but its expression could only be detected by ESTs obtained from an adult blood cell cDNA library (Takatori *et al.*, 2004). This is in striking contrast with the early broad expression in the cardiogenic mesoderm of mouse (Carson *et al.*, 2000), zebrafish (Ahn *et al.*, 2000), and Amphioxus (Belgacem *et al.*, 2011) *Tbx20* orthologs.

Another heterochronic sequence in *Ciona* gene expression compared to flies and vertebrates concerns the expression of *NK4*, the *tin/Nkx2.5* ortholog, and *GATAa*, the *Pnr/GATA4/-5/-6* ortholog. In both *Drosophila* and vertebrates, *tin/Nkx2.5* are expressed broadly in the cardiogenic mesoderm and precede the expression of *Pnr/GATA4/-5/-6*. Careful examination of *GATAa* and *NK4* expression in *Ciona* TVC indicated that *GATAa* is expressed priori to cell migration, before *NK4* (Christiaen *et al.*, 2010; Ragkousi *et al.*, 2011). Consistent with a conserved positive regulatory interaction between *Pnr/GATA4/5/6* and *Tin/Nkx2.5*, overexpression of a constitutive repressor form of GATAa in the TVCs inhibited *NK4* expression (Ragkousi *et al.*, 2011).

Preliminary analyses indicated that both *NK4* and *GATAa* expressions are sensitive to distinct levels of BMP-Smad1/5/8 signaling, which may contribute to determining their timing of expression in the TVCs (Christiaen *et al.*, 2010). Directed binding of the activated Smad complex on a *tin/Nkx2.5* cardiac enhancer mediates the conserved dependency of *NK4* orthologs to BMP signaling in *Drosophila* and vertebrates

(Frasch, 1995). These observations support the notion that the bilaterian heart kernel is conserved but may be deployed in a temporally distinct manner in *Ciona*.

In this regard, it is of note that in ascidians, heart cardiomyocyte differentiation does not occur during embryogenesis, but after metamorphosis (Davidson, 2007; Satou *et al.*, 2004). For instance, following the initial expression in the TVCs, *NK4* expression is no longer detectable in the heart precursors of swimming larvae but reinitiates in the heart primordium of metamorphosing juveniles (Davidson and Levine, 2003). The cellular and molecular mechanisms that underlie this striking pause in the heart developmental program are unknown but warrant further investigation, inasmuch as they may share essential regulatory features with the process of stem-cell maintenance and tissue regeneration, which also requires the reactivation of a paused developmental program.

5.2. Evolutionary diversity of the upstream regulators of the heart kernel factors

Initial activation of the heart kernel regulators appears to be controlled by different upstream factors in chordates as compared to lophotrochozoans and even basal deuterostomes. Whereas in flies, *tinman* is initially activated downstream of the panmesodermal regulator Twist; in vertebrates, a key regulator of early heart specification is the transcription factor Mesp1 (Bondue and Blanpain, 2010; Saga *et al.*, 2000; Satou *et al.*, 2004). In contrast, the proposed *Drosophila* ortholog of Mesp, *Sage*, is expressed in the salivary gland but seemingly not in the cardiac mesoderm (Moore *et al.*, 2000). To the best of our knowledge, the cloning and expression of *Mesp* orthologs has not been reported in other Protostomians, Hemichordates, or Cephalochordates. These data would be required to evaluate whether a role for *Mesp* in the early cardiogenic lineage is a synapomorphy of Deuterostomes or Chordates or whether it was specifically lost in the lineage leading to *Drosophila*.

In *Ciona*, *Mesp* is activated in part by Tbx6 and, unlike the *Mesp* ortholog *Sage*, the Tbx6 orthologs *Dorsocross-1*, *-2*, and *-3* are required early in the cardiogenic mesoderm for expression of *tinman* and subsequent heart development in *Drosophila* (Reim and Frasch, 2005, 2010). *Tbx6/16* is also expressed broadly in the early mesoderm in the European Amphioxus *Branchiostoma lanceolatum* (Belgacem *et al.*, 2011). These observations suggest that an early widespread expression in the mesoderm, including the precardiac mesoderm, is an ancestral feature of *Tbx6* orthologs. In vertebrates, *Tbx6* expression seems to have been restricted to the presomitic mesoderm. Interestingly, Tbx6 directly activates *Mesp2* during somitogenesis, opening the possibility of an ancient Tbx6-Mesp regulatory connection in the early mesoderm of Chordates (Yasuhiko *et al.*, 2006). This connection may have been lost in the precardiac mesoderm of vertebrates, where it is conceivable that another factor may have replaced Tbx6 since the T-box factor Eomesodermin was recently reported to activate *Mesp1* in the precardiac mesoderm by binding to the same conserved T-box sites required for Tbx6-mediated *trans*-activation in the presomitic mesoderm (Costello *et al.*, 2011).

5.3. Chordate-specific evolution of the cardiopharyngeal mesoderm

A striking chordate innovation appears to be the integration of cardiac and pharyngeal mesoderm into a single embryonic lineage. We proposed to name this ontogenetically transient population of mesodermal cells the cardiopharyngeal mesoderm which (1) derives from *Mesp*-expressing precursors, (2) expresses the orthologs of *FoxF* and *Nkx2.5*, and (3) gives birth to both cardiac and noncardiac muscles, the latter of which (3) display sustained expression of *Islet* (Stolfi *et al.*, 2010). In Amniotes, the hypothetical existence of such a cardiopharyngeal mesoderm is supported by evidence that *Islet1* specifically marks the branchiomeric muscle precursors and the progenitors of the second heart field (Cai *et al.*, 2003; Nathan *et al.*, 2008), which derive clonally from common progenitors (Lescroart *et al.*, 2010). These progenitors probably derive from the *Mesp1*-expressing mesoderm and share a common ontogenetic origin with the first heart field (Meilhac *et al.*, 2004; Saga *et al.*, 2000; Fig. 4.6A).

Determining the precise phylogenetic distribution of cardiopharyngeal mesoderm as described above will await further investigation in phyla such as the Cephalochordates, Hemichordates, Echinoderms, and Protostomes other than *Drosophila* but an emerging picture points to specific innovations within the Chordate lineage (Fig. 4.6B).

Detailed analysis of the pharyngeal gills development in the Hemichordate *Saccoglossus kowalevskii* showed that Hemichordates lack the *Tbx1/10*-positive pharyngeal mesoderm within branchial pouches of endodermal origin (Gillis *et al.*, 2012). By contrast, both the Amphioxus (Cephalochordate) *Branchiostoma floridae* and the lampreys *Lethenteron japonicum* and *Lampetra fluviatilis* express *Tbx1/10* orthologs in territories overlapping and/or adjacent to the Nkx2.5-expressing mesoderm, which includes the precardiac and pharyngeal mesoderm (Holland *et al.*, 2003; Kokubo *et al.*, 2010; Mahadevan *et al.*, 2004; Onimaru *et al.*, 2011; Sauka-Spengler *et al.*, 2002; Tiecke *et al.*, 2007). It is not clear whether *FoxF* orthologs are expressed in the precardiac mesoderm of higher vertebrates, but broad expression domains were detected in the ventral mesoderm of Amphioxus and lateral plate mesoderm of the lamprey, in both cases the expression domain appears to encompass the cardiac and pharyngeal mesoderm (Onimaru *et al.*, 2011). Given the basal position of Cephalochordates within the Chordate lineage (Delsuc *et al.*, 2006; Putnam *et al.*, 2008), these observations are consistent with the parsimonious hypothesis that the

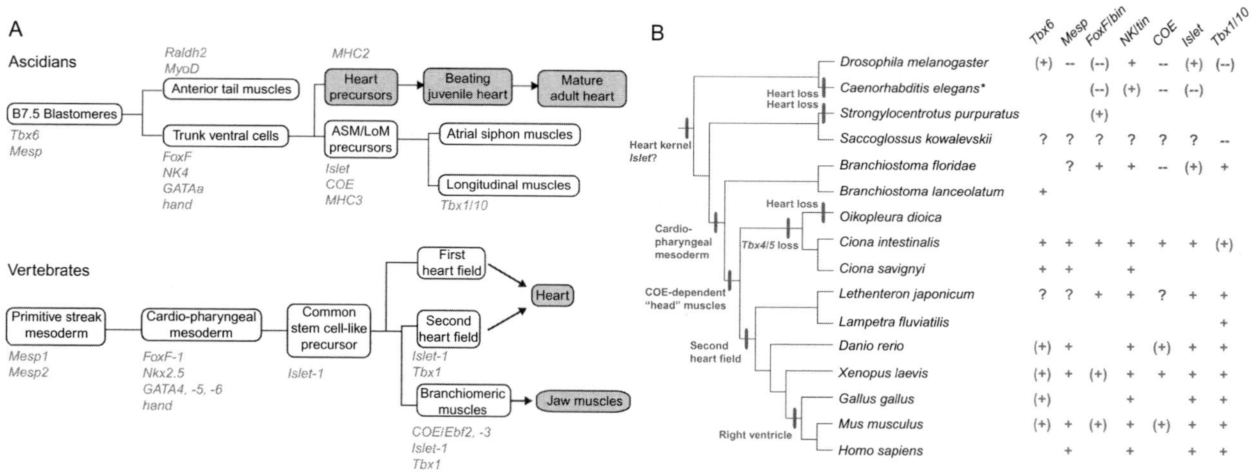

Figure 4.6 Evolutionary origins and diversification of the cardiopharyngeal mesoderm. (A) Ontogeny of the heart and other cardiopharyngeal mesoderm derivatives in ascidians (upper panel) and vertebrates (lower panel; modified from Stolfi *et al.*, 2010). The diagrams represent simplified lineages and key markers expressed at successive stages. Note that genes that are maintained in offspring tissues are not necessarily indicated twice. (B) Phylogenetic distribution of the main markers of the cardiopharyngeal mesoderm in bilaterian. Empty spaces mark the absence of information. Brackets indicate a possible conservation (+) and a possible lack of conservation (–). (?) Mark missing information that would be particularly informative to reconstruct key evolutionary transitions. Phylogenetic distribution partially according to in-text citations and the following references: Bally-Cuif *et al.* (1998), Crozatier and Vincent (1999), Chapman *et al.* (1996), Crozatier *et al.* (1996), Deimling and Drysdale (2011), Dubois *et al.* (1998), David *et al.* (2008), Hug *et al.* (1997), Hitachi *et al.* (2009), Holland *et al.* (2003), Jackman *et al.* (2000), Kokubo *et al.* (2010), Knezevic *et al.* (1997), Koster *et al.* (1999), Moore *et al.* (2000), Mann *et al.* (2009), Mahlapuu *et al.* (1998), Mahadevan *et al.* (2004), Mazet *et al.* (2004), Onimaru *et al.* (2011), Peterson *et al.* (1997), Porsch *et al.* (1998), Prasad *et al.* (1998), Reim and Frasch (2005), Sawada *et al.* (2000), Sauka-Spengler *et al.* (2002), Tu *et al.* (2006), Tiecke *et al.* (2007), Uchiyama *et al.* (2001), Voutev *et al.* (2009), Wotton and Shimeld (2006, 2011), Wotton *et al.* (2007, 2008), and Zaffran *et al.* (2001). (For color version of this figure, the reader is referred to the Web version of this chapter.)

cardiopharyngeal mesoderm was already present in the last common ancestor of Chordates (Fig. 4.5).

Within the cardiopharyngeal mesoderm population, we proposed that sustained *Islet* expression correlates with a pharyngeal muscle fate but this view is somewhat challenged by several observations. First, *tailup* (*tup*), the *Drosophila* ortholog of *Islet*, is expressed in the cardiogenic mesoderm where it has been reported to *trans*-activate *hand* expression, but its prevalent role in upregulating the ectodermal expression of *Dpp*, a BMP ligand coding gene necessary for heart specification, complicates the analysis of *tup* function in the heart precursor cells (Mann *et al.*, 2009; Reim and Frasch, 2010). In Amniote embryos, *Islet1* is specifically maintained in pluripotent cardiopharyngeal muscle progenitors but it is also transiently expressed in the primary heart field (Prall *et al.*, 2007). In both the lamprey (Kokubo *et al.*, 2010) and amphioxus (Jackman *et al.*, 2000), the observed staining was consistent with a possible expression of *Islet* orthologs throughout the cardiopharyngeal mesoderm. These observations suggest that the sustained expression of *Ci-islet* in the ASM reflects evolutionary variability of *Islet* expression and/or function within the cardiopharyngeal linage rather than a specific role in the pharyngeal versus heart muscle specification.

Finally, we found that ASM-specific expression of *COE* in *Ciona* was instrumental for non-heart pharyngeal versus heart muscle specification. The *Drosophila* ortholog *Collier* has been implicated in the specification of DA3 somatic muscles but it does not seem to be expressed in the cardiac lineage (Crozatier and Vincent, 1999; Crozatier *et al.*, 1996, 1999). Recently, it was shown that the *Xenopus* orthologs of COE, *XEBF2* and *XEBF3*, act redundantly and together with MyoD-related myogenic factors to specify non-heart muscles, including jaw muscles (Green and Vetter, 2011). Expression of the Amphioxus ortholog *AmphiCoe* is conserved in the central nervous system but *AmphiCoe* does not seem to be expressed in the pharyngeal mesoderm (Mazet *et al.*, 2004). Most notably, expression of conserved MyoD-related myogenic factors in Amphioxus appeared to be restricted to the somitic mesoderm (Schubert *et al.*, 2003). Taken together, these observations suggest that *COE/EBF* genes and MyoD-related myogenic factors function in the cardiopharyngeal mesoderm of vertebrates to promote jaw muscle specification and restrict heart formation, a key developmental feature that may have contributed to development of novel head structures in the vertebrate lineage.

ACKNOWLEDGMENTS

We thank Alexandra Ketcham for the double fluorescent *in situ* hybridization and immunostaining showed in Fig. 4.4C. We would also like to thank Alberto Stolfi for his thoughtful comments on the chapter. Our work is supported by Grants 10SDG4310061 from the

American Heart Association, R01GM096032 from NIGMS/NIH, R01HL108643 from NHLBI/NIH, by the New York Cardiac Center, and by the New York University College of Arts and Sciences.

REFERENCES

Ahn, D. G., Ruvinsky, I., Oates, A. C., Silver, L. M., and Ho, R. K. (2000). tbx20, a new vertebrate T-box gene expressed in the cranial motor neurons and developing cardiovascular structures in zebrafish. *Mech. Dev.* **95**(1–2), 253–258.

Azumi, K., Takahashi, H., Miki, Y., Fujie, M., Usami, T., Ishikawa, H., Kitayama, A., Satou, Y., Ueno, N., and Satoh, N. (2003). Construction of a cDNA microarray derived from the ascidian Ciona intestinalis. *Zoolog. Sci.* **20**(10), 1223–1229.

Bally-Cuif, L., Dubois, L., and Vincent, A. (1998). Molecular cloning of Zcoe2, the zebrafish homolog of Xenopus Xcoe2 and mouse EBF-2, and its expression during primary neurogenesis. *Mech. Dev.* **77**(1), 85–90.

Beh, J., Shi, W., Levine, M., Davidson, B., and Christiaen, L. (2007). FoxF is essential for FGF-induced migration of heart progenitor cells in the ascidian Ciona intestinalis. *Development* **134**(18), 3297–3305.

Belgacem, M. R., Escande, M. L., Escriva, H., and Bertrand, S. (2011). Amphioxus Tbx6/16 and Tbx20 embryonic expression patterns reveal ancestral functions in chordates. *Gene Expr. Patterns* **11**(3–4), 239–243.

Bertrand, V., Hudson, C., Caillol, D., Popovici, C., and Lemaire, P. (2003). Neural tissue in ascidian embryos is induced by FGF9/16/20, acting via a combination of maternal GATA and Ets transcription factors. *Cell* **115**(5), 615–627.

Bondue, A., and Blanpain, C. (2010). Mesp1: A key regulator of cardiovascular lineage commitment. *Circ. Res.* **107**(12), 1414–1427.

Bondue, A., Lapouge, G., Paulissen, C., Semeraro, C., Iacovino, M., Kyba, M., and Blanpain, C. (2008). Mesp1 acts as a master regulator of multipotent cardiovascular progenitor specification. *Cell Stem Cell* **3**(1), 69–84.

Buckingham, M., Meilhac, S., and Zaffran, S. (2005). Building the mammalian heart from two sources of myocardial cells. *Nat. Rev. Genet.* **6**(11), 826–835.

Cai, C. L., Liang, X., Shi, Y., Chu, P. H., Pfaff, S. L., Chen, J., and Evans, S. (2003). Isl1 identifies a cardiac progenitor population that proliferates prior to differentiation and contributes a majority of cells to the heart. *Dev. Cell* **5**(6), 877–889.

Carson, C. T., Kinzler, E. R., and Parr, B. A. (2000). Tbx12, a novel T-box gene, is expressed during early stages of heart and retinal development. *Mech. Dev.* **96**(1), 137–140.

Chapman, D. L., Agulnik, I., Hancock, S., Silver, L. M., and Papaioannou, V. E. (1996). Tbx6, a mouse T-Box gene implicated in paraxial mesoderm formation at gastrulation. *Dev. Biol.* **180**(2), 534–542.

Christiaen, L., Davidson, B., Kawashima, T., Powell, W., Nolla, H., Vranizan, K., and Levine, M. (2008). The transcription/migration interface in heart precursors of Ciona intestinalis. *Science* **320**(5881), 1349–1352.

Christiaen, L., Stolfi, A., Davidson, B., and Levine, M. (2009a). Spatio-temporal intersection of Lhx3 and Tbx6 defines the cardiac field through synergistic activation of Mesp. *Dev. Biol.* **328**(2), 552–560.

Christiaen, L., Wagner, E., Shi, W., and Levine, M. (2009b). Electroporation of transgenic DNAs in the sea squirt Ciona. *CSH Protoc.* **2009**(12), pdb prot5345.

Christiaen, L., Wagner, E., Shi, W., and Levine, M. (2009c). Isolation of individual cells and tissues from electroporated sea squirt (Ciona) embryos by fluorescence-activated cell sorting (FACS). *CSH Protoc.* **2009**(12), pdb prot5349.

Christiaen, L., Wagner, E., Shi, W., and Levine, M. (2009d). Microinjection of morpholino oligos and RNAs in sea squirt (Ciona) embryos. *CSH Protoc.* **2009**(12), pdb prot5347.

Christiaen, L., Stolfi, A., and Levine, M. (2010). BMP signaling coordinates gene expression and cell migration during precardiac mesoderm development. *Dev. Biol.* **340**(2), 179–187.

Cooley, J., Whitaker, S., Sweeney, S., Fraser, S., and Davidson, B. (2011). Cytoskeletal polarity mediates localized induction of the heart progenitor lineage. *Nat. Cell Biol.* **13**(8), 952–957.

Corbo, J. C., Levine, M., and Zeller, R. W. (1997). Characterization of a notochord-specific enhancer from the Brachyury promoter region of the ascidian, Ciona intestinalis. *Development* **124**(3), 589–602.

Costello, I., Pimeisl, I. M., Drager, S., Bikoff, E. K., Robertson, E. J., and Arnold, S. J. (2011). The T-box transcription factor Eomesodermin acts upstream of Mesp1 to specify cardiac mesoderm during mouse gastrulation. *Nat. Cell Biol.* **13**(9), 1084–1091.

Crozatier, M., and Vincent, A. (1999). Requirement for the Drosophila COE transcription factor Collier in formation of an embryonic muscle: Transcriptional response to notch signalling. *Development* **126**(7), 1495–1504.

Crozatier, M., Valle, D., Dubois, L., Ibnsouda, S., and Vincent, A. (1996). Collier, a novel regulator of Drosophila head development, is expressed in a single mitotic domain. *Curr. Biol.* **6**(6), 707–718.

Crozatier, M., Valle, D., Dubois, L., Ibnsouda, S., and Vincent, A. (1999). Head versus trunk patterning in the Drosophila embryo; collier requirement for formation of the intercalary segment. *Development* **126**(19), 4385–4394.

David, R., Brenner, C., Stieber, J., Schwarz, F., Brunner, S., Vollmer, M., Mentele, E., Muller-Hocker, J., Kitajima, S., Lickert, H., *et al.* (2008). MesP1 drives vertebrate cardiovascular differentiation through Dkk-1-mediated blockade of Wnt-signalling. *Nat. Cell Biol.* **10**(3), 338–345.

Davidson, B. (2007). Ciona intestinalis as a model for cardiac development. *Semin. Cell Dev. Biol.* **18**(1), 16–26.

Davidson, E. H., and Erwin, D. H. (2006). Gene regulatory networks and the evolution of animal body plans. *Science* **311**(5762), 796–800.

Davidson, B., and Levine, M. (2003). Evolutionary origins of the vertebrate heart: Specification of the cardiac lineage in Ciona intestinalis. *Proc. Natl. Acad. Sci. USA* **100**(20), 11469–11473.

Davidson, B., Shi, W., and Levine, M. (2005). Uncoupling heart cell specification and migration in the simple chordate Ciona intestinalis. *Development* **132**(21), 4811–4818.

Davidson, B., Shi, W., Beh, J., Christiaen, L., and Levine, M. (2006). FGF signaling delineates the cardiac progenitor field in the simple chordate, Ciona intestinalis. *Genes Dev.* **20**(19), 2728–2738.

Dehal, P., Satou, Y., Campbell, R. K., Chapman, J., Degnan, B., De Tomaso, A., Davidson, B., Di Gregorio, A., Gelpke, M., Goodstein, D. M., *et al.* (2002). The draft genome of Ciona intestinalis: Insights into chordate and vertebrate origins. *Science* **298**(5601), 2157–2167.

Deimling, S. J., and Drysdale, T. A. (2011). Fgf is required to regulate anterior-posterior patterning in the Xenopus lateral plate mesoderm. *Mech Dev* **128**, 327–341.

Delsuc, F., Brinkmann, H., Chourrout, D., and Philippe, H. (2006). Tunicates and not cephalochordates are the closest living relatives of vertebrates. *Nature* **439**(7079), 965–968.

Dubois, L., Bally-Cuif, L., Crozatier, M., Moreau, J., Paquereau, L., and Vincent, A. (1998). XCoe2, a transcription factor of the Col/Olf-1/EBF family involved in the specification of primary neurons in Xenopus. *Curr. Biol.* **8**(4), 199–209.

Fishman, M. C., and Olson, E. N. (1997). Parsing the heart: Genetic modules for organ assembly. *Cell* **91**(2), 153–156.

Frasch, M. (1995). Induction of visceral and cardiac mesoderm by ectodermal Dpp in the early Drosophila embryo. *Nature* **374**(6521), 464–467.

Fujiwara, S., Maeda, Y., Shin, I. T., Kohara, Y., Takatori, N., Satou, Y., and Satoh, N. (2002). Gene expression profiles in Ciona intestinalis cleavage-stage embryos. *Mech. Dev.* **112**(1–2), 115–127.

Gillis, J. A., Fritzenwanker, J. H., and Lowe, C. J. (2012). A stem-deuterostome origin of the vertebrate pharyngeal transcriptional network. *Proc. Biol. Sci.* **279**(1727), 237–246.

Green, Y. S., and Vetter, M. L. (2011). EBF proteins participate in transcriptional regulation of Xenopus muscle development. *Dev. Biol.* **358**(1), 240–250.

Hill, M. M., Broman, K. W., Stupka, E., Smith, W. C., Jiang, D., and Sidow, A. (2008). The C. savignyi genetic map and its integration with the reference sequence facilitates insights into chordate genome evolution. *Genome Res.* **18**(8), 1369–1379.

Hirano, T., and Nishida, H. (1997). Developmental fates of larval tissues after metamorphosis in ascidian Halocynthia roretzi. I. Origin of mesodermal tissues of the juvenile. *Dev. Biol.* **192**(2), 199–210.

Hitachi, K., Kondow, A., Danno, H., Nishimura, Y., Okabayashi, K., and Asashima, M. (2009). Molecular analyses of Xenopus laevis Mesp-related genes. *Integr. Zool.* **4**(4), 387–394.

Holland, N. D., Venkatesh, T. V., Holland, L. Z., Jacobs, D. K., and Bodmer, R. (2003). AmphiNk2-tin, an amphioxus homeobox gene expressed in myocardial progenitors: Insights into evolution of the vertebrate heart. *Dev. Biol.* **255**(1), 128–137.

Hopkins, D. R., Keles, S., and Greenspan, D. S. (2007). The bone morphogenetic protein 1/Tolloid-like metalloproteinases. *Matrix Biol.* **26**(7), 508–523.

Hotta, K., Mitsuhara, K., Takahashi, H., Inaba, K., Oka, K., Gojobori, T., and Ikeo, K. (2007). A web-based interactive developmental table for the ascidian Ciona intestinalis, including 3D real-image embryo reconstructions: I. From fertilized egg to hatching larva. *Dev. Dyn.* **236**(7), 1790–1805.

Hug, B., Walter, V., and Grunwald, D. J. (1997). tbx6, a Brachyury-related gene expressed by ventral mesendodermal precursors in the zebrafish embryo. *Dev. Biol.* **183**(1), 61–73.

Hunger-Glaser, I., Fan, R. S., Perez-Salazar, E., and Rozengurt, E. (2004). PDGF and FGF induce focal adhesion kinase (FAK) phosphorylation at Ser-910: dissociation from Tyr-397 phosphorylation and requirement for ERK activation. *J. Cell. Physiol.* **200**(2), 213–222.

Imai, K., Takada, N., Satoh, N., and Satou, Y. (2000). (beta)-Catenin mediates the specification of endoderm cells in ascidian embryos. *Development* **127**(14), 3009–3020.

Imai, K. S., Satoh, N., and Satou, Y. (2002). Early embryonic expression of FGF4/6/9 gene and its role in the induction of mesenchyme and notochord in Ciona savignyi embryos. *Development* **129**(7), 1729–1738.

Imai, K. S., Satoh, N., and Satou, Y. (2003). A twist-like bHLH gene is a downstream factor of an endogenous FGF and determines mesenchymal fate in the ascidian embryos. *Development* **130**(18), 4461–4472.

Imai, K. S., Hino, K., Yagi, K., Satoh, N., and Satou, Y. (2004). Gene expression profiles of transcription factors and signaling molecules in the ascidian embryo: Towards a comprehensive understanding of gene networks. *Development* **131**(16), 4047–4058.

Imai, K. S., Levine, M., Satoh, N., and Satou, Y. (2006). Regulatory blueprint for a chordate embryo. *Science* **312**(5777), 1183–1187.

Jackman, W. R., Langeland, J. A., and Kimmel, C. B. (2000). Islet reveals segmentation in the Amphioxus hindbrain homolog. *Dev. Biol.* **220**(1), 16–26.

Keduka, E., Kaiho, A., Hamada, M., Watanabe-Takano, H., Takano, K., Ogasawara, M., Satou, Y., Satoh, N., and Endo, T. (2009). M-Ras evolved independently of R-Ras and its neural function is conserved between mammalian and ascidian, which lacks classical Ras. *Gene* **429**(1–2), 49–58.

Kitajima, S., Takagi, A., Inoue, T., and Saga, Y. (2000). MesP1 and MesP2 are essential for the development of cardiac mesoderm. *Development* **127**(15), 3215–3226.

Knezevic, V., De Santo, R., and Mackem, S. (1997). Two novel chick T-box genes related to mouse Brachyury are expressed in different, non-overlapping mesodermal domains during gastrulation. *Development* **124**(2), 411–419.

Kobayashi, M., Takatori, N., Nakajima, Y., Kumano, G., Nishida, H., and Saiga, H. (2010). Spatial and temporal expression of two transcriptional isoforms of Lhx3, a LIM class homeobox gene, during embryogenesis of two phylogenetically remote ascidians, Halocynthia roretzi and Ciona intestinalis. *Gene Expr. Patterns* **10**(2–3), 98–104.

Kokubo, N., Matsuura, M., Onimaru, K., Tiecke, E., Kuraku, S., Kuratani, S., and Tanaka, M. (2010). Mechanisms of heart development in the Japanese lamprey, Lethenteron japonicum. *Evol. Dev.* **12**(1), 34–44.

Koshiba-Takeuchi, K., Mori, A. D., Kaynak, B. L., Cebra-Thomas, J., Sukonnik, T., Georges, R. O., Latham, S., Beck, L., Henkelman, R. M., Black, B. L., *et al.* (2009). Reptilian heart development and the molecular basis of cardiac chamber evolution. *Nature* **461**(7260), 95–98.

Koster, M., Dillinger, K., and Knochel, W. (1999). Genomic structure and embryonic expression of the Xenopus winged helix factors XFD-13/13′. *Mech. Dev.* **88**(1), 89–93.

Kubo, A., Suzuki, N., Yuan, X., Nakai, K., Satoh, N., Imai, K. S., and Satou, Y. (2010). Genomic cis-regulatory networks in the early Ciona intestinalis embryo. *Development* **137** (10), 1613–1623.

Kumano, G., Takatori, N., Negishi, T., Takada, T., and Nishida, H. (2011). A maternal factor unique to ascidians silences the germline via binding to P-TEFb and RNAP II regulation. *Curr. Biol.* **21**(15), 1308–1313.

Lescroart, F., Kelly, R. G., Le Garrec, J. F., Nicolas, J. F., Meilhac, S. M., and Buckingham, M. (2010). Clonal analysis reveals common lineage relationships between head muscles and second heart field derivatives in the mouse embryo. *Development* **137** (19), 3269–3279.

Mahadevan, N. R., Horton, A. C., and Gibson-Brown, J. J. (2004). Developmental expression of the amphioxus Tbx1/ 10 gene illuminates the evolution of vertebrate branchial arches and sclerotome. *Dev. Genes Evol.* **214**(11), 559–566.

Maher, P. A., Pasquale, E. B., Wang, J. Y., and Singer, S. J. (1985). Phosphotyrosine-containing proteins are concentrated in focal adhesions and intercellular junctions in normal cells. *Proc. Natl. Acad. Sci. USA* **82**(19), 6576–6580.

Mahlapuu, M., Pelto-Huikko, M., Aitola, M., Enerback, S., and Carlsson, P. (1998). FREAC-1 contains a cell-type-specific transcriptional activation domain and is expressed in epithelial-mesenchymal interfaces. *Dev. Biol.* **202**(2), 183–195.

Mann, T., Bodmer, R., and Pandur, P. (2009). The Drosophila homolog of vertebrate Islet1 is a key component in early cardiogenesis. *Development* **136**(2), 317–326.

Mazet, F., Masood, S., Luke, G. N., Holland, N. D., and Shimeld, S. M. (2004). Expression of AmphiCoe, an amphioxus COE/EBF gene, in the developing central nervous system and epidermal sensory neurons. *Genesis* **38**(2), 58–65.

Meilhac, S. M., Esner, M., Kelly, R. G., Nicolas, J. F., and Buckingham, M. E. (2004). The clonal origin of myocardial cells in different regions of the embryonic mouse heart. *Dev. Cell* **6**(5), 685–698.

Miwata, K., Chiba, T., Horii, R., Yamada, L., Kubo, A., Miyamura, D., Satoh, N., and Satou, Y. (2006). Systematic analysis of embryonic expression profiles of zinc finger genes in Ciona intestinalis. *Dev. Biol.* **292**(2), 546–554.

Miya, T., and Nishida, H. (2003). An Ets transcription factor, HrEts, is target of FGF signaling and involved in induction of notochord, mesenchyme, and brain in ascidian embryos. *Dev. Biol.* **261**(1), 25–38.

Moore, A. W., Barbel, S., Jan, L. Y., and Jan, Y. N. (2000). A genomewide survey of basic helix-loop-helix factors in Drosophila. *Proc. Natl. Acad. Sci. USA* **97**(19), 10436–10441.

Nagatomo, K., and Fujiwara, S. (2003). Expression of Raldh2, Cyp26 and Hox-1 in normal and retinoic acid-treated Ciona intestinalis embryos. *Gene Expr. Patterns* **3**(3), 273–277.

Nathan, E., Monovich, A., Tirosh-Finkel, L., Harrelson, Z., Rousso, T., Rinon, A., Harel, I., Evans, S. M., and Tzahor, E. (2008). The contribution of Islet1-expressing splanchnic mesoderm cells to distinct branchiomeric muscles reveals significant heterogeneity in head muscle development. *Development* **135**(4), 647–657.

Nishida, H., and Sawada, K. (2001). Macho-1 encodes a localized mRNA in ascidian eggs that specifies muscle fate during embryogenesis. *Nature* **409**(6821), 724–729.

Olson, E. N. (2006). Gene regulatory networks in the evolution and development of the heart. *Science* **313**(5795), 1922–1927.

Onimaru, K., Shoguchi, E., Kuratani, S., and Tanaka, M. (2011). Development and evolution of the lateral plate mesoderm: Comparative analysis of amphioxus and lamprey with implications for the acquisition of paired fins. *Dev. Biol.* **359**(1), 124–136.

Pasini, A., Amiel, A., Rothbacher, U., Roure, A., Lemaire, P., and Darras, S. (2006). Formation of the ascidian epidermal sensory neurons: Insights into the origin of the chordate peripheral nervous system. *PLoS Biol.* **4**(7), e225.

Peterson, R. S., Lim, L., Ye, H., Zhou, H., Overdier, D. G., and Costa, R. H. (1997). The winged helix transcriptional activator HFH-8 is expressed in the mesoderm of the primitive streak stage of mouse embryos and its cellular derivatives. *Mech. Dev.* **69**(1–2), 53–69.

Porsch, M., Hofmeyer, K., Bausenwein, B. S., Grimm, S., Weber, B. H., Miassod, R., and Pflugfelder, G. O. (1998). Isolation of a Drosophila T-box gene closely related to human TBX1. *Gene* **212**(2), 237–248.

Prall, O. W., Menon, M. K., Solloway, M. J., Watanabe, Y., Zaffran, S., Bajolle, F., Biben, C., McBride, J. J., Robertson, B. R., Chaulet, H., *et al.* (2007). An Nkx2-5/Bmp2/Smad1 negative feedback loop controls heart progenitor specification and proliferation. *Cell* **128**(5), 947–959.

Prasad, B. C., Ye, B., Zackhary, R., Schrader, K., Seydoux, G., and Reed, R. R. (1998). unc-3, a gene required for axonal guidance in Caenorhabditis elegans, encodes a member of the O/E family of transcription factors. *Development* **125**(8), 1561–1568.

Putnam, N. H., Butts, T., Ferrier, D. E., Furlong, R. F., Hellsten, U., Kawashima, T., Robinson-Rechavi, M., Shoguchi, E., Terry, A., Yu, J. K., *et al.* (2008). The amphioxus genome and the evolution of the chordate karyotype. *Nature* **453**(7198), 1064–1071.

Ragkousi, K., Beh, J., Sweeney, S., Starobinska, E., and Davidson, B. (2011). A single GATA factor plays discrete, lineage specific roles in ascidian heart development. *Dev. Biol.* **352**(1), 154–163.

Reim, I., and Frasch, M. (2005). The Dorsocross T-box genes are key components of the regulatory network controlling early cardiogenesis in Drosophila. *Development* **132**(22), 4911–4925.

Reim, I., and Frasch, M. (2010). Genetic and genomic dissection of cardiogenesis in the Drosophila model. *Pediatr. Cardiol.* **31**(3), 325–334.

Saga, Y., Miyagawa-Tomita, S., Takagi, A., Kitajima, S., Miyazaki, J., and Inoue, T. (1999). MesP1 is expressed in the heart precursor cells and required for the formation of a single heart tube. *Development* **126**(15), 3437–3447.

Saga, Y., Kitajima, S., and Miyagawa-Tomita, S. (2000). Mesp1 expression is the earliest sign of cardiovascular development. *Trends Cardiovasc. Med.* **10**(8), 345–352.

Sardet, C., Nishida, H., Prodon, F., and Sawada, K. (2003). Maternal mRNAs of PEM and macho 1, the ascidian muscle determinant, associate and move with a rough endoplasmic reticulum network in the egg cortex. *Development* **130**(23), 5839–5849.

Satoh, N. (2003). The ascidian tadpole larva: Comparative molecular development and genomics. *Nat. Rev. Genet.* **4**(4), 285–295.

Satou, Y., and Satoh, N. (2006). Gene regulatory networks for the development and evolution of the chordate heart. *Genes Dev.* **20**(19), 2634–2638.

Satou, Y., Imai, K. S., and Satoh, N. (2001a). Action of morpholinos in Ciona embryos. *Genesis* **30**(3), 103–106.

Satou, Y., Imai, K. S., and Satoh, N. (2001b). Early embryonic expression of a LIM-homeobox gene Cs-lhx3 is downstream of beta-catenin and responsible for the endoderm differentiation in Ciona savignyi embryos. *Development* **128**(18), 3559–3570.

Satou, Y., Takatori, N., Yamada, L., Mochizuki, Y., Hamaguchi, M., Ishikawa, H., Chiba, S., Imai, K., Kano, S., Murakami, S. D., *et al.* (2001c). Gene expression profiles in Ciona intestinalis tailbud embryos. *Development* **128**(15), 2893–2904.

Satou, Y., Takatori, N., Fujiwara, S., Nishikata, T., Saiga, H., Kusakabe, T., Shin-i, T., Kohara, Y., and Satoh, N. (2002a). Ciona intestinalis cDNA projects: Expressed sequence tag analyses and gene expression profiles during embryogenesis. *Gene* **287** (1–2), 83–96.

Satou, Y., Yamada, L., Mochizuki, Y., Takatori, N., Kawashima, T., Sasaki, A., Hamaguchi, M., Awazu, S., Yagi, K., Sasakura, Y., *et al.* (2002b). A cDNA resource from the basal chordate Ciona intestinalis. *Genesis* **33**(4), 153–154.

Satou, Y., Imai, K. S., Levine, M., Kohara, Y., Rokhsar, D., and Satoh, N. (2003). A genomewide survey of developmentally relevant genes in Ciona intestinalis. I. Genes for bHLH transcription factors. *Dev. Genes Evol.* **213**(5–6), 213–221.

Satou, Y., Imai, K. S., and Satoh, N. (2004). The ascidian Mesp gene specifies heart precursor cells. *Development* **131**(11), 2533–2541.

Satou, Y., Kawashima, T., Shoguchi, E., Nakayama, A., and Satoh, N. (2005). An integrated database of the ascidian, Ciona intestinalis: Towards functional genomics. *Zoolog. Sci.* **22** (8), 837–843.

Satou, Y., Hamaguchi, M., Takeuchi, K., Hastings, K. E., and Satoh, N. (2006). Genomic overview of mRNA 5′-leader trans-splicing in the ascidian Ciona intestinalis. *Nucleic Acids Res.* **34**(11), 3378–3388.

Sauka-Spengler, T., Le Mentec, C., Lepage, M., and Mazan, S. (2002). Embryonic expression of Tbx1, a DiGeorge syndrome candidate gene, in the lamprey Lampetra fluviatilis. *Gene Expr. Patterns* **2**(1–2), 99–103.

Sawada, A., Fritz, A., Jiang, Y. J., Yamamoto, A., Yamasu, K., Kuroiwa, A., Saga, Y., and Takeda, H. (2000). Zebrafish Mesp family genes, mesp-a and mesp-b are segmentally expressed in the presomitic mesoderm, and Mesp-b confers the anterior identity to the developing somites. *Development* **127**(8), 1691–1702.

Schlaepfer, D. D., Hauck, C. R., and Sieg, D. J. (1999). Signaling through focal adhesion kinase. *Prog. Biophys. Mol. Biol.* **71**(3–4), 435–478.

Schubert, M., Meulemans, D., Bronner-Fraser, M., Holland, L. Z., and Holland, N. D. (2003). Differential mesodermal expression of two amphioxus MyoD family members (AmphiMRF1 and AmphiMRF2). *Gene Expr. Patterns* **3**(2), 199–202.

Shi, W., and Levine, M. (2008). Ephrin signaling establishes asymmetric cell fates in an endomesoderm lineage of the Ciona embryo. *Development* **135**(5), 931–940.

Shirae-Kurabayashi, M., Matsuda, K., and Nakamura, A. (2011). Ci-Pem-1 localizes to the nucleus and represses somatic gene transcription in the germline of Ciona intestinalis embryos. *Development* **138**(14), 2871–2881.

Small, K. S., Brudno, M., Hill, M. M., and Sidow, A. (2007). A haplome alignment and reference sequence of the highly polymorphic Ciona savignyi genome. *Genome Biol.* **8** (3), R41.

Stolfi, A., Gainous, T. B., Young, J. J., Mori, A., Levine, M., and Christiaen, L. (2010). Early chordate origins of the vertebrate second heart field. *Science* **329**(5991), 565–568.

Takatori, N., Hotta, K., Mochizuki, Y., Satoh, G., Mitani, Y., Satoh, N., Satou, Y., and Takahashi, H. (2004). T-box genes in the ascidian Ciona intestinalis: Characterization of cDNAs and spatial expression. *Dev. Dyn.* **230**(4), 743–753.

Tassy, O., Daian, F., Hudson, C., Bertrand, V., and Lemaire, P. (2006). A quantitative approach to the study of cell shapes and interactions during early chordate embryogenesis. *Curr. Biol.* **16**(4), 345–358.

Tassy, O., Dauga, D., Daian, F., Sobral, D., Robin, F., Khoueiry, P., Salgado, D., Fox, V., Caillol, D., Schiappa, R., *et al.* (2010). The ANISEED database: Digital representation, formalization, and elucidation of a chordate developmental program. *Genome Res.* **20** (10), 1459–1468.

Thannickal, V. J., Lee, D. Y., White, E. S., Cui, Z., Larios, J. M., Chacon, R., Horowitz, J. C., Day, R. M., and Thomas, P. E. (2003). Myofibroblast differentiation by transforming growth factor-beta1 is dependent on cell adhesion and integrin signaling via focal adhesion kinase. *J. Biol. Chem.* **278**(14), 12384–12389.

Tiecke, E., Matsuura, M., Kokubo, N., Kuraku, S., Kusakabe, R., Kuratani, S., and Tanaka, M. (2007). Identification and developmental expression of two Tbx1/10-related genes in the agnathan Lethenteron japonicum. *Dev. Genes Evol.* **217**(10), 691–697.

Tomioka, M., Miya, T., and Nishida, H. (2002). Repression of zygotic gene expression in the putative germline cells in ascidian embryos. *Zoolog. Sci.* **19**(1), 49–55.

Tu, Q., Brown, C. T., Davidson, E. H., and Oliveri, P. (2006). Sea urchin Forkhead gene family: Phylogeny and embryonic expression. *Dev. Biol.* **300**(1), 49–62.

Uchiyama, H., Kobayashi, T., Yamashita, A., Ohno, S., and Yabe, S. (2001). Cloning and characterization of the T-box gene Tbx6 in Xenopus laevis. *Dev. Growth Differ.* **43**(6), 657–669.

Vincent, S. D., and Buckingham, M. E. (2010). How to make a heart: The origin and regulation of cardiac progenitor cells. *Curr. Top. Dev. Biol.* **90,** 1–41.

Vinson, J. P., Jaffe, D. B., O'Neill, K., Karlsson, E. K., Stange-Thomann, N., Anderson, S., Mesirov, J. P., Satoh, N., Satou, Y., Nusbaum, C., *et al.* (2005). Assembly of polymorphic genomes: Algorithms and application to Ciona savignyi. *Genome Res.* **15**(8), 1127–1135.

Voutev, R., Keating, R., Hubbard, E. J., and Vallier, L. G. (2009). Characterization of the Caenorhabditis elegans Islet LIM-homeodomain ortholog, lim-7. *FEBS Lett.* **583**(2), 456–464.

Wotton, K. R., and Shimeld, S. M. (2006). Comparative genomics of vertebrate Fox cluster loci. *BMC Genomics* **7,** 271.

Wotton, K. R., and Shimeld, S. M. (2011). Analysis of lamprey clustered Fox genes: Insight into Fox gene evolution and expression in vertebrates. *Gene* **489**(1), 30–40.

Wotton, K. R., French, K. E., and Shimeld, S. M. (2007). The developmental expression of foxl2 in the dogfish Scyliorhinus canicula. *Gene Expr. Patterns* **7**(7), 793–797.

Wotton, K. R., Mazet, F., and Shimeld, S. M. (2008). Expression of FoxC, FoxF, FoxL1, and FoxQ1 genes in the dogfish Scyliorhinus canicula defines ancient and derived roles for Fox genes in vertebrate development. *Dev. Dyn.* **237**(6), 1590–1603.

Xavier-Neto, J., Castro, R. A., Sampaio, A. C., Azambuja, A. P., Castillo, H. A., Cravo, R. M., and Simoes-Costa, M. S. (2007). Parallel avenues in the evolution of hearts and pumping organs. *Cell. Mol. Life Sci.* **64**(6), 719–734.

Yagi, K., Satou, Y., Mazet, F., Shimeld, S. M., Degnan, B., Rokhsar, D., Levine, M., Kohara, Y., and Satoh, N. (2003). A genomewide survey of developmentally relevant genes in Ciona intestinalis. III. Genes for Fox, ETS, nuclear receptors and NFkappaB. *Dev. Genes Evol.* **213**(5–6), 235–244.

Yagi, K., Takatori, N., Satou, Y., and Satoh, N. (2005). Ci-Tbx6b and Ci-Tbx6c are key mediators of the maternal effect gene Ci-macho1 in muscle cell differentiation in Ciona intestinalis embryos. *Dev. Biol.* **282**(2), 535–549.

Yasuhiko, Y., Haraguchi, S., Kitajima, S., Takahashi, Y., Kanno, J., and Saga, Y. (2006). Tbx6-mediated Notch signaling controls somite-specific Mesp2 expression. *Proc. Natl. Acad. Sci. USA* **103**(10), 3651–3656.

Yasuo, H., and Hudson, C. (2007). FGF8/17/18 functions together with FGF9/16/20 during formation of the notochord in Ciona embryos. *Dev. Biol.* **302**(1), 92–103.

Yoshida, S., Marikawa, Y., and Satoh, N. (1996). Posterior end mark, a novel maternal gene encoding a localized factor in the ascidian embryo. *Development* **122**(7), 2005–2012.

Zaffran, S., Kuchler, A., Lee, H. H., and Frasch, M. (2001). biniou (FoxF), a central component in a regulatory network controlling visceral mesoderm development and midgut morphogenesis in Drosophila. *Genes Dev.* **15**(21), 2900–2915.

Zaidel-Bar, R., and Geiger, B. (2010). The switchable integrin adhesome. *J. Cell Sci.* **123**(Pt 9), 1385–1388.

Zamir, E., and Geiger, B. (2001). Molecular complexity and dynamics of cell-matrix adhesions. *J. Cell Sci.* **114**(Pt 20), 3583–3590.

CHAPTER FIVE

Regulation of GATA4 Transcriptional Activity in Cardiovascular Development and Disease

Pingzhu Zhou,[*] Aibin He,[†] *and* William T. Pu[*,†]

Contents

Abstract

Transcription factors regulate formation and function of the heart, and perturbation of transcription factor expression and regulation disrupts normal heart structure and function. Multiple mechanisms regulate the level and locus-specific activity of transcription factors, including transcription, translation, subcellular localization, posttranslational modifications, and context-dependent interactions with other transcription factors, chromatin remodeling enzymes, and epigenetic regulators. The zinc finger transcription factor GATA4 is among the best-studied cardiac transcriptional factors. This review focuses

* Department of Cardiology, Children's Hospital Boston, Boston, Massachusetts, USA
† Harvard Stem Cell Institute, Harvard University, Cambridge, Massachusetts, USA

Current Topics in Developmental Biology, Volume 100
ISSN 0070-2153, DOI: 10.1016/B978-0-12-387786-4.00005-1

on molecular mechanisms that regulate GATA4 transcriptional activity in the cardiovascular system, providing a framework to investigate and understand the molecular regulation of cardiac gene transcription by other transcription factors.

1. Introduction: GATA Factors in Cardiac Specification and Development

Development of the mammalian heart is an intricate dance involving multiple cell types that arise from several sources. The initial heart tube, composed of a layer of cardiomyocytes overlying a layer of endocardial cells, grows through the addition of cardiomyocytes and endothelial cells at both poles. The elongating heart tube forms a rightward loop, and the looped heart tube is then divided into four chambers by growth of the muscular interatrial and interventricular septae and by expansion of the endocardial cushions, which form central portions of these septae as well as the developing heart valves. Division of the outflow tract into systemic and pulmonary arterial circuits requires the coordination of cells arising from neural crest, second heart field, and endocardial cushions. Cells arising from the proepicardium migrate onto the surface of the heart, forming an epithelial sheet known as the epicardium. This sheet of cells undergoes epithelial-to-mesenchymal transition, thereby generating mesenchymal cells that contribute to most of the stromal cells of the heart.

A network of transcription factors precisely choreographs this process. Not surprisingly, the transcription factor network is regulated by multiple mechanisms, and disruption of the structure or activity of transcription factors underlies a significant portion of congenital heart disease, the most common type of major congenital malformation (Bruneau, 2008; Srivastava, 2006). The transcription factor GATA4 is positioned high in the cardiac transcriptional network. Proper regulation of GATA4 levels and activity are crucial for normal heart specification and development, and GATA4 regulatory mechanisms have been intensively studied. Here, we review these mechanisms, providing a framework to understand regulation of transcription factor activity in the developing heart.

1.1. Structure and function of GATA factors

The mammalian genome encodes six GATA factors, GATA1–6, which share a highly conserved region spanning 109 amino acid residues, of which 82 residues (75%) are identical across the six proteins (Fig. 5.1). This region encompasses the DNA-binding domain and contains two C-X_2-C-X_{17}-C-X_2-C zinc fingers and adjacent residues (Evans and Felsenfeld, 1989;

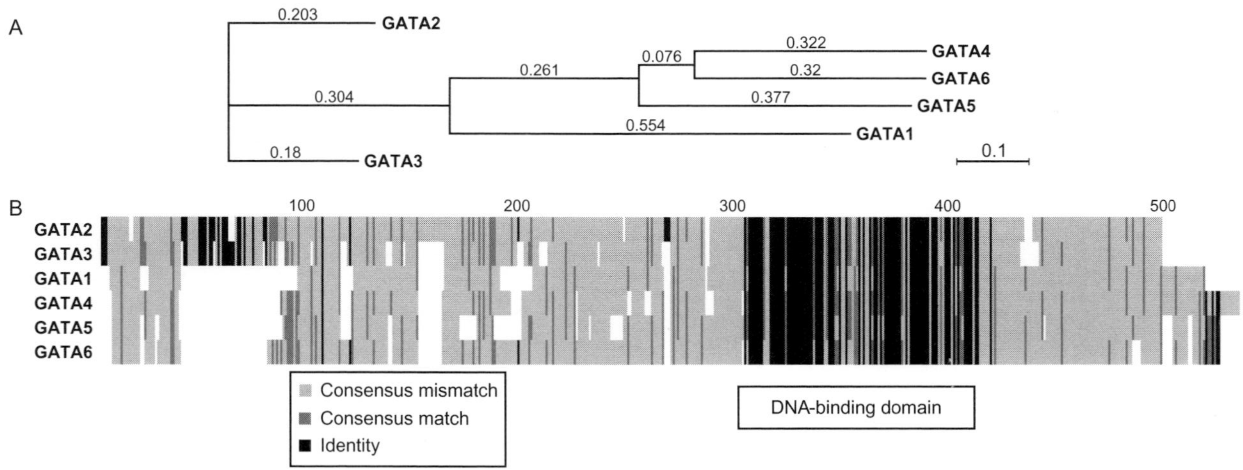

Figure 5.1 Sequence conservation of GATA proteins. (A) Phylogenetic tree shows close relationship of GATA4/5/6. The GATA4/5/6 cluster is more closely related to GATA1 than GATA2/3. (B) Conservation of primary sequence. GATA factors share a highly conserved DNA binding domain. Flanking regions show less sequence conservation. (For color version of this figure, the reader is referred to the Web version of this chapter.)

Molkentin, 2000; Tsai *et al.*, 1989). The C-terminal zinc finger and adjacent basic residues are sufficient for specific DNA binding (Martin and Orkin, 1990; Yang and Evans, 1992). The structure of the C-terminal finger bound to DNA, determined by X-ray crystallography and NMR spectroscopy for GATA3 and GATA1, respectively (Bates *et al.*, 2008; Omichinski *et al.*, 1993), reveals direct base-specific protein-DNA contacts. The N-terminal zinc finger stabilizes DNA–protein interactions and also participates in specific protein–protein interactions. Using protein binding microarrays and other binding site selection approaches (Berger *et al.*, 2006; Merika and Orkin, 1993; Newburger and Bulyk, 2009; Sakai *et al.*, 1998), the DNA sequence recognized by GATA proteins was defined as (a/t) GATAA(g), and the *in vivo* preference of GATA1, GATA2, and GATA4 for this sequence was verified by chromatin immunoprecipitation followed by high-throughput sequencing (ChIP-seq) (Fujiwara *et al.*, 2009; He *et al.*, 2011; Yu *et al.*, 2009). Outside of the DNA-binding domain, the GATA4/5/6 proteins retain the modest similarity to one another but show more distant relationships to GATA1/2/3 (Fig. 5.1).

GATA4/5/6 are predominantly expressed in the cardiovascular system, the gonads, and endodermal derivatives, while GATA1/2/3 are predominantly expressed in the hematopoietic system. However, there are exceptions; for instance, GATA2 is an important transcriptional regulator in endothelial cells (Linnemann *et al.*, 2011). The GATA4/5/6 proteins are closely related, while GATA1 is more distantly related to GATA2 (Fig. 5.1B). Perhaps consistent with this sequence divergence, GATA1 is often functionally dissimilar from GATA2/3, and changes in chromatin occupancy between GATA1 and GATA2 are important for normal progression of hematopoiesis (Bresnick *et al.*, 2010). In contrast, the functional importance of such a "GATA switch" has not been established for GATA4/5/6.

Structure–function relationships for GATA4 have been analyzed by systematic mutagenesis, with the major readouts being activity in DNA-binding and luciferase reporter assays (Fig. 5.2) (Durocher *et al.*, 1997; Lee *et al.*, 1998; Morin *et al.*, 2000; Morrisey *et al.*, 1997). The N-terminal region of GATA4 was necessary and sufficient for transcriptional activation, containing two independent transcriptional activation domains that were moderately conserved in GATA5/6 (Morrisey *et al.*, 1997). While the C-terminal region was not sufficient to activate reporter transcription when fused to a heterologous DNA-binding domain, it was necessary for GATA4 transcriptional activity (but not DNA-binding). Subsequent analysis has revealed that the C-terminus is the target of acetylation and sumoylation (Takaya *et al.*, 2008; Wang *et al.*, 2004) (Fig. 5.2), which increase GATA4 transcriptional activity (see below). In addition, the C-terminal domain was essential for GATA4 transcriptional synergy with a subset of interacting transcription factors, but not their binding (Fig. 5.2). Thus, the C-terminal domain acts through unclear mechanisms to regulate GATA4 transcriptional activity.

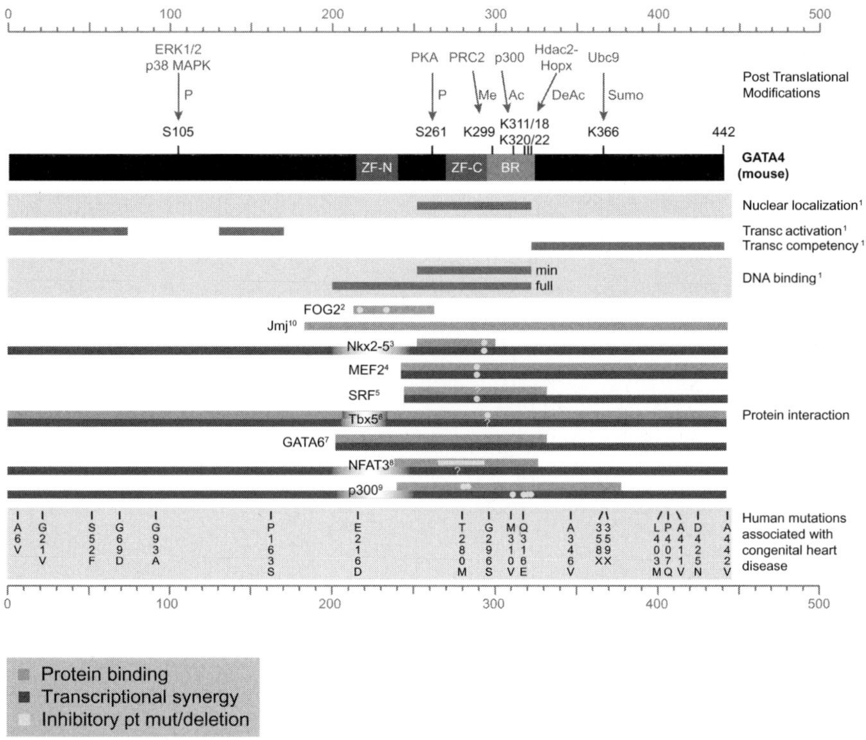

Figure 5.2 GATA4 primary sequence annotated with functional domains, posttranslational modifications, protein–protein interaction domains, and identified pathogenic human mutations. Numbers indicate amino acid residue. The GATA4 schematic shows the N- and C-terminal zinc finger domains (ZF-N and ZF-C) and the basic region (BR). Posttranslational modifications are shown above the GATA4 schematic. Ac, acetylation; DeAc, deacetylation; P, phosphorylation; Sumo, Sumoylation. Domains of biological activity are shown below the GATA4 schematic, with bars representing the minimum region sufficient for the indicated activity. Red and green bars indicate protein binding and transcriptional synergy, respectively. Yellow indicates point mutations or internal deletions that impair the indicated biological activity. Only proteins where mutational analysis of GATA4 interaction has been performed are shown. 1. Morrisey *et al.* (1997). 2. Lu *et al.* (1999), Crispino *et al.* (2001), Svensson *et al.* (2000). 3. Lee *et al.* (1998), Durocher *et al.* (1997). 4. Morin *et al.* (2000). 5. Belaguli *et al.* (2000). 6. Garg *et al.* (2003). 7. Charron *et al.* (1999). 8. Molkentin *et al.* (1998). 9. Dai and Markham (2001), Takaya *et al.* (2008). 10. Kim *et al.* (2004). Human mutations found in patients with congenital heart disease but not controls are indicated. Corresponding references are provided in the text. (For interpretation of the references to color in this figure legend, the reader is referred to the Web version of this chapter.)

1.2. GATA4 function in cardiac specification, development, and function

GATA4 is expressed in cells of the cardiac lineage from the time of their specification through adulthood, and it is a key regulator of gene expression and cellular activity at each of these stages. GATA4 promotes cardiogenesis, probably through both cell autonomous and cell nonautonomous mechanisms. During embryoid body differentiation of embryonic stem (ES) cells, GATA4 overexpression increased formation of cardiomyocytes (Grepin *et al.*, 1997; Holtzinger *et al.*, 2010). However, this effect was through non-cell autonomous mechanisms, as GATA4-overexpressing cells differentiated into Sox17$^+$ endoderm, which then stimulated cardiac differentiation of neighboring cells through secretion of cardiac-inducing paracrine factors such as BMPs and Wnt inhibitors. The cardiogenic effect of GATA4-induced endoderm is reminiscent of similar activity of endoderm observed in developing embryos (Foley *et al.*, 2006).

GATA4 stimulates cardiogenesis through cell autonomous mechanisms as well. Uncommitted mesodermal cells forced to express ectopically GATA4 and the chromatin remodeling factor BAF60C expressed cardiomyocyte markers (Takeuchi and Bruneau, 2009). Similarly, in combination with additional cardiac transcription factors TBX5 and MEF2C, GATA4 was also reported to reprogram fibroblasts into cardiomyocytes (Ieda *et al.*, 2010). Conversely, GATA4/6 deficient ES cells failed to form cardiomyocytes, in both EB differentiation systems and in embryos (rescued from visceral endoderm defects; Narita *et al.*, 1997; through tetraploid complementation; Zhao *et al.*, 2008). These latter experiments did not determine the cell type in which GATA4/6 are required for cardiogenesis, but collectively the data suggest that GATA4/6 act in cardiac progenitors to promote cardiomyocyte differentiation. Consistent with this conclusion, GATA4 participates in the transcriptional regulatory network that regulates second heart field differentiation (Black, 2007). Both *Mef2c* and *Hand2*, important cardiac transcription factors required for development of second heart field derivatives, have been identified as direct downstream targets of GATA4 in this population of heart progenitors (Dodou *et al.*, 2004; McFadden *et al.*, 2000; Zeisberg *et al.*, 2005).

In the developing heart, GATA4 is expressed in proepicardium, epicardium (Watt *et al.*, 2004), myocardium (Arceci *et al.*, 1993), endocardium, and endocardial cushions (Rivera-Feliciano *et al.*, 2006), and its activity is required in each of these compartments. GATA4 deficient embryos, rescued from visceral endoderm defects through tetraploid complementation, showed myocardial abnormalities and absence of the proepicardium (Watt *et al.*, 2004). Selective ablation of GATA4 in endothelial cells caused these cells to fail to undergo EMT, leading to a paucity of valvular mesenchymal cells (Rivera-Feliciano *et al.*, 2006). A point mutation of GATA4 that blocks its interaction with FOG cofactors rescued this defect, but later valve remodeling remained abnormal, indicating continuing requirement for GATA4 in later stages of valve formation (Rivera-Feliciano *et al.*, 2006). GATA4 inactivation in myocardium caused marked myocardial hypoplasia due to decreased cardiomyocyte proliferation (Zeisberg *et al.*, 2005). Broad abnormalities of cell cycle gene expression were identified in GATA4 deficient cardiomyocytes, and Cyclin D2 and Cdk4 were identified as likely direct downstream targets (Rojas *et al.*, 2008). Strikingly, GATA4 inactivation early in heart morphogenesis caused severe, selective defects in right ventricular morphogenesis, due in part to downregulation of *Hand2* (Zeisberg *et al.*, 2005).

GATA4 inactivation in later fetal life or postnatally caused progressive, severe defects in myocardial function, and rapid decompensation with pressure overload stress (Oka *et al.*, 2006). Following pressure overload, GATA4 knockout cardiomyocytes were unable to hypertrophy and underwent increased apoptosis. GATA6 inactivation in adult heart caused similar phenotypes, suggesting that GATA4 and GATA6 function additively in regulating cardiomyocyte function and stress responses (van Berlo *et al.*, 2010). Cardiomyocyte GATA4 was also found to support normal and pressure-overload stimulated increases in myocardial capillary density by upregulating proangiogenic factors including VEGFA and inhibiting anti-angiogenic gene expression (Heineke *et al.*, 2007; Zhou *et al.*, 2009).

The level of GATA4 is critical for normal embryo development and survival, as revealed by a series of alleles that express different levels of GATA4 (Pu *et al.*, 2004). Moreover, *Gata4* heterozygous embryos suffered from a high incidence of cardiac (myocardial hypoplasia and endocardial cushion defects) and extracardiac (diaphragmatic hernia) congenital defects in the pure C57BL6/J strain background (Jay *et al.*, 2007; Rajagopal *et al.*, 2007). These defects were milder forms of phenotypes observed in null mice. In a mixed strain background, *Gata4* heterozygous mice survived to adulthood but were susceptible to pressure-overload induced heart failure due, at least in part, to increased cardiomyocyte apoptosis (Bisping *et al.*, 2006). On the other hand, cardiomyocyte overexpression of GATA4 caused cardiac hypertrophy and progressive heart failure (Liang *et al.*, 2001a). These data underscore the importance of precisely regulating GATA4 level and activity.

Consistent with key dosage sensitive roles of *Gata4* in regulating organ development and function, heterozygous mutation of *GATA4* has been linked to abnormalities of heart, diaphragm, and gonad development or function in humans (Bielinska *et al.*, 2007; Butler *et al.*, 2010; Chen *et al.*, 2010a,b; Garg *et al.*, 2003; Lourenco *et al.*, 2011; Nemer *et al.*, 2006; Peng *et al.*, 2010; Rajagopal *et al.*, 2007; Tomita-Mitchell *et al.*, 2007; Zhang *et al.*, 2008, 2009). Missense and nonsense mutations of *GATA4* cause atrial and ventricular septal defects, often in association with pulmonary stenosis, and endocardial cushion defects (Garg *et al.*, 2003; Rajagopal *et al.*, 2007). GATA4 mutation has also been linked to complex cardiovascular defects involving right ventricular hypoplasia (Rajagopal *et al.*, 2007). Interestingly, human *GATA4* mutation has not been reported to cause ventricular dysfunction, a highly penetrate phenotype in heterozygous mice. Microdeletions involving *GATA4* are also associated with congenital diaphragmatic hernia (Bielinska *et al.*, 2007), and with disorders of sexual development (Lourenco *et al.*, 2011).

2. Regulation of GATA4 Expression Level and Activity

Given the importance of maintaining appropriate levels of GATA4 activity for normal organ development and function, a panoply of mechanisms have evolved to regulate GATA4 level and activity (Fig. 5.3).

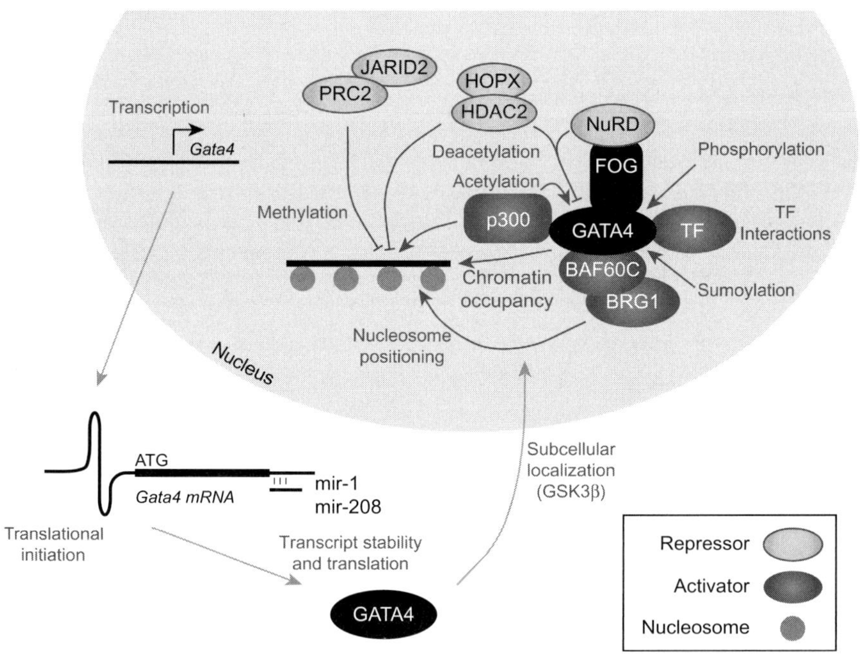

Figure 5.3 Regulation of GATA4 expression and activity. GATA4 is regulated by transcriptional, translational, and posttranslational mechanisms. Locus-specific transcriptional activity is regulated by interactions with other TFs, coactivators, corepressors, covalent modifications, and interactions with chromatin remodeling and modifying enzymes. (See Color Insert.)

The remainder of this review focus on these regulatory mechanisms. These mechanisms undoubtedly pertain to other transcription factors, and thus GATA4 serves as a case study of mechanisms used to regulate transcription factor activity in the developing and adult heart.

2.1. Regulation of GATA4 expression

GATA4 expression level varies between tissues, between developmental stages, and in disease states. Although often used as a marker of cardiomyocytes, in fetal heart GATA4 expression is the highest in the proepicardium followed by the endocardial cushions and then cardiomyocytes. GATA4 expression in the adult heart has been reported to increase by approximately twofold in heart disease (Diedrichs *et al.*, 2004; Hall *et al.*, 2004). However, little is known about the mechanisms that govern GATA4 expression.

The Black lab has identified several *Gata4* enhancers with expression restricted to endoderm or portions of lateral mesoderm including the septum transversum, but not the heart itself (Rojas *et al.*, 2005, 2009, 2010). These enhancers were regulated by GATA factors themselves in combination with Forkhead and homeodomain family transcription factors. We identified two *Gata4* regulatory sequences that drove cardiomyocyte expression in transient transgenic embryos, located immediately upstream and −93 kbp upstream of the GATA4 transcriptional start site (He *et al.*, 2011). However, neither of these sequences exhibited strong cardiac expression compared to the endogenous gene, and the −93 kb site appears to have predominant activity in endocardium and endocardial cushion mesenchyme. Thus, the sequences and mechanisms that regulate GATA4 in heart development and disease remain poorly described.

GATA4 protein levels are also regulated posttranscriptionally. The cardiac-specific microRNA miR-208 appears to regulate GATA4 translation, as GATA4 protein was upregulated in miR-208 knockout mice (Callis *et al.*, 2009). The regulation likely involves direct miR-208 interaction with the *Gata4* 3′untranslated region (UTR), as it contains a putative miR-208 binding site and is sufficient to decrease expression of a linked luciferase reporter in response to miR-208. GATA4 protein translation was also regulated by the cardiac-enriched microRNA miR-1, as GATA4 levels were regulated antithetically to miR-1 in neonatal rat ventricular cardiomyocytes (Ikeda *et al.*, 2009). However, the *Gata4* 3′UTR does not contain a predicted miR-1 binding site, suggesting that miR-1 may regulate GATA4 expression indirectly.

GATA4 translation is also regulated through sequences in the *Gata4* 5′ UTR (Sharma *et al.*, 2007). At 518 nt, this region is longer than the typical 5′UTR. Moreover, it contains 18 upstream ATG sequences and a highly stable predicted secondary structure. The 5′UTR possessed activity of an internal ribosome reentry site (IRES), supporting cap-independent translation and remaining active when cap-dependent translation was inhibited. This sequence stimulated GATA4 translation in a rat embryonic heart-derived cell line (H9c2) treated with the hypertrophic agonist vasopressin. Ca^{2+}- and protein kinase C mediated this effect. Further work is needed to delineate the impact of regulated GATA4 translation on fetal development and adult heart function.

2.2. Regulation of GATA4 activity by posttranslational modifications

A host of posttranscriptional modifications modulate GATA4 transcriptional activity. GATA4 lysine 366 has been reported to be covalently linked to the small ubiquitin-like modifier SUMO-1 (Wang *et al.*, 2004). The reaction was catalyzed by the conjugating enzyme Ubc9, and expedited by the E3 ligase PIAS1, which favored attachment of poly SUMO chains. Sumoylated GATA4 exhibited increased transcriptional activity on GATA4-regulated luciferase reporters. Interestingly, sumoylation of GATA1 and GATA2 had the opposite effect, repressing their transcriptional activity. The combination of GATA4, SUMO-1, and PIAS1 activated transcription of selected cardiac genes in 10T1/2 fibroblasts, suggesting that GATA4 sumoylation may be required for the GATA4 cardiogenic activity.

GATA4 activity is also regulated through phosphorylation. The best studied phosphorylation site is serine 105, which was reported to be phosphorylated by both Erk and p38 MAPK kinases (Charron *et al.*, 2001; Liang *et al.*, 2001b). GATA4 phosphorylation at this site within the N-terminal activation domain increased its DNA-binding affinity and transcriptional activity. Recently, the importance of this posttranslational modification was tested *in vivo* by mutating GATA4 S105 to alanine (van Berlo *et al.*, 2011). This point mutation was compatible with survival to adulthood and normal heart development. However, mutant hearts showed blunted hypertrophy to biomechanical stresses such as phenylephrine infusion and were more susceptible to heart failure and cardiac dilation after pressure overload. Moreover, these mutant hearts were resistant to hypertrophy driven by a MEK1 transgene, indicating that GATA4 is an essential mediator of cardiac hypertrophy downstream of MEK1.

GATA4 phosphorylation has been implicated in its regulation of gene transcription downstream of cyclic adenosine monophosphate (cAMP) signaling. In response to cAMP elevation in MA-10 Leydig tumor cells, GATA4 was phosphorylated by protein kinase A on serine 261, located between the two zinc finger domains (Tremblay and Viger, 2003; Tremblay *et al.*, 2002). S261 phosphorylation enhanced GATA4 activation of cAMP-responsive Leydig cell promoter transcription, at least in part by increasing GATA4 physical interaction with the transcriptional coactivator CBP (CREB (cAMP-response element-binding) binding protein). Interestingly, transcriptional activity of

CBP itself is increased by PKA phosphorylation, suggesting that the GATA4/CBP complex may coordinate regulation of Leydig cell gene transcription downstream of cAMP. cAMP signaling is also important in regulating cardiomyocyte function; however, GATA4 or S261 phosphorylation has yet to be studied in regulation of heart development or function.

GATA4 is an important regulator of cardiac gene expression changes triggered by hypertrophic agonists. The β-adrenergic agonist isoproterenol was reported to increase GATA4 transcriptional activity by enhancing its nuclear accumulation. In unstimulated cells, GATA4 interacted with and was directly phosphorylated by GSK3β at undetermined residue(s), leading to active GATA4 export from the nucleus (Morisco *et al.*, 2001). β-Agonist stimulation inhibited GSK3β, reduced nuclear export, and thereby increased nuclear accumulation. The *in vivo* significance of GSK3β regulation of GATA4 activity and subcellular localization has not been verified *in vivo*.

Protein acetylation is another important form of posttranslational modification. GATA4 physically interacts with p300 and CBP, both acetyltransferases. GATA1 was initially reported to be acetylated by p300, resulting in increased GATA1 DNA binding and transcriptional activity (Boyes *et al.*, 1998). Subsequently, p300 was also found to acetylate GATA4 (Yanazume *et al.*, 2003). As with GATA1, GATA4 acetylation increased its DNA binding affinity and its *in vitro* transcriptional activity, and GATA4 acetylation was implicated as an important mechanism that regulates cardiac hypertrophy (Yanazume *et al.*, 2003). Several residues (K311, K318, K320, K322) were found to be acetylated by p300 and required for p300 stimulation of GATA4 transcriptional activity in luciferase reporter assays (Takaya *et al.*, 2008). Mutation of all four of these residues blocked GATA4 acetylation by p300 and blunted cardiac hypertrophy induced by GATA4 overexpression. Thus, GATA4 acetylation may be an important mechanism to regulate GATA4 transcriptional activity *in vivo*.

Protein acetylation is counterbalanced by histone deacetylases (HDACs). Although initially identified by their ability to deacetylate histones, these enzymes are active on a broad range of proteins, including transcription factors (Glozak *et al.*, 2005). HDACs are key regulators of cardiac development and hypertrophy (Kook *et al.*, 2003; Zhang *et al.*, 2002). The cardiac homeodomain only protein HOPX is selectively expressed in heart and essential for normal cardiac growth and differentiation (Ismat *et al.*, 2005; Shin *et al.*, 2002). This protein, which lacks intrinsic DNA-binding capacity, modulates cardiac gene transcription by recruiting HDAC2 (Kook *et al.*, 2003). Ablation of *Hopx* and *Hdac2* caused severe heart malformation and excessive cardiomyocyte proliferation that was linked to GATA4 hyperacetylation (Trivedi *et al.*, 2010). These data suggest that HOPX serves as an adapter to facilitate HDAC2 recruitment to GATA4, thereby regulating its activity by modulating the balance of GATA4 acetylation/deacetylation.

Recently we reported that GATA4 is also regulated by protein methylation (He *et al.*, 2012). The histone methyltransferase complex PRC2 (polycomb repressive complex 2) specifically binds to and methylates GATA4 at lysine 299. PRC2 methylation of GATA4 inhibited its transcriptional activity. This inhibition of transcriptional potency was due to reduced GATA4 binding to and acetylation by p300.

2.3. Regulation of GATA4 activity by interaction with other transcription factors and cofactors

GATA4 forms protein complexes with several other transcription factors expressed in the heart. Their synergistic interactions have been proposed to regulate cardiac gene transcription, and disruption of these interactions has been reported to underlie some cases of congenital heart disease.

2.3.1. FOG1/2

The N-terminal zinc finger of GATA proteins strongly interacts with Friend of GATA (FOG) cofactors. This protein family contains two members, FOG1 and FOG2, with FOG1 predominantly coexpressed with GATA1/2/3 in hematopoietic cells, and FOG2 predominantly coexpressed in with GATA4/5/6 endodermal and mesodermal derivatives. FOG proteins do not detectably bind DNA directly, but rather associate with other transcription factors, principally GATA factors. FOG2 has been reported to largely repress GATA4/5/6 transcriptional activation. However, FOG1 both facilitates and represses GATA1/2/3-mediated transcription in a context dependent manner, suggesting that FOG2 likely also increases GATA4 transcriptional activity at a subset of targets. Consistent with this idea, Tevosian found that FOG2 stimulates GATA4 activation of *Tnnt1* transcription (Manuylov and Tevosian, 2009).

The function of GATA4–FOG2 interaction has been probed by inactivation of *Fog2* (Svensson *et al.*, 2000; Tevosian *et al.*, 2000; Zhou *et al.*, 2009), and by point mutation of *Gata4* to substitute glycine for valine 217. This substitution selectively ablates GATA4–FOG2 interaction (Crispino *et al.*, 2001; Rivera-Feliciano *et al.*, 2006; Zhou *et al.*, 2009). FOG2 deficient mice died at midgestation with cardiac defects, most notably a marked deficiency of the coronary endothelial plexus (Tevosian *et al.*, 2000). Although FOG2 is expressed in endocardial, myocardial, and epicardial lineages, targeted deletion of FOG2 in cardiomyocytes by Nkx2-5 recapitulated the Fog2$^{-/-}$ phenotype. This result indicates that cardiomyocyte FOG2 promotes coronary plexus formation through cell nonautonomous mechanisms (Zhou *et al.*, 2009). The proangiogenic activity of cardiomyocyte FOG2 continues into adulthood, as adult cardiomyocyte-restricted *Fog2* ablation caused heart failure, decreased myocardial perfusion, and reduced microvascular density (Zhou *et al.*, 2009). The GATA4-V217G mutant largely recapitulated the FOG2 phenotype, suggesting that these activities of FOG2 are mediated through GATA4 interaction (Crispino *et al.*, 2001; Zhou *et al.*, 2009). In keeping with this result, GATA4 and FOG2 ChIP-seq in HL1 cardiomyocyte-like cells showed that nearly all FOG2 binding sites coincide with GATA4 binding sites (P. Zhou and W.T. Pu, unpublished).

Use of the GATA4-V217G mutant also identified an instance of functional GATA4–FOG1 interaction. Embryos with loss of GATA4–FOG interaction restricted to endocardium and endocardial cushion mesenchyme developed atrioventricular canal defects (Rivera-Feliciano *et al.*, 2006). However, FOG1 but not FOG2 is required for atrioventricular valve formation (Katz *et al.*, 2003). Thus, GATA4–FOG1 interaction is essential for atrioventricular valve remodeling.

2.3.2. GATA6

GATA4 and GATA6 proteins physically interact and synergistically activate target gene transcription (Charron *et al.*, 1999). The heterodimeric interaction occurs through the GATA4 DNA-binding domain, but GATA4 DNA-binding activity was dispensible for both GATA6 interaction and GATA6 transcriptional synergy. The crystal structure of GATA3 bound to DNA illustrates how the DNA binding domain mediates dimer formation, and suggests that GATA factors exist as homotypic and heterotypic dimers on DNA (Bates *et al.*, 2008).

GATA4 and GATA6 show similar expression patterns during early murine development, with both expressed in the precardiac mesoderm, the embryonic heart tube, and the developing endoderm (Morrisey *et al.*, 1996). As described in Section 1.2, GATA4 and GATA6 are redundantly required for cardiomyocyte specification. In later stages of heart development, examination of GATA4 and GATA6 doubly heterozygous embryos revealed a genetic interaction between these factors. Doubly heterozygous embryos developed myocardial hypoplasia and defects in ventricular and aortopulmonary septation (Xin *et al.*, 2006). In adult heart, inactivation of either GATA4 or GATA6 caused systolic dysfunction and impaired cardiomyocyte hypertrophy. Inactivation of both GATA4 and GATA6 caused more severe cardiac dysfunction than either alone. Thus, GATA4 and GATA6 function redundantly to maintain adult heart function (van Berlo *et al.*, 2010), consistent with GATA4/6 physical interaction.

Given the similarity of GATA4 and GATA6 loss-of-function phenotypes, it has been difficult to determine whether these proteins function redundantly in a cumulative dosage-dependent manner, or whether there are distinct functions attributable to either protein alone or to interactions between these proteins. Ablation of FOG interaction has revealed at least one qualitative distinction between GATA4 and GATA6. Unlike the dramatic embryonic lethal phenotype of GATA4-V217G mutants, the comparable GATA6 mutants have no discernable cardiovascular phenotype (J. Wang and S. H. Orkin, unpublished). Thus, cardiovascular FOG-dependent functions appear to be predominantly mediated through GATA4 rather than GATA6.

2.3.3. Nkx2-5

The C-terminal zinc finger of GATA4 binds to the cardiac homeodomain protein Nkx2–5, and this interaction leads to synergistic transcriptional activation of reporter genes *in vitro* (Durocher *et al.*, 1997; Lee *et al.*, 1998; Sepulveda *et al.*, 1998). Nkx2-5 DNA-binding activity was required for physical interaction, but GATA4 DNA-binding activity was not (Garg *et al.*, 2003). Consistent with physical interaction between Nkx2-5 and GATA4, the Nkx2-5 consensus binding site was significantly enriched in chromatin regions occupied by GATA4 in HL1 cardiomyocyte-like cells (He *et al.*, 2011). However, *in vivo* evidence of the biological significance of the Nkx2-5 and GATA4 interaction is lacking. Double heterozygosity for GATA4 and Nkx2-5 did not appear to affect fetal survival beyond the effect of heterozygosity for each factor alone (W.T. Pu, unpublished).

2.3.4. TBX5

GATA4 and TBX5 physically interact and synergistically activate reporter gene transcription (Garg *et al.*, 2003), although deletion analysis to identify the interacting domains has not been reported. The human GATA4 G296S mutation ablated both GATA4 binding to both DNA and TBX5 and blocked transcriptional synergy with TBX5 (Garg *et al.*, 2003). Although disruption of TBX5 interaction by this mutation may contribute to malformation of the heart, its abrogation of DNA binding is also likely significant. Further studies in mice showed that double heterozygosity for GATA4 and TBX5 caused more severe cardiovascular defects than single heterozygosity for each factor, thus establishing a genetic interaction between GATA4 and TBX5 (Maitra *et al.*, 2009). GATA4 and TBX5 interact in endocardial cells to promote atrial septation, and this genetic interaction was linked to synergistic activation of endocardial NOS3 expression (Nadeau *et al.*, 2010). Further work is required to define whether this interaction occurs in cardiomyocytes as well, and whether or not the genetic interaction reflects functionally required physical interaction between these factors.

2.3.5. SRF and MEF2

The MEF2 transcription factor family (MEF2A/B/C/D) and SRF contain MADS DNA binding domains. Both muscle and nonmuscle cells express these factors, yet in muscle cells they are key drivers of muscle gene expression (Niu *et al.*, 2007; Potthoff and Olson, 2007). SRF, MEF2A, MEF2C, and MEF2D are each individually required for cardiac development and/or function (Kim *et al.*, 2008; Lin *et al.*, 1997; Naya *et al.*, 2002; Niu *et al.*, 2005). The C-terminal finger of GATA4 and the MADS domain of SRF/MEF2 factors are sufficient for protein–protein interaction

(Belaguli *et al.*, 2000; Morin *et al.*, 2000, 2001). GATA4 and SRF/MEF2 synergistically activated reporter gene transcription *in vitro*, and in both cases, the GATA4 C-terminal region was required in addition to the DNA binding domain for transcriptional synergy. However, *in vivo* evidence that GATA4–SRF or GATA4–MEF2 interaction is functionally significant is lacking currently.

2.4. Chromatin occupancy

The human or mouse genomes contains approximately 7×10^6 GATA motifs, but ChIP-seq experiments show that GATA4 occupies only a small fraction ($10–50\times10^3$) of these potential sites (He *et al.*, 2011). In the case of GATA2, where genome-wide chromatin occupancy has been analyzed in different cell types, the majority of GATA2 occupied sites differed between endothelial and leukocyte cell lines, indicating that sites of chromatin occupancy are regulated in a cell type specific manner that is related to tissue-restricted transcription factor activity (Linnemann *et al.*, 2011). The mechanisms that underlie tissue-restricted transcription factor chromatin occupancy are not well understood but are pivotal to understand the activity of transcription factors like GATA4, which drive tissue-restricted gene expression programs in diverse tissues such as heart, gut, and gonads. Likely, mechanisms involve combinatorial interactions with other tissue-restricted transcription factors, and chromatin structure and accessibility.

Combinatorial interactions between transcription factors have been long proposed to regulate GATA4 and contribute to its regulation of cardiac-specific gene expression. Like GATA4, other major cardiac transcription factors such as SRF, MEF2, TBX5, and NKX2-5 are expressed in multiple tissues. However, combinatorial synergy may enhance specificity for the cardiac gene program. We used ChIP-seq to define genome-wide the chromatin occupancy of GATA4, NKX2-5, TBX5, SRF, and MEF2A in HL1 cardiomyocyte-like cells (He *et al.*, 2011). Our data supported substantial protein–protein interaction between these factors, as over 20% of chromatin regions were bound by two or more factors ($P<10^{-16}$). Four or more of these factors co-occurred within 500bp of one another at 1715 sites, and these sites were highly overrepresented for genes with enriched cardiac expression. In fact, multiple transcription factor binding was an effective criteria to identify enhancers with cardiac activity. The large majority of these multiple transcription factor bound loci were occupied by GATA4, suggesting that it plays an important role in determining cardiac enhancer chromatin occupancy.

Chromatin structure and accessibility are similarly important determinants of transcription factor chromatin occupancy. Chromatin structure is governed by the interaction of epigenetic regulatory machinery with sequence-specific transcription factors. The interaction is likely bidirectional and mutually reinforcing, as chromatin modifying enzymes may induce a more accessible chromatin conformation and thereby recruit transcription factor occupancy. At the same time, sequence specific transcription factors may recruit epigenetic regulators that modify local chromatin structure and accessibility, thereby strengthening transcription factor occupancy and enhancing binding of interacting transcription factors.

The interaction between the histone acetyltransferase p300 and GATA4 is an example of cross talk between transcription factors and epigenetic regulators. p300 is an important transcriptional coactivator in multiple organs including the heart. Embryos lacking p300 or p300 acetyltransferase enzymatic activity die from heart defects (Shikama *et al.*, 2003; Yao *et al.*, 1998), and cardiac hypertrophy in response to pressure overload was quantitatively linked to p300 expression levels (Miyamoto *et al.*, 2006; Wei *et al.*, 2008). p300 is selectively recruited to active transcriptional enhancers, where it acetylates histones and stimulates transcription. Indeed, p300 chromatin occupancy has been shown to identify tissue-restricted transcriptional enhancers, including heart-specific enhancers (Blow *et al.*, 2010; Visel *et al.*, 2009). p300 interacts with a large number of transcriptional regulators including GATA4 (Yanazume *et al.*, 2003), and presumably these interactions govern p300 recruitment to transcriptional enhancers. We measured p300 chromatin occupancy in HL1 cardiomyocyte-like cells by ChIP-seq, and found that nearly 80% of chromatin sites bound by p300 were also bound by GATA4 (He *et al.*, 2011). Thus, GATA4 recruitment appears to be a major mechanism underlying p300 chromatin occupancy in cardiomyocytes. Since p300 also acetylates GATA4 and this increases GATA4 DNA binding affinity (Yanazume *et al.*, 2003), GATA4–p300 interaction may bidirectionally influence both p300 and GATA4 chromatin occupancy.

Specific properties of GATA4 may facilitate its occupancy of compact chromatin inaccessible to many other transcription factors. Zaret and colleagues studied FoxA2 and GATA4 in specification of hepatocytes from early endoderm progenitors (Cirillo *et al.*, 2002; Zaret *et al.*, 2008). FoxA2 and GATA4 occupied regulatory elements of the hepatocyte-specific albumin promoter, even prior to active albumin (*Alb1*) transcription. Unlike other transcription factors, FoxA2 and GATA4 were found to bind efficiently to these regulatory sequences even when compacted in nucleosome arrays. Thus, these factors were proposed to act as "pioneer" factors that could bind chromatinized regulatory regions and facilitate subsequent binding by other factors. In fact, GATA4 and a different FoxA factor (FoxA3) were central components of a set of defined factors that reprogram fibroblasts to hepatocytes (Huang *et al.*, 2011). FoxA2 and GATA4 occupancy of these regulatory elements prior to gene activation may reflect a "bookmarking" or "competence" function by establishing a chromatin state that is compatible with later expression of the gene (Zaret *et al.*, 2008). In fact, in ES cells, the Fox-factor binding site is occupied by the related FoxD3, and this is associated with selective *Alb1* demethylation at the Fox-factor binding site. Thus, occupancy of the *Alb1* site by a series of Fox-factors may preserve competence of this regulatory region for later activity in hepatocytes, in part by promoting focal *Alb1* demethylation at the Fox-factor binding site. GATA factors (GATA4 and GATA6) are redundantly required for cardiomyocyte specification and thus GATA4 may play a pioneer and competence role in cardiac gene expression analogous to its role in hepatocyte specification.

GATA4 chromatin occupancy may be modulated by its interaction with FOG cofactors. Although FOG proteins neither bind DNA intrinsically nor modify GATA factor binding affinity or specificity for naked DNA, GATA1–FOG1 interaction was required for occupancy of specific chromatin loci (Pal *et al.*, 2004). This "chromatin occupancy facilitator" function of FOG1 was hypothesized to account in part for FOG1 modulation of GATA1 transcriptional activity. The mechanisms underlying this chromatin occupancy facilitator activity are poorly understood. The applicability of this paradigm to GATA4–FOG2 interactions has yet to be evaluated by unbiased genome-wide approaches.

Indirect GATA4 binding appears to contribute substantially to GATA4 chromatin occupancy, as a large fraction of genomic regions associated with GATA4 in ChIP-seq experiments lacked recognizable GATA motifs (He *et al.*, 2011). GATA4 was likely cross-linked to these regions as a result of its interaction with other transcriptional complexes rather than through direct DNA binding to cryptic GATA binding sites. Consistent with this, *in vitro* studies show that GATA4 interaction with some transcription factors does not require GATA4 DNA binding activity (Fig. 5.2).

2.5. GATA4 interaction with chromatin remodeling complexes

Chromatin remodeling complexes use the energy of ATP hydrolysis to displace and reposition nucleosomes, thereby altering chromatin accessibility (Ho and Crabtree, 2010). These complexes can be broadly grouped by the ATPase subunit, with the SWI/SNF family being the best studied in the cardiovascular system (Han *et al.*, 2011). SWI/SNF ATPases are large, multisubunit machines that use either BRG1 or BRM1 as the ATPase subunit. The composition of the SWI/SNF complex is under tight spatiotemporal control, and the precise complement of subunits determines the biological activity of the complex (Ho and Crabtree, 2010). The BAF60 subunit of the complex can be one of three distinct proteins, BAF60A–C (encoded by genes *Smarcd1–3*, respectively). Interestingly, BAF60C is specifically expressed in the developing heart from E7.5 to E9.5, and shRNA-mediated BAF60C knockdown caused severe defects in cardiac morphogenesis (Lickert *et al.*, 2004). BAF60C coprecipitated with GATA4 and was required for GATA4 to coprecipitate BRG1, suggesting that BAF60C functions as a cardiac-specific adaptor that facilitates GATA4 interaction with the BRG1 chromatin remodeling complex (Lickert *et al.*, 2004). This interaction was associated with synergistic activation of cardiac reporter genes by BRG1 and GATA4 in the presence of BAF60C.

The importance of BAF60C-mediated interaction of GATA4 with the BRG1 chromatin remodeling complex was reinforced by the discovery that GATA4 and BAF60C are together sufficient to induce ectopic cardiac gene expression in mouse mesoderm (Takeuchi and Bruneau, 2009). Addition of the cardiac transcription factor TBX5 stimulated further cardiac differentiation, inducing mesoderm to form ectopic foci with spontaneously beating, a complex phenotype indicative of maturation of multiple cardiomyocyte-specific functions. At a mechanistic level, coexpression of GATA4 and BAF60C in mesoderm resulted in recruitment of GATA4 and BRG1 to regulatory elements of the cardiac genes *Tnnt2* and *Nppa*. This suggested that the BAF60C–BRG1 complex remodeled chromatin and thereby facilitated GATA4 chromatin occupancy (Takeuchi and Bruneau, 2009). Alternatively, pioneer activity of GATA4 facilitated chromatin occupancy of these regulatory elements, which was subsequently stabilized by the BAF60C/BRG1-containing remodeling complex. BAF60C continues to be expressed in the adult heart, where both BRG1 and GATA4 are essential for normal cardiac function and responses to biomechanical stress (Bisping *et al.*, 2006; Hang *et al.*, 2010; Oka *et al.*, 2006; Takeuchi *et al.*, 2011). This suggests that GATA4–BAF60C–BRG1 continue to regulate adult heart chromatin remodeling, although additional experiments are required to test this hypothesis.

2.6. GATA4 interaction with histone modifying complexes

GATA4 has also been reported to interact with three distinct complexes containing enzymes that covalently modify histones. First, as mentioned above GATA4 coprecipitates with HOPX–HDAC2. While this interaction was studied in the context of GATA4 deacetylation, a GATA4–HOPX–HDAC2 complex may also repress GATA4-bound genes by HDAC2-mediated histone deacetylation. Although GATA4 has primarily been studied as a transcriptional activator, GATA1–3 are active as both transcriptional activators and repressors. The extent to which GATA4 represses transcription via HDAC recruitment requires further study.

GATA4 also forms a complex with the nucleosome remodeling and histone deacetylase (NuRD) complex through its cofactor FOG2. This unique multisubunit complex contains both ATP-dependent chromatin remodeling and histone deacetylase activities, conferred by Mi-2α/β (Denslow and Wade, 2007). The distinct functions of these HDACs in the NuRD complex, compared to HDAC activities in complexes with HOPX or in other corepressor complexes such as mSIN3a, remain to be determined.

The NuRD complex appears to function as a transcriptional repressor, both by decreasing chromatin accessibility through nucleosome repositioning and by removing activating histone acetylation marks. FOG proteins interact with NuRD through the N-terminal 12 amino acids of FOG2 (Roche *et al.*, 2008), which are necessary and sufficient for FOG2-mediated transcriptional repression (Lin *et al.*, 2004). This N-terminal repressor recruitment domain binds the MTA (metastasis associated) protein family members MTA1/2/3, components of the NuRD complex. Interestingly, HDAC inhibitors did not block FOG2-mediated repression *in vitro* (Lin *et al.*, 2004), suggesting that other mechanisms such as nucleosome repositioning mediate gene repression by FOG.

GATA4 also occurs in protein complexes with Jumonji (*Jarid2*), a gene initially identified in a gene trap screen for mutants in heart development. *Jarid2* null mutants developed ventricular septal defect (VSD), ventricular noncompaction, and double-outlet right ventricle. JARID2 and GATA4 proteins interacted at the *Nppa* promoter, where JARID2 repressed GATA4 transcriptional activity (Kim *et al.*, 2004). JARID2 was the founding member of a family of histone demethylases, but unlike other members of this family, JARID2 lacks histone demethylase activity due to substitution of key residues within the conserved catalytic domain (Takeuchi *et al.*, 2006). Recently, JARID2 was found to be an important component of Polycomb Repressive Complex 2 (PRC2), a key repressive complex that silences gene expression by establishing repressive H3K27me3 epigenetic marks (Li *et al.*, 2010; Pasini *et al.*, 2010; Shen *et al.*, 2009). In ES cells, JARID2 colocalized with PRC2 and was required for efficient PRC2 chromatin occupancy. These data are of particular interest because GATA4 interacts directly with the PRC2 complex (He *et al.*, 2012), suggesting that GATA4-JARID2-PRC2 interaction may modulate chromatin occupancy and activity of both GATA4 and PRC2 in cardiac cells.

3. Summary

Achieving normal heart development and function requires precise regulation of transcription factor dose and activity. Every step including transcription factor gene transcription, translation, subcellular localization, chromatin occupancy, and interaction with the transcriptional machinery is regulated, permitting the cell to precisely tune transcription factor activity at each target gene locus. While the work summarized in this review has moved our understanding of transcription factor regulation forward considerably, a number of fundamental questions remain. What determines tissue- and stage-specific transcription factor chromatin occupancy and tissue-specific transcriptional activation? What is the role of three-dimensional chromatin structure in gene expression regulation, and how is the structure initiated, maintained, and regulated? What are the transcription factor-epigenetic factor/chromatin remodeling enzyme interactions that sculpt the chromatin landscape? How do transcription factor occupancy and the chromatin landscape change during development, and in abnormal development and disease? How can the chromatin landscape be manipulated to promote cardiomyocyte expansion and myocardial regeneration? The expanding arsenal of tools to study chromatin structure and occupancy at a genome scale promise to shed light on these fundamental questions in the coming years.

ACKNOWLEDGMENTS

The authors were supported by funding from NIH (U01HL098166 and R01HL095712), the American Heart Association, and the Harvard Stem Cell Institute, and by charitable contributions from Edward Marram, Karen Carpenter, and Gail Federici Smith.

REFERENCES

Arceci, R. J., King, A. A., Simon, M. C., Orkin, S. H., and Wilson, D. B. (1993). Mouse GATA-4: A retinoic acid-inducible GATA-binding transcription factor expressed in endodermally derived tissues and heart. *Mol. Cell Biol.* **13,** 2235–2246.

Bates, D. L., Chen, Y., Kim, G., Guo, L., and Chen, L. (2008). Crystal structures of multiple GATA zinc fingers bound to DNA reveal new insights into DNA recognition and self-association by GATA. *J. Mol. Biol.* **381,** 1292–1306.

Belaguli, N. S., Sepulveda, J. L., Nigam, V., Charron, F., Nemer, M., and Schwartz, R. J. (2000). Cardiac tissue enriched factors serum response factor and GATA-4 are mutual coregulators. *Mol. Cell Biol.* **20,** 7550–7558.

Berger, M. F., Philippakis, A. A., Qureshi, A. M., He, F. S., Estep, P. W.r., and Bulyk, M. L. (2006). Compact, universal DNA microarrays to comprehensively determine transcription-factor binding site specificities. *Nat. Biotechnol.* **24,** 1429–1435.

Bielinska, M., Jay, P. Y., Erlich, J. M., Mannisto, S., Urban, Z., Heikinheimo, M., and Wilson, D. B. (2007). Molecular genetics of congenital diaphragmatic defects. *Ann. Med.* **39,** 261–274.

Bisping, E., Ikeda, S., Kong, S. W., Tarnavski, O., Bodyak, N., McMullen, J. R., Rajagopal, S., Son, J. K., Ma, Q., Springer, Z., Kang, P. M., Izumo, S., *et al.* (2006). Gata4 is required for maintenance of postnatal cardiac function and protection from pressure overload-induced heart failure. *Proc. Natl. Acad. Sci. USA* **103,** 14471–14476.

Black, B. L. (2007). Transcriptional pathways in second heart field development. *Semin. Cell Dev. Biol.* **18,** 67–76.

Blow, M. J., McCulley, D. J., Li, Z., Zhang, T., Akiyama, J. A., Holt, A., Plajzer-Frick, I., Shoukry, M., Wright, C., Chen, F., Afzal, V., Bristow, J., *et al.* (2010). ChIP-Seq identification of weakly conserved heart enhancers. *Nat. Genet.* **42**(9), 806–810.

Boyes, J., Byfield, P., Nakatani, Y., and Ogryzko, V. (1998). Regulation of activity of the transcription factor GATA-1 by acetylation. *Nature* **396,** 594–598.

Bresnick, E. H., Lee, H. Y., Fujiwara, T., Johnson, K. D., and Keles, S. (2010). GATA switches as developmental drivers. *J. Biol. Chem.* **285,** 31087–31093.

Bruneau, B. G. (2008). The developmental genetics of congenital heart disease. *Nature* **451,** 943–948.

Butler, T. L., Esposito, G., Blue, G. M., Cole, A. D., Costa, M. W., Waddell, L. B., Walizada, G., Sholler, G. F., Kirk, E. P., Feneley, M., Harvey, R. P., and Winlaw, D. S. (2010). GATA4 mutations in 357 unrelated patients with congenital heart malformation. *Genet. Test. Mol. Biomarkers* **14,** 797–802.

Callis, T. E., Pandya, K., Seok, H. Y., Tang, R. H., Tatsuguchi, M., Huang, Z. P., Chen, J. F., Deng, Z., Gunn, B., Shumate, J., Willis, M. S., Selzman, C. H., *et al.* (2009). MicroRNA-208a is a regulator of cardiac hypertrophy and conduction in mice. *J. Clin. Invest.* **119,** 2772–2786.

Charron, F., Paradis, P., Bronchain, O., Nemer, G., and Nemer, M. (1999). Cooperative interaction between GATA-4 and GATA-6 regulates myocardial gene expression. *Mol. Cell. Biol.* **19,** 4355–4365.

Charron, F., Tsimiklis, G., Arcand, M., Robitaille, L., Liang, Q., Molkentin, J. D., Meloche, S., and Nemer, M. (2001). Tissue-specific GATA factors are transcriptional effectors of the small GTPase RhoA. *Genes Dev.* **15,** 2702–2719.

Chen, Y., Han, Z. Q., Yan, W. D., Tang, C. Z., Xie, J. Y., Chen, H., and Hu, D. Y. (2010a). A novel mutation in GATA4 gene associated with dominant inherited familial atrial septal defect. *J. Thorac. Cardiovasc. Surg.* **140,** 684–687.

Chen, Y., Mao, J., Sun, Y., Zhang, Q., Cheng, H. B., Yan, W. H., Choy, K. W., and Li, H. (2010b). A novel mutation of GATA4 in a familial atrial septal defect. *Clin. Chim. Acta* **411,** 1741–1745.

Cirillo, L. A., Lin, F. R., Cuesta, I., Friedman, D., Jarnik, M., and Zaret, K. S. (2002). Opening of compacted chromatin by early developmental transcription factors HNF3 (FoxA) and GATA-4. *Mol. Cell* **9,** 279–289.

Crispino, J. D., Lodish, M. B., Thurberg, B. L., Litovsky, S. H., Collins, T., Molkentin, J. D., and Orkin, S. H. (2001). Proper coronary vascular development and heart morphogenesis depend on interaction of GATA-4 with FOG cofactors. *Genes Dev.* **15,** 839–844.

Dai, Y. S., and Markham, B. E. (2001). p300 Functions as a coactivator of transcription factor gata-4. *J. Biol. Chem.* **276,** 37178–37185.

Denslow, S. A., and Wade, P. A. (2007). The human Mi-2/NuRD complex and gene regulation. *Oncogene* **26,** 5433–5438.

Diedrichs, H., Chi, M., Boelck, B., Mehlhorn, U., and Schwinger, R. H. (2004). Increased regulatory activity of the calcineurin/NFAT pathway in human heart failure. *Eur. J. Heart Fail.* **6,** 3–9.

Dodou, E., Verzi, M. P., Anderson, J. P., Xu, S. M., and Black, B. L. (2004). Mef2c is a direct transcriptional target of ISL1 and GATA factors in the anterior heart field during mouse embryonic development. *Development* **131,** 3931–3942.

Durocher, D., Charron, F., Warren, R., Schwartz, R. J., and Nemer, M. (1997). The cardiac transcription factors Nkx2-5 and GATA-4 are mutual cofactors. *EMBO J.* **16,** 5687–5696.

Evans, T., and Felsenfeld, G. (1989). The erythroid-specific transcription factor Eryf1: A new finger protein. *Cell* **58,** 877–885.

Foley, A. C., Gupta, R. W., Guzzo, R. M., Korol, O., and Mercola, M. (2006). Embryonic heart induction. *Ann. N. Y. Acad. Sci.* **1080,** 85–96.

Fujiwara, T., O'Geen, H., Keles, S., Blahnik, K., Linnemann, A. K., Kang, Y. A., Choi, K., Farnham, P. J., and Bresnick, E. H. (2009). Discovering hematopoietic mechanisms through genome-wide analysis of GATA factor chromatin occupancy. *Mol. Cell* **36,** 667–681.

Garg, V., Kathiriya, I. S., Barnes, R., Schluterman, M. K., King, I. N., Butler, C. A., Rothrock, C. R., Eapen, R. S., Hirayama-Yamada, K., Joo, K., Matsuoka, R.,

Cohen, J. C., *et al.* (2003). GATA4 mutations cause human congenital heart defects and reveal an interaction with TBX5. *Nature* **424,** 443–447.

Glozak, M. A., Sengupta, N., Zhang, X., and Seto, E. (2005). Acetylation and deacetylation of non-histone proteins. *Gene* **363,** 15–23.

Grepin, C., Nemer, G., and Nemer, M. (1997). Enhanced cardiogenesis in embryonic stem cells overexpressing the GATA-4 transcription factor. *Development* **124,** 2387–2395.

Hall, J. L., Grindle, S., Han, X., Fermin, D., Park, S., Chen, Y., Bache, R. J., Mariash, A., Guan, Z., Ormaza, S., Thompson, J., Graziano, J., *et al.* (2004). Genomic profiling of the human heart before and after mechanical support with a ventricular assist device reveals alterations in vascular signaling networks. *Physiol. Genomics* **17,** 283–291.

Han, P., Hang, C. T., Yang, J., and Chang, C. P. (2011). Chromatin remodeling in cardiovascular development and physiology. *Circ. Res.* **108,** 378–396.

Hang, C. T., Yang, J., Han, P., Cheng, H. L., Shang, C., Ashley, E., Zhou, B., and Chang, C. P. (2010). Chromatin regulation by Brg1 underlies heart muscle development and disease. *Nature* **466,** 62–67.

He, A., Kong, S. W., Ma, Q., and Pu, W. T. (2011). Co-occupancy by multiple cardiac transcription factors identifies transcriptional enhancers active in heart. *Proc. Natl. Acad. Sci. USA* **108,** 5632–5637.

He, A., Shen, X., Ma, Q., Cao, J., von Gise, A., Zhou, P., Wang, G., Marquez, V. E., Orkin, S. H., and Pu, W. T. (2012). PRC2 directly methylates GATA4 and represses its transcriptional activity. *Genes Dev.* **26,** 37–42.

Heineke, J., Auger-Messier, M., Xu, J., Oka, T., Sargent, M. A., York, A., Klevitsky, R., Vaikunth, S., Duncan, S. A., Aronow, B. J., Robbins, J., Crombleholme, T. M., *et al.* (2007). Cardiomyocyte GATA4 functions as a stress-responsive regulator of angiogenesis in the murine heart. *J. Clin. Invest.* **117,** 3198–3210.

Ho, L., and Crabtree, G. R. (2010). Chromatin remodelling during development. *Nature* **463,** 474–484.

Holtzinger, A., Rosenfeld, G. E., and Evans, T. (2010). Gata4 directs development of cardiac-inducing endoderm from ES cells. *Dev. Biol.* **337,** 63–73.

Huang, P., He, Z., Ji, S., Sun, H., Xiang, D., Liu, C., Hu, Y., Wang, X., and Hui, L. (2011). Induction of functional hepatocyte-like cells from mouse fibroblasts by defined factors. *Nature* **475,** 386–389.

Ieda, M., Fu, J. D., Delgado-Olguin, P., Vedantham, V., Hayashi, Y., Bruneau, B. G., and Srivastava, D. (2010). Direct reprogramming of fibroblasts into functional cardiomyocytes by defined factors. *Cell* **142,** 375–386.

Ikeda, S., He, A., Kong, S. W., Lu, J., Bejar, R., Bodyak, N., Lee, K. H., Ma, Q., Kang, P. M., Golub, T. R., and Pu, W. T. (2009). MicroRNA-1 negatively regulates expression of the hypertrophy-associated calmodulin and Mef2a genes. *Mol. Cell. Biol.* **29,** 2193–2204.

Ismat, F. A., Zhang, M., Kook, H., Huang, B., Zhou, R., Ferrari, V. A., Epstein, J. A., and Patel, V. V. (2005). Homeobox protein Hop functions in the adult cardiac conduction system. *Circ. Res.* **96,** 898–903.

Jay, P. Y., Bielinska, M., Erlich, J. M., Mannisto, S., Pu, W. T., Heikinheimo, M., and Wilson, D. B. (2007). Impaired mesenchymal cell function in Gata4 mutant mice leads to diaphragmatic hernias and primary lung defects. *Dev. Biol.* **301,** 602–614.

Katz, S. G., Williams, A., Yang, J., Fujiwara, Y., Tsang, A. P., Epstein, J. A., and Orkin, S. H. (2003). Endothelial lineage-mediated loss of the GATA cofactor Friend of GATA 1 impairs cardiac development. *Proc. Natl. Acad. Sci. USA* **100,** 14030–14035.

Kim, T. G., Chen, J., Sadoshima, J., and Lee, Y. (2004). Jumonji represses atrial natriuretic factor gene expression by inhibiting transcriptional activities of cardiac transcription factors. *Mol. Cell. Biol.* **24,** 10151–10160.

Kim, Y., Phan, D., van Rooij, E., Wang, D. Z., McAnally, J., Qi, X., Richardson, J. A., Hill, J. A., Bassel-Duby, R., and Olson, E. N. (2008). The MEF2D transcription factor mediates stress-dependent cardiac remodeling in mice. *J. Clin. Invest.* **118**(1), 124–132.

Kook, H., Lepore, J. J., Gitler, A. D., Lu, M. M., Wing-Man Yung, W., Mackay, J., Zhou, R., Ferrari, V., Gruber, P., and Epstein, J. A. (2003). Cardiac hypertrophy and histone deacetylase-dependent transcriptional repression mediated by the atypical homeodomain protein Hop. *J. Clin. Invest.* **112,** 863–871.

Lee, Y., Shioi, T., Kasahara, H., Jobe, S. M., Wiese, R. J., Markham, B. E., and Izumo, S. (1998). The cardiac tissue-restricted homeobox protein Csx/Nkx2.5 physically associates with the zinc finger protein GATA4 and cooperatively activates atrial natriuretic factor gene expression. *Mol. Cell. Biol.* **18,** 3120–3129.

Li, G., Margueron, R., Ku, M., Chambon, P., Bernstein, B. E., and Reinberg, D. (2010). Jarid2 and PRC2, partners in regulating gene expression. *Genes Dev.* **24,** 368–380.

Liang, Q., De Windt, L. J., Witt, S. A., Kimball, T. R., Markham, B. E., and Molkentin, J. D. (2001a). The transcription factors GATA4 and GATA6 regulate cardiomyocyte hypertrophy in vitro and in vivo. *J. Biol. Chem.* **276,** 30245–30253.

Liang, Q., Wiese, R. J., Bueno, O. F., Dai, Y. S., Markham, B. E., and Molkentin, J. D. (2001b). The transcription factor gata4 is activated by extracellular signal-regulated kinase 1- and 2-mediated phosphorylation of serine 105 in cardiomyocytes. *Mol. Cell. Biol.* **21,** 7460–7469.

Lickert, H., Takeuchi, J. K., Von Both, I., Walls, J. R., McAuliffe, F., Adamson, S. L., Henkelman, R. M., Wrana, J. L., Rossant, J., and Bruneau, B. G. (2004). Baf60c is essential for function of BAF chromatin remodelling complexes in heart development. *Nature* **432,** 107–112.

Lin, Q., Schwarz, J., Bucana, C., and Olson, E. N. (1997). Control of mouse cardiac morphogenesis and myogenesis by transcription factor MEF2C. *Science* **276,** 1404–1407.

Lin, A. C., Roche, A. E., Wilk, J., and Svensson, E. C. (2004). The N termini of Friend of GATA (FOG) proteins define a novel transcriptional repression motif and a superfamily of transcriptional repressors. *J. Biol. Chem.* **279,** 55017–55023.

Linnemann, A. K., O'Geen, H., Keles, S., Farnham, P. J., and Bresnick, E. H. (2011). Genetic framework for GATA factor function in vascular biology. *Proc. Natl. Acad. Sci. USA* **108,** 13641–13646.

Lourenco, D., Brauner, R., Rybczynska, M., Nihoul-Fekete, C., McElreavey, K., and Bashamboo, A. (2011). Loss-of-function mutation in GATA4 causes anomalies of human testicular development. *Proc. Natl. Acad. Sci. USA* **108,** 1597–1602.

Lu, J. R., McKinsey, T. A., Xu, H., Wang, D. Z., Richardson, J. A., and Olsen, E. N. (1999). FOG-2, a heart- and brain-enriched cofactor for GATA transcription factors. *Mol. Cell. Biol.* **19,** 4495-4502.

Maitra, M., Schluterman, M. K., Nichols, H. A., Richardson, J. A., Lo, C. W., Srivastava, D., and Garg, V. (2009). Interaction of Gata4 and Gata6 with Tbx5 is critical for normal cardiac development. *Dev. Biol.* **326,** 368–377.

Manuylov, N. L., and Tevosian, S. G. (2009). Cardiac expression of Tnnt1 requires the GATA4-FOG2 transcription complex. *Sci. World J.* **9,** 575–587.

Martin, D. I., and Orkin, S. H. (1990). Transcriptional activation and DNA binding by the erythroid factor GF-1/NF-E1/Eryf 1. *Genes Dev.* **4,** 1886–1898.

McFadden, D. G., Charite, J., Richardson, J. A., Srivastava, D., Firulli, A. B., and Olson, E. N. (2000). A GATA-dependent right ventricular enhancer controls dHAND transcription in the developing heart. *Development* **127,** 5331–5341.

Merika, M., and Orkin, S. H. (1993). DNA-binding specificity of GATA family transcription factors. *Mol. Cell. Biol.* **13,** 3999–4010.

Miyamoto, S., Kawamura, T., Morimoto, T., Ono, K., Wada, H., Kawase, Y., Matsumori, A., Nishio, R., Kita, T., and Hasegawa, K. (2006). Histone acetyltransferase activity of p300 is required for the promotion of left ventricular remodeling after myocardial infarction in adult mice in vivo. *Circulation* **113,** 679–690.

Molkentin, J. D., Lu, J. R., Antos, C. L., Markham, B., Richardson, J., Robbins, J., Grant, S. R., and Olson, E. N. (1998). A calcineurin-dependent transcriptional pathway for cardiac hypertrophy. *Cell* **93,** 215–228.

Molkentin, J. D. (2000). The zinc finger-containing transcription factors GATA-4, -5, and -6 Ubiquitously expressed regulators of tissue-specific gene expression. *J. Biol. Chem.* **275,** 38949–38952.

Morin, S., Charron, F., Robitaille, L., and Nemer, M. (2000). GATA-dependent recruitment of MEF2 proteins to target promoters. *EMBO J.* **19,** 2046–2055.

Morin, S., Paradis, P., Aries, A., and Nemer, M. (2001). Serum response factor-GATA ternary complex required for nuclear signaling by a G-protein-coupled receptor. *Mol. Cell. Biol.* **21,** 1036–1044.

Morisco, C., Seta, K., Hardt, S. E., Lee, Y., Vatner, S. F., and Sadoshima, J. (2001). Glycogen synthase kinase 3beta regulates GATA4 in cardiac myocytes. *J. Biol. Chem.* **276,** 28586–28597.

Morrisey, E. E., Ip, H. S., Lu, M. M., and Parmacek, M. S. (1996). GATA-6: A zinc finger transcription factor that is expressed in multiple cell lineages derived from lateral mesoderm. *Dev. Biol.* **177,** 309–322.

Morrisey, E. E., Ip, H. S., Tang, Z., and Parmacek, M. S. (1997). GATA-4 activates transcription via two novel domains that are conserved within the GATA-4/5/6 subfamily. *J. Biol. Chem.* **272,** 8515–8524.

Nadeau, M., Georges, R. O., Laforest, B., Yamak, A., Lefebvre, C., Beauregard, J., Paradis, P., Bruneau, B. G., Andelfinger, G., and Nemer, M. (2010). An endocardial pathway involving Tbx5, Gata4, and Nos3 required for atrial septum formation. *Proc. Natl. Acad. Sci. USA* **107,** 19356–19361.

Narita, N., Bielinska, M., and Wilson, D. B. (1997). Wild-type endoderm abrogates the ventral developmental defects associated with GATA-4 deficiency in the mouse. *Dev. Biol.* **189,** 270–274.

Naya, F. J., Black, B. L., Wu, H., Bassel-Duby, R., Richardson, J. A., Hill, J. A., and Olson, E. N. (2002). Mitochondrial deficiency and cardiac sudden death in mice lacking the MEF2A transcription factor. *Nat. Med.* **8,** 1303–1309.

Nemer, G., Fadlalah, F., Usta, J., Nemer, M., Dbaibo, G., Obeid, M., and Bitar, F. (2006). A novel mutation in the GATA4 gene in patients with Tetralogy of Fallot. *Hum. Mutat.* **27,** 293–294.

Newburger, D. E., and Bulyk, M. L. (2009). UniPROBE: An online database of protein binding microarray data on protein-DNA interactions. *Nucleic Acids Res.* **37,** D77–D82.

Niu, Z., Yu, W., Zhang, S. X., Barron, M., Belaguli, N. S., Schneider, M. D., Parmacek, M., Nordheim, A., and Schwartz, R. J. (2005). Conditional mutagenesis of the murine serum response factor gene blocks cardiogenesis and the transcription of downstream gene targets. *J. Biol. Chem.* **280,** 32531–32538.

Niu, Z., Li, A., Zhang, S. X., and Schwartz, R. J. (2007). Serum response factor micromanaging cardiogenesis. *Curr. Opin. Cell Biol.* **19,** 618–627.

Oka, T., Maillet, M., Watt, A. J., Schwartz, R. J., Aronow, B. J., Duncan, S. A., and Molkentin, J. D. (2006). Cardiac-specific deletion of Gata4 reveals its requirement for hypertrophy, compensation, and myocyte viability. *Circ. Res.* **98,** 837–845.

Omichinski, J. G., Clore, G. M., Schaad, O., Felsenfeld, G., Trainor, C., Appella, E., Stahl, S. J., and Gronenborn, A. M. (1993). NMR structure of a specific DNA complex of Zn-containing DNA binding domain of GATA-1. *Science* **261,** 438–446.

Pal, S., Cantor, A. B., Johnson, K. D., Moran, T. B., Boyer, M. E., Orkin, S. H., and Bresnick, E. H. (2004). Coregulator-dependent facilitation of chromatin occupancy by GATA-1. *Proc. Natl. Acad. Sci. USA* **101,** 980–985.

Pasini, D., Cloos, P. A., Walfridsson, J., Olsson, L., Bukowski, J. P., Johansen, J. V., Bak, M., Tommerup, N., Rappsilber, J., and Helin, K. (2010). JARID2 regulates binding of the Polycomb repressive complex 2 to target genes in ES cells. *Nature* **464,** 306–310.

Peng, T., Wang, L., Zhou, S. F., and Li, X. (2010). Mutations of the GATA4 and NKX2.5 genes in Chinese pediatric patients with non-familial congenital heart disease. *Genetica* **138,** 1231–1240.

Potthoff, M. J., and Olson, E. N. (2007). MEF2: A central regulator of diverse developmental programs. *Development* **134,** 4131–4140.

Pu, W. T., Ishiwata, T., Juraszek, A. L., Ma, Q., and Izumo, S. (2004). GATA4 is a dosage-sensitive regulator of cardiac morphogenesis. *Dev. Biol.* **275,** 235–244.

Rajagopal, S. K., Ma, Q., Obler, D., Shen, J., Manichaikul, A., Tomita-Mitchell, A., Boardman, K., Briggs, C., Garg, V., Srivastava, D., Goldmuntz, E., Broman, K. W., *et al.* (2007). Spectrum of heart disease associated with murine and human GATA4 mutation. *J. Mol. Cell. Cardiol.* **43,** 677–685.

Rivera-Feliciano, J., Lee, K. H., Kong, S. W., Rajagopal, S., Ma, Q., Springer, Z., Izumo, S., Tabin, C. J., and Pu, W. T. (2006). Development of heart valves requires Gata4 expression in endothelial-derived cells. *Development* **133,** 3607–3618.

Roche, A. E., Bassett, B. J., Samant, S. A., Hong, W., Blobel, G. A., and Svensson, E. C. (2008). The zinc finger and C-terminal domains of MTA proteins are required for FOG-2-mediated transcriptional repression via the NuRD complex. *J. Mol. Cell. Cardiol.* **44,** 352–360.

Rojas, A., De Val, S., Heidt, A. B., Xu, S. M., Bristow, J., and Black, B. L. (2005). Gata4 expression in lateral mesoderm is downstream of BMP4 and is activated directly by Forkhead and GATA transcription factors through a distal enhancer element. *Development* **132,** 3405–3417.

Rojas, A., Kong, S. W., Agarwal, P., Gilliss, B., Pu, W. T., and Black, B. L. (2008). GATA4 is a direct transcriptional activator of Cyclin D2 and Cdk4 and is required for cardiomyocyte proliferation in anterior heart field-derived myocardium. *Mol. Cell. Biol.* **28,** 5420–5431.

Rojas, A., Schachterle, W., Xu, S. M., and Black, B. L. (2009). An endoderm-specific transcriptional enhancer from the mouse Gata4 gene requires GATA and homeodomain protein-binding sites for function in vivo. *Dev. Dyn.* **238,** 2588–2598.

Rojas, A., Schachterle, W., Xu, S. M., Martin, F., and Black, B. L. (2010). Direct transcriptional regulation of Gata4 during early endoderm specification is controlled by FoxA2 binding to an intronic enhancer. *Dev. Biol.* **346,** 346–355.

Sakai, Y., Nakagawa, R., Sato, R., and Maeda, M. (1998). Selection of DNA binding sites for human transcriptional regulator GATA-6. *Biochem. Biophys. Res. Commun.* **250,** 682–688.

Sepulveda, J. L., Belaguli, N., Nigam, V., Chen, C. Y., Nemer, M., and Schwartz, R. J. (1998). GATA-4 and Nkx-2.5 coactivate Nkx-2 DNA binding targets: Role for regulating early cardiac gene expression. *Mol. Cell. Biol.* **18,** 3405–3415.

Sharma, A., Masri, J., Jo, O. D., Bernath, A., Martin, J., Funk, A., and Gera, J. (2007). Protein kinase C regulates internal initiation of translation of the GATA-4 mRNA following vasopressin-induced hypertrophy of cardiac myocytes. *J. Biol. Chem.* **282,** 9505–9516.

Shen, X., Kim, W., Fujiwara, Y., Simon, M. D., Liu, Y., Mysliwiec, M. R., Yuan, G. C., Lee, Y., and Orkin, S. H. (2009). Jumonji modulates polycomb activity and self-renewal versus differentiation of stem cells. *Cell* **139,** 1303–1314.

Shikama, N., Lutz, W., Kretzschmar, R., Sauter, N., Roth, J. F., Marino, S., Wittwer, J., Scheidweiler, A., and Eckner, R. (2003). Essential function of p300 acetyltransferase activity in heart, lung and small intestine formation. *EMBO J.* **22,** 5175–5185.

Shin, C. H., Liu, Z. P., Passier, R., Zhang, C. L., Wang, D. Z., Harris, T. M., Yamagishi, H., Richardson, J. A., Childs, G., and Olson, E. N. (2002). Modulation of cardiac growth and development by HOP, an unusual homeodomain protein. *Cell* **110,** 725–735.

Srivastava, D. (2006). Making or breaking the heart: From lineage determination to morphogenesis. *Cell* **126,** 1037–1048.

Svensson, E. C., Huggins, G. S., Lin, H., Clendenin, C., Jiang, F., Tufts, R., Dardik, F. B., and Leiden, J. M. (2000). A syndrome of tricuspid atresia in mice with a targeted mutation of the gene encoding Fog-2. *Nat. Genet.* **25,** 353–356.

Takaya, T., Kawamura, T., Morimoto, T., Ono, K., Kita, T., Shimatsu, A., and Hasegawa, K. (2008). Identification of p300-targeted acetylated residues in GATA4 during hypertrophic responses in cardiac myocytes. *J. Biol. Chem.* **283,** 9828–9835.

Takeuchi, J. K., and Bruneau, B. G. (2009). Directed transdifferentiation of mouse mesoderm to heart tissue by defined factors. *Nature* **459,** 708–711.

Takeuchi, T., Watanabe, Y., Takano-Shimizu, T., and Kondo, S. (2006). Roles of jumonji and jumonji family genes in chromatin regulation and development. *Dev. Dyn.* **235,** 2449–2459.

Takeuchi, J. K., Lou, X., Alexander, J. M., Sugizaki, H., Delgado-Olguin, P., Holloway, A. K., Mori, A. D., Wylie, J. N., Munson, C., Zhu, Y., Zhou, Y. Q., Yeh, R. F., *et al.* (2011). Chromatin remodelling complex dosage modulates transcription factor function in heart development. *Nat. Commun.* **2,** 187.

Tevosian, S. G., Deconinck, A. E., Tanaka, M., Schinke, M., Litovsky, S. H., Izumo, S., Fujiwara, Y., and Orkin, S. H. (2000). FOG-2, a cofactor for GATA transcription factors, is essential for heart morphogenesis and development of coronary vessels from epicardium. *Cell* **101,** 729–739.

Tomita-Mitchell, A., Maslen, C. L., Morris, C. D., Garg, V., and Goldmuntz, E. (2007). GATA4 sequence variants in patients with congenital heart disease. *J. Med. Genet.* **44,** 779–783.

Tremblay, J. J., and Viger, R. S. (2003). Transcription factor GATA-4 is activated by phosphorylation of serine 261 via the cAMP/protein kinase a signaling pathway in gonadal cells. *J. Biol. Chem.* **278,** 22128–22135.

Tremblay, J. J., Hamel, F., and Viger, R. S. (2002). Protein kinase A-dependent cooperation between GATA and CCAAT/enhancer-binding protein transcription factors regulates steroidogenic acute regulatory protein promoter activity. *Endocrinology* **143,** 3935–3945.

Trivedi, C. M., Zhu, W., Wang, Q., Jia, C., Kee, H. J., Li, L., Hannenhalli, S., and Epstein, J. A. (2010). Hopx and Hdac2 interact to modulate Gata4 acetylation and embryonic cardiac myocyte proliferation. *Dev. Cell* **19,** 450–459.

Tsai, S. F., Martin, D. I., Zon, L. I., D'Andrea, A. D., Wong, G. G., and Orkin, S. H. (1989). Cloning of cDNA for the major DNA-binding protein of the erythroid lineage through expression in mammalian cells. *Nature* **339,** 446–451.

van Berlo, J. H., Elrod, J. W., van den Hoogenhof, M. M., York, A. J., Aronow, B. J., Duncan, S. A., and Molkentin, J. D. (2010). The transcription factor GATA-6 regulates pathological cardiac hypertrophy. *Circ. Res.* **107,** 1032–1040.

van Berlo, J. H., Elrod, J. W., Aronow, B. J., Pu, W. T., and Molkentin, J. D. (2011). Serine 105 phosphorylation of transcription factor GATA4 is necessary for stress-induced cardiac hypertrophy in vivo. *Proc. Natl. Acad. Sci. USA* **108,** 12331–12336.

Visel, A., Blow, M. J., Li, Z., Zhang, T., Akiyama, J. A., Holt, A., Plajzer-Frick, I., Shoukry, M., Wright, C., Chen, F., Afzal, V., Ren, B., *et al.* (2009). ChIP-seq accurately predicts tissue-specific activity of enhancers. *Nature* **457,** 854–858.

Wang, J., Feng, X. H., and Schwartz, R. J. (2004). SUMO-1 modification activated GATA4-dependent cardiogenic gene activity. *J. Biol. Chem.* **279,** 49091–49098.

Watt, A. J., Battle, M. A., Li, J., and Duncan, S. A. (2004). GATA4 is essential for formation of the proepicardium and regulates cardiogenesis. *Proc. Natl. Acad. Sci. USA* **101,** 12573–12578.

Wei, J. Q., Shehadeh, L. A., Mitrani, J. M., Pessanha, M., Slepak, T. I., Webster, K. A., and Bishopric, N. H. (2008). Quantitative control of adaptive cardiac hypertrophy by acetyltransferase p300. *Circulation* **118,** 934–946.

Xin, M., Davis, C. A., Molkentin, J. D., Lien, C. L., Duncan, S. A., Richardson, J. A., and Olson, E. N. (2006). A threshold of GATA4 and GATA6 expression is required for cardiovascular development. *Proc. Natl. Acad. Sci. USA* **103,** 11189–11194.

Yanazume, T., Hasegawa, K., Morimoto, T., Kawamura, T., Wada, H., Matsumori, A., Kawase, Y., Hirai, M., and Kita, T. (2003). Cardiac p300 is involved in myocyte growth with decompensated heart failure. *Mol. Cell. Biol.* **23,** 3593–3606.

Yang, H. Y., and Evans, T. (1992). Distinct roles for the two cGATA-1 finger domains. *Mol. Cell. Biol.* **12,** 4562–4570.

Yao, T. P., Oh, S. P., Fuchs, M., Zhou, N. D., Ch'ng, L. E., Newsome, D., Bronson, R. T., Li, E., Livingston, D. M., and Eckner, R. (1998). Gene dosage-dependent embryonic development and proliferation defects in mice lacking the transcriptional integrator p300. *Cell* **93,** 361–372.

Yu, M., Riva, L., Xie, H., Schindler, Y., Moran, T. B., Cheng, Y., Yu, D., Hardison, R., Weiss, M. J., Orkin, S. H., Bernstein, B. E., Fraenkel, E., *et al.* (2009). Insights into GATA-1-mediated gene activation versus repression via genome-wide chromatin occupancy analysis. *Mol. Cell* **36,** 682–695.

Zaret, K. S., Watts, J., Xu, J., Wandzioch, E., Smale, S. T., and Sekiya, T. (2008). Pioneer factors, genetic competence, and inductive signaling: Programming liver and pancreas progenitors from the endoderm. *Cold Spring Harb. Symp. Quant. Biol.* **73,** 119–126.

Zeisberg, E. M., Ma, Q., Juraszek, A. L., Moses, K., Schwartz, R. J., Izumo, S., and Pu, W. T. (2005). Morphogenesis of the right ventricle requires myocardial expression of Gata4. *J. Clin. Invest.* **115,** 1522–1531.

Zhang, C. L., McKinsey, T. A., Chang, S., Antos, C. L., Hill, J. A., and Olson, E. N. (2002). Class II histone deacetylases act as signal-responsive repressors of cardiac hypertrophy. *Cell* **110,** 479–488.

Zhang, W., Li, X., Shen, A., Jiao, W., Guan, X., and Li, Z. (2008). GATA4 mutations in 486 Chinese patients with congenital heart disease. *Eur. J. Med. Genet.* **51,** 527–535.

Zhang, W. M., Li, X. F., Ma, Z. Y., Zhang, J., Zhou, S. H., Li, T., Shi, L., and Li, Z. Z. (2009). GATA4 and NKX2.5 gene analysis in Chinese Uygur patients with congenital heart disease. *Chin. Med. J. (Engl.)* **122,** 416–419.

Zhao, R., Watt, A. J., Battle, M. A., Li, J., Bondow, B. J., and Duncan, S. A. (2008). Loss of both GATA4 and GATA6 blocks cardiac myocyte differentiation and results in acardia in mice. *Dev. Biol.* **317,** 614–619.

Zhou, B., Ma, Q., Kong, S. W., Hu, Y., Campbell, P. H., McGowan, F. X., Ackerman, K. G., Wu, B., Zhou, B., Tevosian, S. G., and Pu, W. T. (2009). Fog2 is critical for cardiac function and maintenance of coronary vasculature in the adult mouse heart. *J. Clin. Invest.* **119,** 1462–1476.

CHAPTER SIX

Localized and Temporal Gene Regulation in Heart Development

Phil Barnett, Malou van den Boogaard, *and* Vincent Christoffels

Contents

Abstract

The heart is a structurally complex and functionally heterogeneous organ. The repertoire of genes active in a given cardiac cell defines its shapes and function. This process of localized or heterogeneous gene expression is regulated to a large extent at the level of transcription, dictating the degree particular genes in a cell are active. Therefore, errors in the regulation of localized gene expression are at the basis of misregulation of the delicate process of heart development and function. In this review, we provide an overview of the origin of the different components of the vertebrate heart, and discuss our current understanding of the regulation of localized gene expression in the developing heart. We will also discuss where future research may lead to gain more insight into this process, which should provide much needed insight into the dysregulation of heart development and function, and the etiology of congenital defects.

Department of Anatomy, Embryology and Physiology, Heart Failure Research Center, Academic Medical Center, University of Amsterdam, Amsterdam, The Netherlands

Current Topics in Developmental Biology, Volume 100

ISSN 0070-2153, DOI: 10.1016/B978-0-12-387786-4.00004-X

1. Introduction

The heart of vertebrates is a structurally complex conglomerate of multiple components. In one respect, it is different from most other organs. The heart is not built of many identical functional substructures, such as the alveolus of the lung, the lobule of the liver, or the nephron of the kidney, but of substructures each having a unique shape and function. These functions include compartmentalization to separate the blood flows from the lung and the body (septa) and to ensure forward flow (valves), contraction to pump blood (chambers) and a system to generate (sinus node) and propagate (conduction system) the electrical signal that activates the contraction of the muscle (excitation–contraction coupling). While each part has its own specific functions, together they work as an efficient muscle pump that ensures sufficient circulation adapted to the body's requirement. If one of these component parts does not develop correctly, severe dysfunction of the whole heart ensues. In fact, this occurs frequently, as about 1 in 100 children is born with some kind of cardiac defect, and an estimated 10–20% of still births result from heart dysfunction. The spectrum of defects is wide, ranging from minor valve defects (e.g., regurgitation) to holes in the septum and inverted great arteries, requiring surgery, to undeveloped left ventricle not compatible with life. Also, electrical defects occur including bradycardia (too slow heart rate) resulting from hypoplastic pacemaker or atrioventricular (AV) conduction system components and tachycardia (too fast) due to incorrect connections between the atria and ventricles (Hoffman *et al.*, 2004; Pierpont *et al.*, 2007). And there is more epigenetics and predisposition. Variations in our genomes or little errors in our development may predispose us to develop arrhythmias or other dysfunctions in later life. Our understanding of these relations is underdeveloped to say the least.

One important aspect of heart development and function is localization. In the properly working heart, every component is uniquely shaped and positioned at exactly the right spot. This spatial organization is at the basis of a well-functioning heart pump. It is errors in this spatial organization that lead to heart defects. Therefore, researchers across the globe are working to understand the mechanisms responsible for setting up this organization. During embryogenesis, a few progenitor cells will multiply and, in a series of inputs from interactions with other cells in the constantly changing environment of the growing embryo develop into all the substructures. One of the main aims is to understand how the cells of the developing heart react to these inputs, and how they subsequently differentiate to the right cell type at the right location. A particularly relevant process in this regard is that of localized gene expression. The repertoire of genes that is active in a cell defines its phenotype, that is, its shape, behaviors, and function. The spatial and developmental differences in this repertoire between cells of the developing heart are at the basis of heart development. This process of localized or heterogeneous gene expression is regulated to a large extent at the level of transcription, dictating the degree particular genes in a cell are active. In this review, we aim to provide an overview of our current understanding of the regulation of regionalized gene expression in the developing heart. We will also discuss the approaches we need to take to find out what we do not know, and where future research may lead to gain more insight into this process. To set the stage, we will provide an overview of the development of the vertebrate heart and our current insight into the developmental origins of the component parts, which is importantly related to regionalized gene expression during development.

2. Morphogenesis of the Heart: Series of Localized Processes of Differentiation and Growth Stacked on to One Another

The heart is the first organ to form and function during embryonic development (Fig. 6.1). It develops from the cardiac plate that is formed by cells from the mesoderm in the cranial part of the embryonic disk. During the process of gastrulation, early cardiac progenitor cells are located in two symmetrical, bilaterally positioned regions on either side of the embryonic midline. These regions are often referred to as the heart-forming regions. The developing embryo undergoes what is perhaps one of the most difficult descriptive morphological transformations during development that of embryonic folding (see, e.g., van den Berg and Moorman, 2009). By the lateral and cranial folding of the embryo and formation of the foregut, the anterior heart-forming regions come together, deriving a horseshoe like cardiac crescent. Subsequently, the foregut expands and folds further, the lateral walls of the cardiac crescent move toward the central region and fuse at the midline to form the primitive heart tube. In human and chicken, a straight heart tube is seen to form; in mouse, it is nothing more than a small bulb.

The early heart tube itself does not undergo significant proliferation but instead elongates mainly by recruiting cells of the heart-forming regions surrounding the heart tube (Cai *et al.*, 2003; Kelly and Buckingham, 2002; van den Berg *et al.*, 2009). The foregut and the medial cells of the heart-forming region together move dorsally and become positioned dorsal of the heart tube (Fig. 6.1). These cells will form the "anterior second heart field" that contributes cells to the tube through the dorsal mesocardium and at the outflow tract. The caudal cells of the heart-forming regions are referred to as the "posterior second heart field" contributing cells to the caudal dorsal side of the heart (Cai *et al.*, 2003; Galli *et al.*, 2008; Goddeeris *et al.*, 2008; Kelly and Buckingham, 2002; Snarr *et al.*, 2007; van den Berg *et al.*, 2009). As a

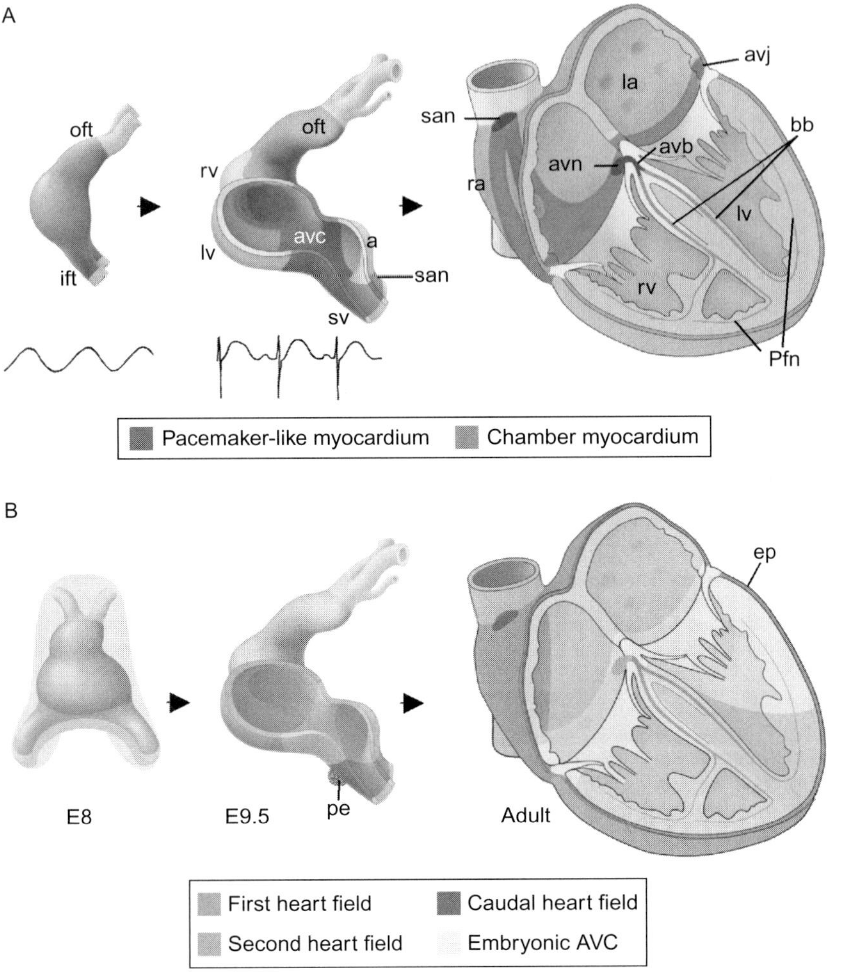

Figure 6.1 Development of the heart. (A) Formation of the atrial and ventricular chambers, and the development of the pacemaker and conduction system. A mature type of ECG can be observed already during chamber development. (B) Contribution of different cell lineages to the heart. At E8.0 the primary heart tube primarily consists of first heart field (FHF) derived cells, shown in green. This region will become restricted to the left part of the interventricular septum and part of the left ventricle in the adult heart. The second heart field (blue) enters at the inflow tract, dorsal mesocardium, and outflow tract and will eventually derive a large part of the atria in the right ventricle. Part of the inflow tract (yellow) will form the embryonic atrioventricular canal (AVC) and thereafter contribute to the ventricle, particularly the left ventricular wall. The caudal heart field will contribute to the formation of the (pro) epicardium and sinus venosus. l/ra, left/right atrium; l/rv, left/right ventricle; avc/n, atrioventricular canal/node; san, sinus node; pe, proepicardium; ep, epicardium. Modified from Moorman and Christoffels (2003). (See Color Insert.)

result of the folding process, the lateral rim of the heart-forming region has become positioned caudal and ventral of the heart tube where it contributes cells. This rim has been referred to as the "caudal heart field" and can be distinguished from the "posterior second heart field" (Mommersteeg *et al.*, 2010).

The primary myocardium of the early heart tube possesses a phenotype that resembles that of the nodal (pacemaker) tissues in the adult heart, having slow contraction, underdeveloped contractile apparatus, slow transmission of electrical signals, and high automaticity (reviewed in Moorman and Christoffels, 2003). Already prior to first contractions are observed, dominant pacemaker activity develops at the caudal venous pole of the heart tube, and remains there even though new cells are being added to the venous pole (Christoffels *et al.*, 2010; van Mierop, 1967). While the heart tube further lengthens, localized regions in the primary heart tube will differentiate into atrial and ventricular chambers, whose myocardium acquires a phenotype that does not resemble that of the early heart tube. The chamber myocardium displays rapid transmission of electrical signals and low automaticity (Moorman and Christoffels, 2003). Formation of the chambers is marked by the differentiation of localized regions of primary myocardium along the embryonic heart tube, which is characterized by a slow propagation of the cardiac impulse, into the fast-conducting working myocardium of the chambers. The inflow, the AV canal, the inner curvature, and the outflow retain their primary myocardial phenotype and initially do not differentiate into the working myocardium phenotype (reviewed in Moorman and Christoffels, 2003). Taken together, cardiac chamber development requires the localized differentiation and proliferation of primary myocardial cells into working myocardial cells. The primary and working myocardium can be distinguished from each other by the expression and repression patterns of numerous genes.

The fully developed, adult, four-chambered heart consists of ventricular and atrial chambers separated by muscular and fibrous septa with inflow (venous) and outflow (arterial) components and valves that ensure a unidirectional blood flow. During development, two blood flows, one from the right atrium draining into the forming right ventricle and the other from the left atrium into the left ventricle, remain separate laminar flows that do not mix laterally, a process guided by the developing endocardial cushions. These cushions and cells from the dorsal mesenchymal protrusion (Snarr *et al.*, 2007) will form the AV septation complex required for formation of the septa of the atria and ventricles and of the AV valves (Lamers and Moorman, 2002). The outflow tract will also septate and remodel and form valves, a process involving the interaction between neural crest-derived cells, mesoderm-derived cells, and foregut cells, to properly connect the aorta to the left ventricle and the pulmonary artery to the right ventricle (Bajolle *et al.*, 2006; Hutson and Kirby, 2007; Rochais *et al.*, 2009).

The heart will also develop a pacemaker and electrical wiring system, the conduction system. These conduction system components gradually emerge from the different heart components. The sinus node, the primary pacemaker, develops within the sinus venosus, the AV node that delays the impulse from the atria to the ventricles develops within the AV canal, which already has this function in the embryo (reviewed in Christoffels *et al.*, 2010). The AV bundle develops from the crest of the ventricular septum and the bundle branches and Purkinje fiber network from trabecular component (reviewed in Christoffels and Moorman, 2009; Miquerol *et al.*, 2011).

3. Lineage Relationships: Which Cells in the Embryo Generate What Component Parts?

We have seen 10 years of discussion about whether the progenitors that form the initial heart tube and the progenitors that are recruited to the rims (outflow, dorsal mesocardium, dorsal, and ventral inflow) are different and justifiably referred to as heart fields with distinct names. The heart-forming field was originally defined as all cells in the early gastrula embryo that ended up in the heart tube of approximately HH12 chicken embryos (slightly beyond straight heart tube stage). At that time, it was already known that cells from outside the heart tube were still added to the tube at a later stage, but it was technically not possible to make a retrospective fate map of the fully formed heart and the "real" heart-forming region in the early embryo (reviewed in Abu-Issa and Kirby, 2007; Moorman *et al.*, 2007). In 2001, the cells that are added after the heart tube has been formed were identified in the embryo using physical and genetic labeling techniques and duly christened the "secondary" or "second heart field" (for a discussion and clarification on this issue, see Buckingham *et al.*, 2005; Dyer and Kirby, 2009). The main arguments in favor of the existence of a different field are that (1) several markers were found to be expressed more or less specifically in the second heart field, or parts thereof, which was further supported by lineage tracing experiments and (2) perturbation of genes for several regulatory factors did not disrupt heart tube formation but did affect the process of elongation (reviewed in Rochais *et al.*, 2009). However, the recruitment process is a gradual temporal process. The border between "first heart field cells" already differentiated to cardiomyocyte and "second heart field" cells not yet differentiated is defined simply by the time point in cardiac development chosen by the observer. Further, no two marker genes distinguish the same cell population, and several key markers for the second heart field, such as Islet1 (ISL1 transcription factor, LIM/homeodomain, required for heart development), turned out to be expressed temporally in all heart progenitors, including the "first heart field" (Brade *et al.*, 2007; Mommersteeg *et al.*, 2010; Prall *et al.*, 2007). Therefore, one could also postulate a large heart field at each lateral side of the embryo that acquires patterns of gene expression and properties during development. The cells more lateral differentiate first while becoming incorporated in the tube during folding, and the cells more medial, caudal, or toward the lateral rim of the field maintain high levels of proliferation, retain a more progenitor like phenotype, and are gradually incorporated in the tube where they will differentiate (Fig. 6.1).

The discussion of heart fields is part of an important aspect of understanding organogenesis. Which cells in the embryo will form particular components in the developing and in the mature heart? This knowledge is instrumental in understanding the history of cells, that is, where the cells were during development, which signals they received from their neighbors that made them behave in a certain way, how much they proliferated, when they differentiated, and which signals they provided to make things happen around them. Classical and more recent genetic fate maps have taught us a great deal about these relationships (reviewed in Abu-Issa and Kirby, 2007; Mikawa, 1999; van den Berg and Moorman, 2009). A key implication of these studies is that during subsequent developmental stages, structures bearing the same name are made of different cells, because the cell lineages "migrate" through structures. Here, we will briefly summarize the origins of the main cardiac components.

The cells of the initial heart tube of the chicken and mouse embryo will form the apex of the left ventricle and the left portion of the interventricular septum (Abu-Issa and Kirby, 2007; Cai *et al.*, 2003; De la Cruz *et al.*, 1989; van den Berg and Moorman, 2009). The "limbs" of the cardiac crescent (first heart field-derived) will contribute to the AV canal (Aanhaanen *et al.*, 2009). The medial and caudal region of the heart-forming fields will be displaced dorsally and caudally of the forming heart tube and becoming recognized as the "second heart field." The cranial portion of this field will form the future right ventricle and outflow tract, the caudal portion the atria and probably contribute to the AV canal (Abu-Issa and Kirby, 2007). The cells of the lateral rim of the heart-forming regions, adjacent to the extra-embryonic mesoderm overlying the yolk sac, will be displaced to the caudal and ventral side of the heart tube. These cells will form the proepicardium and the sinus venosus (Mommersteeg *et al.*, 2010; Moorman *et al.*, 2007; van Wijk *et al.*, 2009). The sinus node will be formed from cells at the border of the Isl1+ caudal and Tbx18+ ventral progenitors at the right side (Mommersteeg *et al.*, 2010).

The embryonic AV canal will form the mature AV node and ring bundles and provide a large contribution to the left ventricular free wall (Aanhaanen *et al.*, 2010). Virtually, all of the original primary outflow tract myocardium differentiates to right ventricular working myocardium, including the right side of the interventricular septum, whereas apoptosis will remove remaining cells (De la Cruz *et al.*, 1977; Kelly *et al.*, 2001; Rana *et al.*, 2007). The AV bundle will develop from the crest of the interventricular septum (Bakker *et al.*, 2008; Virágh and Challice, 1977, 1982) and the ventricular conduction system from the embryonic trabecular component (Christoffels and Moorman, 2009; Miquerol *et al.*, 2010), although it has also been proposed to form relatively late in chicken from subendocardial myocytes by induction (Pennisi *et al.*, 2002). Based on genetic fate maps, lineage boundaries have been observed between the sinus node and atrial myocardium (Hoogaars *et al.*, 2007; Mommersteeg *et al.*, 2007), between the AV node and AV bundle (Aanhaanen *et al.*, 2010) and between the left and right ventricle (Franco *et al.*, 2006). The proepicardium that develops from the caudal ventral progenitors will envelope the heart to form the epicardium. Subsequently, epicardium-derived cells will migrate into the muscle walls and form the interstitial fibroblasts, provide large contributions to the fibrous skeleton, and play a key role in the generation of the coronary vasculature, providing the fibroblasts and smooth muscle surrounding the vessels (Winter and Gittenberger-de Groot, 2007).

In most cases, the boundaries between these progenitor populations and the degree of their contributions to each heart component has not been defined precisely. The reason for this is that clear definitions on the borders between these progenitor populations is lacking and that genetic fate maps have been based on the expression of a particular gene marking a particular progenitor population at a particular stage of development. Moreover, these same genetic fate maps suffer from imprecision and misinterpretation, as the efficiency of labeling, and thereby the relation between level and pattern of marker gene expression and the labeling efficiency differs between Cre drivers and Cre reporters (Fig. 6.2). More efficient drivers or reporters will, in general, indicate a larger contribution of a particular progenitor population as marked by the domain of expression of the corresponding marker gene (Ma *et al.*, 2008). This has already resulted in interesting discrepancies between studies and some erroneous conclusions. For example, the second heart field progenitor population expresses Isl1 and was initially found to form the outflow tract, right ventricle, and part of the atria, but not the initial heart tube forming the left ventricle. This observation has greatly contributed to the idea of a discrete second heart field that was marked by *Isl1* (Cai *et al.*, 2003). However, *Isl1* was found to be expressed briefly during development in all lateral plate mesoderm, including the progenitors of the heart tube (Brade *et al.*, 2007; Mommersteeg *et al.*, 2010; Prall *et al.*, 2007), and the use of more efficient *Isl1*-Cre drivers and Cre reporters confirmed this, as now the entire heart was labeled (Ma *et al.*, 2008; Sun *et al.*, 2006). Similarly, we found that sinus venosus was derived from precursor population not expressing Nkx2-5 (NK2 Homeobox Protein 5, a homeobox transcription factor required for heart development;

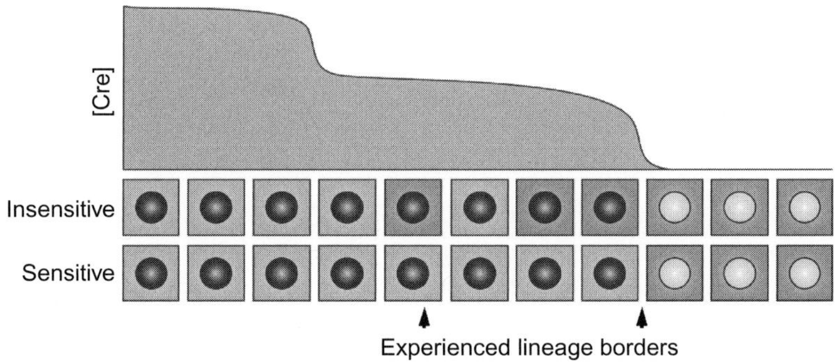

Figure 6.2 Effect of Cre reporter on Cre-based fate map. Gradient showing the expression level of Cre in the nucleus of a row of cells. To analyze the fate of these Cre+cells during development or in the adult, the Cre mouse line is crossed with a Cre reporter, in which a reporter gene is irreversibly activated. The process of recombination is stochastic, the probability depending on Cre concentration and the "sensitivity" of the reporter. An insensitive Cre reporter indicates a smaller Cre-expressing cell population compared to a more sensitive reporter, strongly affecting the genetic lineage analysis. (For interpretation of the references to color in this figure legend, the reader is referred to the Web version of this chapter.)

Christoffels *et al.*, 2006) but was marked by *Nkx2-5*-driven Cre when using a more efficient reporter (Ma *et al.*, 2008; W. Pu, Boston, personal communications). The underlying cause of the discrepancy is the brief expression of *Nkx2-5* in the progenitors of the sinus venosus before it is turned off, in contrast to all other progenitors and derived cardiomyocytes in which *Nkx2-5* expression is maintained. Finally, epicardial cells were suggested to contribute to the myocardium (Cai *et al.*, 2008; Zhou *et al.*, 2008). However, in one case, where *Tbx18* was used to mark the epicardial cells, expression of *Tbx18* in myocardium was missed. This myocardial expression, not transdifferentiation of epicardium into myocardium, was (at least largely) responsible for the observed Cre reporter activity in the myocardium (Christoffels *et al.*, 2009). Further, a technical but important aspect of lineage tracing using Cre, the neomycin resistance cassette driven by its own promoter was not removed from the Cre driver construct, which affected the strictness of the *Tbx18*-driven Cre pattern of expression. Similarly, expression of Wt1 in early heart progenitors prior to its restriction to epicardium has probably caused misinterpretation of the fate of the Wt1-expressing epicardium as well. Nevertheless, when interpreted conservatively, these Cre-mediated lineage analyses have given interesting and sometimes unexpected insights into origins of heart components and cellular movements.

In short, research has taught us many of the physical processes and changes in morphology that occur during heart development. In terms of the regulation of development as a whole it is becoming clear that the genome plays the leading role, governing the timing and cell-specific expression of the genes that will shape the cells and define their ultimate function. Although the basic elements of the genome are the same in every cell, its arrangement and organization within different cells can be vastly different, presenting sets of rules and blueprints that guide cells along their differentiation pathways, instructions that themselves may change as cells progress toward terminal differentiation. We are only now beginning to gain insight into these processes. In the preceding text, we have tried to present some of the essential steps that take place during heart development. In the next sections, we discuss and touch upon some of the elements and mechanisms that are known to play key roles in gene regulation and ultimately regulation of the processes mentioned. We describe some of the basic concepts that must be considered if one is to go on and fully understand how the genome is regulated. We also present the reader with some of the modern techniques and approaches that are presently available for studying gene regulation.

4. Regulatory Modules Control the Spatiotemporal Patterns of Gene Expression, but How?

Development and function of the heart relies on the coordinated and timed expression of the genes that will ultimately dictate cellular function. When considering a single gene downstream of a set of effector proteins that regulate its expression, the molecular mechanism underlying the transcription process seems fairly straightforward. For example, gene expression is often considered to be regulated by its promoter, a sequence of a few hundred to thousand base pairs positioned upstream of the transcriptional start site. In transfection assays, transcription factors are added in greater combinations and expressed at levels many-fold higher than would ever be normally found within the nucleus. Binding sites are verified using a combination of *in silico* site finding programs, alignments, and guess work, all of which is backed up with further *in vitro* binding assays and mutation studies. Although conceptually this approach is useful and may provide some novel mechanistic insights, it represents a very minimalistic view that takes a complex process, freezes its dynamics, and ignores its context.

Models and functional hypotheses that have combined aspects from both *in vitro* and *in vivo* gene regulation studies usually turn out to be much more accurate and robust, often being supported and expanded upon by further data that makes use of the very latest techniques. These studies have indicated that genes are regulated by multiple sequence elements that can sometimes be positioned very far away from the transcription start site. A promoter region alone is almost never sufficient to drive correct endogenous expression, requiring the often autonomous activity of regulatory modules to provide specific aspects of the total spatiotemporal pattern of gene expression (Gray and Levine, 1996; Habets *et al.*, 2003; Visel *et al.*, 2009a).

Using transgenic mouse assays, regulatory DNA regions, or modules, have been identified in several cardiac genes (Habets *et al.*, 2003). In these assays, small putative regulatory fragments are coupled to a minimal promoter-reporter and the constructs are subsequently injected into mouse zygotes to allow their (random) integration into the genome. The pattern of the reporter is subsequently assayed in the developing embryo. These include *Nkx2-5* and *Nppa* (see below); *Tnni3* and *chicken Gata6*, both containing a module close to the transcription start site driving expression in the AV canal (Davis *et al.*, 2001; Di Lisi *et al.*, 2000); *Mef2c*, where an enhancer was found to drive expression in the second heart field and right ventricle (Verzi *et al.*, 2005); *Cx30.2* and *Tbx2*, where enhancers were found that drive expression in the AV canal, similar to the pattern of the endogenous genes (Munshi *et al.*, 2009; Shirai *et al.*, 2009; Singh *et al.*, 2009); *Isl1*, for which a downstream enhancer has been described that drives expression in the lateral plate mesoderm and migrating heart progenitor cells (Kappen and Salbaum, 2009); and *Pitx2*, where an intronic enhancer was shown to be responsible for asymmetric expression during cardiac morphogenesis (Nowotschin *et al.*, 2006). For several of these enhancers, the factors involved in their regulation have been identified. However, only rarely can they explain the spatiotemporal pattern of activity of the enhancer module *in vivo*.

It is important to realize that these transgenic assays only test whether a fragment can drive expression in a particular environment. Thus, they do not test requirement or relative contribution to the overall regulatory system, or even whether the enhancer indeed contributes to the regulation of a particular gene, as the fragments are tested outside their context and the assays is notoriously nonquantitative. Herein also lies another problem. Identification of a proven active enhancer in the vicinity of a gene does not dictate that this gene is in fact the endogenous target of the enhancer. Further, these relationships become even more complicated if one takes into account the sometimes redundant behavior of *cis*-acting primary and secondary (shadow) enhancers, a feature which may underlie the phenotypic robustness of gene expression (Frankel *et al.*, 2010; Hong *et al.*, 2008). *De facto*, we do not understand well the relationship between transcriptional activities and the spatiotemporal response of the modules and their concerted mechanism of action.

One example that highlights these issues is the regulation of the spatiotemporal expression pattern of the aforementioned *Nkx2-5* in the early heart (Schwartz and Olson, 1999). Multiple regulatory fragments were identified using transgenic embryos, each providing unique and overlapping activities (Fig. 6.3). Building on previous studies by other groups showing the importance of Gata4 in the regulation of *Nkx2-5 in vivo* (Searcy *et al.*, 1998), Lien *et al.* (1999) identified an enhancer element 9kbp upstream of *Nkx2-5* that recapitulated the expression of the endogenous gene. Like the previous study, binding of Gata4 to this enhancer was demonstrated both *in vitro* and *in vivo*. This study also clearly demonstrated that although Gata4 was important for the regulation of this element, its presence alone did not dictate expression. It is this modularity, the spatial positioning, the temporal activity, and the cooperative function of enhancers that can often not be inferred from *in vitro* studies alone.

The transgenic mouse assay (nonquantitative testing of relatively small regulatory fragments outside genomic context) may not give an accurate picture of the role of regulatory regions in regulation of gene expression *in vivo* and usually is limited to testing relatively proximal regulatory sequences. One approach which may circumvent these issues is the use of bacterial artificial chromosomes (BACs) containing several hundred thousand base pairs of the region of interest (Yang *et al.*, 1997). Following on from studies into the regulation of *Nkx2-5*, the Schwartz lab expanded the repertoire of regulatory elements following a BAC approach (Chi *et al.*, 2003, 2005). Understanding the regulation of *Nkx2-5* or any important gene to this level will provide insights that may prove key to unlocking the mechanisms underlying lineage decisions made during early heart development.

A second example is *Nppa*, encoding Atrial Natriuretic Factor. Regulation of *Nppa* has been extensively studied, and its expression can be linked to the differentiation of the working myocardium of the atria and ventricles (Christoffels *et al.*, 2000; Houweling *et al.*, 2005), making it suitable and thus commonly used marker for this process. Notably, *Nppa* expression is absent in the primary myocardium (e.g., AV canal; Fig. 6.1). The transcription factors Nkx2-5, Gata4, and Tbx5 activate the expression of *Nppa* in the working myocardium of the chambers, thereby indicating that these factors play a role in chamber differentiation (Bruneau *et al.*, 2001; Garg *et al.*, 2003; Hiroi *et al.*, 2001; Small and Krieg, 2003). As with Gata4 regulation of *Nkx2-5*, the presence of these three factors alone cannot be considered as dominant, since they are also expressed in areas that do not differentiate into working myocardium. Indeed, other T-box factors, Tbx2 and Tbx3, are both expressed in primary myocardium, act as transcriptional repressors, and display high homology in both structure and function (Christoffels *et al.*, 2004; Habets *et al.*, 2002; He *et al.*, 1999). During heart development, *Tbx2* is expressed in the myocardium of the AV canal and outflow tract in mouse (Christoffels *et al.*, 2004; Harrelson *et al.*, 2004) *Tbx3* expression overlaps *Tbx2* expression in the AV canal and is uniquely expressed in precursors of the sinus node and AV bundle. Later in development, *Tbx2* expression

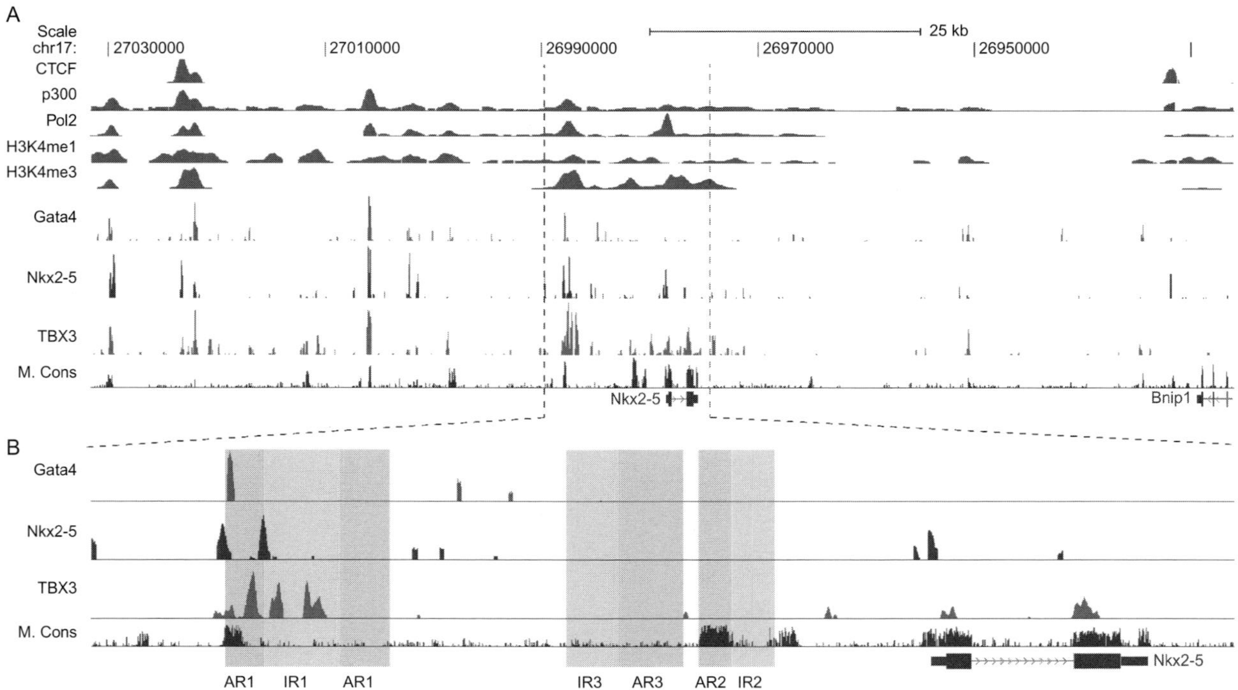

Figure 6.3 ChIP-sequence analysis of the *Nkx2-5* gene region. (A, B) ChIP-seq traces for various transcription factor analyses carried out *in vivo* using mouse hearts. (A) specifically shows ENCODE/UCSC (Rosenbloom *et al.*, 2010) adult mouse heart ChIP-seq traces for insulator associated factor CTCF, polymerase II (Pol2), histone monomethylation and trimethylation of lysine residue 4 (H3K4me1, H3K4me3), and the enhancer associated protein P300 in and around the *Nkx2-5* gene locus. (B) Closeup of the region directly surrounding *Nkx2-5* showing adult mouse heart ChIP-seq data traces (our unpublished results) for Gata4, Nkx2-5, and Tbx3. Both figures also highlight (dotted lines) the relative positions of previously discovered enhancer elements that have been identified and verified *in vivo* (see text). Significant overlap can be seen between the various datasets and the verified enhancer positions, particularly at the most distally located elements, shown to mark expression specifically in the heart. (See Color Insert.)

gradually fades, whereas *Tbx3* expression persists in the heart into adult life in the sinus node, the AV node and rings, and the AV bundle and branches (Bakker *et al.*, 2008; Hoogaars *et al.*, 2004). Both factors turned out to interact with the *Nppa* upstream promoter *in vivo*, and repress its activity in all domains of their own expression. This finding has led to the concept that repression of working differentiation is crucial for sinus node, AV canal, and AV bundle development (Christoffels *et al.*, 2010).

Initial studies of *Nppa* regulation focused on a small upstream proximal promoter that appeared to possess all information necessary to drive correct *in vivo* expression during development (reviewed in Habets *et al.*, 2003). However, a more in-depth study in to the regulation of *Nppa* via these promoter fragments revealed a failing of correct ventricular expression and downregulation of the reporter gene after birth (Horsthuis *et al.*, 2008), contrary to previous reports (Knowlton *et al.*, 1995). Further, this fragment was unable to recapitulate correct reactivation upon hypertrophic stimulation. Using the BAC approach, our group was also able to demonstrate that separable and distinct distal acting genomic elements were responsible for regulating the embryonic activity and the reactivation of *Nppa* during cardiac disease, respectively.

The above examples are also evidence to the fact that although transcription factors remain effectors of expression, the eukaryotic genome itself plays just as big, if not greater, role in gene expression and regulation. This form of regulation probably underlies much of the aforementioned "modular" behavior of enhancer elements in the developing heart. Research within this field of cardiac development is still in its early stages (Bruneau, 2010), but an understanding of its influence on gene expression during heart development will prove crucial to formulating a more accurate picture of cardiogenesis and cardiac disease. Therefore, the following sections give a brief overview of some of the epigenetic aspects that must be considered if one wants to explore cardiac development, how regulatory sequences are being identified and their *in vivo* function determined.

5. *In Vivo* Regulatory Apparatus of the Genome: Identification of Regulatory Sequences and the Genes They Control

The regulatory apparatus that governs the transcriptional profile of any given eukaryotic cell consists in its simplest terms of proximal promoters, enhancers, silencers, and insulators. The basic function of such elements is to provide a localized DNA scaffold for the binding and interaction of effector proteins that will either stimulate or repress gene transcription. Although not discussed here (Dean, 2011; Yang and Corces, 2011), insulators and silencers are high-order protein-DNA complexes that play a role in regulating and restricting enhancer promoter interactions and establishing regions of chromatin regulated gene expression. Silencers basically function by establishing regions for nucleation of stable repression by recruitment of polycomb complexes. These centers of repression may then spread via histone modification and looping of chromatin (Lanzuolo *et al.*, 2007) to silence transcriptional activity at target genes. To control the possible promiscuity of silencers and enhancers, insulators, via recruitment (in vertebrates) of the protein CTCF (Phillips and Corces, 2009), are believed to function as barriers to these activities. They define strict boundaries of regions across which silencers and enhancers may function, regions that may be several hundreds of kbps in length. To illustrate, the enhancer for the sonic hedgehog gene (*Shh*) is positioned at a distance in excess of 1 Mbp from its target promoter (Lettice *et al.*, 2003; Sagai *et al.*, 2005).

Employing the BAC approach and enhancer screens for single gene loci has proven that the modular workings of enhancers regulating the gene of interest can be unlocked, albeit that this is a slow and sometimes labor intensive process. However, with recent advances in gene sequencing technology, it has become possible to hunt for distal regulatory elements on a genome-wide scale. Several unifying features of such elements and the way they function within the context of the genome can be used as markers for their location. Typically, when active, regulatory elements affect the chromatin environment in their direct vicinity. Promoters and notably enhancers are typically found to colocalize with disruptions in chromatin structure, making the DNA more accessible to DNase I treatment. Such disruptions are believed to be caused by the localized association of transcription factor complexes that result in reorganization of the nucleosome and thus chromatin structure (Felsenfeld and Groudine, 2003). As such, this technique, if performed on purified nuclei, can be used to identify active promoters and putative enhancer elements. Huang and colleagues (Huang and Liew, 1998; Huang *et al.*, 1997), using nuclei from neonatal hamster cardiomyocytes, mapped the DNase I hypersensitive sites within the 5′ upstream regions of the β-myosin heavy chain gene. Although the isolated elements, one of which was positioned in the proximal promoter, did not behave as classical enhancers in *in vitro* transient transfection assays, a link could be made between the appearance of the DNase hypersensitive site and promoter activity during development. DNase sensitivity and concordant promoter activity was notably diminished at the adult stage. With the advent of next generation sequencing techniques, mapping DNase I hypersensitive sites can now be carried out genome wide and when combined with data from ChIP-seq studies (see section 8 below), provides a powerful and comprehensive method for identifying regulatory elements (Boyle *et al.*, 2011).

Another criterion that can be used for predicting enhancer position is noncoding sequence conservation. This approach has grown over the past

15 years due to the growing availability of sequenced genomes and user-friendly interfaces for creating and analyzing whole genome alignments. One study, which provides an interesting example for early comparative genomic analysis in cardiac development and disease, set out to map the velocardiofacial syndrome region on chromosome 22 (Lund *et al.*, 2000). Although the gene underlying this disorder (also known as diGeorge syndrome), *TBX1*, was eventually identified (Lindsay *et al.*, 2001; Merscher *et al.*, 2001), the approach taken by Lund and coworkers used comparison of genomic sequences across species as a tool for predicting and identifying gene transcripts via their evolutionary conservation. However, the study also noted that across the 634 kb stretch analyzed, approximately 6% of the sequence showed conservation in regions of "noncoding" DNA, implying that this elements may be conserved because they are of intrinsic importance.

With the availability of full genome sequences from multiple species, prediction accuracy increased. The power of this technique was illustrated by Pennacchio *et al.* (2006) who demonstrated that overall 29% of noncoding regions found to be conserved between humans and fish showed demonstrable enhancer activity *in vivo*. Further, 61% of those which were also ultraconserved between mammals proved to function as enhancers *in vivo*. This analysis was also in agreement with a later study demonstrating that about 30% of functional orthologous conserved noncoding sequence elements between paired distant species (e.g., human–fish) exhibited a similar functional activity (Ritter *et al.*, 2010). Currently, several user-friendly web interfaces are available for such analyses. Using the ECR (Evolutionary Conserved Region) browser (http://ecrbrowser.dcode.org; Miller *et al.*, 2008a) examined 100 kb flanking regions of the *MSX1* gene locus leading to the identification of an evolutionary conserved enhancer element able to support expression of *Msx1* in migrating cardiac neural crest cells and the developing outflow tract of the heart.

Although conservation may represent a useful tool for predicting the presence of enhancers, it is by no means absolute in its capacity. As with genes, enhancers are also open to sequence divergence over evolution, being negatively selected if divergence disrupts function too radically, otherwise evolving in a neutral fashion (Meireles-Filho and Stark, 2009). Moreover, conservation of enhancers may differ between tissue types. Enhancers identified by assessing p300 occupancy (see below) and *in vivo* testing revealed that "heart" enhancers are less evolutionary conserved than "forebrain" enhancers and in fact are not very evolutionary conserved at all (Blow *et al.*, 2010).

Although basal promoters can typically be assigned to genes simply based on their direct upstream location to the transcriptional start site of a gene, the distance of an enhancer from its target gene appears of no constraint to its function in transcription and, although probably rare, interactions may even occur interchromosomally (Noordermeer *et al.*, 2011). This then has been perhaps one of the major limiting factors in the identification of specific enhancers controlling regionalized gene expression. This issue can be resolved using an unbiased analysis such as chromatin conformation capture (Simonis *et al.*, 2007). In these assays, the interaction between one position in the genome (e.g., a promoter) and the surrounding chromatin (including enhancers) can be assessed. Using this approach, Tena and coworkers showed that an evolutionary conserved three-dimensional structure in the Iroquois homeobox gene cluster, which is important for regionalized gene expression and function in the heart (Tena *et al.*, 2011), facilitates enhancer sharing. They identified tissue-specific enhancers and explained how two of the three genes in each cluster are coregulated by these enhancers. Warren and coworkers identified several distal enhancers involved in Nkx2-5-mediated regulation and developmental ventricular activity far upstream of *Nppa* (Warren *et al.*, 2011). These fragments were positioned within the BAC that recapitulates the endogenous developmental expression of *Nppa*, but two of three are outside the BAC sufficient to mediate the hypertrophy response (Horsthuis *et al.*, 2008). Taken together, enhancers and proximal promoters can find and specifically interact with each other even at large distances, that these interactions are subject to context (e.g., tissue-type, developmental stage) dependent regulation and that chromatin structure plays an important functional role. Deep sequence-based variations of the 3C-methods to capture DNA interactions genome wide will greatly facilitate the identification of the mechanism of spatiotemporal gene regulation during heart development.

6. Signposts in a Chromatin Landscape

Histones, the basic subunits of the nucleosomes of chromatin, are characterized by numerous posttranslational modifications (see Fig. 6.4). Such modifications are known to affect transcription in either a positive or negative way, though the rules governing the functional aspects of these modifications either have yet to be discovered or are somewhat flexible. Further, it appears that the code that determines the level of repression or activation may be provided by the combinatorial action of several modifications (Wang *et al.*, 2008b). Of all the modifications, acetylation appears to be the one which is the most consistently associated with active gene expression. Specific patterns of acetylation have also been used to predict enhancer positions within the genome. Notably, p300, perhaps one of the better known acetyltransferases, itself is a clear marker for regulatory elements. In one study, the distribution of H3K9/14 acetylation was used to identify enhancer elements across the genome irrespective of their evolutionary

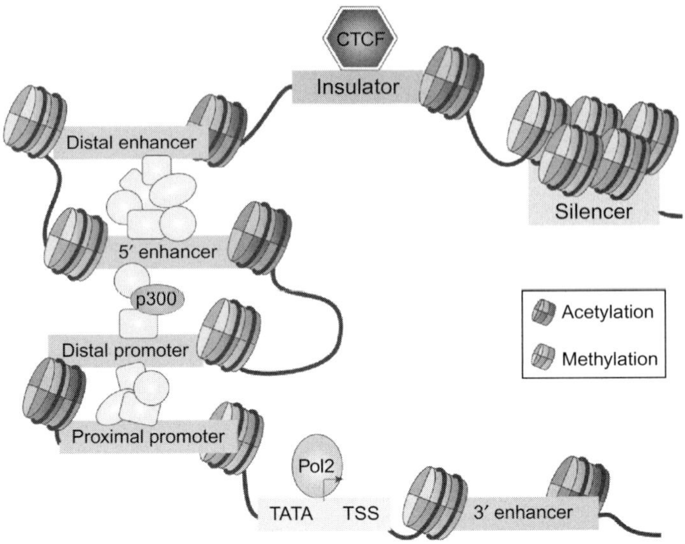

Figure 6.4 Histone modifications mark functional elements in the mammalian genome. Active and inactive transcription is associated with different features on chromatin organization. Here, generally acetylated (red) and methylated (blue) histone states of chromatin are indicated. Regions of gene silencing and condensed chromatin are typically marked by extensive histone methylation (e.g., H3K27me3). Insulators are bound by CTCF and are typically marked by methylation (e.g., H3K4me3) and acetylation of the type associated with active genomic regions. Regulatory elements are bound by DNA-binding proteins, or transcription factors (yellow), and RNA polymerase II. Among these are enhancers, of which the activated state is relatively enriched for p300 and both acetylated and methylated histones (e.g., H3K4me1/3). These various features of chromatin help organize the DNA and distinguish functional elements in the large expanse of the genome. Modified from Zhou *et al.* (2011). (For color version of this figure, the reader is referred to the Web version of this chapter.)

conservation (Roh *et al.*, 2007). Likewise, deacetylation is associated with inactive genes. However, the proteins that regulate these modifications can both be found at sites of active transcription (Wang *et al.*, 2009). Further, binding of HDAC proteins (Histone DeACetylase) was not found to be correlated with silenced genes, suggesting that although HDAC recruitment may be involved in down regulation of gene activity, it is not required for maintenance of gene silence. Interestingly, these studies revealed that cross talk between histone modifications was possible and that H3K4 methylation facilitates acetylation events. H3K4 mono-, di-, and trimethylation (me3) are all typically associated with active transcription marking both the direct promoter region surrounding the TSS of the active gene and also *cis*-acting enhancer elements (Barski *et al.*, 2007). H3K4 monomethylation has recently been directly linked to active enhancers in stem cells

(Rada-Iglesias *et al.*, 2011). As stem cells began to differentiate, in this case, the differentiation of neuroepithelium was followed, "poised" promoters, marked by H3K27me3, begin to display diminished levels of this typically repressive associated chromatin mark and adopt markers of gene activity such as H3K27 acetylation. The maintenance of H3K4me3, though perhaps not surprisingly, has been suggested to play an important role in the transcriptional stability of cardiomyocytes and has been linked to progression of heart failure in rats (Kaneda *et al.*, 2009). Loss of H3K4me3 could also be shown to correlate with a reduction in ion channel expression, upregulation of *Nppa*, a marked alteration in the electrophysiology of cardiomyocytes and the potential for the development of ventricular arrhythmias (Stein *et al.*, 2011).

At this point, it is worth pointing out that histone methylation is different to methylation of DNA directly, a modification that is typically associated with gene silencing. This form of methylation is almost exclusively associated with CpG nucleotide repeats, commonly occurring in or around TSS, and is involved in the recruitment of gene silencing proteins such as MeCP2 (Deaton and Bird, 2011), which in turn alter the chromatin state, affect histone methylation, and physically repress gene expression.

Clearly then, during development, the lineage choices made by cells are greatly influenced by changes in chromatin that modulate its conformation, accessibility, and the signals that mark both poised and active enhancers and promoters. Recent studies show that members of a family of lineage-restricted transcription factors, the T-box family, also involved in cardiac development, have a conserved ability to interact with both H3K4 methyltransferases and H3K27 demethylases (Lewis *et al.*, 2007; Miller *et al.*, 2008b) and may thereby influence the status of both enhancers and promoters at the chromatin level. This interaction was recently further ratified by the finding that Tbx3, although classed as a general repressor of transcription, can maintain the expression of *Gata6* in embryonic stem cells via inhibition of PRC2 recruitment, an effector of H3K27 trimethylation (Lu *et al.*, 2011). Such an observation could offer up a solution to the question of how some transcription factors like Tbx3 can possess dual and seemingly opposing functions. This illustrates then that chromatin structure, modifications, and transcription are closely interwoven and influenced by the transcription factors that read and respond to its status.

The DNA sequence, chromatin organization, and its various modifications provide the active blueprint that is available at a given moment. However, this blueprint remains dynamic, a feature which will be retained to a certain degree after birth. Lineage decisions made during development rely on the correct modular reading and timing of this blueprint to ensure correct patterning of differentiation. Reading, maintenance, and modification of the blueprint are the duty of spatiotemporally regulated transcription factors, coactivators/repressors, and other mediators of transcription.

The association of these factors with their target enhancers, silencers, insulators, and promoters in turn provides us with another tool for the identification of regulatory units that drive gene expression.

7. Transcription Factors Can Be Modified Too

Like histones, transcription factors are also open to modification, which, when it affects the transcriptional capacity of the factor, can be considered a form of epigenetic regulation. Interestingly, some of the factors involved in catalyzing the modifications are the very same factors that also modify histone proteins in the same way.

Posttranslational modifications such as phosphorylation, sumoylation, and acetylation have been reported on numerous transcription factors. Such modifications have been demonstrated to have differing roles and effects on the target protein such as subcellular localization, cofactor binding, and DNA association. Gata4, for example, has been shown to be modified by all three examples given above (Kawamura *et al.*, 2005; Liang *et al.*, 2001; Wang *et al.*, 2004). Like its association with histone modification, the histone acetyltransferase p300 has been shown to bind and acetylated Gata4, thus enhancing its transcriptional activity (Takaya *et al.*, 2008). In a recent study by Trivedi and coworkers, Hdac2 together with the homodomain protein Hopx were shown to regulate acetylation of Gata4. In the absence of these factors, Gata4 becomes hyperacetylated and is associated with an increase in cardiomyocyte proliferation and the upregulation of many known Gata4 target genes (Trivedi *et al.*, 2010). Although the exact mechanism by which the transcriptional activity of Gata4 is modified is not clear, a possible analogy can be drawn from work centering on the structural aspects of DNA recognition by FoxO transcription factors. X-ray data comparing the crystal states of three different DNA elements bound by FoxO1 reveal the effects of Mst1-phosphorylation and p300-acetylation in this case appear to directly alter DNA-binding affinity constants of FoxO1 (Brent *et al.*, 2008). Although the effect of acetylation on Gata4 appears different to that of FoxO1, that is, Gata4 acetylation enhances activity, the crystal structure of Gata reveals the positioning of Lys311, a residue shown to be acetylated (Takaya *et al.*, 2008), to be juxtapositioned within the minor groove of the DNA-binding site. Secondary modification of this residue, therefore, will probably have a direct effect on DNA recognition and binding (personal observation based on Bates *et al.*, 2008). Treatment of ES cells with trichostatin A, an HDAC inhibitor, resulted in an increase in the acetylated form of Gata4 and appeared to augment its DNA-binding capacity. Like Gata4, Nkx2-5 is also subject to secondary modifications that can alter its transcriptional activity. Sumoylation by small ubiquitin-like modifiers has been demonstrated for both Gata4 and Nkx2-5. For Nkx2-5, sumoylation of Lys51, catalyzed by PIAS proteins, appears to alter both DNA-binding and transcription characteristics resulting in enhanced activity (Wang *et al.*, 2008a) possibly by stabilizing protein–protein interactions. Similarly, sumoylation has been shown to function as a molecular switch for myocardin in much the same way as Nkx2-5 and Gata4. If one takes into account that a transcription factor may possess multiple sites for anyone of the mentioned modifications and be modified simultaneously by more than one at a time, this may provide one piece of evidence as to why transcription factor ChIP-seq data often contains many binding sites that appear to diverge considerably from their normal consensus sites as such modifications may alter subcellular localization, DNA-binding characteristics, and protein–protein affinities. This fact obviously greatly complicates *in silico* approaches to finding DNA-binding sites across the genome using pattern searching and may even complicate the interpretation of *in vitro* data provided by routinely used techniques such as electromobility shift assays and promoter cell transfection studies.

8. Transcription Factor Occupancy

TFs are key components of gene regulatory systems and are known to associate with and regulate gene transcription via interaction with gene promoters and enhancers. In this respect, enhancer prediction becomes much more accurate if one combines a number of the features mentioned earlier with transcription factor binding information.

The acquisition of the location and sequence of specific transcription factor binding sites has been radically improved with the introduction of new generation sequencing and its combination with the standard biochemistry of protein precipitation. Following this approach, transcription factors or DNA/chromatin associated proteins or modifications can be specifically immunoprecipitated from a cell-line or tissue of interest and purified with their target DNA still attached. This DNA can then be purified and sequenced. Once aligned to the target genome, a read out is generated of the genome-wide occupancy of the transcription factor of interest (Robertson *et al.*, 2007).

Visel and coworkers were among the first to use this technique to address the issue of identifying tissue-specific enhancers on a genome-wide scale (Blow *et al.*, 2010; Visel *et al.*, 2009b). By targeting the ubiquitous enhancer binding protein p300, several thousands of binding sites across the genome of embryonic mouse brain, limbs, or heart were identified, the majority of which functioned as a regulatory element when tested in a transgenic assay. A strong enrichment of p300 at conserved DNA elements was observed in the tissues tested, although the conservation was much less for p300 occupied sites in heart tissue. This convincingly revealed that considering sequence conservation alone when trying to identify enhancer regions will return a considerable underestimate of the actual numbers of active enhancers in any one tissue. Further, the conservation and functional analysis of enhancers between species, especially if one wants to consider applications that translate to human development and disease, must also take into account variation in transcription factor binding which may occur at both the inter- and intraspecies level (Kasowski *et al.*, 2010).

Although p300 appears a reliable means of identifying enhancers active in a tissue, its ubiquitous nature may make it difficult to identify active enhancers in specific cell populations and lineages. In this respect, a combinatorial approach of p300 with one or more transcription factor of interest, particularly if the transcription factor has a defined function in a particular tissue, could prove the most effective in identifying active enhancers involved in a definitive process such as heart development. Boogerd and coworkers used this approach to identify an evolutionary conserved enhancer in intron 1 of the gap junction protein Cx43 encoding gene (*Gja1*) bound by Tbx3, Sox4, and p300 (Boogerd *et al.*, 2011). In a much broader study, He *et al.* (2011) recently published a genome-wide ChIP-seq study in HL-1 cells (cardiac cell line) using the cardiac developmental transcription factors Tbx5, Nkx2.5, Gata4, Srf, and Mef2a. Analysis of regions, bound by multiple factors, revealed a significant proportion of these sites to be able to drive luciferase expression *in vitro* and a transgene in the heart *in vivo*, establishing their function as *bona fide* enhancer elements. To exemplify the accuracy and thus use of this combinatorial approach, we turn back to the previous description of the regulatory elements that drive expression of *Nkx2-5* in the early heart (Lien *et al.*, 1999). Figure 6.3 shows UCSC genome plot of the region approximating to that studied by the Olson lab. In this figure, four separate sets of data appear, one showing the conservation of this genomic region across vertebrate species and three others displaying cumulative data from three ChIP-seq experiments including our labs own unpublished *in vivo* heart data (our unpublished observations). The identified upstream enhancer region is marked in all data sets; the position of the proposed Gata4 binding site exactly matching that discovered using an *in vivo* ChIP-seq approach in adult mouse heart. However, it is worth pointing out that although positionally we can now identify enhancers across the genome in a given tissue, linking the enhancer to the correct downstream targets remains to be resolved.

The issue of transcription factor co-occupancy also brings a point to light which has been discussed in many articles dealing with transcriptional regulation, the key issue of protein–protein interactions and their combinatorial relevance to the complexity of development. The discovery of several key cardiac developmental protein interactions, such as Nkx2-5 and Tbx5 (Hiroi *et al.*, 2001), prior to ChIP-seq technologies already demonstrated that transcription factors do not function alone, but rather as a complex. Although not discussed here, these interactions are often diverse, themselves being governed by their own spatiotemporal expression patterns and those of their interaction partners. Further, interactions between key cardiac transcription factors and the proteins that govern and modulate chromatin structure, such as the BAF complex (Lickert *et al.*, 2004), demonstrate that we need to embrace and learn to understand all the rules that govern gene regulation if we intend to get to grips with the complexities of not only heart development, but of development as a whole.

9. Conclusions

The past few decades have brought a vast increase into our understanding of heart development. We now know which progenitor cells form which components, when and how they interact and differentiate. Many of the molecular players have been uncovered. However, as our work progresses we find that understanding the mechanisms underlying localized differentiation is becoming extremely difficult due to their complex nature, their propensity to change, and shift during the very processes that drive differentiation. Our insights into the mechanisms of gene regulation have increased dramatically as well, and new genome-wide and high-throughput methods to study gene regulation and epigenetics are developing rapidly. The time is ripe to start using these tools to uncover the mechanisms of cell-specific gene regulation in the developing heart. This will enable us to gradually generate a four-dimensional picture of mechanisms that regulate heart development, and may, in the end, give us very useful clues as to how we can steer this process or understand why and how it sometimes goes wrong.

References

Aanhaanen, W. T., Brons, J. F., Dominguez, J. N., Rana, M. S., Norden, J., Airik, R., Wakker, V., de Gier-de Vries, C., Brown, N. A., Kispert, A., Moorman, A. F., and Christoffels, V. M. (2009). The Tbx2+ primary myocardium of the atrioventricular canal forms the atrioventricular node and the base of the left ventricle. *Circ. Res.* **104,** 1267.

Aanhaanen, W. T., Mommersteeg, M. T., Norden, J., Wakker, V., de Gier-de Vries, C., Anderson, R. H., Kispert, A., Moorman, A. F., and Christoffels, V. M. (2010). Developmental origin, growth, and three-dimensional architecture of the atrioventricular conduction axis of the mouse heart. *Circ. Res.* **107,** 728–736.

Abu-Issa, R., and Kirby, M. L. (2007). Heart field: From mesoderm to heart tube. *Annu. Rev. Cell Dev. Biol.* **23,** 45–68.

Bajolle, F., Zaffran, S., Kelly, R. G., Hadchouel, J., Bonnet, D., Brown, N. A., and Buckingham, M. E. (2006). Rotation of the myocardial wall of the outflow tract is implicated in the normal positioning of the great arteries. *Circ. Res.* **98,** 421–428.

Bakker, M. L., Boukens, B. J., Mommersteeg, M. T. M., Brons, J. F., Wakker, V., Moorman, A. F. M., and Christoffels, V. M. (2008). Transcription factor Tbx3 is required for the specification of the atrioventricular conduction system. *Circ. Res.* **102,** 1340–1349.

Barski, A., Cuddapah, S., Cui, K., Roh, T. Y., Schones, D. E., Wang, Z., Wei, G., Chepelev, I., and Zhao, K. (2007). High-resolution profiling of histone methylations in the human genome. *Cell* **129,** 823–837.

Bates, D. L., Chen, Y., Kim, G., Guo, L., and Chen, L. (2008). Crystal structures of multiple GATA zinc fingers bound to DNA reveal new insights into DNA recognition and self-association by GATA. *J. Mol. Biol.* **381,** 1292–1306.

Blow, M. J., McCulley, D. J., Li, Z., Zhang, T., Akiyama, J. A., Holt, A., Plajzer-Frick, I., Shoukry, M., Wright, C., Chen, F., Afzal, V., Bristow, J., *et al.* (2010). ChIP-Seq identification of weakly conserved heart enhancers. *Nat. Genet.* **42,** 806–810.

Boogerd, C. J., Wong, L. Y., van den Boogaard, M., Bakker, M. L., Tessadori, F., Bakkers, J., 't Hoen, P. A., Moorman, A. F., Christoffels, V. M., and Barnett, P. (2011). Sox4 mediates Tbx3 transcriptional regulation of the gap junction protein Cx43. *Cell. Mol. Life Sci.* **68,** 3949–3961.

Boyle, A. P., Song, L., Lee, B. K., London, D., Keefe, D., Birney, E., Iyer, V. R., Crawford, G. E., and Furey, T. S. (2011). High-resolution genome-wide in vivo footprinting of diverse transcription factors in human cells. *Genome Res.* **21,** 456–464.

Brade, T., Gessert, S., Kuhl, M., and Pandur, P. (2007). The amphibian second heart field: Xenopus islet-1 is required for cardiovascular development. *Dev. Biol.* **311,** 297–310.

Brent, M. M., Anand, R., and Marmorstein, R. (2008). Structural basis for DNA recognition by FoxO1 and its regulation by posttranslational modification. *Structure* **16,** 1407–1416.

Bruneau, B. G. (2010). Chromatin remodeling in heart development. *Curr. Opin. Genet. Dev.* **20,** 505–511.

Bruneau, B. G., Nemer, G., Schmitt, J. P., Charron, F., Robitaille, L., Caron, S., Conner, D. A., Gessler, M., Nemer, M., Seidman, C. E., and Seidman, J. G. (2001). A murine model of Holt-Oram syndrome defines roles of the T-box transcription factor Tbx5 in cardiogenesis and disease. *Cell* **106,** 709–721.

Buckingham, M., Meilhac, S., and Zaffran, S. (2005). Building the mammalian heart from two sources of myocardial cells. *Nat. Rev. Genet.* **6,** 826–837.

Cai, C. L., Liang, X., Shi, Y., Chu, P. H., Pfaff, S. L., Chen, J., and Evans, S. (2003). Isl1 identifies a cardiac progenitor population that proliferates prior to differentiation and contributes a majority of cells to the heart. *Dev. Cell* **5,** 877–889.

Cai, C. L., Martin, J. C., Sun, Y., Cui, L., Wang, L., Ouyang, K., Yang, L., Bu, L., Liang, X., Zhang, X., Stallcup, W. B., Denton, C. P., *et al.* (2008). A myocardial lineage derives from Tbx18 epicardial cells. *Nature* **454,** 104–108.

Chi, X., Zhang, S. X., Yu, W., DeMayo, F. J., Rosenberg, S. M., and Schwartz, R. J. (2003). Expression of Nkx2-5-GFP bacterial artificial chromosome transgenic mice closely resembles endogenous Nkx2-5 gene activity. *Genesis* **35,** 220–226.

Chi, X., Chatterjee, P. K., Wilson, W., III, Zhang, S. X., DeMayo, F. J., and Schwartz, R. J. (2005). Complex cardiac Nkx2-5 gene expression activated by noggin-sensitive enhancers followed by chamber-specific modules. *Proc. Natl. Acad. Sci. USA* **102,** 13490–13495.

Christoffels, V. M., and Moorman, A. F. (2009). Development of the cardiac conduction system: Why are some regions of the heart more arrhythmogenic than others? *Circ. Arrhythm. Electrophysiol.* **2,** 195–207.

Christoffels, V. M., Habets, P. E. M. H., Franco, D., Campione, M., de Jong, F., Lamers, W. H., Bao, Z. Z., Palmer, S., Biben, C., Harvey, R. P., and Moorman, A. F. M. (2000). Chamber formation and morphogenesis in the developing mammalian heart. *Dev. Biol.* **223,** 266–278.

Christoffels, V. M., Hoogaars, W. M. H., Tessari, A., Clout, D. E. W., Moorman, A. F. M., and Campione, M. (2004). T-box transcription factor Tbx2 represses differentiation and formation of the cardiac chambers. *Dev. Dyn.* **229,** 763–770.

Christoffels, V. M., Mommersteeg, M. T. M., Trowe, M. O., Prall, O. W. J., de Gier-de Vries, C., Soufan, A. T., Bussen, M., Schuster-Gossler, K., Harvey, R. P., Moorman, A. F. M., and Kispert, A. (2006). Formation of the venous pole of the heart from an *Nkx2-5*-negative precursor population requires Tbx18. *Circ. Res.* **98,** 1555–1563.

Christoffels, V. M., Grieskamp, T., Norden, J., Mommersteeg, M. T., Rudat, C., and Kispert, A. (2009). Tbx18 and the fate of epicardial progenitors. *Nature* **458,** E8–E9.

Christoffels, V. M., Smits, G. J., Kispert, A., and Moorman, A. F. (2010). Development of the pacemaker tissues of the heart. *Circ. Res.* **106,** 240–254.

Davis, D. L., Edwards, A. V., Juraszek, A. L., Phelps, A., Wessels, A., and Burch, J. B. (2001). A GATA-6 gene heart-region-specific enhancer provides a novel means to mark and probe a discrete component of the mouse cardiac conduction system. *Mech. Dev.* **108,** 105–119.

De la Cruz, M. V., Sanchez Gomez, C., Arteaga, M. M., and Arguëllo, C. (1977). Experimental study of the development of the truncus and the conus in the chick embryo. *J. Anat.* **123,** 661–686.

De la Cruz, M. V., Sánchez-Gómez, C., and Palomino, M. (1989). The primitive cardiac regions in the straight tube heart (stage 9) and their anatomical expression in the mature heart: An experimental study in the chick embryo. *J. Anat.* **165,** 121–131.

Dean, A. (2011). In the loop: Long range chromatin interactions and gene regulation. *Brief. Funct. Genomics* **10,** 3–10.

Deaton, A. M., and Bird, A. (2011). CpG islands and the regulation of transcription. *Genes Dev.* **25,** 1010–1022.

Di Lisi, R., Sandri, C., Franco, D., Ausoni, S., Moorman, A. F. M., and Schiaffino, S. (2000). An atrioventricular canal domain defined by cardiac troponin I transgene expression in the embryonic myocardium. *Anat. Embryol. (Berl.)* **202,** 95–101.

Dyer, L. A., and Kirby, M. L. (2009). The role of secondary heart field in cardiac development. *Dev. Biol.* **336,** 137–144.

Felsenfeld, G., and Groudine, M. (2003). Controlling the double helix. *Nature* **421,** 448–453.

Franco, D., Meilhac, S. M., Christoffels, V. M., Kispert, A., Buckingham, M., and Kelly, R. G. (2006). Left and right ventricular contributions to the formation of the interventricular septum in the mouse heart. *Dev. Biol.* **294,** 366–375.

Frankel, N., Davis, G. K., Vargas, D., Wang, S., Payre, F., and Stern, D. L. (2010). Phenotypic robustness conferred by apparently redundant transcriptional enhancers. *Nature* **466,** 490–493.

Galli, D., Dominguez, J. N., Zaffran, S., Munk, A., Brown, N. A., and Buckingham, M. E. (2008). Atrial myocardium derives from the posterior region of the second heart field, which acquires left-right identity as Pitx2c is expressed. *Development* **135,** 1157–1167.

Garg, V., Kathiriya, I. S., Barnes, R., Schluterman, M. K., King, I. N., Butler, C. A., Rothrock, C. R., Eapen, R. S., Hirayama-Yamada, K., Joo, K., Matsuoka, R., Cohen, J. C., *et al.* (2003). GATA4 mutations cause human congenital heart defects and reveal an interaction with TBX5. *Nature* **424,** 443–447.

Goddeeris, M. M., Rho, S., Petiet, A., Davenport, C. L., Johnson, G. A., Meyers, E. N., and Klingensmith, J. (2008). Intracardiac septation requires hedgehog-dependent cellular contributions from outside the heart. *Development* **135,** 1887–1895.

Gray, S., and Levine, M. (1996). Transcriptional repression in development. *Curr. Opin. Cell Biol.* **8,** 358–364.

Habets, P. E. M. H., Moorman, A. F. M., Clout, D. E. W., van Roon, M. A., Lingbeek, M., Lohuizen, M., Campione, M., and Christoffels, V. M. (2002). Cooperative action of Tbx2 and Nkx2.5 inhibits ANF expression in the atrioventricular canal: Implications for cardiac chamber formation. *Genes Dev.* **16,** 1234–1246.

Habets, P. E. M. H., Moorman, A. F. M., and Christoffels, V. M. (2003). Regulatory modules in the developing heart. *Cardiovasc. Res.* **58,** 246–263.

Harrelson, Z., Kelly, R. G., Goldin, S. N., Gibson-Brown, J. J., Bollag, R. J., Silver, L. M., and Papaioannou, V. E. (2004). Tbx2 is essential for patterning the atrioventricular canal and for morphogenesis of the outflow tract during heart development. *Development* **131,** 5041–5052.

He, M.l, Wen, L., Campbell, C. E., Wu, J. Y., and Rao, Y. (1999). Transcription repression by Xenopus ET and its human ortholog TBX3, a gene involved in ulnar-mammary syndrome. *Proc. Natl. Acad. Sci. USA* **96,** 10212–10217.

He, A., Kong, S. W., Ma, Q., and Pu, W. T. (2011). Co-occupancy by multiple cardiac transcription factors identifies transcriptional enhancers active in heart. *Proc. Natl. Acad. Sci. USA* **108,** 5632–5637.

Hiroi, Y., Kudoh, S., Monzen, K., Ikeda, Y., Yazaki, Y., Nagai, R., and Komuro, I. (2001). Tbx5 associates with Nkx2-5 and synergistically promotes cardiomyocyte differentiation. *Nat. Genet.* **28,** 276–280.

Hoffman, J. I., Kaplan, S., and Liberthson, R. R. (2004). Prevalence of congenital heart disease. *Am. Heart J.* **147,** 425–439.

Hong, J. W., Hendrix, D. A., and Levine, M. S. (2008). Shadow enhancers as a source of evolutionary novelty. *Science* **321,** 1314.

Hoogaars, W. M. H., Tessari, A., Moorman, A. F. M., de Boer, P. A. J., Hagoort, J., Soufan, A. T., Campione, M., and Christoffels, V. M. (2004). The transcriptional repressor Tbx3 delineates the developing central conduction system of the heart. *Cardiovasc. Res.* **62,** 489–499.

Hoogaars, W. M., Engel, A., Brons, J. F., Verkerk, A. O., de Lange, F. J., Wong, L. Y., Bakker, M. L., Clout, D. E., Wakker, V., Barnett, P., Ravesloot, J. H., Moorman, A. F., *et al.* (2007). Tbx3 controls the sinoatrial node gene program and imposes pacemaker function on the atria. *Genes Dev.* **21,** 1098–1112.

Horsthuis, T., Houweling, A. C., Habets, P. E. M. H., de Lange, F. J., el Azzouzi, H., Clout, D. E. W., Moorman, A. F. M., and Christoffels, V. M. (2008). Distinct regulation of developmental and heart disease induced atrial natriuretic factor expression by two separate distal sequence. *Circ. Res.* **102,** 849–859.

Houweling, A. C., van Borren, M. M., Moorman, A. F. M., and Christoffels, V. M. (2005). Expression and regulation of the atrial natriuretic factor encoding gene *Nppa* during development and disease. *Cardiovasc. Res.* **67,** 583–593.

Huang, W. Y., and Liew, C. C. (1998). Chromatin remodelling of the cardiac beta-myosin heavy chain gene. *Biochem. J.* **330**(Pt 2), 871–876.

Huang, W. Y., Chen, J. J., Shih, N., and Liew, C. C. (1997). Multiple muscle-specific regulatory elements are associated with a DNase I hypersensitive site of the cardiac beta-myosin heavy-chain gene. *Biochem. J.* **327**(Pt 2), 507–512.

Hutson, M. R., and Kirby, M. L. (2007). Model systems for the study of heart development and disease. Cardiac neural crest and conotruncal malformations. *Semin. Cell Dev. Biol.* **18,** 101–110.

Kaneda, R., Takada, S., Yamashita, Y., Choi, Y. L., Nonaka-Sarukawa, M., Soda, M., Misawa, Y., Isomura, T., Shimada, K., and Mano, H. (2009). Genome-wide histone methylation profile for heart failure. *Genes Cells* **14,** 69–77.

Kappen, C., and Salbaum, J. M. (2009). Identification of regulatory elements in the Isl1 gene locus. *Int. J. Dev. Biol.* **53,** 935–946.

Kasowski, M., Grubert, F., Heffelfinger, C., Hariharan, M., Asabere, A., Waszak, S. M., Habegger, L., Rozowsky, J., Shi, M., Urban, A. E., Hong, M. Y., Karczewski, K. J., *et al.* (2010). Variation in transcription factor binding among humans. *Science* **328,** 232–235.

Kawamura, T., Ono, K., Morimoto, T., Wada, H., Hirai, M., Hidaka, K., Morisaki, T., Heike, T., Nakahata, T., Kita, T., and Hasegawa, K. (2005). Acetylation of GATA-4 is involved in the differentiation of embryonic stem cells into cardiac myocytes. *J. Biol. Chem.* **280,** 19682–19688.

Kelly, R. G., and Buckingham, M. E. (2002). The anterior heart-forming field: Voyage to the arterial pole of the heart. *Trends Genet.* **18,** 210–216.

Kelly, R. G., Brown, N. A., and Buckingham, M. E. (2001). The arterial pole of the mouse heart forms from Fgf10-expressing cells in pharyngeal mesoderm. *Dev. Cell* **1,** 435–440.

Knowlton, K. U., Rockman, H. A., Itani, M., Vovan, A., Seidman, C. E., and Chien, K. R. (1995). Divergent pathways mediate the induction of ANF transgenes in neonatal and hypertrophic ventricular myocardium. *J. Clin. Invest.* **96,** 1311–1318.

Lamers, W. H., and Moorman, A. F. M. (2002). Cardiac septation: A late contribution of the embryonic primary myocardium to heart morphogenesis. *Circ. Res.* **91,** 93–103.

Lanzuolo, C., Roure, V., Dekker, J., Bantignies, F., and Orlando, V. (2007). Polycomb response elements mediate the formation of chromosome higher-order structures in the bithorax complex. *Nat. Cell Biol.* **9,** 1167–1174.

Lettice, L. A., Heaney, S. J., Purdie, L. A., Li, L., de, B. P., Oostra, B. A., Goode, D., Elgar, G., Hill, R. E., and de, G. E. (2003). A long-range Shh enhancer regulates expression in the developing limb and fin and is associated with preaxial polydactyly. *Hum. Mol. Genet.* **12,** 1725–1735.

Lewis, M. D., Miller, S. A., Miazgowicz, M. M., Beima, K. M., and Weinmann, A. S. (2007). T-bet's ability to regulate individual target genes requires the conserved T-box domain to recruit histone methyltransferase activity and a separate family member-specific transactivation domain. *Mol. Cell. Biol.* **27,** 8510–8521.

Liang, Q., Wiese, R. J., Bueno, O. F., Dai, Y. S., Markham, B. E., and Molkentin, J. D. (2001). The transcription factor GATA4 is activated by extracellular signal-regulated kinase 1- and 2-mediated phosphorylation of serine 105 in cardiomyocytes. *Mol. Cell. Biol.* **21,** 7460–7469.

Lickert, H., Takeuchi, J. K., Von I, B., Walls, J. R., McAuliffe, F., Von, B., Adamson, S. L., Henkelman, R. M., Wrana, J. L., Rossant, J., and Bruneau, B. G. (2004). Baf60c is essential for function of BAF chromatin remodelling complexes in heart development. *Nature* **432,** 107–112.

Lien, C. L., Wu, C., Mercer, B., Webb, R., Richardson, J. A., and Olson, E. N. (1999). Control of early cardiac-specific transcription of Nkx2-5 by a GATA-dependent enhancer. *Development* **126,** 75–84.

Lindsay, E. A., Vitelli, F., Su, H., Morishima, M., Huynh, T., Pramparo, T., Jurecic, V., Ogunrinu, G., Sutherland, H. F., Scambler, P. J., Bradley, A., and Baldini, A. (2001). Tbx1 haploinsufficieny in the DiGeorge syndrome region causes aortic arch defects in mice. *Nature* **410,** 97–101.

Lu, R., Yang, A., and Jin, Y. (2011). Dual functions of T-box 3 (Tbx3) in the control of self-renewal and extraembryonic endoderm differentiation in mouse embryonic stem cells. *J. Biol. Chem.* **286,** 8425–8436.

Lund, J., Chen, F., Hua, A., Roe, B., Budarf, M., Emanuel, B. S., and Reeves, R. H. (2000). Comparative sequence analysis of 634 kb of the mouse chromosome 16 region of conserved synteny with the human velocardiofacial syndrome region on chromosome 22q11.2. *Genomics* **63,** 374–383.

Ma, Q., Zhou, B., and Pu, W. T. (2008). Reassessment of Isl1 and Nkx2-5 cardiac fate maps using a Gata4-based reporter of Cre activity. *Dev. Biol.* **323,** 98–104.

Meireles-Filho, A. C., and Stark, A. (2009). Comparative genomics of gene regulation-conservation and divergence of cis-regulatory information. *Curr. Opin. Genet. Dev.* **19,** 565–570.

Merscher, S., Funke, B., Epstein, J. A., Heyer, J., Puech, A., Lu, M. M., Xavier, R. J., Demay, M. B., Russell, R. G., Factor, S., Tokooya, K., Jore, B. S., *et al.* (2001). TBX1 is responsible for cardiovascular defects in velo-cardio-facial/DiGeorge syndrome. *Cell* **104,** 619–629.

Mikawa, T. (1999). Cardiac lineages. *In* "Heart Development," (R. P. Harvey and N. Rosenthal, eds.), pp. 19–33. Sydney en Boston, Academic Press.

Miller, K. A., Davidson, S., Liaros, A., Barrow, J., Lear, M., Heine, D., Hoppler, S., and MacKenzie, A. (2008a). Prediction and characterisation of a highly conserved, remote and cAMP responsive enhancer that regulates Msx1 gene expression in cardiac neural crest and outflow tract. *Dev. Biol.* **317,** 686–694.

Miller, S. A., Huang, A. C., Miazgowicz, M. M., Brassil, M. M., and Weinmann, A. S. (2008b). Coordinated but physically separable interaction with H3K27-demethylase and H3K4-methyltransferase activities are required for T-box protein-mediated activation of developmental gene expression. *Genes Dev.* **22,** 2980–2993.

Miquerol, L., Moreno-Rascon, N., Beyer, S., Dupays, L., Meilhac, S. M., Buckingham, M. E., Franco, D., and Kelly, R. G. (2010). Biphasic development of the mammalian ventricular conduction system. *Circ. Res.* **107,** 153–161.

Miquerol, L., Beyer, S., and Kelly, R. G. (2011). Establishment of the mouse ventricular conduction system. *Cardiovasc. Res.* **91,** 232–242.

Mommersteeg, M. T. M., Hoogaars, W. M. H., Prall, O. W. J., de Gier-de Vries, C., Wiese, C., Clout, D. E. W., Papaioannou, V. E., Brown, N. A., Harvey, R. P., Moorman, A. F. M., and Christoffels, V. M. (2007). Molecular pathway for the localized formation of the sinoatrial node. *Circ. Res.* **100,** 354–362.

Mommersteeg, M. T., Dominguez, J. N., Wiese, C., Norden, J., de Gier-de, V. C., Burch, J. B., Kispert, A., Brown, N. A., Moorman, A. F., and Christoffels, V. M. (2010). The sinus venosus progenitors separate and diversify from the first and second heart fields early in development. *Cardiovasc. Res.* **87,** 92–101.

Moorman, A. F. M., and Christoffels, V. M. (2003). Cardiac chamber formation: Development, genes and evolution. *Physiol. Rev.* **83,** 1223–1267.

Moorman, A. F. M., Christoffels, V. M., Anderson, R. H., and van den Hoff, M. J. B. (2007). The heart-forming fields: One or multiple? *Philos. Trans. R. Soc. Lond. B Biol. Sci.* **362,** 1257–1265.

Munshi, N. V., McAnally, J., Bezprozvannaya, S., Berry, J. M., Richardson, J. A., Hill, J. A., and Olson, E. N. (2009). Cx30.2 enhancer analysis identifies Gata4 as a novel regulator of atrioventricular delay. *Development* **136,** 2665–2674.

Noordermeer, D., de, W. E., Klous, P., van de Werken, H., Simonis, M., Lopez-Jones, M., Eussen, B., de, K. A., Singer, R. H., and de, L. W. (2011). Variegated gene expression caused by cell-specific long-range DNA interactions. *Nat. Cell Biol.* **13,** 944–951.

Nowotschin, S., Liao, J., Gage, P. J., Epstein, J. A., Campione, M., and Morrow, B. E. (2006). Tbx1 affects asymmetric cardiac morphogenesis by regulating Pitx2 in the secondary heart field. *Development* **133,** 1565–1573.

Pennacchio, L. A., Ahituv, N., Moses, A. M., Prabhakar, S., Nobrega, M. A., Shoukry, M., Minovitsky, S., Dubchak, I., Holt, A., Lewis, K. D., Plajzer-Frick, I., Akiyama, J., *et al.* (2006). In vivo enhancer analysis of human conserved non-coding sequences. *Nature* **444,** 499–502.

Pennisi, D. J., Rentschler, S., Gourdie, R. G., Fishman, G. I., and Mikawa, T. (2002). Induction and patterning of the cardiac conduction system. *Int. J. Dev. Biol.* **46,** 765–775.

Phillips, J. E., and Corces, V. G. (2009). CTCF: Master weaver of the genome. *Cell* **137,** 1194–1211.

Pierpont, M. E., Basson, C. T., Benson, D. W., Jr., Gelb, B. D., Giglia, T. M., Goldmuntz, E., McGee, G., Sable, C. A., Srivastava, D., and Webb, C. L. (2007). Genetic basis for congenital heart defects: Current knowledge: A scientific statement from the American Heart Association Congenital Cardiac Defects Committee, Council on Cardiovascular Disease in the Young: Endorsed by the American Academy of Pediatrics. *Circulation* **115,** 3015–3038.

Prall, O. W. J., Menon, M. K., Solloway, M. J., Watanabe, Y., Zaffran, S., Bajolle, F., Biben, C., McBride, J. J., Robertson, B. R., Chaulet, H., Stennard, F. A., Wise, N., *et al.* (2007). An Nkx2-5/Bmp2/Smad1 negative feedback loop controls heart progenitor specification and proliferation. *Cell* **128,** 947–959.

Rada-Iglesias, A., Bajpai, R., Swigut, T., Brugmann, S. A., Flynn, R. A., and Wysocka, J. (2011). A unique chromatin signature uncovers early developmental enhancers in humans. *Nature* **470,** 279–283.

Rana, M. S., Horsten, N. C. A., Tesink-Taekema, S., Lamers, W. H., Moorman, A. F. M., and van den Hoff, M. J. B. (2007). Trabeculated right ventricular free wall in the chicken heart forms by ventricularization of the myocardium initially forming the outflow tract. *Circ. Res.* **100,** 1000–1007.

Ritter, D. I., Li, Q., Kostka, D., Pollard, K. S., Guo, S., and Chuang, J. H. (2010). The importance of being cis: Evolution of orthologous fish and mammalian enhancer activity. *Mol. Biol. Evol.* **27,** 2322–2332.

Robertson, G., Hirst, M., Bainbridge, M., Bilenky, M., Zhao, Y., Zeng, T., Euskirchen, G., Bernier, B., Varhol, R., Delaney, A., Thiessen, N., Griffith, O. L., *et al.* (2007). Genome-wide profiles of STAT1 DNA association using chromatin immunoprecipitation and massively parallel sequencing. *Nat. Methods* **4,** 651–657.

Rochais, F., Mesbah, K., and Kelly, R. G. (2009). Signaling pathways controlling second heart field development. *Circ. Res.* **104,** 933–942.

Roh, T. Y., Wei, G., Farrell, C. M., and Zhao, K. (2007). Genome-wide prediction of conserved and nonconserved enhancers by histone acetylation patterns. *Genome Res.* **17,** 74–81.

Rosenbloom, K. R., Dreszer, T. R., Pheasant, M., Barber, G. P., Meyer, L. R., Pohl, A., Raney, B. J., Wang, T., Hinrichs, A. S., Zweig, A. S., Fujita, P. A., Learned, K., Rhead, B., Smith, K. E., Kuhn, R. M., Karolchik, D., Haussler, D., and Kent, W. J. (2010). ENCODE whole-genome data in the UCSC Genome Browser. *Nucleic Acids Res.* **38,** D620–D625.

Sagai, T., Hosoya, M., Mizushina, Y., Tamura, M., and Shiroishi, T. (2005). Elimination of a long-range cis-regulatory module causes complete loss of limb-specific Shh expression and truncation of the mouse limb. *Development* **132,** 797–803.

Schwartz, R. J., and Olson, E. N. (1999). Building the heart piece by piece: Modularity of cis-elements regulating Nkx2-5 transcription. *Development* **126,** 4187–4192.

Searcy, R. D., Vincent, E. B., Liberatore, C. M., and Yutzey, K. E. (1998). A GATA-dependent *nkx-2.5* regulatory element activates early cardiac gene expression in transgenic mice. *Development* **125,** 4461–4470.

Shirai, M., Imanaka-Yoshida, K., Schneider, M. D., Schwartz, R. J., and Morisaki, T. (2009). T-box 2, a mediator of Bmp-Smad signaling, induced hyaluronan synthase 2 and Tgfbeta2 expression and endocardial cushion formation. *Proc. Natl. Acad. Sci. USA* **106,** 18604–18609.

Simonis, M., Kooren, J., and de, L. W. (2007). An evaluation of 3C-based methods to capture DNA interactions. *Nat. Methods* **4,** 895–901.

Singh, R., Horsthuis, T., Farin, H. F., Grieskamp, T., Norden, J., Petry, M., Wakker, V., Moorman, A. F., Christoffels, V. M., and Kispert, A. (2009). Tbx20 interacts with smads to confine tbx2 expression to the atrioventricular canal. *Circ. Res.* **105,** 442–452.

Small, E. M., and Krieg, P. A. (2003). Transgenic analysis of the atrialnatriuretic factor (ANF) promoter: Nkx2-5 and GATA-4 binding sites are required for atrial specific expression of ANF. *Dev. Biol.* **261,** 116–131.

Snarr, B. S., O'Neal, J. L., Chintalapudi, M. R., Wirrig, E. E., Phelps, A. L., Kubalak, S. W., and Wessels, A. (2007). Isl1 expression at the venous pole identifies a novel role for the second heart field in cardiac development. *Circ. Res.* **101,** 971–974.

Stein, A. B., Jones, T. A., Herron, T. J., Patel, S. R., Day, S. M., Noujaim, S. F., Milstein, M. L., Klos, M., Furspan, P. B., Jalife, J., and Dressler, G. R. (2011). Loss of H3K4 methylation destabilizes gene expression patterns and physiological functions in adult murine cardiomyocytes. *J. Clin. Invest.* **121,** 2641–2650.

Sun, Y., Liang, Y., Najafi, N., Cass, M., Lin, L., Cai, C., Chen, J., and Evans, S. (2006). Islet 1 is expressed in distinct cardiovascular lineages, including pacemaker and coronary vascular cells. *Dev. Biol.* **304**(1), 286–296.

Takaya, T., Kawamura, T., Morimoto, T., Ono, K., Kita, T., Shimatsu, A., and Hasegawa, K. (2008). Identification of p300-targeted acetylated residues in GATA4 during hypertrophic responses in cardiac myocytes. *J. Biol. Chem.* **283,** 9828–9835.

Tena, J. J., Alonso, M. E., Calle-Mustienes, E., Splinter, E., de, L. W., Manzanares, M., and Gomez-Skarmeta, J. L. (2011). An evolutionarily conserved three-dimensional structure in the vertebrate Irx clusters facilitates enhancer sharing and coregulation. *Nat. Commun.* **2,** 310.

Trivedi, C. M., Zhu, W., Wang, Q., Jia, C., Kee, H. J., Li, L., Hannenhalli, S., and Epstein, J. A. (2010). Hopx and Hdac2 interact to modulate Gata4 acetylation and embryonic cardiac myocyte proliferation. *Dev. Cell* **19,** 450–459.

van den Berg, G., and Moorman, A. F. (2009). Concepts of cardiac development in retrospect. *Pediatr. Cardiol.* **30,** 580–587.

van den Berg, G., Abu-Issa, R., de Boer, B. A., Hutson, M. R., de Boer, P. A., Soufan, A. T., Ruijter, J. M., Kirby, M. L., van den Hoff, M. J., and Moorman, A. F. (2009). A caudal proliferating growth center contributes to both poles of the forming heart tube. *Circ. Res.* **104,** 179–188.

van Mierop, L. H. S. (1967). Localization of pacemaker in chick embryo heart at the time of initiation of heartbeat. *Am. J. Physiol.* **212,** 407–415.

van Wijk, B., van den Berg, G., Abu-Issa, R., Barnett, P., van der Velden, S., Schmidt, M., Ruijter, J. M., Kirby, M. L., Moorman, A. F., and van den Hoff, M. J. (2009). Epicardium and myocardium separate from a common precursor pool by crosstalk between bone morphogenetic protein- and fibroblast growth factor-signaling pathways. *Circ. Res.* **105,** 431–441.

Verzi, M. P., McCulley, D. J., De, V. S., Dodou, E., and Black, B. L. (2005). The right ventricle, outflow tract, and ventricular septum comprise a restricted expression domain within the secondary/anterior heart field. *Dev. Biol.* **287,** 134–145.

Virágh, Sz, and Challice, C. E. (1977). The development of the conduction system in the mouse embryo heart. II. Histogenesis of the atrioventricular node and bundle. *Dev. Biol.* **56,** 397–411.

Virágh, Sz, and Challice, C. E. (1982). The development of the conduction system in the mouse embryo heart. IV. Differentiation of the atrioventricular conduction system. *Dev. Biol.* **89,** 25–40.

Visel, A., Akiyama, J. A., Shoukry, M., Afzal, V., Rubin, E. M., and Pennacchio, L. A. (2009a). Functional autonomy of distant-acting human enhancers. *Genomics* **93,** 509–513.

Visel, A., Blow, M. J., Li, Z., Zhang, T., Akiyama, J. A., Holt, A., Plajzer-Frick, I., Shoukry, M., Wright, C., Chen, F., Afzal, V., Ren, B., *et al.* (2009b). ChIP-seq accurately predicts tissue-specific activity of enhancers. *Nature* **457,** 854–858.

Wang, J., Feng, X. H., and Schwartz, R. J. (2004). SUMO-1 modification activated GATA4-dependent cardiogenic gene activity. *J. Biol. Chem.* **279,** 49091–49098.

Wang, J., Zhang, H., Iyer, D., Feng, X. H., and Schwartz, R. J. (2008a). Regulation of cardiac specific Nkx2.5 gene activity by small ubiquitin-like modifier. *J. Biol. Chem.* **283,** 23235–23243.

Wang, Z., Zang, C., Rosenfeld, J. A., Schones, D. E., Barski, A., Cuddapah, S., Cui, K., Roh, T. Y., Peng, W., Zhang, M. Q., and Zhao, K. (2008b). Combinatorial patterns of histone acetylations and methylations in the human genome. *Nat. Genet.* **40,** 897–903.

Wang, Z., Zang, C., Cui, K., Schones, D. E., Barski, A., Peng, W., and Zhao, K. (2009). Genome-wide mapping of HATs and HDACs reveals distinct functions in active and inactive genes. *Cell* **138,** 1019–1031.

Warren, S. A., Terada, R., Briggs, L. E., Cole-Jeffrey, C. T., Chien, W. M., Seki, T., Weinberg, E. O., Yang, T. P., Chin, M. T., Bungert, J., and Kasahara, H. (2011). Differential role of Nkx2–5 in activation of the ANF gene in developing vs. failing heart. *Mol. Cell. Biol.* **31**(22), 4633–4645.

Winter, E. M., and Gittenberger-de Groot, A. C. (2007). Epicardium-derived cells in cardiogenesis and cardiac regeneration. *Cell. Mol. Life Sci.* **64,** 692–703.

Yang, J., and Corces, V. G. (2011). Chromatin insulators: A role in nuclear organization and gene expression. *Adv. Cancer Res.* **110,** 43–76.

Yang, X. W., Model, P., and Heintz, N. (1997). Homologous recombination based modification in Escherichia coli and germline transmission in transgenic mice of a bacterial artificial chromosome. *Nat. Biotechnol.* **15,** 859–865.

Zhou, B., Ma, Q., Rajagopal, S., Wu, S. M., Domian, I., Rivera-Feliciano, J., Jiang, D., von, G. A., Ikeda, S., Chien, K. R., and Pu, W. T. (2008). Epicardial progenitors contribute to the cardiomyocyte lineage in the developing heart. *Nature* **454,** 109–113.

Zhou, v.w., Goren, v.w., and Bernstein, v.w. (2011). Charting histone modifications and the functional organization of mammalian genomes. *Nat. Rev. Genet.* **12,** 7–18.

CHAPTER SEVEN

Heart Valve Development, Maintenance, and Disease: The Role of Endothelial Cells

Ge Tao,* James D. Kotick,† *and* Joy Lincoln‡

Contents

* Molecular Cell and Developmental Biology Graduate Program, Leonard M. Miller School of Medicine, University of Miami, Miami, Florida, USA

† MD Program, Leonard M. Miller School of Medicine, University of Miami, Miami, Florida, USA

‡ Center for Cardiovascular and Pulmonary Research, The Research Institute at Nationwide Children's Hospital, Department of Pediatrics, The Ohio State University, Columbus, Ohio, USA

Current Topics in Developmental Biology, Volume 100

ISSN 0070-2153, DOI: 10.1016/B978-0-12-387786-4.00006-3

Abstract

Heart valves are dynamic structures that open and close during the cardiac cycle to maintain unidirectional blood flow throughout life. Insufficient valve function, commonly due to congenital malformations leads to disruptions in hemodynamics and eventual heart failure. Mature valve leaflets are composed of a heterogeneous population of interstitial cells and stratified extracellular matrix, surrounded by a layer of endothelial cells. This defined connective tissue "architecture" provides the valve with all the necessary biomechanical properties required to efficiently function while withstanding constant cyclic shear stress. Valvular endothelial cells (VECs) play essential roles in establishing the valve structures during embryonic development and are important for maintaining lifelong valve integrity and function. In contrast to a continuous endothelium over the surface of healthy valve leaflets, VEC disruption is commonly observed in malfunctioning valves and is associated with pathological processes that promote valve sclerosis and calcification. Increasing our understanding of the roles of VECs in development and disease has lead to promising advances in the development of endothelial cell-based therapies for treating valve disease.

Abbreviations

AV	atrioventricular
AVC	atrioventricular canal
BAV	bicuspid aortic valve
BMP	bone morphogenetic protein
CJ	cardiac jelly
ECM	extracellular matrix
EMT	endothelial-to-mesenchymal transformation
MVP	mitral valve prolapse
OFT	outflow tract
TGF	transforming growth factor
VEC	valve endothelial cells
VEGF	vascular endothelial growth factor
VIC	valve interstitial cell

1. Introduction

Heart valve disease is a significant contributor of mortality, resulting in approximately 20,000 deaths a year, and the incidence is increasing (Lloyd-Jones *et al.*, 2010). Despite the clinical significance, therapies are limited and the only effective treatment is surgical intervention. In the United States, the direct cost for valve disease has been estimated at $1 billion per year (Rajamannan *et al.*, 2003a). Therefore, the need to develop alternative effective therapies is imperative. Normal heart valve function is largely facilitated by the organization of valve interstitial cells (VICs) and extracellular matrices (ECM) that collectively provide the leaflet with the necessary biomechanical properties to open and close during the cardiac cycle. This structure–function relationship is supported by the controlled interaction of VICs and matrix with a peripheral layer of valve endothelial cells (VECs) that respond to the surrounding hemodynamic environment to establish and maintain valve homeostasis in the embryo and throughout adult life (Balachandran *et al.*, 2011; Hinton and Yutzey, 2011). In contrast to normal valves, disease in the form of valve thickening (sclerosis) and/or calcification is associated with disruption to the valve endothelium resulting in attenuated responses to molecular or hemodynamic cues (Butcher and Nerem, 2007; Ge and Sotiropoulos, 2010; Poggianti *et al.*, 2003). This review will focus on the importance of VEC function for valve development, maturation, and maintenance through embryogenesis and adult life. In addition, the potential of VEC-based therapies as an alternative approach in the treatment of valve disease will be discussed.

2. Heart Valve Structure and Function

2.1. Heart valve function

As the average adult's heart beats around 70 times per minute, heart valves must open and close over 100,000 times a day to maintain unidirectional blood flow through the heart. In healthy individuals, this is achieved by structure–function relationships between the cellular and extracellular components of the mature valve and the hemodynamic environment. There are two sets of cardiac valves: the atrioventricular (AV) valves including the mitral and tricuspid that separate the atria from the ventricles on the left and right sides respectively, and the aortic and pulmonic semilunar valves that separate the ventricles from the great arteries (Hinton and Yutzey, 2011). Although the functional demand of each valve set is similar, their anatomies are different. The AV valves consist of the mitral and tricuspid valve sets made up of two (mitral) or three (tricuspid) valve leaflets and external supporting chordae tendineae that attach the underside of the valve leaflet to the papillary muscles within the ventricle (Anderson *et al.*, 2000). The three leaflets of the semilunar valves (aortic, pulmonic) are referred to as cusps, and lack external support, although a unique supporting structure is apparent within the aortic roots in the form of a fibrous annulus (Anderson, 2000). It is the coordinated movement of these valvular structures that maintain unidirectional blood flow during the cardiac cycle. In diastole, the papillary muscles are relaxed and high pressure in the atrium causes opening of the mitral (left) and tricuspid (right) valve leaflets to promote blood flow into the respective ventricle. Once ventricular pressure increases during diastole, the chordae "pull" the AV valve leaflets closed and maintain coaptation to prevent eversion of the valve into the atria. As the ventricle contracts, blood exits through the now open semilunar valves and the ventricle relaxes to begin the cycle again. Therefore, throughout the cardiac cycle the heart valve structures are exposed to constant changes in hemodynamic force as a result of pressure differences between systole to diastole. To withstand this complex mechanical environment, the valve leaflets/cusps develop and maintain an intricate and highly organized connective tissue system that provides all the necessary biomechanical properties for efficient function.

2.2. Heart valve structure

The mature valve structures are primarily composed of an outer layer of VECs that surround three stratified layers of specialized ECM, interspersed with differentiated VICs (Fig. 7.1) (Garcia-Martinez *et al.*, 1991; Gross and Kugel, 1931; Lincoln *et al.*, 2006b). Each ECM layer is organized according to blood flow and collectively they work together to withstand the continual

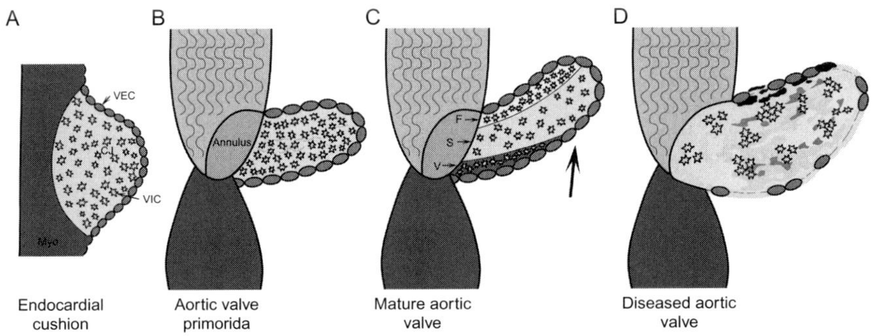

Figure 7.1 Overview of aortic heart valve maturation and disease. (A) Heart valve development is initiated with formation of endocardial cushions following EMT. These primitive valve structures are composed of cardiac jelly (CJ) containing a population of mesenchymal VICs surrounded by a layer of VECs. (B) As embryogenesis proceeds, endocardial cushions elongate to form aortic valve primorida. At this time, VICs begin to differentiate and secrete specialized ECM to give rise to the trilaminar mature aortic valve structure (C) consisting of the fibrosa (F), spongiosa (S), and ventricularis (V) layers. Arrow indicates blood flow. Throughout this process, the valve endothelium is maintained. In contrast to healthy valves, the valve endothelium is disrupted, associated with alterations in VIC and ECM organization. These histopathological often lead to sclerosis and calcification (black lesions) on the fibrosa layer away from blood flow. (D) Myo, myocardium. (For color version of this figure, the reader is referred to the Web version of this chapter.)

changes in hemodynamic flow during the cardiac cycle (Fig. 7.1C) (Balachandran *et al.*, 2011; Lincoln *et al.*, 2006b). The fibrosa layer is located on the ventricular side of the AV valve leaflets and atrial side of the semilunar valves, away from blood flow. This layer is predominantly composed of bundles of collagen fibers aligned along the circumferential direction of the leaflets (Icardo and Colvee, 1995; Kunzelman *et al.*, 1993; Lincoln *et al.*, 2006b; Rabkin-Aikawa *et al.*, 2005). This arrangement provides tensile strength to the valve leaflet during opening, while transmitting forces to promote coaptation of the leaflets in the closed position (Aldous *et al.*, 2009; Balachandran *et al.*, 2011; Grande-Allen and Liao, 2011). Adjacent to the fibrosa is the spongiosa layer, with a lower abundance of collagens and high prevalence of proteoglycans. This composition provides a more compressible matrix, allowing the valve to geometrically "flex" and absorb high force (Lincoln *et al.*, 2006b; Sacks *et al.*, 2009). Finally, the layer adjacent to blood flow is termed the atrialis (AV) or ventricularis (semilunar) and largely consists of radially orientated elastin fibers that facilitate tissue movement by extending as the valve leaflet opens and recoiling during closure (Hinton and Yutzey, 2011; Schoen, 1997; Scott and Vesely, 1995). Further support is provided by the fibrous annulus structure at the connection between the leaflets/cusps and the myocardium (Hinton and Yutzey, 2011; Misfeld and Sievers, 2007), and the external chordae tendineae in the AV position. In addition to the ECM, the VIC and VEC populations play essential roles in maintaining connective tissue homeostasis in the functional valve leaflet. Although originally thought of as a homogenous population of fibroblast-like cells, VICs are now considered highly heterogeneous with quiescent, activated, and progenitor-like phenotypes reported (Liu *et al.*, 2007). Studies have shown that *in vitro*, ECM secretion by activated VICs in response to flow is dependent upon their location within the valve leaflet. For example, VICs isolated from the center of the anterior mitral valve leaflet secrete less collagen, but more proteoglycans than VICs from the posterior leaflet free edge (Blevins *et al.*, 2008). Considering the dynamic organization of ECM of the mature valve, the importance of this function by VICs *in vivo* can be easily appreciated. In addition to VICs, VECs are also highly responsive to changes in the hemodynamic environment, and this will be discussed in greater detail later in this review. As normal valve function is dependent on the complex arrangement of connective tissue and overall valve morphology, it is not surprising that alterations in localization, and/or contribution of matrix components leads to inefficiencies and valve disease.

2.3. Heart valve disease

In contrast to healthy valves, diseased valves are characterized by disruptions in ECM layer stratification, VIC disarray, and VEC denudation (Fig. 7.1D; Hinton *et al.*, 2006). These histopathological changes have detrimental effects on the overall structure and biomechanical properties of the valve, leading to primary insufficiency and secondary ventricular compensation (Hinton and Yutzey, 2011). Valve disease is a significant public health problem and results in approximately 20,000 deaths annually (Lloyd-Jones *et al.*, 2010). The current treatment for heart valve disease is replacement, and approximately 95,000 valve replacement surgeries are performed in the United States annually (Lloyd-Jones *et al.*, 2010). Clinically, valve disease is the most commonly diagnosed as regurgitation, caused by inefficient closing of the valve leaflets leading to backward flow, or stenosis, resulting from outflow obstruction. Both scenarios are caused by disruptions in the structure–function relationship as a result of valve malformations. Clinical phenotypes resulting in structural valve defects are often diagnosed at birth or become apparent later in life.

Congenital valve disease occurs in approximately 2% of live births (Hoffman and Kaplan, 2002). Bicuspid aortic valve (BAV) is the most common with a reported incidence of 1–2% of the adult population (Hoffman and Kaplan, 2002) and is characterized by the presence of two, rather than the normal three cusps, leading to stenosis or narrowing of the aortic valve opening (Ward, 2000). In human patients with BAV, this structural defect in cusp morphology present at birth leads to lifelong dysfunction and abnormal biomechanics. In addition, the bicuspid cusps are susceptible to calcification: a pathological process mediated by an "osteoblast-like" VIC phenotype and matrix mineralization (Rajamannan *et al.*, 2003b). This pathological disruption to the valve's normal connective tissue homeostasis leads to valve thickening (sclerosis) and mechanical stiffness, resulting in stenosis or regurgitation that can ultimately require surgical valve replacement (Freeman and Otto, 2005; Garg, 2006). In contrast to BAV and associated calcification, mitral valve prolapse (MVP) affects more than 2% of the U.S. population and is characterized by increased deposition of collagens and proteoglycans. These changes in ECM contribution and distribution weaken valve biomechanics leading to "floppy" and regurgitant leaflets that "bulge" or prolapse into the atrium, preventing valve closure (Grande-Allen *et al.*, 2003; Gupta *et al.*, 2009). While BAV is apparent at birth and considered congenital, MVP phenotypes are largely latent until adulthood. As MVP is heritable and linked to connective tissue disorders, it could be considered a congenital malformation (Devereux *et al.*, 1982). Although end-stage valve insufficiency could also be the result of predisposing genotypes and structural malformations that increase susceptibility to degenerative "wear and tear" caused by hemodynamic exposure over time. While the underlying genotypes of congenital and latent valve disease remain largely unknown, there is increasing evidence to suggest that structural abnormalities are the result of defects in the developmental pathways required for valve formation in the embryo.

3. Heart Valve Development

3.1. Endocardial cushion formation

The primitive vertebrate heart tube consists of an outer myocardial cell layer and a continual inner layer of specialized endothelial cells known as the endocardium (Perez-Pomares *et al.*, 2009). The myocardial and endocardial cell layers are separated by specialized ECM, referred to as cardiac jelly (CJ) (Person *et al.*, 2005). Soon after rightward looping, myocardial cells localized within the atrioventricular canal (AVC) and outflow tract (OFT) regions increase deposition of ECM components including proteoglycans and hyaluronan and give rise to CJ "swellings." Concurrently, a subpopulation of endocardial cells overlying these "swellings" lose contact with neighboring cells and undergo endothelial-to-mesenchymal transformation (EMT) (Eisenberg and Markwald, 1995; Person *et al.*, 2005). As a result, newly transformed mesenchyme cells invade the underlying CJ and proliferate to populate endocardial cushions (Figs. 7.1A and 7.2; Combs and Yutzey, 2009a; Eisenberg and Markwald, 1995; Person *et al.*, 2005). This essential role of VECs during early stages of valvulogenesis is initiated in the chick around embryonic day (E) 3, in the mouse at E9.5 and between E31 and E35 in human development (Lopez-Sanchez and Garcia-Martinez, 2011). In the AVC, four endocardial cushions form: the superior, inferior, and left and right lateral cushions; while in the OFT, two cushions form in proximal and distal locations (Markwald *et al.*, 2010). These cushion structures serve as physical barriers to prevent backflow of blood through the primitive heart tube (Schroeder *et al.*, 2003), in addition to hosting a population of undifferentiated VICs that serve as precursors to the mature valve structures (de Lange *et al.*, 2004; Lincoln *et al.*, 2004). Fused inferior and superior endocardial cushions in the AVC give rise to the septal valve leaflets whereas the lateral cushions form the mural leaflets (de Lange *et al.*, 2004; Lincoln *et al.*, 2004). However, the precise origins of the right, left, and anterior (aortic)/posterior (pulmonic) valve cusps from the proximal and distal endocardial cushions in the OFT following EMT are less well defined but are likely important in BAV pathogenesis.

3.1.1. Growth factor signaling during endocardial cushion EMT

EMT, epithelial-to-mesenchymal transition, and its reverse process MET (mesenchymal-to-endothelial transformation), occurs several times successively during heart development. It features in the specification of cardiac mesodermal cells during gastrulation, the formation of the endothelial cell layer following folding of the two cardiogenic areas around the primitive foregut, and eventually, as a crucial step for endocardial cushion formation (Nieto, 2009). With a focus on heart valve development, this review will discuss the molecular mechanisms of endocardial cushion EMT.

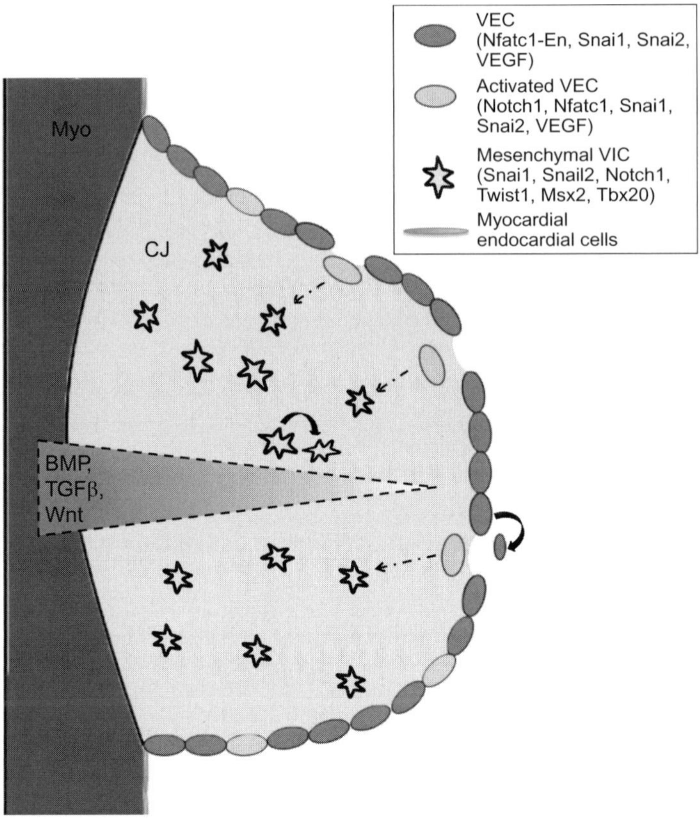

Figure 7.2 EMT and endocardial cushion formation. Schematic diagram to indicate molecular and cellular processes involved in endocardial cushion endothelial-to-mesenchymal transformation (EMT). Growth factor signals emanating from the myocardium (myo) diffuse through the cardiac jelly (CJ) to activate VECs and initiate EMT. During this process, VECs lose cell–cell contacts with neighboring cells, transform, and migrate into the underlying CJ where they proliferate to expand the population size (curved arrow). (See Color Insert.)

The initiation of endocardial cushion EMT has been well studied due to its significance for proper valve development. The onset is marked by activation of VECs in response to signals emanating from the adjacent nonworking myocardium (Harrelson *et al.*, 2004; Lyons *et al.*, 1990; Ma and Martin, 2005; Plageman and Yutzey, 2004). This leads to the morphological appearance of cobble stone-like VECs that harbor progenitor potential (Butcher *et al.*, 2004; Person *et al.*, 2005). Many signaling pathways critical for early embryogenesis have been shown to play roles initiating endocardial cushion EMT. Most of them, if not all, are secreted signaling molecules, including transforming growth factor (TGF) β, Wnt, and bone morphogenetic protein (BMP) ligands capable of diffusing through the CJ to reach the endothelium and trigger a series of downstream effects (Fig. 7.2; Combs and Yutzey, 2009a).

TGFβs and their downstream signaling mediators are among the pioneer molecules derived from the myocardium that regulate several EMT-related processes. During avian endocardial cushion formation, Tgfβ2 and Tgfβ3 are required for VECs to initiate EMT (Mercado-Pimentel and Runyan, 2007; Nakajima *et al.*, 2000). In particular, Tgfβ2 mediates VEC cell–cell activation and separation, while Tgfβ3-activating TgfβRII promotes invasion of transformed mesenchyme cells into the CJ (Mercado-Pimentel and Runyan, 2007). In mice deficient for *Tgfβ2,* EMT is initiated although developing cushions are variably hypoplastic (Bartram *et al.*, 2001; Sanford *et al.*, 1997). Similar phenotypes are also observed in other mouse models with targeted downregulation of Tgfβ signaling including endocardial-deletion of *TgfβRI* (*Alk5*) (Sridurongrit *et al.*, 2008) and global deletion of the long form of latent Tgfβ binding protein 1, in which EMT initiation is impaired (Todorovic *et al.*, 2011).

TGFβ-mediated activation of the downstream effectors Smad2/3 is sufficient to induce expression of the zinc finger transcription factors Snai1 (Snail) and Snai2 (Slug) (Cho *et al.*, 2007; Romano and Runyan, 2000). Both these family members are highly expressed in VECs undergoing EMT, as well as in newly transformed mesenchyme cells within the developing cushion (Niessen *et al.*, 2008; Oram *et al.*, 2003; Tao *et al.*, 2011). Studies in chicken and mice have shown that Snai1 and Snai2 are essential for endocardial cushion EMT (Romano and Runyan, 2000; Tao *et al.*, 2011). Mice deficient for *Snai2* display hypocellular cushions at E9.5, although this phenotype is compensated by E10.5 due to increased *Snai1* (Niessen *et al.*, 2008). However, *Snai2* cannot compensate for *Snai1* deficiency and therefore *Snai1*$^{-/-}$ mice die early during development due to defective gastrulation and mesoderm formation prior to endocardial cushion formation (Carver *et al.*, 2001). Using a conditional approach, our group has recently shown that similar to *TGFβ* mutants, mice with reduced *Snai1* function in endothelial-derived cells using the *Tie2cre* model (*Tie2cre;Snai1*$^{fl/+}$) also develop hypoplastic cushions as a result of impaired EMT (Tao *et al.*, 2011). As expected, *Snai1* overexpression in avian AVC explants is sufficient to induce EMT at the level of VEC transformation and mesenchyme cell migration and invasion (Tao *et al.*, 2011). In other noncardiac systems, Snai1 is known to regulate EMT by directly repressing expression of cell adhesion molecules including *E-cadherin* (Cano *et al.*, 2000). The downregulation of cell adhesion genes in *Tie2cre;Snai1*$^{fl/+}$ mice suggests similar mechanisms in the developing heart, therefore attenuating the breakdown of cell–cell contacts between VECs in *Snai1* mutants. We have also shown that *Snai1* directly regulates *mmp15:* a matrix metalloproteinase family member important for cell motility. Therefore, in addition to maintained VEC contacts, the hypoplastic endocardial cushion phenotype observed in *Tie2cre;Snai1*$^{fl/+}$ mice may also be the result of impaired cell invasion into the CJ (Tao *et al.*, 2011). This collection of studies indicate that Tgfβ signaling pathways are important for several biological processes required for endocardial cushion formation including EMT initiation, breakdown of cell–cell contacts, mesenchymal cell motility and proliferation.

In addition to targeting downstream transcription factors, Tgfβ cross talks with other signaling pathways including β-catenin: a downstream mediator of canonical Wnt signaling. This cross talk between Tgfβ and β-catenin signaling suggests that the Wnt pathway may also be involved in endocardial cushion EMT. *Tie2*-cre deletion of β-catenin inhibits Tgfβ-mediated induction of EMT in mice (Liebner *et al.*, 2004), suggesting a requirement of the canonical Wnt signaling in this process. In zebrafish, overexpression of a secreted Wnt inhibitor (*Dickkopf1*) blocks cushion formation, while truncation mutants develop hyperplastic cushions as a result of increased cell proliferation (Hurlstone *et al.*, 2003). The notion that Wnt signaling regulates VIC proliferation is continued in the avian system where the Wnt receptor Frzb and Wnt9a promote mesenchyme cell number in the AV cushions (Person *et al.*, 2005). Consistently, expression patterns of other Wnt pathway genes including *Wnt2a, LEF1*, and *Fzd2* in murine mesenchyme cells of the AVC and OFT cushions suggest conserved mechanisms across species.

As members of the TGF superfamily, BMPs also signal through the canonical SMAD pathway, however, predominantly through activated Smad1/5/8. There is a wealth of evidence from studies in primary cell cultures and animal model to show that similar to Tgfβ ligands, BMPs are also a major source of myocardial-derived signals for EMT initiation (Delot, 2003). From the extensive number of BMP family members, Bmp2 and 4 are most predominantly expressed in the AV and OFT myocardium adjacent to developing cushions, respectively (Nakajima *et al.*, 2000; Somi *et al.*, 2004). As with Tgfβ and Wnts, mutations that target decreased BMP signaling in mice commonly result in hypoplastic endocardial cushions as a result of impaired EMT (Delot, 2003). Myocardial-specific knockdown of *Bmp2* severely disrupts VEC transformation, particularly in the AVC (Ma *et al.*, 2005). Further studies have shown that targeted *Bmp2* function *in vivo* leads to downregulation of *Tbx2*, which is necessary for suppression of chamber-specific gene expression and generation of the CJ (Ma *et al.*, 2005; Rivera-Feliciano and Tabin, 2006). In addition to BMP ligands, endocardial cushion EMT defects have also been reported in mice with targeted deletions of BMP receptors including *BmpRIa* and *BmpRII* (Delot *et al.*, 2003), as well as inhibitors of BMP signaling (Bachiller *et al.*, 2003; Brunet *et al.*, 1998; Galvin *et al.*, 2000; Matzuk *et al.*, 1995). The significance of endocardial cushion defects observed in BMP mutant mice is likely the result of compromised phospho-Smad1/5/8 activity which in healthy hearts is required to sustain BMP signaling from the myocardium to VECs during EMT (Combs and Yutzey, 2009a). Further, proper BMP signaling is also required for sufficient functioning of TGFβ, as well as the expression of transcription factors *Twist1* and *Msx2*, which are all well-known positive regulators of EMT (Combs and Yutzey, 2009a). These collective studies highlight the complexity of the transcriptional networks activated from growth factor signaling during initial stages of endocardial cushion formation for progressive steps of the EMT process.

3.1.2. VEC signal transduction during endocardial cushion EMT

Studies have shown that many of the myocardial-derived growth factors important for EMT signal to downstream target genes highly expressed in VECs, suggesting that the valve endothelium serves as a signaling center during cushion formation. Integrated with TGFβ and BMP signaling (Luna-Zurita *et al.*, 2010; Niessen *et al.*, 2008), Notch signaling is an important activator of EMT, through the receptor Notch1 predominantly expressed in VECs. With both the ligands (Delta and Jagged) and receptors (Notch) as membrane proteins, Notch sets up an efficient intercellular local signaling system that affects multiple downstream targets (Kopan and Ilagan, 2009). Notch regulates Tgfβ2 production through translocation of its cytoplasmic domain to the nucleus, where it binds to the RBPjk/CSL cotranscription factor (Kopan and Ilagan, 2009). The detrimental effects of altered Tgfβ signaling have been discussed previous; however, the additional significance of the Notch pathway is appreciated in RBPjk- and Notch1-targeted mutants where hypoplastic endocardial cushions lack significant numbers of transformed mesenchymal cells (Macgrogan *et al.*, 2011; Timmerman *et al.*, 2004). Critical target genes affected by reduced Notch signaling in developing cushions include Snai2 (Niessen *et al.*, 2008) (discussed above), and members of the bHLH Hey family of transcription factors (Macgrogan *et al.*, 2011). Studies in chick and mouse have shown that *Hey1* and *Hey2* are highly expressed in the AVC VECs and are important for restricting expression of *Bmp2* and *Tbx2* in the AVC region (Kokubo *et al.*, 2007; Rutenberg *et al.*, 2006). In contrast to reduced Notch signaling, overexpression of constitutively active Notch1 in the endocardium delays endocardial cushion formation (Venkatesh *et al.*, 2008) and ectopic EMT is observed in the ventricular chambers (Luna-Zurita *et al.*, 2010). In humans, *NOTCH1* mutations have been linked with BAV (Garg *et al.*, 2005), while mutations in *JAG1* and *NOTCH2* cause Alagille syndrome, an autosomal-dominant disease that affects the cardiac OFT (Li *et al.*, 1997; McDaniell *et al.*, 2006; Oda *et al.*, 1997). *JAG1* mutations have also been identified in other forms of congenital heart disease with endocardial cushion involvement (Eldadah *et al.*, 2001; Krantz *et al.*, 1999); therefore suggesting that defects in Notch signaling pathways during valvulogenesis may have detrimental effects on heart valve structure and function after birth.

In addition to Notch, vascular endothelial growth factor A (VEGF) is a potent cytokine highly expressed by the myocardium and VECs before cushion formation. However, expression of the ligand and its receptors (VEGFR) become restricted to the endocardium once cushion formation has been initiated (Dor *et al.*, 2001; Miquerol *et al.*, 1999). Given the role that VEGF plays in response to hypoxia in the vascular system, it has been suggested that this temporal and spatial change in VEGF expression during EMT correlates with oxygen levels in the developing embryonic heart (Dor *et al.*, 2003). Studies in chick, mouse, and zebrafish demonstrate that endocardial VEGF signaling contributes to early valvulogenesis by promoting the VEC phenotype and proliferation, thereby maintaining a proliferative population of endothelial cells throughout EMT and cushion maturation (Carmeliet *et al.*, 1996; Combs and Yutzey, 2009b; Dor *et al.*, 2001; Ferrara *et al.*, 1996; Miquerol *et al.*, 2000). From these studies, it is apparent that VEGF levels of expression must be tightly controlled during endocardial cushion formation: too much VEGF inhibits EMT, while too little attenuates VEC proliferation, thereby decreasing VEC availability for transformation (Dor *et al.*, 2001; Miquerol *et al.*, 2000). Interestingly, it appears that VEGF signaling is required for complete EMT in the OFT but has less essential roles in AVC endocardial cushion formation (Stankunas *et al.*, 2010). This observation has been attributed to the differential mediation of VEGF signaling through VEGFR1 in the OFT and VEGFRII in the AVC (Stankunas *et al.*, 2010). Further downstream, VEGF regulates the transcription factor NFATc1 (nuclear factor of activated T-cells cytoplasmic 1) in both developing and mature VECs to promote VEC cell proliferation (Combs and Yutzey, 2009b; Johnson *et al.*, 2003). While *Nfatc1*$^{-/-}$ mice undergo successful endocardial cushion EMT (de la Pompa *et al.*, 1998; Lange and Yutzey, 2006; Ranger *et al.*, 1998), a new role has recently emerged in fate decisions of VEC during transformation stages.

During a tightly regulated temporal window, VECs break cell–cell contacts, transform, and migrate into the underlying CJ. However, after around E13 in the mouse when cushion formation is complete, VECs no longer transform, despite maintained expression of many EMT-inducers. For many years, it has remained a challenge to identify the molecular cues that determine the fate of individual VECs during this window of events: which VECs transform and which VECs remain in the endothelium? In the heart, Nfatc1 is expressed in VECs during endocardial cushion formation (de la Pompa *et al.*, 1998; Lange and Yutzey, 2006; Ranger *et al.*, 1998); however, a role during early stages of valve formation has remained elusive. A recent study (Wu *et al.*, 2011) identified a transcriptional enhancer region (En) within *Nfatc1* that may play an important role in regulating VEC fate during EMT. In contrast to *Nfatc1* expression in all VECs, the study used reporter mice to reveal a small subpopulation of VECs that express the *Nfatc1*-En. VECs positive for *Nfatc1-En* expression do not appear to undergo EMT and remain part of the proliferative endothelial cell population surrounding the developing valve (Fig. 7.2). Further, they showed that Nfatc1 inhibits EMT in a cell-autonomous manner by suppressing transcription of *Snai1* and *Snai2* (Wu *et al.*, 2011). This regulation of VECs by Nfatc1 is the first to provide a mechanism for cell fate decisions of endothelial cells during endocardial cushion EMT.

3.2. Contribution of nonendothelial cell lineages to the heart valve structures

As discussed and supported by cell fate lineage analysis using *Tie2cre; ROSA26R* mice, the majority of mesenchymal VICs that form the endocardial cushions and later give rise to the mature valve structures are derived from VECs as a result of EMT (de Lange *et al.*, 2004; Lincoln *et al.*, 2004). However, extensive cell mapping studies have shown that in the mature AV valves, some VICs are derived from alternative sources. Using the chick-quail chimera system, it is apparent that epicardium-derived cells significantly contribute to the avian AV valves (Gittenberger-de Groot *et al.*, 1998). However, studies in mice using *Wilms Tumor 1 (WT1)-Cre* reporters suggest little contribution of this cell population to murine valves (Zhou *et al.*, 2010). In OFT endocardial cushions, findings from lineage tracing using *Wnt1-Cre* suggest that neural crest-derived cells contribute to these structures in addition to the aorticopulmonary septum (de Lange *et al.*, 2004; Jiang *et al.*, 2000). These studies demonstrate that although the VIC population is derived from multiple sources, the endothelial cell lineage is the major provider and plays essential roles throughout valve development.

3.3. Post-EMT and maturation stages of valve development

Once endocardial cushion formation is complete, overlying VECs stop undergoing EMT, regain cell–cell contacts, and form an uninterrupted endothelium (Person *et al.*, 2005). Around this time, the individual endocardial cushions fuse and elongate into mitral and tricuspid valve primordia in the AVC region, and primitive aortic and pulmonic structures in the OFT (Fig. 7.1B) (Lincoln *et al.*, 2004). Concurrently, cells within the developing valve structures continue to proliferate and the connective tissue undergoes extensive remodeling. Compared to the EMT, much less is known about the regulation of valve remodeling, particularly at the level of the VECs as more recent work has focused on VICs during this process. Work by our group and others has shown that during mid-stages of valvulogenesis, VICs lose expression of mesenchyme genes including

Twist1, Tbx20, Msx1, and *Msx2* and begin to express differentiated lineage markers including those characteristic of chondrocyte and tendon cells (Chakraborty *et al.*, 2010; Hurle *et al.*, 1994; Lincoln *et al.*, 2006a,b; Montero *et al.*, 2002). In the mature avian heart, there appears to be definitive compartmentalization between localization of these two lineages with cartilaginous genes including *Sox9* in the maturing valve leaflets and tendon-associated genes (*Scleraxis*) in supporting structures (Lincoln *et al.*, 2006a). In the comparatively smaller mouse valve, there is less distinction and both *Sox9* and *Scleraxis* are highly expressed in both leaflets and supporting structures (Levay *et al.*, 2008; Lincoln *et al.*, 2007). Using an established avian culture system, BMP2 and FGF4 treatments promote differentiation of mesenchymal VICs into *Sox9*- and *Scleraxis*-positive lineages, respectively. However, this is only achievable in VICs isolated from prefused endocardial cushions. Following fusion, cultured VICs are not responsive to exogenous BMP or FGF treatment (Lincoln *et al.*, 2006a). Therefore suggesting that the differentiation potential of embryonic VICs decreases as development progresses. In addition to VICs, the valve ECM also undergoes extensive changes during stages of valve remodeling. This is associated with degradation of the primitive CJ by matrix metalloproteinases, and deposition of collagens and proteoglycans by differentiating VICs that will later form the fibrosa and spongiosa layers (Dreger *et al.*, 2002; Rabkin *et al.*, 2001). It is not until postnatal stages that the elastin fibers are laid down in the atrialis/ventricularis layers and the valve becomes tristratified (Hinton *et al.*, 2006). A significant reduction in cell proliferation is another feature of late embryonic valve development, while normally little, if any, cell proliferation occurs after birth (Aikawa *et al.*, 2006; Hinton *et al.*, 2006). In addition to the regulation of cellular processes by complex transcriptional networks during valve maturation, the influence of mechanical flow also plays an important role in mediating valve cell function.

4. Endothelial Cells and the Hemodynamic Environment

4.1. Endothelial cells and blood flow

Several studies have shown the importance of biomechanical stimuli exerted by blood flow, blood pressure, and cyclic strain for normal cardiac development (Santhanakrishnan and Miller, 2011). In particular, valvulogenesis appears especially sensitive to changes in flow-induced shear stress and pressure-related stretch (Butcher *et al.*, 2007; Hove *et al.*, 2003). Extensive studies in the cardiovascular system have shown that biomechanical forces exerted by blood flow and blood pressure are sensed by vascular endothelial and endocardial cells within the myocardium (Davies *et al.*, 1997;

Santhanakrishnan and Miller, 2011). This is largely regulated through several mechanisms including activation of cell adhesion molecules, surface glycoproteins, ions channels, phosphorylation of surface receptors, secondary messengers, and cilium (Hierck *et al.*, 2007; Lehoux *et al.*, 2006). In response to these stimuli, vascular endothelial cells alter their morphology, growth rate, metabolism, and gene expression profile to ensure proper morphogenesis of the embryonic heart, and maintenance of the adult cardiovascular system (Davies, 1995; Girard and Nerem, 1995; Groenendijk *et al.*, 2005, 2007; Hierck *et al.*, 2008; Passerini *et al.*, 2003; Sorescu *et al.*, 2003; Stamatas and McIntire, 2001). While the available literature on vascular endothelial cell behavior is extensive, comparatively little is known about the "mechanosensory" properties of VECs in response to the hemodynamic environment. Despite the common morphological phenotype of vascular and valvular endothelial cells, emerging recent studies have revealed striking differences between these two cell populations.

4.2. VECs and flow dynamics

Electron micrograph studies of aortic valves indicate that on both ventricularis and fibrosa surfaces of open or closed cusps, cobblestone-like VECs are aligned perpendicular to blood flow and follow the orientation of collagen fibers (Deck, 1986). This alignment of VECs is in contrast to vascular endothelial cells that align parallel in response to flow (Girard and Nerem, 1995; Stamatas and McIntire, 2001). This discrepancy in shear stress response between endothelial cells of the valves and vasculature is likely due to the observed differential molecular profiles and functional responses reported for these two cell populations (Butcher and Nerem, 2006; Farivar *et al.*, 2003). Although blood flow does not influence VEC orientation, changes in the hemodynamic environment can affect other VEC-mediated processes in the valves. Obstructive blood flow studies in zebrafish have shown that flow is important for endothelial cell signal transduction and required for valve formation (Bartman *et al.*, 2004; Hove *et al.*, 2003; Sehnert *et al.*, 2002; Vermot *et al.*, 2009). Further, blood flow is only influential on valvulogenesis after initial patterning of the endothelial cell ring by genetic cues (Vermot *et al.*, 2009). Contradictory to these observations, zebrafish mutants with cardiac contractility defects and abnormal blood flow have normal valves (Sultana *et al.*, 2008), therefore, highlighting the multifactorial regulation of valvulogenesis. In addition to morphogenesis, Butcher *et al.* reported that ECM properties of endocardial cushions change with mechanical flow: with high shear stress correlating with increased elastic rigidity (Butcher *et al.*, 2007). While these studies highlight the importance of flow during early stages of valvulogenesis, little is known about the potential role of flow in establishing the stratified trilaminar layers of the elongated mature valve leaflets. In addition to VECs playing important roles in heart valve development, VEC dysfunction and denudation are largely associated with valve leaflet failure.

5. Endothelial Cell Function in Heart Valve Disease

Valvular heart disease is a major cause of morbidity requiring surgical intervention in the form of repair or replacement (O'Brien, 2007). Degenerative valve disease is commonly manifested by thickening (sclerosis) of the valve leaflets associated with histopathological changes in VEC morphology, VIC distribution and proliferation, in addition to defects in ECM organization (Butcher and Nerem, 2006; Hinton *et al.*, 2006; Leask *et al.*, 2003; Lincoln *et al.*, 2006b). Overall, these alterations have detrimental effects on valve biomechanics leading to impaired leaflet movement, increased resistance to blood flow and eventual heart failure (Freeman and Otto, 2005; Lincoln *et al.*, 2006b).

5.1. VEC dysfunction in valve disease: Comparisons to vascular pathology

Although the role of VICs in valve disease has been an active area of research, the contribution of VECs to valve pathology is less well understood. Atherosclerosis, thrombosis, stroke, and hypertension are vascular disorders associated with endothelial cell dysfunction and pathological changes in endothelial cell-mediated responses (Leask *et al.*, 2003). Similarly, endothelial cell disruption is also characteristic of early stage degenerative valve disease (Butcher and Nerem, 2006; Stein *et al.*, 1989) and associated phenotypes including neo-vascularization, infiltration of inflammatory cells, and lipid deposition are common to sclerotic valves and atherosclerotic lesions (Mohler, 2000; Otto *et al.*, 1994; Pastor-Perez and Marin, 2009). The invasion of macrophage foam cells and accumulation of oxidized low density lipoproteins within the fibrosa layer immediately beneath the endothelium are thought to indirectly facilitate in the passive formation of calcific, bone-like lesions during later stages of pathogenesis (Helske *et al.*, 2007). However, studies in mice have also shown that calcification of valve tissue can be an intrinsic process independent of significant inflammatory responses (Lincoln *et al.*, 2007; Peacock *et al.*, 2010). In addition to ECM mineralization, VEC dysfunction correlates with VIC activation, proliferation, and deposition of ECM, leading to further valve thickening (Hinton *et al.*, 2006). Therefore consistent with previous studies, communication between VECs and VICs is important during valve disease processes, in addition to valvulogenesis (Butcher and

Nerem, 2006). However, it remains unclear if alterations in VEC responses directly cause pathogenic processes or are the result of an alternative primary insult. Nonetheless, it is clear that disruption to the valve endothelium is an important initial step in the pathogenesis of heart valve disease.

5.2. Hemodynamics and valve disease

Despite the clinical prevalence of valve disease, the initiating events that lead to VEC dysfunction are not clear. There is emerging evidence to suggest a genetic component with *NOTCH1* mutations identified in human BAV as well as mouse models of BAV caused by endothelial-specific loss of *Gata5* and global knockout of *endothelial-nitric oxide synthase (eNOS or NOS3)* (Laforest *et al.*, 2011; Lee *et al.*, 2000). In addition to genes, a number of environmental risk factors shared with atherosclerosis including hypercholesterolemia, ageing, smoking, hypertension, and diabetes have been implicated in valve pathologies (Stewart *et al.*, 1997). Indeed older and/or hypercholesterolemic mice develop premature valve calcification (Nus *et al.*, 2011; Tanaka *et al.*, 2005; Weiss *et al.*, 2006), with noted VEC dysfunction (Wang *et al.*, 2010). However, hemodynamic flow across the endothelium is also thought to be an influential factor of physical injury. In the healthy heart, blood flows from low (atria) to high (ventricles) pressure regions; therefore, the left side of the heart is subject to greater pressure levels than the right. As a result, the impact of hemodynamic forces on mitral/aortic versus tricuspid valves is dependent upon position (Hinton and Yutzey, 2011) and the endothelium is subjected to distinctly different fluid dynamics on each side of the leaflet/cusp (Davies *et al.*, 2004; Ge and Sotiropoulos, 2010; Nicosia *et al.*, 2003). In calcific aortic valve disease (and atherosclerosis), patterns of focal calcification on the aortic (fibrosa) side of the valve correspond to sites of increased mechanical stress (Otto *et al.*, 1994; Thubrikar *et al.*, 1986a,b). Further, congenital malformations resulting in valves exposed to higher stress, for example BAV, have accelerated pathogenesis (Beppu *et al.*, 1993). *In vitro*, hemodynamic forces such as fluid shear stress, cyclic strain, and hydrostatic pressure have been shown to activate fibro-calcific endothelial cell responses (Frangos *et al.*, 1999). While at the molecular level, studies have shown that in the absence of flow, VECs isolated from the aortic side of nondiseased valves are permissive to calcification and generally protected against inflammatory processes compared to VECs isolated from the ventricular surface (Simmons *et al.*, 2005). Similarly, the transcriptional profiles of VECs under shear stress conditions also suggest protective roles against pathogenic processes (Butcher *et al.*, 2006). These studies highlight the heterogeneity of the VEC population and suggest a correlative link between VEC localization, hemodynamic cues, and molecular responses in the susceptibility of valve disease.

6. Therapeutics of VECs

The prevalence of valve disease is increasing, yet effective treatments remain limited. Over 95,000 valve replacement surgeries are performed in the United States each year (Supino *et al.*, 2004). However, surgical procedures to repair or replace malformed valves using mechanical or biological valves are inefficient and often temporary due to inappropriate replacements that fail to grow, remodel, repair, or withstand life-long mechanical demands *in vivo* (Mann and West, 2001; Rabkin-Aikawa *et al.*, 2005). In addition, bioprosthetic valve tissue is highly susceptible to calcification, particularly in younger recipients, as well as progressive collagen damage and thrombosis (Schoen, 2008). Therefore, alternative approaches are being developed as a proposed solution to these problems by replacing damaged valve tissue with living cells. As initiation of valve disease is concomitant with damaged endothelia, therapeutic strategies have recently focused on stimulating autologous endothelial cells to replenish injured lesions, and replacing diseased tissue with bioprosthetic valves seeded with healthy endothelial cells. The long-term goal is to generate a therapeutic alternative that will replace damaged VECs, VICs, and ECM, as well as adapt to the changing hemodynamic environment in the absence of secondary pathologies, throughout life.

As previously discussed, transformation of a subset of endothelial cells within the developing heart is essential for establishing the pool of mesenchymal VICs that later give rise to the mature heart valve structures (Lincoln *et al.*, 2004). However, the potential of adult VECs to undergo EMT to replace damaged endothelia and replenish VICs in response to pathological stimuli has not been extensively examined. Previous studies have shown that in response to molecular cues including VEGF stimulation, clonal VECs from adult human pulmonary valves transition into a mesenchymal phenotype through molecular events similar to developmental EMT (Paranya *et al.*, 2001; Paruchuri *et al.*, 2006). Unpublished data from our group suggests that postnatal and adult murine VECs may also have EMT potential, as indicated by increased transformation in response to *Snai1* overexpression (Tao and Lincoln, Unpublished data). In addition to molecular stimuli, sheer stress has also been shown to promote expression of mesenchyme cell markers, cell polarity, and proliferation in endothelial cells (Butcher *et al.*, 2006; Egorova *et al.*, 2011). However, based upon the limited regeneration capacity of mature endothelial cells compared to embryonic endothelial cells or endothelial progenitor cells (Dimmeler and Zeiher, 2004), it is considered that VEC "plasticity" may be greater in the embryo than in neonatal or adult valves: although VECs from postnatal staged mice maintain expression of endothelial progenitor cell markers including CD133 (*prominin1*) (Fig. 7.3) (Kotick and Lincoln, Unpublished data). With plasticity differences in mind, studies have shown that circulating endothelial progenitor cells and bone marrow-derived mesenchyme precursor cells may play a role in maintaining endothelial cell integrity and function in the adult vascular system (Deanfield *et al.*, 2007; Schoen, 2008; Sorrentino *et al.*, 2007). Further, reduced number and function of endothelial progenitor cells have been observed in patients with aortic valve stenosis (and vascular disease) (Hill *et al.*, 2003; Matsumoto *et al.*, 2009; Werner *et al.*, 2005). Therefore, initiation and progression of valve pathogenesis could be advanced if endothelial cell repair is impaired as a result of insufficient endothelial progenitor cell number, immobilization, homing, or increased senescence. Although further studies are required, these data suggest that in response to molecular and environmental stimuli, autologous endothelial cells have the potential as a source for self-repair in the diseased valve.

The generation of heart valve tissue substitutes in the form of bioprosthetic valves has been an active area of research for a number of decades. Recent advances have been made by replacing diseased valves with synthetic scaffolds of decellularized xenogenic tissues seeded with autologous (and exogenous) cell types, including endothelial progenitor cells (Fang *et al.*, 2007; Sales *et al.*, 2010). A recent study by Sales *et al.* showed that subpopulations of endothelial progenitor-derived cells have the potential to transdifferentiate into myofibroblast cells and express markers characteristic of activated VICs, including α-SMA. In addition, seeded endothelial progenitor cells demonstrate ECM production and dynamic remodeling (Sales *et al.*, 2010). Therefore, highlighting the potential of circulatory endothelial

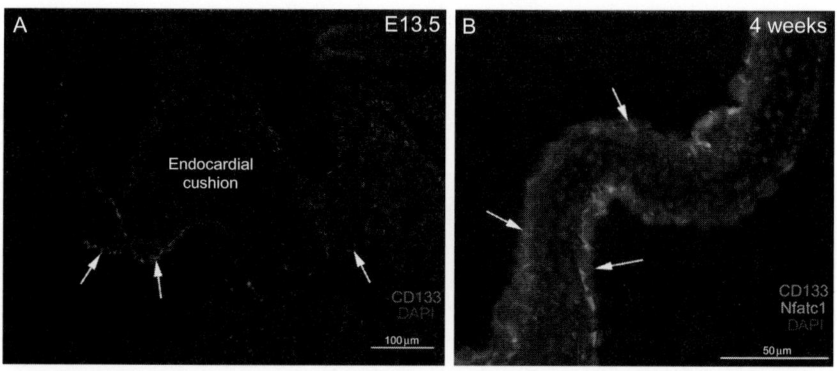

Figure 7.3 CD133 expression in developing and mature murine heart valve structures. Immunohistochemistry to show expression of the endothelial progenitor cell marker CD133 in VECs of the developing endocardial cushion at E13.5 (A) and mature mitral valve leaflets at 4 weeks of age (B). VECs are highlighted by Nfatc1 immunoreactivity in (B) and DAPI indicates nuclei in (A) and (B). (See Color Insert.)

progenitor cells as a single-source for construction of autologous heart valves. Although promising, further work is required to examine the capabilities of autologous endothelial progenitor cells to respond to pathological stimuli *in vivo*.

7. Conclusions and Perspectives

Over the past several decades, the field of heart valve biology has made significant advancements in understanding the structure–function relationship of heart valves in both health and disease. Early studies have extensively characterized the role of VECs in establishing the pool of valve precursor cells that later give rise to the mature valve structures. The molecular regulation of endothelial cell function during the EMT process is not limited to the cardiovascular system, and therefore findings have proven applicable to other development and pathological systems, including cancer metastasis. However, questions remain unanswered as to the role of VECs during post-EMT stages of valve development, particularly as expression of many EMT-initiating genes is maintained in VECs of remodeling valves. More recent studies have identified the interaction of VECs with underlying VICs and ECM components in response to changes in the hemodynamic environment to maintain valve homeostasis in the adult. With the emerging concept that valve disease processes recapitulate developmental gene programs, there are opportunities to focus future studies on the potential of VECs to reactivate the EMT program to repair or replace damaged lesions in diseased valves. To add another level of complexity, it is becoming clear that VECs are heterogeneous: Therefore, understanding the mechanisms that govern differential responses of VECs based upon their circumferential location is going to require further investigation. Finally, there are currently no therapies available to prevent or treat valve disease. Addressing these currently unanswered questions will improve our understanding of valve pathogenesis and provide insights into therapeutic targeting of endothelial cells and allow for intervention prior to end-stage heart failure.

Acknowledgments

We thank William Escobar for illustrations and Harriet Hammond and Danielle Huk for editorial assistance. This work is supported by NIH/NHLBI R01HL091878 (J. L.), American Heart Association Predoctoral Fellowship 10PRE4360052 (G. T.), American Heart Association Student Scholarship in Cardiovascular Disease and Stroke (J. D. K.).

REFERENCES

Aikawa, E., Whittaker, P., Farber, M., Mendelson, K., Padera, R. F., Aikawa, M., and Schoen, F. J. (2006). Human semilunar cardiac valve remodeling by activated cells from fetus to adult: Implications for postnatal adaptation, pathology, and tissue engineering. *Circulation* **113,** 1344–1352.

Aldous, I. G., Veres, S. P., Jahangir, A., and Lee, J. M. (2009). Differences in collagen cross-linking between the four valves of the bovine heart: A possible role in adaptation to mechanical fatigue. *Am. J. Physiol. Heart Circ. Physiol.* **296,** H1898–H1906.

Anderson, R. H. (2000). Clinical anatomy of the aortic root. *Heart* **84,** 670–673.

Anderson, R. H., Ho, S. Y., and Becker, A. E. (2000). Anatomy of the human atrioventricular junctions revisited. *Anat. Rec.* **260,** 81–91.

Bachiller, D., Klingensmith, J., Shneyder, N., Tran, U., Anderson, R., Rossant, J., and De Robertis, E. M. (2003). The role of chordin/Bmp signals in mammalian pharyngeal development and DiGeorge syndrome. *Development* **130,** 3567–3578.

Balachandran, K., Sucosky, P., and Yoganathan, A. P. (2011). Hemodynamics and mechanobiology of aortic valve inflammation and calcification. *Int. J. Inflam.* **2011,** 263870.

Bartman, T., Walsh, E. C., Wen, K. K., McKane, M., Ren, J., Alexander, J., Rubenstein, P. A., and Stainier, D. Y. (2004). Early myocardial function affects endocardial cushion development in zebrafish. *PLoS Biol.* **2,** E129.

Bartram, U., Molin, D. G., Wisse, L. J., Mohamad, A., Sanford, L. P., Doetschman, T., Speer, C. P., Poelmann, R. E., and Gittenberger-de Groot, A. C. (2001). Double-outlet right ventricle and overriding tricuspid valve reflect disturbances of looping, myocardialization, endocardial cushion differentiation, and apoptosis in TGF-beta(2)-knockout mice. *Circulation* **103,** 2745–2752.

Beppu, S., Suzuki, S., Matsuda, H., Ohmori, F., Nagata, S., and Miyatake, K. (1993). Rapidity of progression of aortic stenosis in patients with congenital bicuspid aortic valves. *Am. J. Cardiol.* **71,** 322–327.

Blevins, T. L., Peterson, S. B., Lee, E. L., Bailey, A. M., Frederick, J. D., Huynh, T. N., Gupta, V., and Grande-Allen, K. J. (2008). Mitral valvular interstitial cells demonstrate regional, adhesional, and synthetic heterogeneity. *Cells Tissues Organs* **187,** 113–122.

Brunet, L. J., McMahon, J. A., McMahon, A. P., and Harland, R. M. (1998). Noggin, cartilage morphogenesis, and joint formation in the mammalian skeleton. *Science* **280,** 1455–1457.

Butcher, J. T., and Nerem, R. M. (2006). Valvular endothelial cells regulate the phenotype of interstitial cells in co-culture: Effects of steady shear stress. *Tissue Eng.* **12,** 905–915.

Butcher, J. T., and Nerem, R. M. (2007). Valvular endothelial cells and the mechanoregulation of valvular pathology. *Philos. Trans. R. Soc. Lond. B Biol. Sci.* **362,** 1445–1457.

Butcher, J. T., Penrod, A. M., Garcia, A. J., and Nerem, R. M. (2004). Unique morphology and focal adhesion development of valvular endothelial cells in static and fluid flow environments. *Arterioscler. Thromb. Vasc. Biol.* **24,** 1429–1434.

Butcher, J. T., Tressel, S., Johnson, T., Turner, D., Sorescu, G., Jo, H., and Nerem, R. M. (2006). Transcriptional profiles of valvular and vascular endothelial cells reveal phenotypic differences: Influence of shear stress. *Arterioscler. Thromb. Vasc. Biol.* **26,** 69–77.

Butcher, J. T., McQuinn, T. C., Sedmera, D., Turner, D., and Markwald, R. R. (2007). Transitions in early embryonic atrioventricular valvular function correspond with changes in cushion biomechanics that are predictable by tissue composition. *Circ. Res.* **100,** 1503–1511.

Cano, A., Perez-Moreno, M. A., Rodrigo, I., Locascio, A., Blanco, M. J., del Barrio, M. G., Portillo, F., and Nieto, M. A. (2000). The transcription factor snail controls epithelial-mesenchymal transitions by repressing E-cadherin expression. *Nat. Cell Biol.* **2,** 76–83.

Carmeliet, P., Ferreira, V., Breier, G., Pollefeyt, S., Kieckens, L., Gertsenstein, M., Fahrig, M., Vandenhoeck, A., Harpal, K., Eberhardt, C., Declercq, C., Pawling, J., *et al.* (1996). Abnormal blood vessel development and lethality in embryos lacking a single VEGF allele. *Nature* **380,** 435–439.

Carver, E. A., Jiang, R., Lan, Y., Oram, K. F., and Gridley, T. (2001). The mouse snail gene encodes a key regulator of the epithelial-mesenchymal transition. *Mol. Cell. Biol.* **21,** 8184–8188.

Chakraborty, S., Combs, M. D., and Yutzey, K. E. (2010). Transcriptional regulation of heart valve progenitor cells. *Pediatr. Cardiol.* **31,** 414–421.

Cho, H. J., Baek, K. E., Saika, S., Jeong, M. J., and Yoo, J. (2007). Snail is required for transforming growth factor-beta-induced epithelial-mesenchymal transition by activating PI3 kinase/Akt signal pathway. *Biochem. Biophys. Res. Commun.* **353,** 337–343.

Combs, M. D., and Yutzey, K. E. (2009a). Heart valve development: Regulatory networks in development and disease. *Circ. Res.* **105,** 408–421.

Combs, M. D., and Yutzey, K. E. (2009b). VEGF and RANKL regulation of NFATc1 in heart valve development. *Circ. Res.* **105,** 565–574.

Davies, P. F. (1995). Flow-mediated endothelial mechanotransduction. *Physiol. Rev.* **75,** 519–560.

Davies, P. F., Barbee, K. A., Volin, M. V., Robotewskyj, A., Chen, J., Joseph, L., Griem, M. L., Wernick, M. N., Jacobs, E., Polacek, D. C., dePaola, N., and Barakat, A. I. (1997). Spatial relationships in early signaling events of flow-mediated endothelial mechanotransduction. *Annu. Rev. Physiol.* **59,** 527–549.

Davies, P. F., Passerini, A. G., and Simmons, C. A. (2004). Aortic valve: Turning over a new leaf(let) in endothelial phenotypic heterogeneity. *Arterioscler. Thromb. Vasc. Biol.* **24,** 1331–1333.

de la Pompa, J. L., Timmerman, L. A., Takimoto, H., Yoshida, H., Elia, A. J., Samper, E., Potter, J., Wakeham, A., Marengere, L., Langille, B. L., Crabtree, G. R., and Mak, T. W. (1998). Role of the NF-ATc transcription factor in morphogenesis of cardiac valves and septum. *Nature* **392,** 182–186.

de Lange, F. J., Moorman, A. F., Anderson, R. H., Manner, J., Soufan, A. T., de Gier-de Vries, C., Schneider, M. D., Webb, S., van den Hoff, M. J., and Christoffels, V. M. (2004). Lineage and morphogenetic analysis of the cardiac valves. *Circ. Res.* **95,** 645–654.

Deanfield, J. E., Halcox, J. P., and Rabelink, T. J. (2007). Endothelial function and dysfunction: Testing and clinical relevance. *Circulation* **115,** 1285–1295.

Deck, J. D. (1986). Endothelial cell orientation on aortic valve leaflets. *Cardiovasc. Res.* **20,** 760–767.

Delot, E. C. (2003). Control of endocardial cushion and cardiac valve maturation by BMP signaling pathways. *Mol. Genet. Metab.* **80,** 27–35.

Delot, E. C., Bahamonde, M. E., Zhao, M., and Lyons, K. M. (2003). BMP signaling is required for septation of the outflow tract of the mammalian heart. *Development* **130,** 209–220.

Devereux, R. B., Brown, W. T., Kramer-Fox, R., and Sachs, I. (1982). Inheritance of mitral valve prolapse: Effect of age and sex on gene expression. *Ann. Intern. Med.* **97,** 826–832.

Dimmeler, S., and Zeiher, A. M. (2004). Vascular repair by circulating endothelial progenitor cells: The missing link in atherosclerosis? *J. Mol. Med.* **82,** 671–677.

Dor, Y., Camenisch, T. D., Itin, A., Fishman, G. I., McDonald, J. A., Carmeliet, P., and Keshet, E. (2001). A novel role for VEGF in endocardial cushion formation and its potential contribution to congenital heart defects. *Development* **128,** 1531–1538.

Dor, Y., Klewer, S. E., McDonald, J. A., Keshet, E., and Camenisch, T. D. (2003). VEGF modulates early heart valve formation. *Anat. Rec. A Discov. Mol. Cell. Evol. Biol.* **271,** 202–208.

Dreger, S. A., Taylor, P. M., Allen, S. P., and Yacoub, M. H. (2002). Profile and localization of matrix metalloproteinases (MMPs) and their tissue inhibitors (TIMPs) in human heart valves. *J. Heart Valve Dis.* **11,** 875–880:Discussion 880.

Egorova, A. D., Van der Heiden, K., Van de Pas, S., Vennemann, P., Poelma, C., Deruiter, M. C., Goumans, M. J., Gittenberger-de Groot, A. C., Ten Dijke, P., Poelmann, R. E., and Hierck, B. P. (2011). Tgfbeta/Alk5 signaling is required for shear stress induced klf2 expression in embryonic endothelial cells. *Dev. Dyn.* **240,** 1670–1680.

Eisenberg, L. M., and Markwald, R. R. (1995). Molecular regulation of atrioventricular valvuloseptal morphogenesis. *Circ. Res.* **77,** 1–6.

Eldadah, Z. A., Hamosh, A., Biery, N. J., Montgomery, R. A., Duke, M., Elkins, R., and Dietz, H. C. (2001). Familial Tetralogy of Fallot caused by mutation in the jagged1 gene. *Hum. Mol. Genet.* **10,** 163–169.

Fang, N. T., Xie, S. Z., Wang, S. M., Gao, H. Y., Wu, C. G., and Pan, L. F. (2007). Construction of tissue-engineered heart valves by using decellularized scaffolds and endothelial progenitor cells. *Chin. Med. J.* **120,** 696–702.

Farivar, R. S., Cohn, L. H., Soltesz, E. G., Mihaljevic, T., Rawn, J. D., and Byrne, J. G. (2003). Transcriptional profiling and growth kinetics of endothelium reveals differences between cells derived from porcine aorta versus aortic valve. *Eur. J. Cardiothorac. Surg.* **24,** 527–534.

Ferrara, N., Carver-Moore, K., Chen, H., Dowd, M., Lu, L., O'Shea, K. S., Powell-Braxton, L., Hillan, K. J., and Moore, M. W. (1996). Heterozygous embryonic lethality induced by targeted inactivation of the VEGF gene. *Nature* **380,** 439–442.

Frangos, S. G., Gahtan, V., and Sumpio, B. (1999). Localization of atherosclerosis: Role of hemodynamics. *Arch. Surg.* **134,** 1142–1149.

Freeman, R. V., and Otto, C. M. (2005). Spectrum of calcific aortic valve disease: Pathogenesis, disease progression, and treatment strategies. *Circulation* **111,** 3316–3326.

Galvin, K. M., Donovan, M. J., Lynch, C. A., Meyer, R. I., Paul, R. J., Lorenz, J. N., Fairchild-Huntress, V., Dixon, K. L., Dunmore, J. H., Gimbrone, M. A., Jr., Falb, D., and Huszar, D. (2000). A role for smad6 in development and homeostasis of the cardiovascular system. *Nat. Genet.* **24,** 171–174.

Garcia-Martinez, V., Sanchez-Quintana, D., and Hurle, J. M. (1991). Histochemical and ultrastructural changes in the extracellular matrix of the developing chick semilunar heart valves. *Acta Anat.* **142,** 87–96.

Garg, V. (2006). Insights into the genetic basis of congenital heart disease. *Cell. Mol. Life. Sci.* **63,** 1141–1148.

Garg, V., Muth, A. N., Ransom, J. F., Schluterman, M. K., Barnes, R., King, I. N., Grossfeld, P. D., and Srivastava, D. (2005). Mutations in NOTCH1 cause aortic valve disease. *Nature* **437,** 270–274.

Ge, L., and Sotiropoulos, F. (2010). Direction and magnitude of blood flow shear stresses on the leaflets of aortic valves: Is there a link with valve calcification? *J. Biomech. Eng.* **132,** 014505.

Girard, P. R., and Nerem, R. M. (1995). Shear stress modulates endothelial cell morphology and F-actin organization through the regulation of focal adhesion-associated proteins. *J. Cell. Physiol.* **163,** 179–193.

Gittenberger-de Groot, A. C., Vrancken Peeters, M. P., Mentink, M. M., Gourdie, R. G., and Poelmann, R. E. (1998). Epicardium-derived cells contribute a novel population to the myocardial wall and the atrioventricular cushions. *Circ. Res.* **82,** 1043–1052.

Grande-Allen, K. J., and Liao, J. (2011). The heterogeneous biomechanics and mechanobiology of the mitral valve: Implications for tissue engineering. *Curr. Cardiol. Rep.* **13,** 113–120.

Grande-Allen, K. J., Griffin, B. P., Ratliff, N. B., Cosgrove, D. M., and Vesely, I. (2003). Glycosaminoglycan profiles of myxomatous mitral leaflets and chordae parallel the severity of mechanical alterations. *J. Am. Coll. Cardiol.* **42,** 271–277.

Groenendijk, B. C., Hierck, B. P., Vrolijk, J., Baiker, M., Pourquie, M. J., Gittenberger-de Groot, A. C., and Poelmann, R. E. (2005). Changes in shear stress-related gene expression after experimentally altered venous return in the chicken embryo. *Circ. Res.* **96,** 1291–1298.

Groenendijk, B. C., Van der Heiden, K., Hierck, B. P., and Poelmann, R. E. (2007). The role of shear stress on ET-1, KLF2, and NOS-3 expression in the developing cardiovascular system of chicken embryos in a venous ligation model. *Physiology* **22,** 380–389.

Gross, L., and Kugel, M. A. (1931). Topographic anatomy and histology of the valves in the human heart. *Am. J. Pathol.* **7,** 445–474.

Gupta, V., Barzilla, J. E., Mendez, J. S., Stephens, E. H., Lee, E. L., Collard, C. D., Laucirica, R., Weigel, P. H., and Grande-Allen, K. J. (2009). Abundance and location of proteoglycans and hyaluronan within normal and myxomatous mitral valves. *Cardiovasc. Pathol.* **18,** 191–197.

Harrelson, Z., Kelly, R. G., Goldin, S. N., Gibson-Brown, J. J., Bollag, R. J., Silver, L. M., and Papaioannou, V. E. (2004). Tbx2 is essential for patterning the atrioventricular canal and for morphogenesis of the outflow tract during heart development. *Development* **131,** 5041–5052.

Helske, S., Kupari, M., Lindstedt, K. A., and Kovanen, P. T. (2007). Aortic valve stenosis: An active atheroinflammatory process. *Curr. Opin. Lipidol.* **18,** 483–491.

Hierck, B. P., van der Heiden, K., DeRuiter, M. C., Gittenberger-de Groot, A. C., and Poelmann, R. E. (2007). Fluid shear stress controls cardiovascular development. A functionomic approach. *Wien. Klin. Wochenschr.* **119,** 10–13.

Hierck, B. P., Van der Heiden, K., Poelma, C., Westerweel, J., and Poelmann, R. E. (2008). Fluid shear stress and inner curvature remodeling of the embryonic heart. Choosing the right lane!. *The Scientific World Journal* **8,** 212–222.

Hill, J. M., Zalos, G., Halcox, J. P., Schenke, W. H., Waclawiw, M. A., Quyyumi, A. A., and Finkel, T. (2003). Circulating endothelial progenitor cells, vascular function, and cardiovascular risk. *N. Engl. J. Med.* **348,** 593–600.

Hinton, R. B., and Yutzey, K. E. (2011). Heart valve structure and function in development and disease. *Annu. Rev. Physiol.* **73,** 29–46.

Hinton, R. B., Jr., Lincoln, J., Deutsch, G. H., Osinska, H., Manning, P. B., Benson, D. W., and Yutzey, K. E. (2006). Extracellular matrix remodeling and organization in developing and diseased aortic valves. *Circ. Res.* **98,** 1431–1438.

Hoffman, J. I., and Kaplan, S. (2002). The incidence of congenital heart disease. *J. Am. Coll. Cardiol.* **39,** 1890–1900.

Hove, J. R., Koster, R. W., Forouhar, A. S., Acevedo-Bolton, G., Fraser, S. E., and Gharib, M. (2003). Intracardiac fluid forces are an essential epigenetic factor for embryonic cardiogenesis. *Nature* **421,** 172–177.

Hurle, J. M., Kitten, G. T., Sakai, L. Y., Volpin, D., and Solursh, M. (1994). Elastic extracellular matrix of the embryonic chick heart: An immunohistological study using laser confocal microscopy. *Dev. Dyn.* **200,** 321–332.

Hurlstone, A. F., Haramis, A. P., Wienholds, E., Begthel, H., Korving, J., Van Eeden, F., Cuppen, E., Zivkovic, D., Plasterk, R. H., and Clevers, H. (2003). The Wnt/beta-catenin pathway regulates cardiac valve formation. *Nature* **425,** 633–637.

Icardo, J. M., and Colvee, E. (1995). Atrioventricular valves of the mouse: III. Collagenous skeleton and myotendinous junction. *Anat. Rec.* **243,** 367–375.

Jiang, X., Rowitch, D. H., Soriano, P., McMahon, A. P., and Sucov, H. M. (2000). Fate of the mammalian cardiac neural crest. *Development* **127,** 1607–1616.

Johnson, E. N., Lee, Y. M., Sander, T. L., Rabkin, E., Schoen, F. J., Kaushal, S., and Bischoff, J. (2003). NFATc1 mediates vascular endothelial growth factor-induced proliferation of human pulmonary valve endothelial cells. *J. Biol. Chem.* **278,** 1686–1692.

Kokubo, H., Tomita-Miyagawa, S., Hamada, Y., and Saga, Y. (2007). Hesr1 and Hesr2 regulate atrioventricular boundary formation in the developing heart through the repression of Tbx2. *Development* **134,** 747–755.

Kopan, R., and Ilagan, M. X. (2009). The canonical Notch signaling pathway: Unfolding the activation mechanism. *Cell* **137,** 216–233.

Krantz, I. D., Piccoli, D. A., and Spinner, N. B. (1999). Clinical and molecular genetics of Alagille syndrome. *Curr. Opin. Pediatr.* **11,** 558–564.

Kunzelman, K. S., Cochran, R. P., Murphree, S. S., Ring, W. S., Verrier, E. D., and Eberhart, R. C. (1993). Differential collagen distribution in the mitral valve and its influence on biomechanical behaviour. *J. Heart Valve Dis.* **2,** 236–244.

Laforest, B., Andelfinger, G., and Nemer, M. (2011). Loss of Gata5 in mice leads to bicuspid aortic valve. *J. Clin. Invest.* **121,** 2876–2887.

Lange, A. W., and Yutzey, K. E. (2006). NFATc1 expression in the developing heart valves is responsive to the RANKL pathway and is required for endocardial expression of cathepsin K. *Dev. Biol.* **292,** 407–417.

Leask, R. L., Jain, N., and Butany, J. (2003). Endothelium and valvular diseases of the heart. *Microsc. Res. Tech.* **60,** 129–137.

Lee, T. C., Zhao, Y. D., Courtman, D. W., and Stewart, D. J. (2000). Abnormal aortic valve development in mice lacking endothelial nitric oxide synthase. *Circulation* **101,** 2345–2348.

Lehoux, S., Castier, Y., and Tedgui, A. (2006). Molecular mechanisms of the vascular responses to haemodynamic forces. *J. Intern. Med.* **259,** 381–392.

Levay, A. K., Peacock, J. D., Lu, Y., Koch, M., Hinton, R. B., Jr., Kadler, K. E., and Lincoln, J. (2008). Scleraxis is required for cell lineage differentiation and extracellular matrix remodeling during murine heart valve formation in vivo. *Circ. Res.* **103,** 948–956.

Li, L., Krantz, I. D., Deng, Y., Genin, A., Banta, A. B., Collins, C. C., Qi, M., Trask, B. J., Kuo, W. L., Cochran, J., Costa, T., Pierpont, M. E., *et al.* (1997). Alagille syndrome is caused by mutations in human Jagged1, which encodes a ligand for Notch1. *Nat. Genet.* **16,** 243–251.

Liebner, S., Cattelino, A., Gallini, R., Rudini, N., Iurlaro, M., Piccolo, S., and Dejana, E. (2004). Beta-catenin is required for endothelial-mesenchymal transformation during heart cushion development in the mouse. *J. Cell Biol.* **166,** 359–367.

Lincoln, J., Alfieri, C. M., and Yutzey, K. E. (2004). Development of heart valve leaflets and supporting apparatus in chicken and mouse embryos. *Dev. Dyn.* **230,** 239–250.

Lincoln, J., Alfieri, C. M., and Yutzey, K. E. (2006a). BMP and FGF regulatory pathways control cell lineage diversification of heart valve precursor cells. *Dev. Biol.* **292,** 292–302.

Lincoln, J., Lange, A. W., and Yutzey, K. E. (2006b). Hearts and bones: Shared regulatory mechanisms in heart valve, cartilage, tendon, and bone development. *Dev. Biol.* **294,** 292–302.

Lincoln, J., Kist, R., Scherer, G., and Yutzey, K. E. (2007). Sox9 is required for precursor cell expansion and extracellular matrix organization during mouse heart valve development. *Dev. Biol.* **305,** 120–132.

Liu, A. C., Joag, V. R., and Gotlieb, A. I. (2007). The emerging role of valve interstitial cell phenotypes in regulating heart valve pathobiology. *Am. J. Pathol.* **171,** 1407–1418.

Lloyd-Jones, D., Adams, R. J., Brown, T. M., Carnethon, M., Dai, S., De Simone, G., Ferguson, T. B., Ford, E., Furie, K., Gillespie, C., Go, A., Greenlund, K., *et al.* (2010). Heart disease and stroke statistics—2010 update: A report from the American Heart Association. *Circulation* **121,** e46–e215.

Lopez-Sanchez, C., and Garcia-Martinez, V. (2011). Molecular determinants of cardiac specification. *Cardiovasc. Res.* **91,** 185–195.

Luna-Zurita, L., Prados, B., Grego-Bessa, J., Luxan, G., del Monte, G., Benguria, A., Adams, R. H., Perez-Pomares, J. M., and de la Pompa, J. L. (2010). Integration of a Notch-dependent mesenchymal gene program and Bmp2-driven cell invasiveness regulates murine cardiac valve formation. *J. Clin. Invest.* **120,** 3493–3507.

Lyons, K. M., Pelton, R. W., and Hogan, B. L. (1990). Organogenesis and pattern formation in the mouse: RNA distribution patterns suggest a role for bone morphogenetic protein-2A (BMP-2A). *Development* **109,** 833–844.

Ma, L., and Martin, J. F. (2005). Generation of a Bmp2 conditional null allele. *Genesis* **42,** 203–206.

Ma, L., Lu, M. F., Schwartz, R. J., and Martin, J. F. (2005). Bmp2 is essential for cardiac cushion epithelial-mesenchymal transition and myocardial patterning. *Development* **132,** 5601–5611.

Macgrogan, D., Luna-Zurita, L., and de la Pompa, J. L. (2011). Notch signaling in cardiac valve development and disease. *Birth Defects Res. A Clin. Mol. Teratol.* **91,** 449–459.

Mann, B. K., and West, J. L. (2001). Tissue engineering in the cardiovascular system: Progress toward a tissue engineered heart. *Anat. Rec.* **263,** 367–371.

Markwald, R. R., Norris, R. A., Moreno-Rodriguez, R., and Levine, R. A. (2010). Developmental basis of adult cardiovascular diseases: Valvular heart diseases. *Ann. N. Y. Acad. Sci.* **1188,** 177–183.

Matsumoto, Y., Adams, V., Walther, C., Kleinecke, C., Brugger, P., Linke, A., Walther, T., Mohr, F. W., and Schuler, G. (2009). Reduced number and function of endothelial progenitor cells in patients with aortic valve stenosis: A novel concept for valvular endothelial cell repair. *Eur. Heart J.* **30,** 346–355.

Matzuk, M. M., Lu, N., Vogel, H., Sellheyer, K., Roop, D. R., and Bradley, A. (1995). Multiple defects and perinatal death in mice deficient in follistatin. *Nature* **374,** 360–363.

McDaniell, R., Warthen, D. M., Sanchez-Lara, P. A., Pai, A., Krantz, I. D., Piccoli, D. A., and Spinner, N. B. (2006). NOTCH2 mutations cause Alagille syndrome, a heterogeneous disorder of the notch signaling pathway. *Am. J. Hum. Genet.* **79,** 169–173.

Mercado-Pimentel, M. E., and Runyan, R. B. (2007). Multiple transforming growth factor-beta isoforms and receptors function during epithelial-mesenchymal cell transformation in the embryonic heart. *Cells Tissues Organs* **185,** 146–156.

Miquerol, L., Gertsenstein, M., Harpal, K., Rossant, J., and Nagy, A. (1999). Multiple developmental roles of VEGF suggested by a LacZ-tagged allele. *Dev. Biol.* **212,** 307–322.

Miquerol, L., Langille, B. L., and Nagy, A. (2000). Embryonic development is disrupted by modest increases in vascular endothelial growth factor gene expression. *Development* **127,** 3941–3946.

Misfeld, M., and Sievers, H. H. (2007). Heart valve macro- and microstructure. *Philos. Trans. R. Soc. Lond. B Biol. Sci.* **362,** 1421–1436.

Mohler, E. R., 3rd (2000). Are atherosclerotic processes involved in aortic-valve calcification? *Lancet* **356,** 524–525.

Montero, J. A., Giron, B., Arrechedera, H., Cheng, Y. C., Scotting, P., Chimal-Monroy, J., Garcia-Porrero, J. A., and Hurle, J. M. (2002). Expression of Sox8, Sox9 and Sox10 in the developing valves and autonomic nerves of the embryonic heart. *Mech. Dev.* **118,** 199–202.

Nakajima, Y., Yamagishi, T., Hokari, S., and Nakamura, H. (2000). Mechanisms involved in valvuloseptal endocardial cushion formation in early cardiogenesis: Roles of transforming growth factor (TGF)-beta and bone morphogenetic protein (BMP). *Anat. Rec.* **258,** 119–127.

Nicosia, M. A., Cochran, R. P., Einstein, D. R., Rutland, C. J., and Kunzelman, K. S. (2003). A coupled fluid-structure finite element model of the aortic valve and root. *J. Heart Valve Dis.* **12,** 781–789.

Niessen, K., Fu, Y., Chang, L., Hoodless, P. A., McFadden, D., and Karsan, A. (2008). Slug is a direct Notch target required for initiation of cardiac cushion cellularization. *J. Cell Biol.* **182,** 315–325.

Nieto, M. A. (2009). Epithelial-Mesenchymal Transitions in development and disease: Old views and new perspectives. *Int. J. Dev. Biol.* **53,** 1541–1547.

Nus, M., Macgrogan, D., Martinez-Poveda, B., Benito, Y., Casanova, J. C., Fernandez-Aviles, F., Bermejo, J., and de la Pompa, J. L. (2011). Diet-Induced Aortic Valve Disease in Mice Haploinsufficient for the Notch Pathway Effector RBPJK/CSL. *Arterioscler. Thromb. Vasc. Biol.* **31,** 1580–1588.

O'Brien, K. D. (2007). Epidemiology and genetics of calcific aortic valve disease. *J. Investig. Med.* **55,** 284–291.

Oda, T., Elkahloun, A. G., Pike, B. L., Okajima, K., Krantz, I. D., Genin, A., Piccoli, D. A., Meltzer, P. S., Spinner, N. B., Collins, F. S., and Chandrasekharappa, S. C. (1997). Mutations in the human Jagged1 gene are responsible for Alagille syndrome. *Nat. Genet.* **16,** 235–242.

Oram, K. F., Carver, E. A., and Gridley, T. (2003). Slug expression during organogenesis in mice. *Anat. Rec. A Discov. Mol. Cell. Evol. Biol.* **271,** 189–191.

Otto, C. M., Kuusisto, J., Reichenbach, D. D., Gown, A. M., and O'Brien, K. D. (1994). Characterization of the early lesion of 'degenerative' valvular aortic stenosis. Histological and immunohistochemical studies. *Circulation* **90,** 844–853.

Paranya, G., Vineberg, S., Dvorin, E., Kaushal, S., Roth, S. J., Rabkin, E., Schoen, F. J., and Bischoff, J. (2001). Aortic valve endothelial cells undergo transforming growth factor-beta-mediated and non-transforming growth factor-beta-mediated transdifferentiation in vitro. *Am. J. Pathol.* **159,** 1335–1343.

Paruchuri, S., Yang, J. H., Aikawa, E., Melero-Martin, J. M., Khan, Z. A., Loukogeorgakis, S., Schoen, F. J., and Bischoff, J. (2006). Human pulmonary valve progenitor cells exhibit endothelial/mesenchymal plasticity in response to vascular endothelial growth factor-A and transforming growth factor-beta2. *Circ. Res.* **99,** 861–869.

Passerini, A. G., Milsted, A., and Rittgers, S. E. (2003). Shear stress magnitude and directionality modulate growth factor gene expression in preconditioned vascular endothelial cells. *J. Vasc. Surg.* **37,** 182–190.

Pastor-Perez, F., and Marin, F. (2009). Hypertension, aortic sclerosis and the prothrombotic state: Understanding the complex interaction. *J. Hum. Hypertens.* **23,** 287–288.

Peacock, J. D., Levay, A. K., Gillaspie, D. B., Tao, G., and Lincoln, J. (2010). Reduced sox9 function promotes heart valve calcification phenotypes in vivo. *Circ. Res.* **106,** 712–719.

Perez-Pomares, J. M., Gonzalez-Rosa, J. M., and Munoz-Chapuli, R. (2009). Building the vertebrate heart - an evolutionary approach to cardiac development. *Int. J. Dev. Biol.* **53,** 1427–1443.

Person, A. D., Klewer, S. E., and Runyan, R. B. (2005). Cell biology of cardiac cushion development. *Int. Rev. Cytol.* **243,** 287–335.

Plageman, T. F., Jr., and Yutzey, K. E. (2004). Differential expression and function of Tbx5 and Tbx20 in cardiac development. *J. Biol. Chem.* **279,** 19026–19034.

Poggianti, E., Venneri, L., Chubuchny, V., Jambrik, Z., Baroncini, L. A., and Picano, E. (2003). Aortic valve sclerosis is associated with systemic endothelial dysfunction. *J. Am. Coll. Cardiol.* **41,** 136–141.

Rabkin, E., Aikawa, M., Stone, J. R., Fukumoto, Y., Libby, P., and Schoen, F. J. (2001). Activated interstitial myofibroblasts express catabolic enzymes and mediate matrix remodeling in myxomatous heart valves. *Circulation* **104,** 2525–2532.

Rabkin-Aikawa, E., Mayer, J. E., Jr., and Schoen, F. J. (2005). Heart valve regeneration. *Adv. Biochem. Eng. Biotechnol.* **94,** 141–179.

Rajamannan, N. M., Gersh, B., and Bonow, R. O. (2003a). Calcific aortic stenosis: From bench to the bedside—Emerging clinical and cellular concepts. *Heart* **89,** 801–805.

Rajamannan, N. M., Subramaniam, M., Rickard, D., Stock, S. R., Donovan, J., Springett, M., Orszulak, T., Fullerton, D. A., Tajik, A. J., Bonow, R. O., and Spelsberg, T. (2003b). Human aortic valve calcification is associated with an osteoblast phenotype. *Circulation* **107,** 2181–2184.

Ranger, A. M., Grusby, M. J., Hodge, M. R., Gravallese, E. M., de la Brousse, F. C., Hoey, T., Mickanin, C., Baldwin, H. S., and Glimcher, L. H. (1998). The transcription factor NF-ATc is essential for cardiac valve formation. *Nature* **392,** 186–190.

Rivera-Feliciano, J., and Tabin, C. J. (2006). Bmp2 instructs cardiac progenitors to form the heart-valve-inducing field. *Dev. Biol.* **295,** 580–588.

Romano, L. A., and Runyan, R. B. (2000). Slug is an essential target of TGFbeta2 signaling in the developing chicken heart. *Dev. Biol.* **223,** 91–102.

Rutenberg, J. B., Fischer, A., Jia, H., Gessler, M., Zhong, T. P., and Mercola, M. (2006). Developmental patterning of the cardiac atrioventricular canal by Notch and Hairy-related transcription factors. *Development* **133,** 4381–4390.

Sacks, M. S., David Merryman, W., and Schmidt, D. E. (2009). On the biomechanics of heart valve function. *J. Biomech.* **42,** 1804–1824.

Sales, V. L., Mettler, B. A., Engelmayr, G. C., Jr., Aikawa, E., Bischoff, J., Martin, D. P., Exarhopoulos, A., Moses, M. A., Schoen, F. J., Sacks, M. S., and Mayer, J. E., Jr. (2010). Endothelial progenitor cells as a sole source for ex vivo seeding of tissue-engineered heart valves. *Tissue Eng. Part A* **16,** 257–267.

Sanford, L. P., Ormsby, I., Gittenberger-de Groot, A. C., Sariola, H., Friedman, R., Boivin, G. P., Cardell, E. L., and Doetschman, T. (1997). TGFbeta2 knockout mice have multiple developmental defects that are non-overlapping with other TGFbeta knockout phenotypes. *Development* **124,** 2659–2670.

Santhanakrishnan, A., and Miller, L. A. (2011). Fluid Dynamics of Heart Development. *Cell Biochem Biophys.* **61**(1), 1–22.

Schoen, F. J. (1997). Aortic valve structure-function correlations: Role of elastic fibers no longer a stretch of the imagination. *J. Heart Valve Dis.* **6,** 1–6.

Schoen, F. J. (2008). Evolving concepts of cardiac valve dynamics: The continuum of development, functional structure, pathobiology, and tissue engineering. *Circulation* **118,** 1864–1880.

Schroeder, J. A., Jackson, L. F., Lee, D. C., and Camenisch, T. D. (2003). Form and function of developing heart valves: Coordination by extracellular matrix and growth factor signaling. *J. Mol. Med.* **81,** 392–403.

Scott, M., and Vesely, I. (1995). Aortic valve cusp microstructure: The role of elastin. *Ann. Thorac. Surg.* **60,** S391–S394.

Sehnert, A. J., Huq, A., Weinstein, B. M., Walker, C., Fishman, M., and Stainier, D. Y. (2002). Cardiac troponin T is essential in sarcomere assembly and cardiac contractility. *Nat. Genet.* **31,** 106–110.

Simmons, C. A., Grant, G. R., Manduchi, E., and Davies, P. F. (2005). Spatial heterogeneity of endothelial phenotypes correlates with side-specific vulnerability to calcification in normal porcine aortic valves. *Circ. Res.* **96,** 792–799.

Somi, S., Buffing, A. A., Moorman, A. F., and Van Den Hoff, M. J. (2004). Dynamic patterns of expression of BMP isoforms 2, 4, 5, 6, and 7 during chicken heart development. *Anat. Rec. A Discov. Mol. Cell. Evol. Biol.* **279,** 636–651.

Sorescu, G. P., Sykes, M., Weiss, D., Platt, M. O., Saha, A., Hwang, J., Boyd, N., Boo, Y. C., Vega, J. D., Taylor, W. R., and Jo, H. (2003). Bone morphogenic protein

4 produced in endothelial cells by oscillatory shear stress stimulates an inflammatory response. *J. Biol. Chem.* **278,** 31128–31135.

Sorrentino, S. A., Bahlmann, F. H., Besler, C., Muller, M., Schulz, S., Kirchhoff, N., Doerries, C., Horvath, T., Limbourg, A., Limbourg, F., Fliser, D., Haller, H., *et al.* (2007). Oxidant stress impairs in vivo reendothelialization capacity of endothelial progenitor cells from patients with type 2 diabetes mellitus: Restoration by the peroxisome proliferator-activated receptor-gamma agonist rosiglitazone. *Circulation* **116,** 163–173.

Sridurongrit, S., Larsson, J., Schwartz, R., Ruiz-Lozano, P., and Kaartinen, V. (2008). Signaling via the Tgf-beta type I receptor Alk5 in heart development. *Dev. Biol.* **322,** 208–218.

Stamatas, G. N., and McIntire, L. V. (2001). Rapid flow-induced responses in endothelial cells. *Biotechnol. Prog.* **17,** 383–402.

Stankunas, K., Ma, G. K., Kuhnert, F. J., Kuo, C. J., and Chang, C. P. (2010). VEGF signaling has distinct spatiotemporal roles during heart valve development. *Dev. Biol.* **347,** 325–336.

Stein, P. D., Wang, C. H., Riddle, J. M., Sabbah, H. N., Magilligan, D. J., Jr., and Hawkins, E. T. (1989). Scanning electron microscopy of operatively excised severely regurgitant floppy mitral valves. *Am. J. Cardiol.* **64,** 392–394.

Stewart, B. F., Siscovick, D., Lind, B. K., Gardin, J. M., Gottdiener, J. S., Smith, V. E., Kitzman, D. W., and Otto, C. M. (1997). Clinical factors associated with calcific aortic valve disease. Cardiovascular health study. *J. Am. Coll. Cardiol.* **29,** 630–634.

Sultana, N., Nag, K., Hoshijima, K., Laird, D. W., Kawakami, A., and Hirose, S. (2008). Zebrafish early cardiac connexin, Cx36.7/Ecx, regulates myofibril orientation and heart morphogenesis by establishing Nkx2.5 expression. *Proc. Natl. Acad. Sci. USA* **105,** 4763–4768.

Supino, P. G., Borer, J. S., Yin, A., Dillingham, E., and McClymont, W. (2004). The epidemiology of valvular heart diseases: The problem is growing. *Adv. Cardiol.* **41,** 9–15.

Tanaka, K., Sata, M., Fukuda, D., Suematsu, Y., Motomura, N., Takamoto, S., Hirata, Y., and Nagai, R. (2005). Age-associated aortic stenosis in apolipoprotein E-deficient mice. *J. Am. Coll. Cardiol.* **46,** 134–141.

Tao, G., Levay, A. K., Gridley, T., and Lincoln, J. (2011). Mmp15 is a direct target of Snai1 during endothelial to mesenchymal transformation and endocardial cushion development. *Dev. Biol.* **359,** 209–221.

Thubrikar, M. J., Aouad, J., and Nolan, S. P. (1986a). Patterns of calcific deposits in operatively excised stenotic or purely regurgitant aortic valves and their relation to mechanical stress. *Am. J. Cardiol.* **58,** 304–308.

Thubrikar, M. J., Nolan, S. P., Aouad, J., and Deck, J. D. (1986b). Stress sharing between the sinus and leaflets of canine aortic valve. *Ann. Thorac. Surg.* **42,** 434–440.

Timmerman, L. A., Grego-Bessa, J., Raya, A., Bertran, E., Perez-Pomares, J. M., Diez, J., Aranda, S., Palomo, S., McCormick, F., Izpisua-Belmonte, J. C., and de la Pompa, J. L. (2004). Notch promotes epithelial-mesenchymal transition during cardiac development and oncogenic transformation. *Genes Dev.* **18,** 99–115.

Todorovic, V., Finnegan, E., Freyer, L., Zilberberg, L., Ota, M., and Rifkin, D. B. (2011). Long form of latent TGF-beta binding protein 1 (Ltbp1L) regulates cardiac valve development. *Dev. Dyn.* **240,** 176–187.

Venkatesh, D. A., Park, K. S., Harrington, A., Miceli-Libby, L., Yoon, J. K., and Liaw, L. (2008). Cardiovascular and hematopoietic defects associated with Notch1 activation in embryonic Tie2-expressing populations. *Circ. Res.* **103,** 423–431.

Vermot, J., Forouhar, A. S., Liebling, M., Wu, D., Plummer, D., Gharib, M., and Fraser, S. E. (2009). Reversing blood flows act through klf2a to ensure normal valvulogenesis in the developing heart. *PLoS Biol.* **7,** e1000246.

Wang, H., Goland, S., Burton, K., Czer, L. S., Schwarz, E. R., and Trento, A. (2010). An agonist of liver X receptor slows valvular disease in a hypercholesterolemia mouse model. *J. Heart Valve Dis.* **19,** 653–664.

Ward, C. (2000). Clinical significance of the bicuspid aortic valve. *Heart* **83,** 81–85.

Weiss, R. M., Ohashi, M., Miller, J. D., Young, S. G., and Heistad, D. D. (2006). Calcific aortic valve stenosis in old hypercholesterolemic mice. *Circulation* **114,** 2065–2069.

Werner, N., Kosiol, S., Schiegl, T., Ahlers, P., Walenta, K., Link, A., Bohm, M., and Nickenig, G. (2005). Circulating endothelial progenitor cells and cardiovascular outcomes. *N. Engl. J. Med.* **353,** 999–1007.

Wu, B., Wang, Y., Lui, W., Langworthy, M., Tompkins, K. L., Hatzopoulos, A. K., Baldwin, H. S., and Zhou, B. (2011). Nfatc1 coordinates valve endocardial cell lineage development required for heart valve formation. *Circ. Res.* **109,** 183–192.

Zhou, B., von Gise, A., Ma, Q., Hu, Y. W., and Pu, W. T. (2010). Genetic fate mapping demonstrates contribution of epicardium-derived cells to the annulus fibrosis of the mammalian heart. *Dev. Biol.* **338,** 251–261.

CHAPTER EIGHT

An Epicardial Floor Plan for Building and Rebuilding the Mammalian Heart

Paul R. Riley

Contents

Abstract

The epicardium is a mesothelial cell layer which contributes to the coronary vessels and myocardium and acts as an important source of trophic signals to maintain continued growth and differentiation of the developing heart. The precise lineage potential of the embryonic epicardium has come under recent scrutiny with notable questions around its capacity to give rise to derivative vascular endothelial cells and cardiomyocytes. The importance of the epicardium is not restricted to heart formation. Recent studies in the adult heart have highlighted a paracrine role in modulating injury and have begun to realize its potential as a source of progenitor cells (EPDCs) which may be reactivated toward facilitating neovascularization and myocardial repair after ischemic injury. Thus, the adult epicardium has an embryological origin and emerges as a prime exemplar of the paradigm of activated resident stem cell therapy in regenerative medicine, whereupon a major goal is to restore embryonic plasticity to otherwise dormant adult progenitors and facilitate organ repair. In this review, we will

Department of Physiology, Anatomy and Genetics, University of Oxford, Oxford, OX1 3PT, United Kingdom

Current Topics in Developmental Biology, Volume 100
ISSN 0070-2153, DOI: 10.1016/B978-0-12-387786-4.00007-5

explore current thinking on the origins of the epicardium, its role as a signaling center, lineage heterogeneity, and controversy around epicardial potential within the developing heart. We will extrapolate to the adult injury setting, drawing on key studies in zebrafish and mouse which establish the basis for the adult epicardium as a target for cardiovascular regeneration. Finally, we will consider translation of this potential to the human lineage alongside the prospects for discovery of target cell-based therapeutics.

1. Introduction

Epicardium describes the outer layer of heart tissue (from Greek; *epi*—outer, *cardium*—heart). Historically we have known about the epicardium for over 150 years with the first description of its origin as the proepicardium in 1855 by Robert Remak the Polish–German embryologist and anatomist (1815–1865; Remak, 1855). Subsequently, there followed a serious embryological misunderstanding, with Mollier describing it as a *myoepicardialer mantle*, implying it to be an anlage of both the myocardium and the epicardium (Mollier, 1906). This was challenged by Kurkiewicz in 1909, who established a primarily extracardiac origin of the epicardium (Kurkiewicz, 1909) and later verified by the work of Manasek, who used TEM to establish that the epicardium is a simple epithelium, formed by the spreading of mesothelial cells from the dorsal wall of the sinus venosus region (Manasek, 1969).

Thus, from its earliest inception, the epicardium seems to have been somewhat misinterpreted and, to some extent, this persists in contemporary studies with debate as to the precise origins of the epicardium (or proepicardial organ), its lineage potential and role in the adult heart. Studies on epicardial ontogeny, moving from dogfish through chick-quail to mammals, identified differences around the emergence of the earliest mesothelial protrusions and whether the proepicardial anlage persist unilaterally or encompasses the ventral walls of both horns of the sinus venous. Here the debate is subtle, whereas the discussion around lineage contribution is more polarized. A number of studies present definitive evidence for epicardium-derived cells (EPDCs) contributing vascular smooth muscle cells (vSMCs) and interstitial fibroblasts in the developing heart; recently reviewed in Vincent and Buckingham (2010), but the picture is less clear on a putative contribution, however minor, to coronary endothelial cells (ECs) and cardiomyocytes (Cai *et al.*, 2008; Christoffels *et al.*, 2009; Red-Horse *et al.*, 2010; Zhou *et al.*, 2008).

In the adult zebrafish heart, there appears to be a continued requirement for epicardial function in maintaining homeostatic organ growth but, since the mammalian heart does not continue to grow postnatally, this may be a property largely restricted to teleosts. Consequently, in adult mammals questions remain as to whether the epicardium has any homeostatic role, either via signaling to maintain cardiac function and/or through contribution to (low-level) cell turnover and replenishment. What has become apparent in recent times is that following ischemic injury the adult epicardium is not as quiescent as once thought. This has been demonstrated in both mouse and zebrafish. In the latter, organ-wide activation underpins the inherent regenerative potential of the adult zebrafish heart (Lepilina *et al.*, 2006), whereas in mammals adult epicardial cells proliferate and establish signaling cues to modulate injury (Zhou *et al.*, 2011). If ectopically stimulated, preinjury, they rediscover lost embryonic potential to contribute vascular cells and cardiomyocytes for heart repair (Bock-Marquette *et al.*, 2009; Smart *et al.*, 2007, 2010, 2011).

There have been many insightful reviews on the developing epicardium (refer to Olivey and Svensson, 2010 and references therein); hence, the focus here will be on the areas highlighted which court some degree of controversy and for which there remain unanswered questions with respect to both the embryonic and the adult lineage. The emphasis will be on mouse, as a model organism with potential to more seamlessly translate to humans, but important cross-species comparisons will be drawn where necessary.

2. The Origins, Lineage Checkpoints, and Heterogeneity of the Developing Epicardium

In mammals, the proepicardial organ (PEO) arises from clusters of mesothelial protrusions, in a bilaterally symmetrical fashion (similar to that found in dogfish) which form at the pericardial surface of the septum transversum in close proximity to the sinus venosus. Initially, the protrusions consist exclusively of rounded mesothelial cells but remodel to form villi containing mesenchyme rich in extracellular matrix (reviewed in Manner *et al.*, 2001). The genetic factors which induce the mesothelial protrusions in mammals are not known, but since they develop bilaterally, as opposed to exclusively on the right side as in avians, they are not thought to be subject to L–R signaling. Advances made in understanding determinants of the PEO and early cell fate have largely been made in the chick. Here there are two principal schools of thought centered on the relative importance of FGF and BMP signaling. On the one hand, development of the PEO anlage on the right was shown to be dependent upon FGF8 and Snai1, downstream of the L–R axis, inducing BMP2 in both sinus horns but maintained only on the right by BMP4 in the mesenchyme precursors of the PEO. Subsequent cellular outgrowth and survival of the PEO was dependent upon further FGF signaling, specifically on FGF2 and FGF10 acting in both an autocrine and a paracrine manner (Torlopp *et al.*, 2010). On the other hand, FGF signaling was reported to be dominant over BMP/Smad signaling, to ensure separation of the PEO lineage from precardiac mesoderm. Here, contrary to

BMP being a proepicardial signal, it was shown to cause developmental arrest of the epicardium and enhance myocardium formation at the inflow of the heart (van *et al.*, 2009). While not easily reconciled, these two studies demonstrate cross-talk between FGF and BMP signaling as a critical checkpoint in determining epicardial fate. Equivalent studies have yet to be reported in the mouse, since specific lineage tracing at these early stages remains elusive and the early mouse embryo does not lend itself as well to DiI tracing. A study in zebrafish, however, has established a genetic framework for PEO development. Here tcf21 (also known as epicardin/capsulin/pod1 in mammals) and Wt1 (the orthologue of the Wilm's Tumor transcriptional repressor) were used to trace epicardial development following morpholino knockdown of *miles apart* (a sphingosine-1-phosphate receptor) or *casanova* (a Sox-related factor). Both knockdowns resulted in cardiabifida and each bilateral heart was associated with its own PEO, suggesting the earliest progenitors of the epicardium lie within the lateral plate mesoderm. In the same study, Wt1 was shown to be essential for the PEO lineage and the cell polarity genes *heart and soul* (an atypical protein kinase C iota) and *nagie oko* (a membrane protein palmitoylated 5) required for PEO morphogenesis (Serluca, 2008). In mammals, functional redundancy precludes such early essential roles for Wt1 and cell polarity genes in PEO formation: Wt1, while not required for establishing the PEO, is essential for repression of the epithelial phenotype in epicardial cells to promote epithelial-to-mesenchyme transformation (EMT; Martinez-Estrada *et al.*, 2010; Moore *et al.*, 1999) and loss of function of the planar cell polarity gene, Vangl2, results in non-cell-autonomous epicardial defects and impaired coronary vessel formation (Phillips *et al.*, 2008).

Aside from understanding how the PEO is formed, a significant question is the extent of heterogeneity of the PEO and resulting epicardium in terms of lineage potential and cell fate. It remains to be unequivocally determined whether all PEO/epicardial cells have the same potential or whether they respond differently to different signals instructing them down alternate differentiation pathways, notably vSMCs and/or fibroblasts or, to a lesser extent, ECs and/or cardiomyocytes. The variation in abundance of these derivatives would suggest that epicardial progenitors are stratified to alternate fates and respond differently to developmental signals. Further evidence for this arises via careful expression analyses of the cells within the PEO which reveals that, while Tbx18-expressing and Wt1-expressing cells partially overlap, there is evidence of genetically distinct compartments which are positive for either Tbx18 or Wt1 (Cai *et al.*, 2008; Kraus *et al.*, 2001; Zhou *et al.*, 2008). Moreover, there is at least one other subpopulation which appears to express neither marker (Paul Riley, unpublished observations). As such, it is entirely plausible that this double negative, Tbx18-/Wt1-, population may contribute ECs of the coronary vasculature and/or myocardial cells.

3. The Importance of Epicardial Signaling

A number of canonical signaling pathways, implicated throughout numerous developmental events, have been shown to play a key role within the epicardium. Signaling orchestrates epicardial contribution to the developing coronary vasculature and establishes trophic support to the developing myocardium. What is not so well established is the hierarchy of signaling or how the various pathways intersect with respect to epicardial cell behavior (migration and proliferation) and subsequent lineage contributions to the developing vasculature and myocardium. Moreover, the distinction between signaling requirements for the development and maintenance of the epicardium versus the role of equivalent signals from the epicardium itself is not well defined. That said, undoubtedly a major contributor to coronary development is communication between the epicardium, subepicardial mesenchyme, and the myocardium, which is mediated in part by complex exchange of mitogenic factors (reviewed in Olivey and Svensson, 2010). In terms of coronary angiogenesis, the most documented effect of signaling pathways on the epicardium is vSMC formation. There are two distinct levels of regulation here: either through an early effect on epicardial EMT and/or cell migration or through a direct effect on vSMC differentiation *per se*. Both processes if perturbed result in reduced smooth muscle cell coverage of the developing coronary arteries. Canonical Wnt signaling is required for expansion of the subepicardial space and invasion of the myocardium, such that epicardial-specific loss of β-catenin impaired migration and ultimately vSMC differentiation (Zamora *et al.*, 2007). Intriguingly, the source of Wnt remains unknown. Similar phenotypes have been recorded for aberrant FGFR1 and PDGFRβ signaling. Knockdown of FGFR1 in the PEO in chick compromised the ability of epicardial progeny to invade the myocardium, although this was independent of EMT (Pennisi and Mikawa, 2009). Loss of PDGFRβ and downstream PI3K signaling affected the cytoskeletal arrangement of epicardial cells leading to aberrant migration *in vivo* and a complete lack of coronary vSMCs (Mellgren *et al.*, 2008). To date, only the Notch pathway has formally been shown to have a direct affecting on the differentiation of epicardial cells into vSMCs without effecting EMT or migration. Epicardial inactivation of Rbpj (the intracellular mediator of Notch) led to failed differentiation of vSMCs, whereas conditional activation of Notch induced premature SMC differentiation. Moreover, Notch signaling was shown to cooperate with the TGFβ pathway, a well-established mediator of SMC differentiation (Grieskamp *et al.*, 2011). In addition to establishing the vSMC support of the developing coronaries, the proliferation and morphogenesis of the myocardium is directly regulated by signaling from epicardial cells. A major player here is the retinoic acid (RA) pathway. In RA receptor (RXRα)-deficient embryos, proliferation and morphogenesis of the compact myocardium was perturbed (Kang and Sucov, 2005) and epicardial-specific knockdown of RXRα resulted in impaired myocardial growth and coronary artery formation (Merki *et al.*, 2005). Here, cooperation of RA with an FGF2/Wnt9b feedback loop was shown to be instrumental in EMT (Merki *et al.*, 2005), as a rare example of cross-talk between distinct pathways. RALDH2, which controls the synthesis of RA, is expressed in the developing epicardium, and its spatiotemporal localization directly reflects RA responsiveness (Moss *et al.*, 1998). Studies in chick-quail chimeras revealed dual roles for the Wilms' Tumor transcription factor Wt1 and RALDH2, whereby Wt1+/RALHD2+ epicardial cells were shown to be important for proper development of the ventricular myocardium (Perez-Pomares *et al.*, 2002b). Subsequently Chip analyses in Wt1-null and transgenic mice demonstrated direct binding of RALDH2 by Wt1 (Guadix *et al.*, 2011) establishing a signaling hierarchy and extending the role of Wt1 beyond promoting epicardial EMT (Martinez-Estrada *et al.*, 2010). It is now well established that the developing epicardium secretes RA-responsive trophic factors which drive cardiomyocyte proliferation and promote ventricular growth. The identity of these downstream factors, however, remains elusive. Attempts to draw on parallel RA signaling in the endocardium, which has an instructive role in directing myocardial trabeculation (Lee *et al.*, 1995), proved unsuccessful: while RA responses in endocardium and epicardium were shown to converge via common proliferative components, such as PI3K and Erk, they subsequently diverged downstream at the level of neuregulin signaling (Kang and Sucov, 2005). Therefore, the elucidation of (RA-induced) secreted factors from the developing epicardium and an interrogation of their overlapping and unique roles remains a priority within the field (Sucov *et al.*, 2009). This is a significant goal restricted to not only informing on heart development and congenital heart disease but also identifying candidate mitogens which may reactivate epicardial cells toward repair in the injured adult heart.

4. Do EPDCs Give Rise to Coronary Artery Endothelial Cells?

Studies carried out in chick with quail PEO explants documented a mesothelial origin for endothelial and smooth muscle cells (Perez-Pomares *et al.*, 2002a; Poelmann *et al.*, 1993). This contribution did not account for all of the ECs in the chicken, but was instrumental in establishing a view that propepicardial cells undergo an EMT and then differentiate into isolated endothelial progenitors that assemble *de novo* vasculogenesis into endothelial tubes. However, not all studies in chick are consistent with this model

(Poelmann *et al.*, 1993) and lineage-tracing experiments in mouse show that either Tbx18+ or Wt1+ labeled cells from the proepicardium give rise to few if any coronary ECs (Cai *et al.*, 2008; Zhou *et al.*, 2008). Recently, an alternate source was proposed following the demonstration that coronary vessels arise from angiogenic sprouts of the sinus venosus, the cavity at the inflow region which returns venous blood to the embryonic heart. Local signals were postulated to induce angiogenic outgrowth, dedifferentiation from a venous fate, and stepwise conversion into coronary arteries, capillaries, and veins (Red-Horse *et al.*, 2010). This study established a new paradigm in coronary vessel development and lent weight to the idea that differentiated venous cells retain developmental plasticity as well as shifted the emphasis away from the epicardium as the major source of the coronaries. That said, the authors were unable to categorically rule out a minor epicardial contribution of ECs. Moreover, it was apparent that clonal analysis revealed a secondary source of endocardium-derived coronary ECs. A significant issue now remains in determining the relative contribution from multiple sources, sinus venosus, endocardium, and epicardium, currently difficult to reconcile with existing cre-lox tools available, as reviewed in Riley and Smart (2011). What is also not fully resolved is the earliest spatial contribution and, in particular, whether there is a common progenitor with EC potential which accounts for sources in close anatomical proximity (Fig. 8.1). The sinus venosus and PEO are almost physically associated at the inflow region of the developing heart (Fig. 8.1A) and it is not inconceivable that angioblasts from the liver sinuses may contribute to both tissues. Extending this one step further, it is also plausible that proepicardial cells contribute to the early sinus venosus and cardiac endocardium, respectively, reflecting both a common progenitor contribution to coronary vessel formation and the diversity of progenitor populations that comprise the proepicardium (Fig 8.1B). Temporally regulated contribution is another important yet unresolved issue in coronary vasculogenesis. The timing of addition of discrete progenitors to the developing coronaries is undoubtedly important. For example, as speculated by Red-Horse *et al.* (2010), the endocardial source of coronary ECs may contribute later via endocardial budding from the myocardium to form endothelial blood islands which are subsequently added to the sinus-venosus-derived plexus.

5. Do EPDCs Contribute to the Myocardial Lineage?

Uncovering the origins of myocardial cells is important for understanding and treating congenital and acquired heart disease. The rationale for investigating whether the embryonic epicardium contained cardiogenic

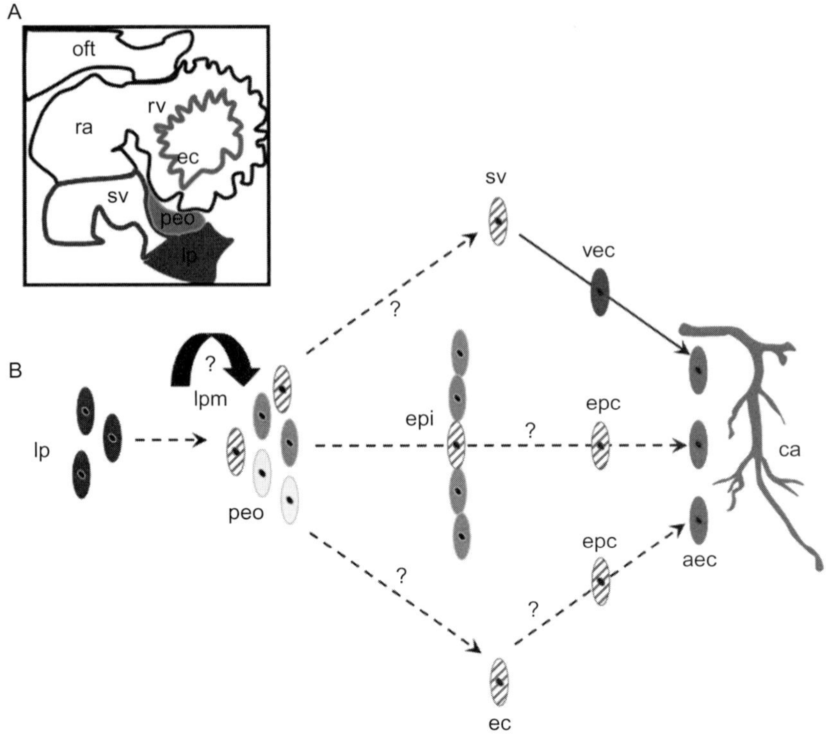

Figure 8.1 The proepicardial organ may contribute discrete progenitors to the three lineages implicated in coronary artery formation. A schematic of the mouse heart at embryonic day (E) 9.5 (A) illustrates the close anatomical proximity of the developing liver plexus (lp), proepicardial organ (peo), and sinus venosus (sv) at the inflow of the heart, relative to the endocardium (ec) forming within the right ventricle (rv); ra, right atrium. Potential contribution of peo progenitors to distinct sources of coronary artery endothelial cells (B). A common progenitor emerging from the lp may contribute to the peo during its formation by unknown signal induction in the lateral plate mesoderm (lpm). Diverse progenitor populations comprise the peo of which, a discrete subtype, with endothelial potential, may contribute to the sv, the epicardium (epi), and the ec. Ultimately this lineage may contribute to the sv-derived venous endothelial cells (vec) which can dedifferentiate and/or endothelial progenitors (epc) from either epi or ec to contribute arterial endothelial cells (aec) to the developing coronary arteries (ca). (For color version of this figure, the reader is referred to the Web version of this chapter.)

precursors stemmed, in the first instance, from a need to genetically trace all epicardium-derived progenitors to accurately establish their fate and, second, from the finding in the adult zebrafish, following resection injury, that Tbx18-expressing (epicardial) cells clustered around the wound site potentially contributing a source of cardiomyocytes during regeneration (Lepilina *et al.*, 2006). Two studies lineage-traced Tbx18+ and Wt1+ expressing cells from the PEO to establish a significant contribution of cardiomyocytes in the ventricles, interventricular septum, and atria (Cai *et al.*, 2008; Zhou *et al.*, 2008). Thus, in combination, these studies identified a previously unrecognized source of cardiomyocyte progenitors and extended our insight into the pluripotency of the proepicardium. In the case of the Tbx18+ population, the findings were subsequently challenged on the basis that the lineage trace was flawed. Tbx18 is itself expressed in the myocardium, most notably in the outer left ventricular wall and throughout the interventricular septum (Christoffels *et al.*, 2009). This raised a critical point regarding interpretation, established a need for careful and appropriately sensitive expression analysis of the gene employed as the Cre driver, prior to genetic lineage tracing, and also highlighted a requirement for independent approaches to assess a specific progenitor to differentiated-cell-type relationship. The Wt1+ lineage trace was more rigorous in employing a tamoxifen-inducible Cre approach for pulse labeling of cells, but there remained the possibility that Wt1 might be active in preexisting cardiomyocytes via the GFPCre and CreERT2 knock-in alleles. On balance the two studies appear to provide evidence of epicardial progenitors contributing to the cardiomyocyte lineage, but it is likely the extent of contribution is overestimated. However, within the field the fundamental finding remains controversial and will only be resolved by combining further rigorous lineage tracing, using alternate cre drivers, with support from clonal analyses and cell transplantation.

6. Is There a Role for the Adult Epicardium in Cardiac Homeostasis?

The epicardium in mouse dramatically loses its migratory and differentiation potential from midgestation through to early postnatal stages (Chen *et al.*, 2002). It is an intriguing question, therefore, as to whether during adulthood the epicardium contributes to maintaining cardiac function either via direct replenishment of "exhausted" cardiovascular cells or through paracrine signaling to sustain the underlying myocardium. In the adult zebrafish, there is evidence of epicardial cell addition to the ventricle in an FGF-dependent fashion to support homeostasis (Wills *et al.*, 2008). These findings, while undoubtedly of interest, have to be set against the unique indeterminate growth of the adult heart in the fish, which can occur under conditions of low population density, altered social interactions, and increased resources. Thus the zebrafish epicardium facilitates growth by acting as a supplement to cardiomyocyte hyperplasia; whether it can also act to maintain homeostasis in the absence of growth has yet to be determined.

7. What, if Any, Is the Role of the Epicardium Following Injury?

It is now apparent that the adult epicardium responds to injury, countering the idea that it is an entirely quiescent lineage. In the resected zebrafish heart, the earliest documented event is reactivation of embryonic epicardial genes and organ-wide activation, such that epicardial cells migrate into the underlying myocardium to stimulate resident cardiac progenitors (Lepilina *et al.*, 2006). In an alternate cryo-injury model, the initial upregulation of epicardial genes is coincident with significant epicardial proliferation (Gonzalez-Rosa *et al.*, 2011). During development, the epicardium has been shown to stimulate myocardial cell proliferation through the secretion of trophic factors (reviewed above and in Sucov *et al.*, 2009). Therefore, following adult injury the increased epicardial population might act as a source of signaling molecules driving regeneration of the underlying myocardium. This has recently been demonstrated in zebrafish where there is a requirement for epicardial RA production and the targeting of RA synthesis to the damaged heart tissue to promote cardiomyocyte proliferation (Kikuchi *et al.*, 2011b). A role for epicardial signals modulating injury has also recently been extrapolated to mammals, including man. *In vitro* and *in vivo* assays demonstrated that EPDCs produced paracrine factors, chief amongst them VEGFA and FGF2, that strongly promoted angiogenesis. In the same study, EPDC-conditioned medium significantly reduced infarct size and improved cardiac function (Zhou *et al.*, 2011). Epicardial thickening and relative sparing of proximal myocardium in receipt of paracrine factors were observed in ischemic canine and human myocardium, suggesting protective roles of the epicardium in larger mammals (Zhou *et al.*, 2011). Extrapolating from the known cardiogenic signaling during development may provide a means to understand how adult epicardium can sustain the immediate subepicardial myocardium. In turn, this may lead to interventional strategies to establish a gradient of protection which encompasses the more distal regions of injured heart muscle.

8. Restoration of Embryonic Plasticity to Adult EPDCs to Effect Cardiovascular Repair

In the mouse, migration of embryonic epicardial cells to give rise to vSMCs, fibroblasts, and putatively a small number of ECs was shown to be dependent upon a myocardial source of the actin binding protein, thymosin β4 (Tβ4; Fig. 8.2A; Smart *et al.*, 2007). Subsequently, it was revealed that not only was Tβ4 necessary for epicardial-derived coronary vessel

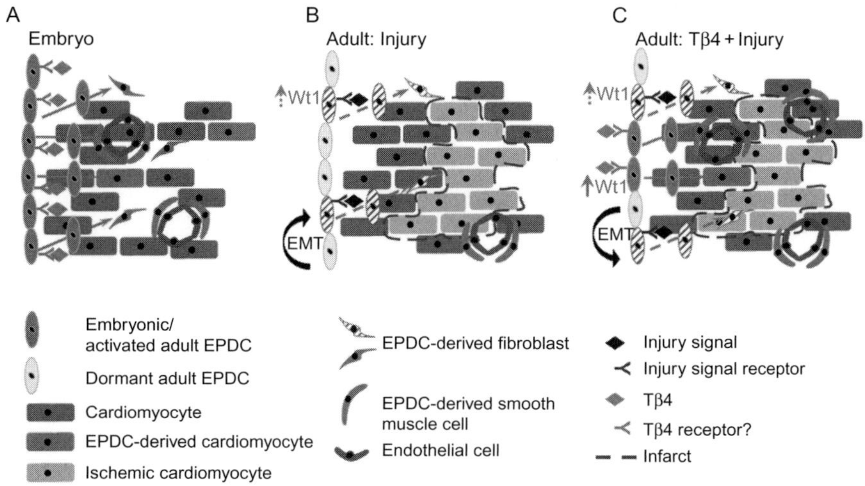

Figure 8.2 Model for how Thymosin β4 and injury might restore embryonic epicardial potential in adult heart repair. (A) During heart development, a myocardial source of thymosin β4 (Tβ4) acts in a paracrine fashion to stimulate migration of embryonic epicardium-derived cells (EPDCs) into the underlying myocardium where they can contribute vSMCs, fibroblasts, and cardiomyocytes. (B) Following ischemic injury in the adult heart, the epicardium can respond to, as-yet-unknown (?), injury-induced signals to partially upregulate the embryonic epicardial gene Wt1, proliferate, and undergo an epithelial-to-mesenchmye transformation (EMT) followed by restricted migration and contribution of fibroblasts to partially recapturing embryonic potential. (C) Pretreatment (priming) with ectopic Tβ4 prior to injury is proposed to recruit additional epicardial progenitor subpopulation(s), via significant and precocious upregulation of Wt1 (relative to injury alone). Tβ4-dependent subpopulations of activated epicardial progenitors can contribute vSMCs and cardiomyocytes. Tβ4 priming plus injury thus restores complete embryonic potential to the adult epicardium to establish the basis for neovascular and myocardial repair. (See Color Insert.)

development, but when applied ectopically to explants, it was sufficient to activate adult epicardial cells to give rise to vascular derivatives analogous to its embryonic potential (Smart *et al.*, 2007). Subsequent studies confirmed a role for Tβ4 in stimulating neovascularization *in vivo*, such that coincident with epicardial thickening, stable new blood vessels incorporating epicardium-derived vSMCs were tracked in adult mouse hearts following Tβ4-treatment and myocardial infarction (Fig. 8.1B; Bock-Marquette *et al.*, 2009; Smart *et al.*, 2010). The source of ECs underpinning new vessel growth was unclear from these studies consistent with the embryonic potential and the apparent lack, or at best minimal contribution, of epicardium-derived ECs during development (Cai *et al.*, 2008; Zhou *et al.*, 2008). The ability of embryonic EPDCs to contribute vSMCs to the developing coronaries, however, is well established, and an adult contribution, to enable collateral support of developing arterioles, with Tβ4 treatment is highly significant. Stable neovascularization addresses a major shortcoming of recent angiogenic therapies in clinical trials which have thus far been limited to short-term capillary growth (Tomanek *et al.*, 2004).

A key observation regarding the neovascular response to Tβ4 was reactivation of the embryonic epicardial gene program, as defined by upregulation of Tbx18, Wt1, and Raldh2 (Bock-Marquette *et al.*, 2009; Smart *et al.*, 2010). This was also documented in the mouse for Tbx18 and Wt1 after injury alone (Limana *et al.*, 2010; Wagner *et al.*, 2002), suggesting that tissue damage in adult animals reactivates the developmental gene regulatory network. The increase in Wt1 was subsequently exploited by Smart and colleagues to translate the myocardial potential observed for embryonic epicardium into the injured adult heart (Smart *et al.*, 2011). The Wt1 knock-in mice, employed by Zhou *et al.* (2008), were inactive in the intact adult heart in accordance with the downregulation of Wt1 during late stages of development (Smart *et al.*, 2011). Smart and colleagues reactivated Wt1, and with it a fluorescent cell label, by priming with Tβ4 followed by myocardial infarction, enabling determination of fate. Tβ4 pretreatment not only significantly and precociously increased Wt1 expression, compared to injury alone (Fig. 8.2B), but was instrumental in the differentiation of Wt1-expressing cells into cardiac muscle in the injured heart (Fig. 8.2C). *De novo* cardiomyocytes arising from transplantation of donor-labeled progenitors, into unlabelled host hearts, excluded the possibility that Wt1 was simply activated in existing myocytes and FISH analyses, following sex mismatched donor–host transplantation, ruled out cell fusion. Importantly, the new cardiomyocytes were fully integrated and contributed to improved cardiac function (Smart *et al.*, 2011).

Interestingly, in zebrafish inducible cre targeting of the tcf21 locus, to specifically trace adult epicardial cells following ventricular resection injury, suggested epicardial fates are limited to nonmyocardial cell types (Kikuchi *et al.*, 2011a). This may be because the zebrafish does not require a progenitor source of new myocytes, since it is clear from two recent studies that myocardial repair is instigated by dedifferentiation and proliferation of existing cardiomyocytes in the adult zebrafish (Jopling *et al.*, 2010; Kikuchi *et al.*, 2010). While a nonmyocardial fate for zebrafish epicardial cells may reflect the natural state, this does not preclude the possibility of retained plasticity, such that with ectopic stimulation they may be persuaded to differentiate into cardiomyocytes. Stimulation of adult zebrafish epicardium during injury, with a molecule such as Tβ4, remains to be determined.

In a recent study, early postnatal mice were subjected to ventricular resection, paralleling the zebrafish model. One day after birth (P1) resected hearts could fully regenerate, in the absence of scarring, via existing myocyte proliferation, whereas by 7 days (P7) regenerative capacity was lost and replaced by fibrosis (Porrello *et al.*, 2011). This transient regeneration was attributed to cardiomyocyte transition from a primitive mononuclear state, as exists throughout adulthood in zebrafish, to a more mature, binucleated state, incapable of sarcomere disassembly, and dedifferentiating to facilitate subsequent proliferation. Interestingly, the authors of this study were unable to rule out a progenitor contribution to the process, and it is of interest to place this conclusion in the context of the postnatal epicardium. As previously noted, the epicardium loses its potential to outgrow and contribute to cardiovascular cell types during development from midgestation through to postnatal stages (Chen *et al.*, 2002). Here, the critical cut off for epicardial activity was 4 days after birth (P4), right in the middle of the 7th day window for transient myocardial regeneration in the postnatal mouse. This raises the intriguing possibility that some of the differences between regenerative capacities in P1 versus P7 mice might be explained by loss of epicardial potential.

Irrespective of whether postnatal epicardial cells ordinarily contribute cardiovascular cells during homeostasis or regeneration, it would appear that, at least in the mouse, they retain plasticity and can be ectopically activated by trophic molecules to contribute to both neovascular and myocardial repair of the injured adult heart.

9. Human Epicardial Cells: Potential and Putative Role in Disease

A significant step toward realizing any therapeutic potential of activating resident epicardial cells is translating their plasticity into man. In this regard, a number of studies have isolated and begun to characterize human primary epicardial cells (Bax *et al.*, 2011; vanTuyn *et al.*, 2006; Winter *et al.*, 2007). Notably, cultures were established from right atrial appendages, taken from human patients undergoing right coronary artery bypass. While this includes a caveat that atrial epicardial cells may differ from their ventricular counterparts, with the latter being the more optimal subpopulation with respect to stimulation postventricular infarction, they are consistent with the ultimate target population of resident cells from diseased and aged individuals. Study of human adult atrial epicardial cells *in vitro* has revealed that they express the epicardial marker WT1 and can spontaneously undergo EMT, further promoted by TGFβ1 and BMP2 to give rise to a smooth muscle cell phenotype (Bax *et al.*, 2011; vanTuyn *et al.*, 2006). Further, zoonotic transplantation of human EPDCs into infarcted mouse hearts resulted in increased vascularization and improved functional parameters over a 6-week duration (Winter *et al.*, 2007). It should be noted here that at the final time point few EPDCs were observed within the murine host hearts, and all were negative for cardiomyocyte markers. As for the adult zebrafish, it remains to be determined whether human EPDCs can be ectopically stimulated (with molecules such as Tβ4) to ensure optimal activation, elevated numbers, and expansion of their potential to a myocardial fate. A study examining cell types within the human epicardium from archived adult autopsies revealed c-kit+ and CD34+ cardiovascular precursors, a subpopulation of which expressed the early cardiomyocyte markers Nkx2.5 and Gata4 (Limana *et al.*, 2007). Equivalent cells were identified within mouse epicardium and analogous c-kit+/Gata4+ cells were shown to expand following myocardial infarction (Limana *et al.*, 2007). In studies looking directly at human hearts with ischemic and hypertrophic cardiomyopathy (HCM), the epicardium appears to be implicated in both forms of the disease. In the ischemic setting, epicardial cells were absent from the heart surface and accumulated in the subepicardium, suggesting activated EMT, epicardial pool exhaustion, and loss of regenerative potential (Di Meglio *et al.*, 2009, 1436 /id). In HCM, the cross-talk between healthy EPDCs and abnormally contracting cardiomyocytes was suggested to account for the diverse manifestations of HCM via mechanotransduction leading to abnormal gene expression and differentiation (Olivotto *et al.*, 2009). While largely correlative observations at this stage, the possibility that the epicardium may contribute to chronic heart disease (in addition to acute injury) is intriguing with respect to therapeutic reactivation of the epicardium and reversal of the underlying pathology and progression to heart failure.

10. Future Perspectives

The identification of signals that instruct epicardial cells is a key objective, given their inherent potential in the embryo and their apparent retained plasticity in the adult heart. There is a clear need to better understand the potential heterogeneity of the developing and adult epicardium and identify subpopulations which respond to distinct signals to instruct cell fate. New candidate signals, for stimulating the adult lineage, may well be extrapolated directly from developmental factors (active metabolites) which induce embryonic epicardium to contribute extensively to formation of the heart. Other signals might be identified via less conventional routes; for example, it is evident that the pericardial fluid contains trophic stimuli which have the potential to activate a subpopulation (c-kit+) of epicardial cells following injury (Limana *et al.*, 2010). Thus far this has only been demonstrated in mouse, but in humans pericardial fluid from patients with ischemic and nonischemic heart disease has biological effects and can modulate growth and survival of cardiomyocytes (Corda *et al.*, 1997) and endothelial and smooth muscle cells (Iwakura *et al.*, 2000; Yoneda *et al.*, 2000). Consequently, the pericardium may be an important source of signals to restore activity to the human adult epicardium following ischemic injury.

As reviewed herein, controversy surrounds some of the attributed epicardial lineage potential during development. Equally, in the adult heart it is unclear whether the epicardium can contribute ECs for neovascularization or whether at any point during adulthood the epicardium acts as a source of either vascular or myocardial cell replenishment. What is evident is the adult epicardium can be activated to switch on embryonic epicardial genes, rediscover embryonic potential, and contribute vSMCs and cardiomyocytes. Thymosin β4 is an epicardial activator which has received much attention, and studies are ongoing to determine the precise mechanism by which this trophic molecule can reactivate dormant adult epicardial cells. While Tβ4 has provided a window into adult epicardial cell potential, the differentiation to cardiomyocytes was inefficient. This may reflect restricted recruitment of a discrete subpopulation of dormant progenitors with myocardial potential from within the epicardium (Fig. 8.2C). Consequently, the door is open to identify other molecules as potential drug candidates which can stimulate the various subpopulations of adult epicardial cells to adopt either vascular or myocardial fates for cardiac repair. This is likely to be more than one factor and may result in combinatorial treatments to achieve what is certainly an ambitious goal. The demonstration that adult epicardial cells can contribute cardiomyocytes was dependent upon priming with Tβ4, followed by injury (Smart *et al.*, 2011). The issue of priming presents a clinical conundrum, whereby treatment before the actual injury event may be required. As such, any therapy will need to be targeted to either "at risk" individuals, identified based on genetic predisposition (family history) and monitoring of risk factors such as blood pressure and circulating LDL cholesterol or those "patients" who present at hospital with chest pain (coronary angina) as the first indications of an infarction and who, therefore, may offer a brief window of opportunity for resident cell stimulation as a preemptive strike against the disease.

The epicardium, after a slightly checkered start has emerged as a tractable lineage for resident cell-based cardiac repair. Developmental biology is critical here in facilitating a back-to-basics approach to understand how the embryo employs the epicardium to assist in building a heart, thus providing invaluable insight into how the adult epicardium might be exploited, following injury, to rebuild a diseased heart.

ACKNOWLEDGMENTS

Paul R.Riley is funded by a British Heart Foundation Personal Chair Award in Regenerative Medicine.

REFERENCES

Bax, N. A., Pijnappels, D. A., van Oorschot, A. A., Winter, E. M., de Vries, A. A., van, T. J., Braun, J., Maas, S., Schalij, M. J., Atsma, D. E., Goumans, M. J., and Gittenberger-de Groot, A. C. (2011). Epithelial-to-mesenchymal transformation alters electrical conductivity of human epicardial cells. *J. Cell. Mol. Med.* **15**(12), 2675–2683. DOI:10.1111/j.1582-4934.2011.01266.x. Epub ahead of print.

Bock-Marquette, I., Shrivastava, S., Pipes, G. C., Thatcher, J. E., Blystone, A., Shelton, J. M., Galindo, C. L., Melegh, B., Srivastava, D., Olson, E. N., and Dimaio, J. M. (2009). Thymosin beta4 mediated PKC activation is essential to initiate the embryonic coronary developmental program and epicardial progenitor cell activation in adult mice in vivo. *J. Cell Mol. Cardiol.* **46,** 728–738.

Cai, C. L., Martin, J. C., Sun, Y., Cui, L., Wang, L., Ouyang, K., Yang, L., Bu, L., Liang, X., Zhang, X., Stallcup, W. B., Denton, C. P., *et al.* (2008). A myocardial lineage derives from Tbx18 epicardial cells. *Nature* **454,** 104–108.

Chen, T. H., Chang, T. C., Kang, J. O., Choudhary, B., Makita, T., Tran, C. M., Burch, J. B., Eid, H., and Sucov, H. M. (2002). Epicardial induction of fetal cardiomyocyte proliferation via a retinoic acid-inducible trophic factor. *Dev. Biol.* **250,** 198–207.

Christoffels, V. M., Grieskamp, T., Norden, J., Mommersteeg, M. T., Rudat, C., and Kispert, A. (2009). Tbx18 and the fate of epicardial progenitors. *Nature* **458,** E8–E9.

Corda, S., Mebazaa, A., Gandolfini, M. P., Fitting, C., Marotte, F., Peynet, J., Charlemagne, D., Cavaillon, J. M., Payen, D., Rappaport, L., and Samuel, J. L. (1997). Trophic effect of human pericardial fluid on adult cardiac myocytes. Differential role of fibroblast growth factor-2 and factors related to ventricular hypertrophy. *Circ. Res.* **81,** 679–687.

Di Meglio, F., Castaldo, C., Nurzynska, D., Romano, V., Miraglia, R., and Montagnani, S. (2009). Epicardial cells are missing from the surface of hearts with ischemic cardiomyopathy: A useful clue about the self-renewal potential of the adult human heart? *Int. J. Cardiol.* **145,** e44–e46.

Gonzalez-Rosa, J. M., Martin, V., Peralta, M., Torres, M., and Mercader, N. (2011). Extensive scar formation and regression during heart regeneration after cryoinjury in zebrafish. *Development* **138,** 1663–1674.

Grieskamp, T., Rudat, C., Ludtke, T. H., Norden, J., and Kispert, A. (2011). Notch signaling regulates smooth muscle differentiation of epicardium-derived cells. *Circ. Res.* **108,** 813–823.

Guadix, J. A., Ruiz-Villalba, A., Lettice, L., Velecela, V., Munoz-Chapuli, R., Hastie, N. D., Perez-Pomares, J. M., and Martinez-Estrada, O. M. (2011). Wt1 controls retinoic acid signalling in embryonic epicardium through transcriptional activation of Raldh2. *Development* **138,** 1093–1097.

Iwakura, A., Fujita, M., Hasegawa, K., Sawamura, T., Nohara, R., Sasayama, S., and Komeda, M. (2000). Pericardial fluid from patients with unstable angina induces vascular endothelial cell apoptosis. *J. Am. Coll. Cardiol.* **35,** 1785–1790.

Jopling, C., Sleep, E., Raya, M., Marti, M., Raya, A., and Belmonte, J. C. (2010). Zebrafish heart regeneration occurs by cardiomyocyte dedifferentiation and proliferation. *Nature* **464,** 606–609.

Kang, J. O., and Sucov, H. M. (2005). Convergent proliferative response and divergent morphogenic pathways induced by epicardial and endocardial signaling in fetal heart development. *Mech. Dev.* **122,** 57–65.

Kikuchi, K., Holdway, J. E., Werdich, A. A., Anderson, R. M., Fang, Y., Egnaczyk, G. F., Evans, T., Macrae, C. A., Stainier, D. Y., and Poss, K. D. (2010). Primary contribution to zebrafish heart regeneration by gata4(+) cardiomyocytes. *Nature* **464,** 601–605.

Kikuchi, K., Gupta, V., Wang, J., Holdway, J. E., Wills, A. A., Fang, Y., and Poss, K. D. (2011a). tcf21+ epicardial cells adopt non-myocardial fates during zebrafish heart development and regeneration. *Development* **138,** 2895–2902.

Kikuchi, K., Holdway, J. E., Major, R. J., Blum, N., Dahn, R. D., Begemann, G., and Poss, K. D. (2011b). Retinoic acid production by endocardium and epicardium is an injury response essential for zebrafish heart regeneration. *Dev. Cell* **20,** 397–404.

Kraus, F., Haenig, B., and Kispert, A. (2001). Cloning and expression analysis of the mouse T-box gene Tbx18. *Mech. Dev.* **100,** 83–86.

Kurkiewicz, T. (1909). O histogenezie miesna sercowego zwierzat kregowych—Zur Histogenese des Herzmuskels der Wirbeltiere. *Bull. Int. Acad. Sci. Cracovie* 148–191.

Lee, K. F., Simon, H., Chen, H., Bates, B., Hung, M. C., and Hauser, C. (1995). Requirement for neuregulin receptor erbB2 in neural and cardiac development. *Nature* **378,** 394–398.

Lepilina, A., Coon, A. N., Kikuchi, K., Holdway, J. E., Roberts, R. W., Burns, C. G., and Poss, K. D. (2006). A dynamic epicardial injury response supports progenitor cell activity during zebrafish heart regeneration. *Cell* **127,** 607–619.

Limana, F., Zacheo, A., Mocini, D., Mangoni, A., Borsellino, G., Diamantini, A., De, M. R., Battistini, L., Vigna, E., Santini, M., Loiaconi, V., Pompilio, G., *et al.* (2007). Identification of myocardial and vascular precursor cells in human and mouse epicardium. *Circ. Res.* **101,** 1255–1265.

Limana, F., Bertolami, C., Mangoni, A., Di, C. A., Avitabile, D., Mocini, D., Iannelli, P., De, M. R., Marchetti, C., Pozzoli, O., Gentili, C., Zacheo, A., *et al.* (2010). Myocardial infarction induces embryonic reprogramming of epicardial c-kit(+) cells: Role of the pericardial fluid. *J. Mol. Cell. Cardiol.* **48,** 609–618.

Manasek, F. J. (1969). Embryonic development of the heart. II. Formation of the epicardium. *J. Embryol. Exp. Morphol.* **22,** 333–348.

Manner, J., Perez-Pomares, J. M., Macias, D., and Munoz-Chapuli, R. (2001). The origin, formation and developmental significance of the epicardium: A review. *Cells Tissues Organs* **169,** 89–103.

Martinez-Estrada, O. M., Lettice, L. A., Essafi, A., Guadix, J. A., Slight, J., Velecela, V., Hall, E., Reichmann, J., Devenney, P. S., Hohenstein, P., Hosen, N., Hill, R. E., *et al.* (2010). Wt1 is required for cardiovascular progenitor cell formation through transcriptional control of Snail and E-cadherin. *Nat. Genet.* **42,** 89–93.

Mellgren, A. M., Smith, C. L., Olsen, G. S., Eskiocak, B., Zhou, B., Kazi, M. N., Ruiz, F. R., Pu, W. T., and Tallquist, M. D. (2008). Platelet-derived growth factor receptor beta signaling is required for efficient epicardial cell migration and development of two distinct coronary vascular smooth muscle cell populations. *Circ. Res.* **103,** 1393–1401.

Merki, E., Zamora, M., Raya, A., Kawakami, Y., Wang, J., Zhang, X., Burch, J., Kubalak, S. W., Kaliman, P., Belmonte, J. C., Chien, K. R., and Ruiz-Lozano, P. (2005). Epicardial retinoid X receptor alpha is required for myocardial growth and coronary artery formation. *Proc. Natl. Acad. Sci. USA* **102,** 18455–18460.

Mollier, S. (1906). Die erste anlage des herzens bei den wirbeltieren. *In* "Handbuch der vergleichenden und experimentellen Entwicklungslehre der Wirbeltiere," (O. Hertwig, ed.) **1,** pp. 1026–1051. Gustav Fischer, JenaSect 1.

Moore, A. W., McInnes, L., Kreidberg, J., Hastie, N. D., and Schedl, A. (1999). YAC complementation shows a requirement for Wt1 in the development of epicardium, adrenal gland and throughout nephrogenesis. *Development* **126,** 1845–1857.

Moss, J. B., Xavier-Neto, J., Shapiro, M. D., Nayeem, S. M., McCaffery, P., Drager, U. C., and Rosenthal, N. (1998). Dynamic patterns of retinoic acid synthesis and response in the developing mammalian heart. *Dev. Biol.* **199,** 55–71.

Olivey, H. E., and Svensson, E. C. (2010). Epicardial-myocardial signaling directing coronary vasculogenesis. *Circ. Res.* **106,** 818–832.

Olivotto, I., Cecchi, F., Poggesi, C., and Yacoub, M. H. (2009). Developmental origins of hypertrophic cardiomyopathy phenotypes: A unifying hypothesis. *Nat. Rev. Cardiol.* **6,** 317–321.

Pennisi, D. J., and Mikawa, T. (2009). FGFR-1 is required by epicardium-derived cells for myocardial invasion and correct coronary vascular lineage differentiation. *Dev. Biol.* **328,** 148–159.

Perez-Pomares, J. M., Carmona, R., Gonzalez-Iriarte, M., Atencia, G., Wessels, A., and Munoz-Chapuli, R. (2002a). Origin of coronary endothelial cells from epicardial mesothelium in avian embryos. *Int. J. Dev. Biol.* **46,** 1005–1013.

Perez-Pomares, J. M., Phelps, A., Sedmerova, M., Carmona, R., Gonzalez-Iriarte, M., Munoz-Chapuli, R., and Wessels, A. (2002b). Experimental studies on the spatiotemporal expression of WT1 and RALDH2 in the embryonic avian heart: A model for the regulation of myocardial and valvuloseptal development by epicardially derived cells (EPDCs). *Dev. Biol.* **247,** 307–326.

Phillips, H. M., Hildreth, V., Peat, J. D., Murdoch, J. N., Kobayashi, K., Chaudhry, B., and Henderson, D. J. (2008). Non-cell-autonomous roles for the planar cell polarity gene Vangl2 in development of the coronary circulation. *Circ. Res.* **102,** 615–623.

Poelmann, R. E., Gittenberger-de Groot, A. C., Mentink, M. M., Bokenkamp, R., and Hogers, B. (1993). Development of the cardiac coronary vascular endothelium, studied with antiendothelial antibodies, in chicken-quail chimeras. *Circ. Res.* **73,** 559–568.

Porrello, E. R., Mahmoud, A. I., Simpson, E., Hill, J. A., Richardson, J. A., Olson, E. N., and Sadek, H. A. (2011). Transient regenerative potential of the neonatal mouse heart. *Science* **331,** 1078–1080.

Red-Horse, K., Ueno, H., Weissman, I. L., and Krasnow, M. A. (2010). Coronary arteries form by developmental reprogramming of venous cells. *Nature* **464,** 549–553.

Remak, R. (1855). Untersuchungen über die entwicklung der wirbelthiere. Reimer, Berlin.

Riley, P. R., and Smart, N. (2011). Vascularizing the heart. *Cardiovasc. Res.* **91,** 260–268.

Serluca, F. C. (2008). Development of the proepicardial organ in the zebrafish. *Dev. Biol.* **315,** 18–27.

Smart, N., Risebro, C. A., Melville, A. A., Moses, K., Schwartz, R. J., Chien, K. R., and Riley, P. R. (2007). Thymosin beta4 induces adult epicardial progenitor mobilization and neovascularization. *Nature* **445,** 177–182.

Smart, N., Risebro, C. A., Clark, J. E., Ehler, E., Miquerol, L., Rossdeutsch, A., Marber, M. S., and Riley, P. R. (2010). Thymosin β4 facilitates epicardial neovascularization of the injured adult heart. *Ann. N. Y. Acad. Sci.* **1194,** 97–104.

Smart, N., Bollini, S., Dube, K. N., Vieira, J. M., Zhou, B., Davidson, S., Yellon, D., Riegler, J., Price, A. N., Lythgoe, M. F., Pu, W. T., and Riley, P. R. (2011). De novo cardiomyocytes from within the activated adult heart after injury. *Nature* **474,** 640–644.

Sucov, H. M., Gu, Y., Thomas, S., Li, P., and Pashmforoush, M. (2009). Epicardial control of myocardial proliferation and morphogenesis. *Pediatr. Cardiol.* **30,** 617–625.

Tomanek, R., Zheng, W., and Yue, X. (2004). Growth factor activation in myocardial vascularization: Therapeutic implications. *Mol. Cell. Biochem.* **264,** 3–11.

Torlopp, A., Schlueter, J., and Brand, T. (2010). Role of fibroblast growth factor signaling during proepicardium formation in the chick embryo. *Dev. Dyn.* **239,** 2393–2403.

vanTuyn, J., Atsma, D. E., Winter, E. M., van der Velde-van Dijke, I., Pijnappels, D. A., Bax, N. A., Knaan-Shanzer, S., Gittenberger-de Groot, A. C., Poelmann, R. E., van der, L. A., van der Wall, E. E., Schalij, M. J., and de Vries, A. A. (2006). Epicardial cells of human adults can undergo an epithelial-to-mesenchymal transition and obtain characteristics of smooth muscle cells in vitro. *Stem Cells* **25,** 271–278.

van, W. B., van den, B. G., Bu-Issa, R., Barnett, P., van der Velden, S., Schmidt, M., Ruijter, J. M., Kirby, M. L., Moorman, A. F., and van den Hoff, M. J. (2009). Epicardium and myocardium separate from a common precursor pool by crosstalk between bone morphogenetic protein- and fibroblast growth factor-signaling pathways. *Circ. Res.* **105,** 431–441.

Vincent, S. D., and Buckingham, M. E. (2010). How to make a heart: The origin and regulation of cardiac progenitor cells. *Curr. Top. Dev. Biol.* **90,** 1–41.

Wagner, K. D., Wagner, N., Bondke, A., Nafz, B., Flemming, B., Theres, H., and Scholz, H. (2002). The Wilms' tumor suppressor Wt1 is expressed in the coronary vasculature after myocardial infarction. *FASEB J.* **16,** 1117–1119.

Wills, A. A., Holdway, J. E., Major, R. J., and Poss, K. D. (2008). Regulated addition of new myocardial and epicardial cells fosters homeostatic cardiac growth and maintenance in adult zebrafish. *Development* **135,** 183–192.

Winter, E. M., Grauss, R. W., Hogers, B., Van, T. J., van der, G. R., Lie-Venema, H., Steijn, R. V., Maas, S., DeRuiter, M. C., Devries, A. A., Steendijk, P., Doevendans, P. A., *et al.* (2007). Preservation of left ventricular function and attenuation of remodeling after transplantation of human epicardium-derived cells into the infarcted mouse heart. *Circulation* **116,** 917–927.

Yoneda, T., Fujita, M., Kihara, Y., Hasegawa, K., Sawamura, T., Tanaka, T., Inanami, M., Nohara, R., and Sasayama, S. (2000). Pericardial fluid from patients with ischemic heart disease accelerates the growth of human vascular smooth muscle cells. *Jpn. Circ. J.* **64,** 495–498.

Zamora, M., Manner, J., and Ruiz-Lozano, P. (2007). Epicardium-derived progenitor cells require beta-catenin for coronary artery formation. *Proc. Natl. Acad. Sci. USA* **104,** 18109–18114.

Zhou, B., Ma, Q., Rajagopal, S., Wu, S. M., Domian, I., Rivera-Feliciano, J., Jiang, D., von, G. A., Ikeda, S., Chien, K. R., and Pu, W. T. (2008). Epicardial progenitors contribute to the cardiomyocyte lineage in the developing heart. *Nature* **454,** 109–113.

Zhou, B., Honor, L. B., He, H., Ma, Q., Oh, J. H., Butterfield, C., Lin, R. Z., Melero-Martin, J. M., Dolmatova, E., Duffy, H. S., Gise, A., Zhou, P., *et al.* (2011). Adult mouse epicardium modulates myocardial injury by secreting paracrine factors. *J. Clin. Invest.* **121,** 1894–1904.

CHAPTER NINE

Transcription Factor Pathways and Congenital Heart Disease

David J. McCulley *and* Brian L. Black

Contents

Abstract

Congenital heart disease is a major cause of morbidity and mortality throughout life. Mutations in numerous transcription factors have been identified in patients and families with some of the most common forms of cardiac malformations and arrhythmias. This review discusses transcription factor pathways known to be important for normal heart development and how abnormalities in these pathways have been linked to morphological and functional forms of congenital heart defects. A comprehensive, current list of known transcription factor mutations associated with congenital heart disease is provided, but the review focuses primarily on three key transcription factors, Nkx2-5, GATA4, and Tbx5, and their known biochemical and genetic partners. By understanding the interaction partners, transcriptional targets, and upstream activators of these core cardiac transcription factors, additional information about normal heart formation and further insight into genes and pathways affected in congenital heart disease should result.

Cardiovascular Research Institute and Department of Biochemistry and Biophysics, University of California, San Francisco, California, USA

Current Topics in Developmental Biology, Volume 100
ISSN 0070-2153, DOI: 10.1016/B978-0-12-387786-4.00008-7

1. Introduction

Defects in heart formation are a significant cause of neonatal morbidity and mortality in humans, while defects in the formation of the cardiac conduction system present a significant mortality risk throughout life (Hoffman *et al.*, 2004; Rubart and Zipes, 2005). Congenital heart defects may be broadly grouped into two major categories: (1) morphological abnormalities, including developmental defects resulting in structural malformations and (2) functional abnormalities, including cardiac rhythm disturbances and cardiomyopathies. While the underlying genetic basis for many of these defects remains elusive, mutations in genes encoding core cardiac transcription factors have emerged as major contributors to many forms of congenital heart disease.

As more has been learned about the transcriptional circuits controlling normal heart development, the number of transcription factors implicated in human congenital heart disease has grown concomitantly. In this review, we highlight transcription factor pathways known to be important for normal heart development and discuss how abnormalities in these pathways have been linked to morphological and functional forms of congenital heart defects. We focus primarily on three key transcription factors, Nkx2-5, GATA4, and Tbx5, and their known biochemical and genetic partners. All three of these transcription factors play central roles in cardiac development, and mutations in each have been implicated in congenital heart disease. We discuss the hypothesis that the important genetic or biochemical interactions between these transcription factors during normal development may be important causes or modifiers of congenital heart defects in humans.

2. Basic Heart Development

Mammalian heart development begins with the specification of cardiac progenitor cells within the anterior lateral plate mesoderm (Brand, 2003; Evans *et al.*, 2010; Harvey, 2002). These progenitor cells condense into two lateral heart primordia around day 15 of human embryonic development and embryonic day (E) 7.5 of mouse development and include both myocardial and endocardial lineage precursors (Brand, 2003; Harvey, 2002; Srivastava, 2006). By E8.5 of mouse development and 3 weeks of human development, the bilaterally paired condensations of cardiac precursors move medially to form a primitive heart tube (Harvey, 2002). The linear heart tube undergoes rightward looping as cells from the second heart field are added to both the inflow and outflow poles (Black, 2007; Vincent and Buckingham, 2010).

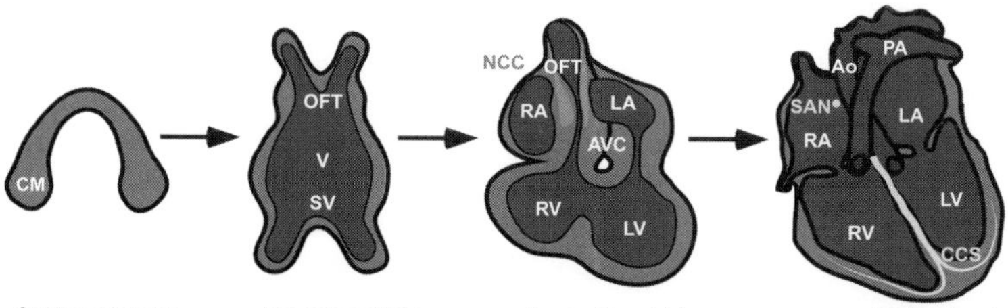

Stage	Cardiac crescent	Linear heart tube	Looped heart tube	Mature heart
Human development	Day 15	Day 20	Day 30	Day 40-birth
Mouse development	E7.5	E8.0	E10–12	E14–18
Key events	Myocyte differentiation Migration to the midline	Bidirection contractions	Looping morphogenesis Addition of SHF progenitors Cushion formation Neural crest addition Ventricular septation begins	Conduction system OFT septation Cushion remodeling

Figure 9.1 Schematic representations of several stages of mammalian heart development with key events in mouse and human heart development denoted. Mouse development is denoted in embryonic days (E). Ao, aorta; AVC, atrioventricular canal; CCS, cardiac conduction system; CM, cardiac mesoderm; LA, left atrium; LV, left ventricle; NCC, neural crest cells; OFT, outflow tract; PA, pulmonary artery; RA, right atrium; RV, right ventricle; SAN, sinoatrial node; SV, sinus venosus; V, ventricle. (See Color Insert.)

Following cardiac looping, endocardial cushions begin to form within the outflow tract (OFT) and at the common atrioventricular canal (AVC) during the sixth and seventh weeks of human development (Person *et al.*, 2005). The endocardial cushions help to septate the heart into the four cardiac chambers and divide the OFT into the aorta and pulmonary artery (Combs and Yutzey, 2009; Person *et al.*, 2005). The early cardiac conduction system also begins to be specified at this time, while the developing heart receives important contributions from the neural crest and proepicardium (Brand, 2003; Hutson and Kirby, 2007; Vincent and Buckingham, 2010). Toward the later stages of development, the heart undergoes extensive remodeling before it assumes the mature four-chambered structure with divided inflow and outflow, finely formed valve leaflets, and functional conduction network (Combs and Yutzey, 2009; Evans *et al.*, 2010; Fig. 9.1).

3. A Group of "Core Cardiac Transcription Factors" Controls Heart Development

A group of transcription factors, including the homeodomain protein Nkx2-5; GATA family zinc finger proteins GATA4, 5, and 6; MEF2 factors and SRF (MADS box proteins); T-box factors, including Tbx1, Tbx2,

Tbx3, Tbx5, Tbx18, and Tbx20; and the Lim-homeodomain protein Isl1, is critical for heart development (Black and Cripps, 2010; Greulich *et al.*, 2011; Harvey *et al.*, 2002; He *et al.*, 2011; Peterkin *et al.*, 2005). These core transcription factors interact with each other and with an array of other transcription factors to control heart development. Later, many of the same transcription factors are reutilized to control cardiac chamber maturation, conduction system development, and endocardial cushion remodeling (Oka *et al.*, 2007; Olson, 2006).

The core cardiac transcription factors function in a mutually reinforcing transcriptional network in which each of the factors regulate the expression of the others (Black, 2007; He *et al.*, 2011; Olson, 2006). Several of the core factors involved in heart development also function as biochemical partners for each other, reflecting a complex molecular and genetic interplay controlling multiple stages of heart and conduction system development (Figs. 9.2 and 9.3). Therefore, it is not surprising that mutations in several of

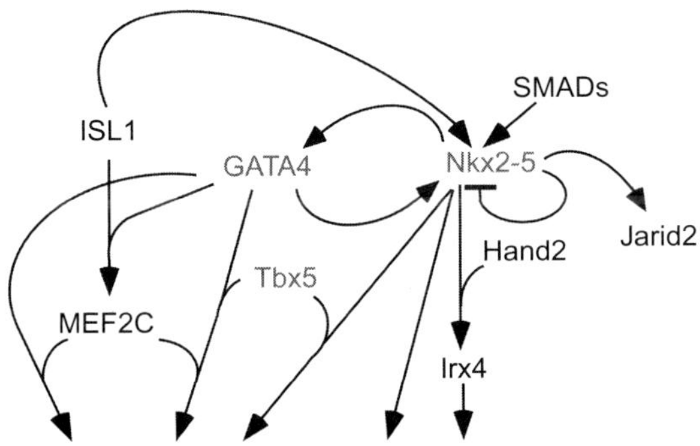

Figure 9.2 Transcription factor pathways discussed in this review that are involved in myocardial development and heart morphogenesis. (For color version of this figure, the reader is referred to the Web version of this chapter.)

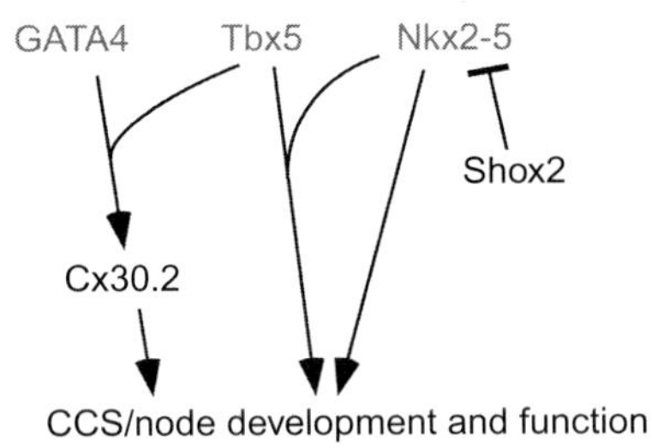

Figure 9.3 Transcription factor pathways discussed in this review that are involved in cardiac conduction system development, maturation, and function. (For color version of this figure, the reader is referred to the Web version of this chapter.)

the genes encoding the core cardiac transcription factors are associated with congenital heart disease (Table 9.1). Nkx2-5, GATA4, and Tbx5 are perhaps the most studied and well characterized of the cardiac transcription factors implicated in patients with congenital heart disease, and all three are critical for normal heart development.

4. Nkx2-5 Is a Central Regulator of Heart Development and Is Affected in Congenital Heart Disease

The homeodomain protein Nkx2-5 controls many aspects of cardiac development, beginning with specification and proliferation of cardiac precursors to later events, such as the formation of the OFT (Harvey *et al.*, 2002). Nkx2-5 acts near the top of a large transcriptional cascade controlling multiple cardiac genes. Expression of *Nkx2-5* is regulated by GATA factors and SMAD proteins and is also controlled by Nkx2-5 itself in an autoregulatory loop (Liberatore *et al.*, 2002; Lien *et al.*, 1999, 2002; Searcy *et al.*, 1998). More recently, *Nkx2-5* expression in the second heart field was shown to be dependent on Isl1, while interactions between Nkx2-5, BMP2, and SMAD1 control cardiac progenitor specification and proliferation (Prall *et al.*, 2007; Takeuchi *et al.*, 2005).

Nkx2-5 acts combinatorially with other core cardiac transcription factors to promote cardiomyocyte differentiation and chamber identity. For example, Nkx2-5 and the basic helix-loop-helix protein Hand2 cooperate to activate *Irx4*, which is necessary for determining ventricular identity (Yamagishi *et al.*, 2001). Ventricular identity is also controlled by Nkx2-5 through a direct physical interaction with the MADS box transcription factor MEF2C (Vincentz *et al.*, 2008). Nkx2-5 interacts with another MADS box transcription factor, SRF, and GATA4 to promote cardiac sarcomeric protein gene expression (Sepulveda *et al.*, 2002). Nkx2-5 also plays a role in OFT development through regulation of the transcriptional repressor *Jarid2* (Barth *et al.*, 2010).

In mice, loss of *Nkx2-5* results in embryonic lethality with failure of cardiac looping and deficient myocardial differentiation (Lyons *et al.*, 1995; Tanaka *et al.*, 1999). In humans, mutations in *NKX2-5* have been found in patients with a variety of structural malformations including septation defects, alignment defects, compaction defects, and cardiac conduction defects (Benson, 2010; Reamon-Buettner and Borlak, 2010). Patients with *NKX2-5* mutations also have cardiac rhythm abnormalities (Benson, 2010). Consistent with those observations, Nkx2-5 functions in the development and maintenance of the cardiac conduction system in the mouse in a transcriptional cascade that involves interactions with Tbx5 and Id2 (Jay

Table 9.1 Transcription factors with mutations found in human patients with congenital heart disease

Transcription factor	Associated CHD phenotype	References
ANKRD1	TAPVR	Cinquetti *et al.* (2008)
CITED2	TOF, ASD, VSD, AS, PS, SI, TGA, RVOTO, TAPVR	Sperling *et al.* (2005)
ETS1	DORV, HLHS, ASD	Ye *et al.* (2010)
FOG2/ZFPM2	TOF, DORV	De Luca *et al.* (2011), Finelli *et al.* (2007), Pizzuti *et al.* (2003)
FOXC1 (18p11.2 del)/*FOXC2*	HLHS, PVA, PAH, VSD, OA, ASD, PDA, BSVC	Brice *et al.* (2002), Finegold *et al.* (2001), Maclean *et al.* (2005), Yu *et al.* (2010)
FOXH1	TGA	De Luca *et al.* (2010)
GATA4	ASD, AVSD, VSD, PS, AR, VPS, PDA, TOF, AF	Butler *et al.* (2010), Chen *et al.* (2010a–c), De Luca *et al.* (2010), Dinesh *et al.* (2010b), Garg *et al.* (2003), Giglio *et al.* (2000), Guida *et al.* (2010), Hirayama-Yamada *et al.* (2005), Liu *et al.* (2011), Nemer *et al.* (2006), Pehlivan *et al.* (1999), Peng *et al.* (2010), Posch *et al.* (2010a), Rajagopal *et al.* (2007), Reamon-Buettner *et al.* (2007), Sarkozy *et al.* (2005), Schluterman *et al.* (2007), Tomita-Mitchell *et al.* (2007), Wang *et al.* (2010, 2011a), Zhang *et al.* (2008, 2009b)
GATA6	PTA, TOF, ASD	Kodo *et al.* (2009), Lin *et al.* (2010), Maitra *et al.* (2010)
HAND1	VSD, ASD, AVSD	Cheng *et al.* (2011a), Goldmuntz *et al.* (2011), Reamon-Buettner *et al.* (2008, 2009), Wang *et al.* (2011a)
HAND2	TOF, DORV, PS	Shen *et al.* (2010)
HOXA1	DORV, TOF, VSD, TAPVR, IAA, PDA	Bosley *et al.* (2008)
IRX4	VSD	Cheng *et al.* (2011b)
JAG1	TOF, VSD, PPS, Ao dextroposition	Eldadah *et al.* (2001), Greenway *et al.* (2009), Heritage *et al.* (2000, 2002), Krantz *et al.* (1999), Li *et al.* (1997a, 2010), Oda *et al.* (1997), Rauch *et al.* (2010)
MYOCD	PVD	Ransom *et al.* (2008)
NFATC1	VSD	Yehya *et al.* (2006)

NKX2-5	ASD, VSD, AVSD, TOF, SVAS, LVNC, PA, PS, PDA, MV anomalies, conduction defects, DORV, PAPVR, TAPVR, heterotaxy, TGA	Baekvad-Hansen *et al.* (2006), Benson *et al.* (1999), De Luca *et al.* (2010), Dinesh *et al.* (2010a), Elliott *et al.* (2003), Gioli-Pereira *et al.* (2010), Gutierrez-Roelens *et al.* (2002), Ikeda *et al.* (2002), Konig *et al.* (2006), McElhinney *et al.* (2003b), Ouyang *et al.* (2011), Pabst *et al.* (2008), Pauli *et al.* (1999), Rauch *et al.* (2010), Reamon-Buettner and Borlak (2004a), Reamon-Buettner *et al.* (2004), Schott *et al.* (1998), Stallmeyer *et al.* (2010), Wang *et al.* (2011b), Zhang *et al.* (2009a), Zhu *et al.* (2000)
NKX2-6	PTA	Heathcote *et al.* (2005)
PITX2	AF	Franco *et al.* (2011)
SALL4	VSD, PTA, TOF	Paradisi and Arias (2007), Wang *et al.* (2010)
TBX1/ 22q11del	IAA, PTA, AAA, TOF, malaligned VSD	Akcakus *et al.* (2003), Alikasifoglu *et al.* (2000), Beauchesne *et al.* (2005), Borgmann *et al.* (1999), Botto *et al.* (2003), Brunet *et al.* (2009), Calderon *et al.* (2009), Devriendt *et al.* (1996), Gawde *et al.* (2006), Gioli-Pereira *et al.* (2008), Goldmuntz *et al.* (1998), Goodship *et al.* (1998), Henwood *et al.* (2001), Hokanson *et al.* (2001), Hu *et al.* (2009), Iserin *et al.* (1998), Ito *et al.* (2002), Jiang *et al.* (2005, 2010), Lindsay *et al.* (1995), Lupski *et al.* (1991), Matsuoka *et al.* (1994), McElhinney *et al.* (2001, 2003a), Oh *et al.* (2002), Poon *et al.* (2007), Rauch *et al.* (2004, 2010), Raymond *et al.* (1997), Repetto *et al.* (2009), Rope *et al.* (2009), Ryan *et al.* (1997), Sorensen *et al.* (2010), Swaby *et al.* (2011), Tomita-Mitchell *et al.* (2010), van Engelen *et al.* (2010), Verhoeven *et al.* (2011), Wilson *et al.* (1992), Worthington *et al.* (1998), Wozniak *et al.* (2010), Yakut *et al.* (2006), Yates *et al.* (1996), Yong *et al.* (1999)
22q11dup	VSD, HLHS, TAPVR, conotruncal anomalies	Ensenauer *et al.* (2003), Ou *et al.* (2008), Sparkes *et al.* (2005)

(*Continued*)

Table 9.1 (*Continued*)

Transcription factor	Associated CHD phenotype	References
TBX5	ASD, VSD, AVSD, rhythm abnormalities	Basson *et al.* (1997), Li *et al.* (1997b), Liu *et al.* (2009), McDermott *et al.* (2008), Postma *et al.* (2008), Reamon-Buettner and Borlak (2004b), Zhu *et al.* (2008)
TBX20	ASD, dilated cardiomyopathy, valve defects	Hammer *et al.* (2008), Kirk *et al.* (2007), Liu *et al.* (2008), Posch *et al.* (2010b), Qian *et al.* (2008)
TFAP2B	Char syndrome, PDA	Chen *et al.* (2011), Mani *et al.* (2005), Satoda *et al.* (2000), Vaughan and Basson (2000), Zhao *et al.* (2001)
ZEB2	Mowat–Wilson syndrome	Adam *et al.* (2006), Balasubramaniam *et al.* (2010), Dastot-Le Moal *et al.* (2007), Garavelli *et al.* (2009), Heinritz *et al.* (2006), McGaughran *et al.* (2005), Sasongko *et al.* (2007), Saunders *et al.* (2009), Wilson *et al.* (2003), Zweier *et al.* (2005)
ZIC3	TGA, DORV, PS, TAPVR, ASD, AVSD, heterotaxy	Bedard *et al.* (2010), De Luca *et al.* (2010), Ware *et al.* (2004)

Abbreviations used: AAA, aortic arch anomalies; AF, atrial fibrillation; AR, aortic regurgitation; AS, aortic stenosis; ASD, atrial septal defect; AVC, atrioventricular canal; AVSD, atrioventricular septal defect; BAV, bicuspid aortic valve; BPV, bicuspid pulmonary valve; BSVC, bilateral superior vena cava; DORV, double outlet right ventricle; HAA, hypoplastic aortic arch; HCM, hypertrophic cardiomyopathy; HLHS, hypoplastic left heart syndrome; HOCM, hypertrophic obstructive cardiomyopathy; IAA, interrupted aortic arch; LVH, left ventricular hypertrophy; LVNC, left ventricular noncompaction; MS, mitral stenosis; MV, mitral valve; MVD, mitral valve dysplasia; MVP, mitral valve prolapse; OA, overriding aorta; PA, pulmonary atresia; PAH, pulmonary artery hypoplasia; PAPVR, partial anomalous pulmonary venous return; PDA, patent ductus arteriosus; PPAS, peripheral pulmonary artery stenosis; PPS, peripheral pulmonic stenosis; PS, pulmonary stenosis; PTA, persistent truncus arteriosus; PVA, pulmonary valve atresia; PVD, pulmonary valve dysplasia; PVS, pulmonary valve stenosis; RAA, right aortic arch; RVOTO, right ventricular outflow tract obstruction; SI, situs inversus; SVAS, supravalvar aortic stenosis; SVPS, supravalvar pulmonary stenosis; SVT, supraventricular tachycardia; TAPVR, total anomalous venous return; TGA, transposition of the great arteries; TOF, tetralogy of Fallot; TVP, tricuspid valve prolapse; VAS, valvar aortic stenosis; VPS, valvar pulmonary stenosis; VSD, ventricular septal defect.

et al., 2004; Moskowitz *et al.*, 2007). *Nkx2-5* heterozygous mice have progressive atrioventricular (AV) block, a phenotype similar to part of the spectrum of congenital heart disease phenotypes described in human patients with *NKX2-5* mutations (Benson, 2010; Biben *et al.*, 2000). In addition, mutation of Shox2, an upstream transcriptional repressor of *Nkx2-5* expressed in the sinoatrial node (SAN), results in mid-gestational embryonic lethality in mice with severe hypoplasia of the SAN and associated abnormal heart rate (Blaschke *et al.*, 2007; Espinoza-Lewis *et al.*, 2009). The essential role of Nkx2-5 in multiple distinct aspects of heart development and congenital heart disease suggests that modifiers, targets, or partners of Nkx2-5 are also likely to be associated with congenital heart disease.

5. GATA4 Regulates Cardiac Morphogenesis and Myoctye Proliferation and Is Frequently Associated with Congenital Heart Disease

The zinc finger transcription factor GATA4 plays a central role in cardiac development and is critical for survival of the embryo (Peterkin *et al.*, 2005; Pikkarainen *et al.*, 2004). Mice that lack GATA4 have abnormal ventral folding, failure of midline fusion of the heart primordia, and extensive endoderm defects with embryonic lethality by E10.5 (Kuo *et al.*, 1997; Molkentin *et al.*, 1997). Dosage of GATA4 is responsible for regulation of cardiac morphogenesis, and graded reduction of GATA4 leads to abnormal cardiac development with a common AVC, double outlet right ventricle, and hypoplasia of the ventricular myocardium (Pu *et al.*, 2004). GATA4 is also an important regulator of cardiomyocyte proliferation through direct transcriptional activation of cell cycle regulators, including *cyclin D2* and *cdk4*, which may explain the lack of sufficient cardiac septation and chamber hypoplasia observed in patients with *GATA4* mutations (Rajagopal *et al.*, 2007; Rojas *et al.*, 2008).

Like Nkx2-5, GATA4 acts combinatorially with other core cardiac transcription factors to regulate heart development, including the previously discussed interactions with Nkx2-5 (Liberatore *et al.*, 2002; Lien *et al.*, 1999, 2002; Searcy *et al.*, 1998). GATA4 also functions as an important partner for the T-box transcription factor Tbx5; mice doubly heterozygous for *Gata4* and *Tbx5* die at E15.5 due to insufficient development of the ventricular myocardium and have atrial septal defects (Maitra *et al.*, 2009). Further, GATA4 and Tbx5 directly interact and activate expression of *Cx30.2*, which encodes a gap-junction protein enriched in the AV node and required for normal AV node delay (Munshi *et al.*, 2009).

In addition to interactions with Nkx2-5 and Tbx5, GATA4 also functions as a transcriptional partner or in transcriptional pathways with several other important cardiac transcription factors, including MEF2C. MEF2 physically interacts with GATA factors, including GATA4 and 6 to synergistically activate the expression of *Nppa*, *α-MHC*, *α-CA*, and *B-type natriuretic peptide* (*BNP*; Morin *et al.*, 2000). Additionally, GATA transcription factors, including GATA4, directly activate *Mef2c* transcription in the second heart field in combination with Isl1 (Dodou *et al.*, 2004). GATA4 also interacts with the BMP signaling effector Smad4 within the developing endocardium, and importantly, *GATA4* mutations identified in human patients with septation defects were found to cause disrupted interaction between GATA4 and Smad4 (Moskowitz *et al.*, 2011). Other patients with deletions in the *GATA4* locus have more severe forms of congenital heart disease, including septation defects, OFT alignment defects, dextrocardia, and pulmonary stenosis (Hirayama-Yamada *et al.*, 2005; Okubo *et al.*, 2004; Pehlivan *et al.*, 1999; Sarkozy *et al.*, 2005). Linkage analyses were used to demonstrate associations between *GATA4* mutations and multiple cardiac defects including ASD and VSD (Garg *et al.*, 2003; Hirayama-Yamada *et al.*, 2005; Okubo *et al.*, 2004; Sarkozy *et al.*, 2005). Interestingly, one of the *GATA4* mutations described in these family studies resulted in a missense mutation within the protein–protein interaction domain of GATA4. This alteration led to a disruption of the interaction of GATA4 and Tbx5, while it maintained the ability of GATA4 to interact with Nkx2-5 (Garg *et al.*, 2003). These data suggest a cooperative role for GATA4 and Tbx5 in cardiac septation, and further support the notion that transcription factor interactions with their partners underscore congenital heart defects.

6. Involvement of Tbx5 in Heart Development and Congenital Heart Disease

Genetic studies in mice have established that Tbx5 plays an important role in cardiac morphogenesis and development of the conduction system. Embryonic mice that lack Tbx5 have abnormal heart tube formation with hypoplastic atria, whereas overexpression of Tbx5 results in inhibition of ventricular maturation (Bruneau *et al.*, 2001). Mice that lack one copy of *Tbx5* have ASD, occasional VSD, and AV block, similar to mice lacking a single copy of *Nkx2-5* (Bruneau *et al.*, 2001). Like the other core cardiac transcription factors, Tbx5 is involved in multiple transcription factor pathways and combinatorial interactions involved in morphological development of the heart and in the development of the cardiac conduction system (Hatcher and Basson, 2009). Mice doubly heterozygous for *Tbx5* and *Gata4* have growth retardation and early neonatal lethality with AVSD and myocardial thinning, similar to the human *GATA4* mutant phenotype (Maitra *et al.*, 2009). Tbx5 also physically interacts with Nkx2-5 to control gene expression in cells of the cardiac conduction system (Bruneau *et al.*, 2001; Hiroi *et al.*, 2001), and Tbx5 and MEF2C physically interact and form a transcriptional complex resulting in synergistic activation of *Myh6* expression (Ghosh *et al.*, 2009).

Given its interactions with GATA4, Nkx2-5, and other core cardiac transcription factors, and its extensive role in cardiac morphological and conduction system development, it is not surprising that *TBX5* is associated with human congenital heart disease. *TBX5* mutations were first identified in human patients with Holt–Oram syndrome (Basson *et al.*, 1997; Li *et al.*, 1997b). These patients have defects in cardiac septation with secundum ASD and VSD as well as defects in cardiac conduction system and upper limb formation (Basson *et al.*, 1994).

7. Summary and Future Directions

In addition to genetic, transcriptional, and biochemical interactions among themselves, Tbx5, Nkx2-5, and GATA4 each interact extensively with other cardiac transcription factors during normal heart development. In a few cases, mutations in these transcription factors have been implicated in congenital heart disease (Table 9.1), but overall, the role that these interacting partners play in human heart malformations has remained largely undefined. Interestingly, Nkx2-5, GATA4, and Tbx5 each interact independently with MEF2C (Ghosh *et al.*, 2009; Morin *et al.*, 2000; Vincentz *et al.*, 2008). MEF2C is critical for normal cardiac development in mice. Mice that lack MEF2C die early in development with severe cardiovascular defects, including failure of normal cardiac looping (Lin *et al.*, 1997), suggesting that *MEF2C* might be associated with congenital heart disease in humans, although this has not been demonstrated to date.

An excellent example of a transcriptional partner that has been associated with congenital heart disease is the friend of GATA (FOG) 2. FOG2 is a cofactor for GATA4, and mutations in *FOG2* have been found in a group of patients with congenital heart disease (Finelli *et al.*, 2007; Svensson *et al.*, 1999). *FOG2* mutations were also identified using a candidate-based approach in patients with tetralogy of Fallot (Pizzuti *et al.*, 2003). Targeted mutation of *Fog2* in a mouse model resulted in embryonic lethality due to congestive heart failure with morphological abnormalities resembling those seen in human tetralogy of Fallot patients and, importantly, those seen in *Gata4* loss-of-function mutants (Svensson *et al.*, 2000; Tevosian *et al.*, 2000). Knowing the importance of GATA family members in directing normal heart formation and the function of FOG2 in modulating GATA activity,

FOG2 was considered as a possible candidate and was subsequently found to be responsible for a common form of human congenital heart disease.

Using multiple model systems, combined with human genetic studies, we are gaining an increased appreciation for the intricate networks of transcriptional modifiers and partners that are required for normal heart development. Importantly, this list of transcriptional modifiers, targets, and partners is very likely to contain genes with mutations that are associated with congenital heart disease. By further studying the interactions of the core cardiac transcription factors, we will more clearly understand how these transcription factors interact to promote normal heart development and how mutations affecting these transcription factor pathways result in congenital heart disease.

ACKNOWLEDGMENTS

We thank members of the Black lab for helpful discussions and comments. The Black lab is supported by funding from the NIH and a grant from Atlantic Philanthropies to BLB.

REFERENCES

Adam, M. P., Schelley, S., Gallagher, R., Brady, A. N., Barr, K., Blumberg, B., Shieh, J. T., Graham, J., Slavotinek, A., Martin, M., Keppler-Noreuil, K., Storm, A. L., *et al.* (2006). Clinical features and management issues in Mowat-Wilson syndrome. *Am. J. Med. Genet. A* **140,** 2730–2741.

Akcakus, M., Ozkul, Y., Gunes, T., Kurtoglu, S., Cetin, N., Kisaarslan, A. P., and Dundar, M. (2003). Associated anomalies in asymmetric crying facies and 22q11 deletion. *Genet. Couns.* **14,** 325–330.

Alikasifoglu, M., Malkoc, N., Ceviz, N., Ozme, S., Uludogan, S., and Tuncbilek, E. (2000). Microdeletion of 22q11 (CATCH 22) in children with conotruncal heart defect and extracardiac malformations. *Turk. J. Pediatr.* **42,** 215–218.

Baekvad-Hansen, M., Tumer, Z., Delicado, A., Erdogan, F., Tommerup, N., and Larsen, L. A. (2006). Delineation of a 2.2 Mb microdeletion at 5q35 associated with microcephaly and congenital heart disease. *Am. J. Med. Genet. A* **140,** 427–433.

Balasubramaniam, S., Keng, W. T., Ngu, L. H., Michel, L. G., and Irina, G. (2010). Mowat-Wilson syndrome: The first two Malaysian cases. *Singapore Med. J.* **51,** e54–e57.

Barth, J. L., Clark, C. D., Fresco, V. M., Knoll, E. P., Lee, B., Argraves, W. S., and Lee, K. H. (2010). Jarid2 is among a set of genes differentially regulated by Nkx2.5 during outflow tract morphogenesis. *Dev. Dyn.* **239,** 2024–2033.

Basson, C. T., Cowley, G. S., Solomon, S. D., Weissman, B., Poznanski, A. K., Traill, T. A., Seidman, J. G., and Seidman, C. E. (1994). The clinical and genetic spectrum of the Holt-Oram syndrome (heart-hand syndrome). *N. Engl. J. Med.* **330,** 885–891.

Basson, C. T., Bachinsky, D. R., Lin, R. C., Levi, T., Elkins, J. A., Soults, J., Grayzel, D., Kroumpouzou, E., Traill, T. A., Leblanc-Straceski, J., Renault, B., Kucherlapati, R., *et al.* (1997). Mutations in human TBX5 cause limb and cardiac malformation in Holt-Oram syndrome. *Nat. Genet.* **15,** 30–35.

Beauchesne, L. M., Warnes, C. A., Connolly, H. M., Ammash, N. M., Grogan, M., Jalal, S. M., and Michels, V. V. (2005). Prevalence and clinical manifestations of

22q11.2 microdeletion in adults with selected conotruncal anomalies. *J. Am. Coll. Cardiol.* **45,** 595–598.

Bedard, J. E., Haaning, A. M., and Ware, S. M. (2010). Identification of a novel ZIC3 isoform and mutation screening in patients with heterotaxy and congenital heart disease. *PLoS One* **6,** e23755.

Benson, D. W. (2010). Genetic origins of pediatric heart disease. *Pediatr. Cardiol.* **31,** 422–429.

Benson, D. W., Silberbach, G. M., Kavanaugh-McHugh, A., Cottrill, C., Zhang, Y., Riggs, S., Smalls, O., Johnson, M. C., Watson, M. S., Seidman, J. G., Seidman, C. E., Plowden, J., *et al.* (1999). Mutations in the cardiac transcription factor NKX2.5 affect diverse cardiac developmental pathways. *J. Clin. Invest.* **104,** 1567–1573.

Biben, C., Weber, R., Kesteven, S., Stanley, E., McDonald, L., Elliott, D. A., Barnett, L., Koentgen, F., Robb, L., Feneley, M., and Harvey, R. P. (2000). Cardiac septal and valvular dysmorphogenesis in mice heterozygous for mutations in the homeobox gene Nkx2-5. *Circ. Res.* **87,** 888–895.

Black, B. L. (2007). Transcriptional pathways in second heart field development. *Semin. Cell Dev. Biol.* **18,** 67–76.

Black, B. L., and Cripps, R. M. (2010). Myocyte enhancer factor 2 transcription factors in heart development and disease. *In* "Heart Development and Regeneration," (N. Rosenthal and R. P. Harvey, eds.), pp. 673–699. Academic Press, Oxford.

Blaschke, R. J., Hahurij, N. D., Kuijper, S., Just, S., Wisse, L. J., Deissler, K., Maxelon, T., Anastassiadis, K., Spitzer, J., Hardt, S. E., Scholer, H., Feitsma, H., *et al.* (2007). Targeted mutation reveals essential functions of the homeodomain transcription factor Shox2 in sinoatrial and pacemaking development. *Circulation* **115,** 1830–1838.

Borgmann, S., Luhmer, I., Arslan-Kirchner, M., Kallfelz, H. C., and Schmidtke, J. (1999). A search for chromosome 22q11.2 deletions in a series of 176 consecutively catheterized patients with congenital heart disease: No evidence for deletions in non-syndromic patients. *Eur. J. Pediatr.* **158,** 958–963.

Bosley, T. M., Alorainy, I. A., Salih, M. A., Aldhalaan, H. M., Abu-Amero, K. K., Oystreck, D. T., Tischfield, M. A., Engle, E. C., and Erickson, R. P. (2008). The clinical spectrum of homozygous HOXA1 mutations. *Am. J. Med. Genet. A* **146A,** 1235–1240.

Botto, L. D., May, K., Fernhoff, P. M., Correa, A., Coleman, K., Rasmussen, S. A., Merritt, R. K., O'Leary, L. A., Wong, L. Y., Elixson, E. M., Mahle, W. T., and Campbell, R. M. (2003). A population-based study of the 22q11.2 deletion: Phenotype, incidence, and contribution to major birth defects in the population. *Pediatrics* **112,** 101–107.

Brand, T. (2003). Heart development: Molecular insights into cardiac specification and early morphogenesis. *Dev. Biol.* **258,** 1–19.

Brice, G., Mansour, S., Bell, R., Collin, J. R., Child, A. H., Brady, A. F., Sarfarazi, M., Burnand, K. G., Jeffery, S., Mortimer, P., and Murday, V. A. (2002). Analysis of the phenotypic abnormalities in lymphoedema-distichiasis syndrome in 74 patients with FOXC2 mutations or linkage to 16q24. *J. Med. Genet.* **39,** 478–483.

Bruneau, B. G., Nemer, G., Schmitt, J. P., Charron, F., Robitaille, L., Caron, S., Conner, D. A., Gessler, M., Nemer, M., Seidman, C. E., and Seidman, J. G. (2001). A murine model of Holt-Oram syndrome defines roles of the T-box transcription factor Tbx5 in cardiogenesis and disease. *Cell* **106,** 709–721.

Brunet, A., Armengol, L., Heine, D., Rosell, J., Garcia-Aragones, M., Gabau, E., Estivill, X., and Guitart, M. (2009). BAC array CGH in patients with Velocardiofacial syndrome-like features reveals genomic aberrations on chromosome region 1q21.1. *BMC Med. Genet.* **10,** 144.

Butler, T. L., Esposito, G., Blue, G. M., Cole, A. D., Costa, M. W., Waddell, L. B., Walizada, G., Sholler, G. F., Kirk, E. P., Feneley, M., Harvey, R. P., and Winlaw, D. S.

(2010). GATA4 mutations in 357 unrelated patients with congenital heart malformation. *Genet. Test. Mol. Biomarkers* **14,** 797–802.

Calderon, J. F., Puga, A. R., Guzman, M. L., Astete, C. P., Arriaza, M., Aracena, M., Aravena, T., Sanz, P., and Repetto, G. M. (2009). VEGFA polymorphisms and cardiovascular anomalies in 22q11 microdeletion syndrome: A case-control and family-based study. *Biol. Res.* **42,** 461–468.

Chen, M. W., Pang, Y. S., Guo, Y., Pan, J. H., Liu, B. L., Shen, J., and Liu, T. W. (2010a). GATA4 mutations in Chinese patients with congenital cardiac septal defects. *Pediatr. Cardiol.* **31,** 85–89.

Chen, Y., Han, Z. Q., Yan, W. D., Tang, C. Z., Xie, J. Y., Chen, H., and Hu, D. Y. (2010b). A novel mutation in GATA4 gene associated with dominant inherited familial atrial septal defect. *J. Thorac. Cardiovasc. Surg.* **140,** 684–687.

Chen, Y., Mao, J., Sun, Y., Zhang, Q., Cheng, H. B., Yan, W. H., Choy, K. W., and Li, H. (2010c). A novel mutation of GATA4 in a familial atrial septal defect. *Clin. Chim. Acta* **411,** 1741–1745.

Chen, Y. W., Zhao, W., Zhang, Z. F., Fu, Q., Shen, J., Zhang, Z., Ji, W., Wang, J., and Li, F. (2011). Familial nonsyndromic patent ductus arteriosus caused by mutations in TFAP2B. *Pediatr. Cardiol.* **32,** 958–965.

Cheng, Z., Lib, L., Li, Z., Liu, M., Yan, J., Wang, B., and Ma, X. (2011b). Two novel HAND1 mutations in Chinese patients with ventricular septal defect. *Clin. Chim. Acta* (in press). doi:http://dx.doi.org/10.1016/j.cca.2011.10.014.

Cheng, Z., Wang, J., Su, D., Pan, H., Huang, G., Li, X., Li, Z., Shen, A., Xie, X., Wang, B., and Ma, X. (2011a). Two novel mutations of the IRX4 gene in patients with congenital heart disease. *Hum. Genet.* **130,** 657–662.

Cinquetti, R., Badi, I., Campione, M., Bortoletto, E., Chiesa, G., Parolini, C., Camesasca, C., Russo, A., Taramelli, R., and Acquati, F. (2008). Transcriptional deregulation and a missense mutation define ANKRD1 as a candidate gene for total anomalous pulmonary venous return. *Hum. Mutat.* **29,** 468–474.

Combs, M. D., and Yutzey, K. E. (2009). Heart valve development: Regulatory networks in development and disease. *Circ. Res.* **105,** 408–421.

Dastot-Le Moal, F., Wilson, M., Mowat, D., Collot, N., Niel, F., and Goossens, M. (2007). ZFHX1B mutations in patients with Mowat-Wilson syndrome. *Hum. Mutat.* **28,** 313–321.

De Luca, A., Sarkozy, A., Consoli, F., Ferese, R., Guida, V., Dentici, M. L., Mingarelli, R., Bellacchio, E., Tuo, G., Limongelli, G., Digilio, M. C., Marino, B., *et al.* (2010). Familial transposition of the great arteries caused by multiple mutations in laterality genes. *Heart* **96,** 673–677.

De Luca, A., Sarkozy, A., Ferese, R., Consoli, F., Lepri, F., Dentici, M. L., Vergara, P., De Zorzi, A., Versacci, P., Digilio, M. C., Marino, B., and Dallapiccola, B. (2011). New mutations in ZFPM2/FOG2 gene in tetralogy of Fallot and double outlet right ventricle. *Clin. Genet.* **80,** 184–190.

Devriendt, K., Eyskens, B., Swillen, A., Dumoulin, M., Gewillig, M., and Fryns, J. P. (1996). The incidence of a deletion in chromosome 22Q11 in sporadic and familial conotruncal heart disease. *Eur. J. Pediatr.* **155,** 721.

Dinesh, S. M., Kusuma, L., Smitha, R., Savitha, M. R., Krishnamurthy, B., Narayanappa, D., and Ramachandra, N. B. (2010a). Single-nucleotide polymorphisms of NKX2.5 found in congenital heart disease patients of Mysore, South India. *Genet. Test. Mol. Biomarkers* **14,** 873–879.

Dinesh, S. M., Lingaiah, K., Savitha, M. R., Krishnamurthy, B., Narayanappa, D., and Ramachandra, N. B. (2010b). GATA4 Specific nonsynonymous single-nucleotide polymorphisms in congenital heart disease patients of Mysore, India. *Genet. Test. Mol. Biomarkers* **15,** 715–720.

Dodou, E., Verzi, M. P., Anderson, J. P., Xu, S. M., and Black, B. L. (2004). Mef2c is a direct transcriptional target of ISL1 and GATA factors in the anterior heart field during mouse embryonic development. *Development* **131,** 3931–3942.

Eldadah, Z. A., Hamosh, A., Biery, N. J., Montgomery, R. A., Duke, M., Elkins, R., and Dietz, H. C. (2001). Familial Tetralogy of Fallot caused by mutation in the jagged1 gene. *Hum. Mol. Genet.* **10,** 163–169.

Elliott, D. A., Kirk, E. P., Yeoh, T., Chandar, S., McKenzie, F., Taylor, P., Grossfeld, P., Fatkin, D., Jones, O., Hayes, P., Feneley, M., and Harvey, R. P. (2003). Cardiac homeobox gene NKX2-5 mutations and congenital heart disease: Associations with atrial septal defect and hypoplastic left heart syndrome. *J. Am. Coll. Cardiol.* **41,** 2072–2076.

Ensenauer, R. E., Adeyinka, A., Flynn, H. C., Michels, V. V., Lindor, N. M., Dawson, D. B., Thorland, E. C., Lorentz, C. P., Goldstein, J. L., McDonald, M. T., Smith, W. E., Simon-Fayard, E., *et al.* (2003). Microduplication 22q11.2, an emerging syndrome: Clinical, cytogenetic, and molecular analysis of thirteen patients. *Am. J. Hum. Genet.* **73,** 1027–1040.

Espinoza-Lewis, R. A., Yu, L., He, F., Liu, H., Tang, R., Shi, J., Sun, X., Martin, J. F., Wang, D., Yang, J., and Chen, Y. (2009). Shox2 is essential for the differentiation of cardiac pacemaker cells by repressing Nkx2-5. *Dev. Biol.* **327,** 376–385.

Evans, S. M., Yelon, D., Conlon, F. L., and Kirby, M. L. (2010). Myocardial lineage development. *Circ. Res.* **107,** 1428–1444.

Finegold, D. N., Kimak, M. A., Lawrence, E. C., Levinson, K. L., Cherniske, E. M., Pober, B. R., Dunlap, J. W., and Ferrell, R. E. (2001). Truncating mutations in FOXC2 cause multiple lymphedema syndromes. *Hum. Mol. Genet.* **10,** 1185–1189.

Finelli, P., Pincelli, A. I., Russo, S., Bonati, M. T., Recalcati, M. P., Masciadri, M., Giardino, D., Cavagnini, F., and Larizza, L. (2007). Disruption of friend of GATA 2 gene (FOG-2) by a de novo t(8;10) chromosomal translocation is associated with heart defects and gonadal dysgenesis. *Clin. Genet.* **71,** 195–204.

Franco, D., Chinchilla, A., Daimi, H., Dominguez, J. N., and Aranega, A. (2011). Modulation of conductive elements by Pitx2 and their impact on atrial arrhythmogenesis. *Cardiovasc. Res.* **91,** 223–231.

Garavelli, L., Zollino, M., Mainardi, P. C., Gurrieri, F., Rivieri, F., Soli, F., Verri, R., Albertini, E., Favaron, E., Zignani, M., Orteschi, D., Bianchi, P., *et al.* (2009). Mowat-Wilson syndrome: Facial phenotype changing with age: Study of 19 Italian patients and review of the literature. *Am. J. Med. Genet. A* **149A,** 417–426.

Garg, V., Kathiriya, I. S., Barnes, R., Schluterman, M. K., King, I. N., Butler, C. A., Rothrock, C. R., Eapen, R. S., Hirayama-Yamada, K., Joo, K., Matsuoka, R., Cohen, J. C., *et al.* (2003). GATA4 mutations cause human congenital heart defects and reveal an interaction with TBX5. *Nature* **424,** 443–447.

Gawde, H., Patel, Z. M., Khatkhatey, M. I., D'Souza, A., Babu, S., Adhia, R., and Kerkar, P. (2006). Chromosome 22 microdeletion by F.I.S.H. in isolated congenital heart disease. *Indian J. Pediatr.* **73,** 885–888.

Ghosh, T. K., Song, F. F., Packham, E. A., Buxton, S., Robinson, T. E., Ronksley, J., Self, T., Bonser, A. J., and Brook, J. D. (2009). Physical interaction between TBX5 and MEF2C is required for early heart development. *Mol. Cell. Biol.* **29,** 2205–2218.

Giglio, S., Graw, S. L., Gimelli, G., Pirola, B., Varone, P., Voullaire, L., Lerzo, F., Rossi, E., Dellavecchia, C., Bonaglia, M. C., Digilio, M. C., Giannotti, A., *et al.* (2000). Deletion of a 5-cM region at chromosome 8p23 is associated with a spectrum of congenital heart defects. *Circulation* **102,** 432–437.

Gioli-Pereira, L., Pereira, A. C., Bergara, D., Mesquita, S., Lopes, A. A., and Krieger, J. E. (2008). Frequency of 22q11.2 microdeletion in sporadic non-syndromic tetralogy of Fallot cases. *Int. J. Cardiol.* **126,** 374–378.

Gioli-Pereira, L., Pereira, A. C., Mesquita, S. M., Xavier-Neto, J., Lopes, A. A., and Krieger, J. E. (2010). NKX2.5 mutations in patients with non-syndromic congenital heart disease. *Int. J. Cardiol.* **138,** 261–265.

Goldmuntz, E., Clark, B. J., Mitchell, L. E., Jawad, A. F., Cuneo, B. F., Reed, L., McDonald-McGinn, D., Chien, P., Feuer, J., Zackai, E. H., Emanuel, B. S., and Driscoll, D. A. (1998). Frequency of 22q11 deletions in patients with conotruncal defects. *J. Am. Coll. Cardiol.* **32,** 492–498.

Goldmuntz, E., Paluru, P., Glessner, J., Hakonarson, H., Biegel, J. A., White, P. S., Gai, X., and Shaikh, T. H. (2011). Microdeletions and microduplications in patients with congenital heart disease and multiple congenital anomalies. *Congenit. Heart Dis.* **6,** 592–602.

Goodship, J., Cross, I., LiLing, J., and Wren, C. (1998). A population study of chromosome 22q11 deletions in infancy. *Arch. Dis. Child.* **79,** 348–351.

Greenway, S. C., Pereira, A. C., Lin, J. C., DePalma, S. R., Israel, S. J., Mesquita, S. M., Ergul, E., Conta, J. H., Korn, J. M., McCarroll, S. A., Gorham, J. M., Gabriel, S., *et al.* (2009). De novo copy number variants identify new genes and loci in isolated sporadic tetralogy of Fallot. *Nat. Genet.* **41,** 931–935.

Greulich, F., Rudat, C., and Kispert, A. (2011). Mechanisms of T-box gene function in the developing heart. *Cardiovasc. Res.* **91,** 212–222.

Guida, V., Lepri, F., Vijzelaar, R., De Zorzi, A., Versacci, P., Digilio, M. C., Marino, B., De Luca, A., and Dallapiccola, B. (2010). Multiplex ligation-dependent probe amplification analysis of GATA4 gene copy number variations in patients with isolated congenital heart disease. *Dis. Markers* **28,** 287–292.

Gutierrez-Roelens, I., Sluysmans, T., Gewillig, M., Devriendt, K., and Vikkula, M. (2002). Progressive AV-block and anomalous venous return among cardiac anomalies associated with two novel missense mutations in the CSX/NKX2-5 gene. *Hum. Mutat.* **20,** 75–76.

Hammer, S., Toenjes, M., Lange, M., Fischer, J. J., Dunkel, I., Mebus, S., Grimm, C. H., Hetzer, R., Berger, F., and Sperling, S. (2008). Characterization of TBX20 in human hearts and its regulation by TFAP2. *J. Cell. Biochem.* **104,** 1022–1033.

Harvey, R. P. (2002). Patterning the vertebrate heart. *Nat. Rev. Genet.* **3,** 544–556.

Harvey, R. P., Lai, D., Elliott, D., Biben, C., Solloway, M., Prall, O., Stennard, F., Schindeler, A., Groves, N., Lavulo, L., Hyun, C., Yeoh, T., *et al.* (2002). Homeodomain factor Nkx2-5 in heart development and disease. *Cold Spring Harb. Symp. Quant. Biol.* **67,** 107–114.

Hatcher, C. J., and Basson, C. T. (2009). Specification of the cardiac conduction system by transcription factors. *Circ. Res.* **105,** 620–630.

He, A., Kong, S. W., Ma, Q., and Pu, W. T. (2011). Co-occupancy by multiple cardiac transcription factors identifies transcriptional enhancers active in heart. *Proc. Natl. Acad. Sci. USA* **108,** 5632–5637.

Heathcote, K., Braybrook, C., Abushaban, L., Guy, M., Khetyar, M. E., Patton, M. A., Carter, N. D., Scambler, P. J., and Syrris, P. (2005). Common arterial trunk associated with a homeodomain mutation of NKX2.6. *Hum. Mol. Genet.* **14,** 585–593.

Heinritz, W., Zweier, C., Froster, U. G., Strenge, S., Kujat, A., Syrbe, S., Rauch, A., and Schuster, V. (2006). A missense mutation in the ZFHX1B gene associated with an atypical Mowat-Wilson syndrome phenotype. *Am. J. Med. Genet. A* **140,** 1223–1227.

Henwood, J., Pickard, C., Leek, J. P., Bennett, C. P., Crow, Y. J., Thompson, J. D., Ahmed, M., Watterson, K. G., Parsons, J. M., Roberts, E., and Lench, N. J. (2001). A region of homozygosity within 22q11.2 associated with congenital heart disease: Recessive DiGeorge/velocardiofacial syndrome? *J. Med. Genet.* **38,** 533–536.

Heritage, M. L., MacMillan, J. C., Colliton, R. P., Genin, A., Spinner, N. B., and Anderson, G. J. (2000). Jagged1 (JAG1) mutation detection in an Australian Alagille syndrome population. *Hum. Mutat.* **16,** 408–416.

Heritage, M. L., MacMillan, J. C., and Anderson, G. J. (2002). DHPLC mutation analysis of Jagged1 (JAG1) reveals six novel mutations in Australian alagille syndrome patients. *Hum. Mutat.* **20,** 481.

Hirayama-Yamada, K., Kamisago, M., Akimoto, K., Aotsuka, H., Nakamura, Y., Tomita, H., Furutani, M., Imamura, S., Takao, A., Nakazawa, M., and Matsuoka, R. (2005). Phenotypes with GATA4 or NKX2.5 mutations in familial atrial septal defect. *Am. J. Med. Genet. A* **135,** 47–52.

Hiroi, Y., Kudoh, S., Monzen, K., Ikeda, Y., Yazaki, Y., Nagai, R., and Komuro, I. (2001). Tbx5 associates with Nkx2-5 and synergistically promotes cardiomyocyte differentiation. *Nat. Genet.* **28,** 276–280.

Hoffman, J. I., Kaplan, S., and Liberthson, R. R. (2004). Prevalence of congenital heart disease. *Am. Heart J.* **147,** 425–439.

Hokanson, J. S., Pierpont, E., Hirsch, B., and Moller, J. H. (2001). 22q11.2 microdeletions in adults with familial tetralogy of Fallot. *Genet. Med.* **3,** 61–64.

Hu, Y., Zhu, X., Yang, Y., Mo, X., Sheng, M., Yao, J., and Wang, D. (2009). Incidences of micro-deletion/duplication 22q11.2 detected by multiplex ligation-dependent probe amplification in patients with congenital cardiac disease who are scheduled for cardiac surgery. *Cardiol. Young* **19,** 179–184.

Hutson, M. R., and Kirby, M. L. (2007). Model systems for the study of heart development and disease. Cardiac neural crest and conotruncal malformations. *Semin. Cell Dev. Biol.* **18,** 101–110.

Ikeda, Y., Hiroi, Y., Hosoda, T., Utsunomiya, T., Matsuo, S., Ito, T., Inoue, J., Sumiyoshi, T., Takano, H., Nagai, R., and Komuro, I. (2002). Novel point mutation in the cardiac transcription factor CSX/NKX2.5 associated with congenital heart disease. *Circ. J.* **66,** 561–563.

Iserin, L., de Lonlay, P., Viot, G., Sidi, D., Kachaner, J., Munnich, A., Lyonnet, S., Vekemans, M., and Bonnet, D. (1998). Prevalence of the microdeletion 22q11 in newborn infants with congenital conotruncal cardiac anomalies. *Eur. J. Pediatr.* **157,** 881–884.

Ito, T., Okubo, T., and Sato, H. (2002). Familial 22q11.2 deletion: An infant with interrupted aortic arch and DiGeorge syndrome delivered from by a mother with tetralogy of Fallot. *Eur. J. Pediatr.* **161,** 173–174.

Jay, P. Y., Harris, B. S., Maguire, C. T., Buerger, A., Wakimoto, H., Tanaka, M., Kupershmidt, S., Roden, D. M., Schultheiss, T. M., O'Brien, T. X., Gourdie, R. G., Berul, C. I., *et al.* (2004). Nkx2-5 mutation causes anatomic hypoplasia of the cardiac conduction system. *J. Clin. Invest.* **113,** 1130–1137.

Jiang, L., Duan, C., Chen, B., Hou, Z., Chen, Z., Li, Y., Huan, Y., and Wu, K. K. (2005). Association of 22q11 deletion with isolated congenital heart disease in three Chinese ethnic groups. *Int. J. Cardiol.* **105,** 216–223.

Jiang, L., Hou, Z., Duan, C., Chen, B., Chen, Z., Li, Y., Huan, Y., and Wu, K. K. (2010). Isolated congenital heart disease is associated with the 22q11 deletion even though it is rare. *Int. J. Cardiol.* **145,** 284–285.

Kirk, E. P., Sunde, M., Costa, M. W., Rankin, S. A., Wolstein, O., Castro, M. L., Butler, T. L., Hyun, C., Guo, G., Otway, R., Mackay, J. P., Waddell, L. B., *et al.* (2007). Mutations in cardiac T-box factor gene TBX20 are associated with diverse cardiac pathologies, including defects of septation and valvulogenesis and cardiomyopathy. *Am. J. Hum. Genet.* **81,** 280–291.

Kodo, K., Nishizawa, T., Furutani, M., Arai, S., Yamamura, E., Joo, K., Takahashi, T., Matsuoka, R., and Yamagishi, H. (2009). GATA6 mutations cause human cardiac outflow tract defects by disrupting semaphorin-plexin signaling. *Proc. Natl. Acad. Sci. USA* **106,** 13933–13938.

Konig, K., Will, J. C., Berger, F., Muller, D., and Benson, D. W. (2006). Familial congenital heart disease, progressive atrioventricular block and the cardiac homeobox transcription factor gene NKX2.5: Identification of a novel mutation. *Clin. Res. Cardiol.* **95,** 499–503.

Krantz, I. D., Smith, R., Colliton, R. P., Tinkel, H., Zackai, E. H., Piccoli, D. A., Goldmuntz, E., and Spinner, N. B. (1999). Jagged1 mutations in patients ascertained with isolated congenital heart defects. *Am. J. Med. Genet.* **84,** 56–60.

Kuo, C. T., Morrisey, E. E., Anandappa, R., Sigrist, K., Lu, M. M., Parmacek, M. S., Soudais, C., and Leiden, J. M. (1997). GATA4 transcription factor is required for ventral morphogenesis and heart tube formation. *Genes Dev.* **11,** 1048–1060.

Li, L., Krantz, I. D., Deng, Y., Genin, A., Banta, A. B., Collins, C. C., Qi, M., Trask, B. J., Kuo, W. L., Cochran, J., Costa, T., Pierpont, M. E., *et al.* (1997a). Alagille syndrome is caused by mutations in human Jagged1, which encodes a ligand for Notch1. *Nat. Genet.* **16,** 243–251.

Li, Q. Y., Newbury-Ecob, R. A., Terrett, J. A., Wilson, D. I., Curtis, A. R., Yi, C. H., Gebuhr, T., Bullen, P. J., Robson, S. C., Strachan, T., Bonnet, D., Lyonnet, S., *et al.* (1997b). Holt-Oram syndrome is caused by mutations in TBX5, a member of the Brachyury (T) gene family. *Nat. Genet.* **15,** 21–29.

Li, F. B., Chen, J., Yu, J. D., Gao, H., and Qi, M. (2010). A Chinese girl molecularly diagnosed with Alagille syndrome. *World J. Pediatr.* **6,** 278–280.

Liberatore, C. M., Searcy-Schrick, R. D., Vincent, E. B., and Yutzey, K. E. (2002). Nkx-2.5 gene induction in mice is mediated by a Smad consensus regulatory region. *Dev. Biol.* **244,** 243–256.

Lien, C. L., Wu, C., Mercer, B., Webb, R., Richardson, J. A., and Olson, E. N. (1999). Control of early cardiac-specific transcription of Nkx2-5 by a GATA-dependent enhancer. *Development* **126,** 75–84.

Lien, C. L., McAnally, J., Richardson, J. A., and Olson, E. N. (2002). Cardiac-specific activity of an Nkx2-5 enhancer requires an evolutionarily conserved Smad binding site. *Dev. Biol.* **244,** 257–266.

Lin, Q., Schwarz, J., Bucana, C., and Olson, E. N. (1997). Control of mouse cardiac morphogenesis and myogenesis by transcription factor MEF2C. *Science* **276,** 1404–1407.

Lin, X., Huo, Z., Liu, X., Zhang, Y., Li, L., Zhao, H., Yan, B., Liu, Y., Yang, Y., and Chen, Y. H. (2010). A novel GATA6 mutation in patients with tetralogy of Fallot or atrial septal defect. *J. Hum. Genet.* **55,** 662–667.

Lindsay, E. A., Goldberg, R., Jurecic, V., Morrow, B., Carlson, C., Kucherlapati, R. S., Shprintzen, R. J., and Baldini, A. (1995). Velo-cardio-facial syndrome: Frequency and extent of 22q11 deletions. *Am. J. Med. Genet.* **57,** 514–522.

Liu, C., Shen, A., Li, X., Jiao, W., Zhang, X., and Li, Z. (2008). T-box transcription factor TBX20 mutations in Chinese patients with congenital heart disease. *Eur. J. Med. Genet.* **51,** 580–587.

Liu, C. X., Shen, A. D., Li, X. F., Jiao, W. W., Bai, S., Yuan, F., Guan, X. L., Zhang, X. G., Zhang, G. R., and Li, Z. Z. (2009). Association of TBX5 gene polymorphism with ventricular septal defect in the Chinese Han population. *Chin. Med. J. (Engl.)* **122,** 30–34.

Liu, X. Y., Wang, J., Zheng, J. H., Bai, K., Liu, Z. M., Wang, X. Z., Liu, X., Fang, W. Y., and Yang, Y. Q. (2011). Involvement of a novel GATA4 mutation in atrial septal defects. *Int. J. Mol. Med.* **28,** 17–23.

Lupski, J. R., Langston, C., Friedman, R., Ledbetter, D. H., and Greenberg, F. (1991). Di George anomaly associated with a de novo Y;22 translocation resulting in monosomy del (22)(q11.2). *Am. J. Med. Genet.* **40,** 196–198.

Lyons, I., Parsons, L. M., Hartley, L., Li, R., Andrews, J. E., Robb, L., and Harvey, R. P. (1995). Myogenic and morphogenetic defects in the heart tubes of murine embryos lacking the homeo box gene Nkx2-5. *Genes Dev.* **9,** 1654–1666.

Maclean, K., Smith, J., St Heaps, L., Chia, N., Williams, R., Peters, G. B., Onikul, E., McCrossin, T., Lehmann, O. J., and Ades, L. C. (2005). Axenfeld-Rieger malformation and distinctive facial features: Clues to a recognizable 6p25 microdeletion syndrome. *Am. J. Med. Genet. A* **132,** 381–385.

Maitra, M., Schluterman, M. K., Nichols, H. A., Richardson, J. A., Lo, C. W., Srivastava, D., and Garg, V. (2009). Interaction of Gata4 and Gata6 with Tbx5 is critical for normal cardiac development. *Dev. Biol.* **326,** 368–377.

Maitra, M., Koenig, S. N., Srivastava, D., and Garg, V. (2010). Identification of GATA6 sequence variants in patients with congenital heart defects. *Pediatr. Res.* **68,** 281–285.

Mani, A., Radhakrishnan, J., Farhi, A., Carew, K. S., Warnes, C. A., Nelson-Williams, C., Day, R. W., Pober, B., State, M. W., and Lifton, R. P. (2005). Syndromic patent ductus arteriosus: Evidence for haploinsufficient TFAP2B mutations and identification of a linked sleep disorder. *Proc. Natl. Acad. Sci. USA* **102,** 2975–2979.

Matsuoka, R., Takao, A., Kimura, M., Imamura, S., Kondo, C., Joh-o, K., Ikeda, K., Nishibatake, M., Ando, M., and Momma, K. (1994). Confirmation that the conotruncal anomaly face syndrome is associated with a deletion within 22q11.2. *Am. J. Med. Genet.* **53,** 285–289.

McDermott, D. A., Hatcher, C. J., and Basson, C. T. (2008). Atrial fibrillation and other clinical manifestations of altered TBX5 dosage in typical Holt-Oram syndrome. *Circ. Res.* **103,** e96.

McElhinney, D. B., Clark, B. J., 3rd, Weinberg, P. M., Kenton, M. L., McDonald-McGinn, D., Driscoll, D. A., Zackai, E. H., and Goldmuntz, E. (2001). Association of chromosome 22q11 deletion with isolated anomalies of aortic arch laterality and branching. *J. Am. Coll. Cardiol.* **37,** 2114–2119.

McElhinney, D. B., Driscoll, D. A., Levin, E. R., Jawad, A. F., Emanuel, B. S., and Goldmuntz, E. (2003a). Chromosome 22q11 deletion in patients with ventricular septal defect: Frequency and associated cardiovascular anomalies. *Pediatrics* **112,** e472.

McElhinney, D. B., Geiger, E., Blinder, J., Benson, D. W., and Goldmuntz, E. (2003b). NKX2.5 mutations in patients with congenital heart disease. *J. Am. Coll. Cardiol.* **42,** 1650–1655.

McGaughran, J., Sinnott, S., Dastot-Le Moal, F., Wilson, M., Mowat, D., Sutton, B., and Goossens, M. (2005). Recurrence of Mowat-Wilson syndrome in siblings with the same proven mutation. *Am. J. Med. Genet. A* **137A,** 302–304.

Molkentin, J. D., Lin, Q., Duncan, S. A., and Olson, E. N. (1997). Requirement of the transcription factor GATA4 for heart tube formation and ventral morphogenesis. *Genes Dev.* **11,** 1061–1072.

Morin, S., Charron, F., Robitaille, L., and Nemer, M. (2000). GATA-dependent recruitment of MEF2 proteins to target promoters. *EMBO J.* **19,** 2046–2055.

Moskowitz, I. P., Kim, J. B., Moore, M. L., Wolf, C. M., Peterson, M. A., Shendure, J., Nobrega, M. A., Yokota, Y., Berul, C., Izumo, S., Seidman, J. G., and Seidman, C. E. (2007). A molecular pathway including Id2, Tbx5, and Nkx2-5 required for cardiac conduction system development. *Cell* **129,** 1365–1376.

Moskowitz, I. P., Wang, J., Peterson, M. A., Pu, W. T., Mackinnon, A. C., Oxburgh, L., Chu, G. C., Sarkar, M., Berul, C., Smoot, L., Robertson, E. J., Schwartz, R., *et al.* (2011). Transcription factor genes Smad4 and Gata4 cooperatively regulate cardiac valve development. *Proc. Natl. Acad. Sci. USA* **108,** 4006–4011.

Munshi, N. V., McAnally, J., Bezprozvannaya, S., Berry, J. M., Richardson, J. A., Hill, J. A., and Olson, E. N. (2009). Cx30.2 enhancer analysis identifies Gata4 as a novel regulator of atrioventricular delay. *Development* **136,** 2665–2674.

Nemer, G., Fadlalah, F., Usta, J., Nemer, M., Dbaibo, G., Obeid, M., and Bitar, F. (2006). A novel mutation in the GATA4 gene in patients with Tetralogy of Fallot. *Hum. Mutat.* **27,** 293–294.

Oda, T., Elkahloun, A. G., Pike, B. L., Okajima, K., Krantz, I. D., Genin, A., Piccoli, D. A., Meltzer, P. S., Spinner, N. B., Collins, F. S., and Chandrasekharappa, S. C. (1997). Mutations in the human Jagged1 gene are responsible for Alagille syndrome. *Nat. Genet.* **16,** 235–242.

Oh, D. C., Min, J. Y., Lee, M. H., Kim, Y. M., Park, S. Y., Won, H. S., Kim, I. K., Lee, Y. H., Yoo, S. J., and Ryu, H. M. (2002). Prenatal diagnosis of tetralogy of Fallot associated with chromosome 22q11 deletion. *J. Korean Med. Sci.* **17,** 125–128.

Oka, T., Xu, J., and Molkentin, J. D. (2007). Re-employment of developmental transcription factors in adult heart disease. *Semin. Cell Dev. Biol.* **18,** 117–131.

Okubo, A., Miyoshi, O., Baba, K., Takagi, M., Tsukamoto, K., Kinoshita, A., Yoshiura, K., Kishino, T., Ohta, T., Niikawa, N., and Matsumoto, N. (2004). A novel GATA4 mutation completely segregated with atrial septal defect in a large Japanese family. *J. Med. Genet.* **41,** e97.

Olson, E. N. (2006). Gene regulatory networks in the evolution and development of the heart. *Science* **313,** 1922–1927.

Ou, Z., Berg, J. S., Yonath, H., Enciso, V. B., Miller, D. T., Picker, J., Lenzi, T., Keegan, C. E., Sutton, V. R., Belmont, J., Chinault, A. C., Lupski, J. R., *et al.* (2008). Microduplications of 22q11.2 are frequently inherited and are associated with variable phenotypes. *Genet. Med.* **10,** 267–277.

Ouyang, P., Saarel, E., Bai, Y., Luo, C., Lv, Q., Xu, Y., Wang, F., Fan, C., Younoszai, A., Chen, Q., Tu, X., and Wang, Q. K. (2011). A de novo mutation in NKX2.5 associated with atrial septal defects, ventricular noncompaction, syncope and sudden death. *Clin. Chim. Acta* **412,** 170–175.

Pabst, S., Wollnik, B., Rohmann, E., Hintz, Y., Glanzer, K., Vetter, H., Nickenig, G., and Grohe, C. (2008). A novel stop mutation truncating critical regions of the cardiac transcription factor NKX2-5 in a large family with autosomal-dominant inherited congenital heart disease. *Clin. Res. Cardiol.* **97,** 39–42.

Paradisi, I., and Arias, S. (2007). IVIC syndrome is caused by a c.2607delA mutation in the SALL4 locus. *Am. J. Med. Genet. A* **143,** 326–332.

Pauli, R. M., Scheib-Wixted, S., Cripe, L., Izumo, S., and Sekhon, G. S. (1999). Ventricular noncompaction and distal chromosome 5q deletion. *Am. J. Med. Genet.* **85,** 419–423.

Pehlivan, T., Pober, B. R., Brueckner, M., Garrett, S., Slaugh, R., Van Rheeden, R., Wilson, D. B., Watson, M. S., and Hing, A. V. (1999). GATA4 haploinsufficiency in patients with interstitial deletion of chromosome region 8p23.1 and congenital heart disease. *Am. J. Med. Genet.* **83,** 201–206.

Peng, T., Wang, L., Zhou, S. F., and Li, X. (2010). Mutations of the GATA4 and NKX2.5 genes in Chinese pediatric patients with non-familial congenital heart disease. *Genetica* **138,** 1231–1240.

Person, A. D., Klewer, S. E., and Runyan, R. B. (2005). Cell biology of cardiac cushion development. *Int. Rev. Cytol.* **243,** 287–335.

Peterkin, T., Gibson, A., Loose, M., and Patient, R. (2005). The roles of GATA-4, -5 and -6 in vertebrate heart development. *Semin. Cell Dev. Biol.* **16,** 83–94.

Pikkarainen, S., Tokola, H., Kerkela, R., and Ruskoaho, H. (2004). GATA transcription factors in the developing and adult heart. *Cardiovasc. Res.* **63,** 196–207.

Pizzuti, A., Sarkozy, A., Newton, A. L., Conti, E., Flex, E., Digilio, M. C., Amati, F., Gianni, D., Tandoi, C., Marino, B., Crossley, M., and Dallapiccola, B. (2003). Mutations of ZFPM2/FOG2 gene in sporadic cases of tetralogy of Fallot. *Hum. Mutat.* **22,** 372–377.

Poon, L. C., Huggon, I. C., Zidere, V., and Allan, L. D. (2007). Tetralogy of Fallot in the fetus in the current era. *Ultrasound Obstet. Gynecol.* **29,** 625–627.

Posch, M. G., Boldt, L. H., Polotzki, M., Richter, S., Rolf, S., Perrot, A., Dietz, R., Ozcelik, C., and Haverkamp, W. (2010a). Mutations in the cardiac transcription factor GATA4 in patients with lone atrial fibrillation. *Eur. J. Med. Genet.* **53,** 201–203.

Posch, M. G., Gramlich, M., Sunde, M., Schmitt, K. R., Lee, S. H., Richter, S., Kersten, A., Perrot, A., Panek, A. N., Al Khatib, I. H., Nemer, G., Megarbane, A., *et al.* (2010b). A gain-of-function TBX20 mutation causes congenital atrial septal defects, patent foramen ovale and cardiac valve defects. *J. Med. Genet.* **47,** 230–235.

Postma, A. V., van de Meerakker, J. B., Mathijssen, I. B., Barnett, P., Christoffels, V. M., Ilgun, A., Lam, J., Wilde, A. A., Lekanne Deprez, R. H., and Moorman, A. F. (2008). A gain-of-function TBX5 mutation is associated with atypical Holt-Oram syndrome and paroxysmal atrial fibrillation. *Circ. Res.* **102,** 1433–1442.

Prall, O. W., Menon, M. K., Solloway, M. J., Watanabe, Y., Zaffran, S., Bajolle, F., Biben, C., McBride, J. J., Robertson, B. R., Chaulet, H., Stennard, F. A., Wise, N., *et al.* (2007). An Nkx2-5/Bmp2/Smad1 negative feedback loop controls heart progenitor specification and proliferation. *Cell* **128,** 947–959.

Pu, W. T., Ishiwata, T., Juraszek, A. L., Ma, Q., and Izumo, S. (2004). GATA4 is a dosage-sensitive regulator of cardiac morphogenesis. *Dev. Biol.* **275,** 235–244.

Qian, L., Mohapatra, B., Akasaka, T., Liu, J., Ocorr, K., Towbin, J. A., and Bodmer, R. (2008). Transcription factor neuromancer/TBX20 is required for cardiac function in Drosophila with implications for human heart disease. *Proc. Natl. Acad. Sci. USA* **105,** 19833–19838.

Rajagopal, S. K., Ma, Q., Obler, D., Shen, J., Manichaikul, A., Tomita-Mitchell, A., Boardman, K., Briggs, C., Garg, V., Srivastava, D., Goldmuntz, E., Broman, K. W., *et al.* (2007). Spectrum of heart disease associated with murine and human GATA4 mutation. *J. Mol. Cell. Cardiol.* **43,** 677–685.

Ransom, J. F., King, I. N., Garg, V., and Srivastava, D. (2008). A rare human sequence variant reveals myocardin autoinhibition. *J. Biol. Chem.* **283,** 35845–35852.

Rauch, A., Devriendt, K., Koch, A., Rauch, R., Gewillig, M., Kraus, C., Weyand, M., Singer, H., Reis, A., and Hofbeck, M. (2004). Assessment of association between variants and haplotypes of the remaining TBX1 gene and manifestations of congenital heart defects in 22q11.2 deletion patients. *J. Med. Genet.* **41,** e40.

Rauch, R., Hofbeck, M., Zweier, C., Koch, A., Zink, S., Trautmann, U., Hoyer, J., Kaulitz, R., Singer, H., and Rauch, A. (2010). Comprehensive genotype-phenotype analysis in 230 patients with tetralogy of Fallot. *J. Med. Genet.* **47,** 321–331.

Raymond, F. L., Simpson, J. M., Mackie, C. M., and Sharland, G. K. (1997). Prenatal diagnosis of 22q11 deletions: A series of five cases with congenital heart defects. *J. Med. Genet.* **34,** 679–682.

Reamon-Buettner, S. M., and Borlak, J. (2004a). Somatic NKX2-5 mutations as a novel mechanism of disease in complex congenital heart disease. *J. Med. Genet.* **41,** 684–690.

Reamon-Buettner, S. M., and Borlak, J. (2004b). TBX5 mutations in non-Holt-Oram syndrome (HOS) malformed hearts. *Hum. Mutat.* **24,** 104.

Reamon-Buettner, S. M., and Borlak, J. (2010). NKX2-5: An update on this hypermutable homeodomain protein and its role in human congenital heart disease (CHD). *Hum. Mutat.* **31,** 1185–1194.

Reamon-Buettner, S. M., Hecker, H., Spanel-Borowski, K., Craatz, S., Kuenzel, E., and Borlak, J. (2004). Novel NKX2-5 mutations in diseased heart tissues of patients with cardiac malformations. *Am. J. Pathol.* **164,** 2117–2125.

Reamon-Buettner, S. M., Cho, S. H., and Borlak, J. (2007). Mutations in the 3'-untranslated region of GATA4 as molecular hotspots for congenital heart disease (CHD). *BMC Med. Genet.* **8,** 38.

Reamon-Buettner, S. M., Ciribilli, Y., Inga, A., and Borlak, J. (2008). A loss-of-function mutation in the binding domain of HAND1 predicts hypoplasia of the human hearts. *Hum. Mol. Genet.* **17,** 1397–1405.

Reamon-Buettner, S. M., Ciribilli, Y., Traverso, I., Kuhls, B., Inga, A., and Borlak, J. (2009). A functional genetic study identifies HAND1 mutations in septation defects of the human heart. *Hum. Mol. Genet.* **18,** 3567–3578.

Repetto, G. M., Guzman, M. L., Puga, A., Calderon, J. F., Astete, C. P., Aracena, M., Arriaza, M., Aravena, T., and Sanz, P. (2009). Clinical features of chromosome 22q11.2 microdeletion syndrome in 208 Chilean patients. *Clin. Genet.* **76,** 465–470.

Rojas, A., Kong, S. W., Agarwal, P., Gilliss, B., Pu, W. T., and Black, B. L. (2008). GATA4 is a direct transcriptional activator of cyclin D2 and Cdk4 and is required for cardiomyocyte proliferation in anterior heart field-derived myocardium. *Mol. Cell. Biol.* **28,** 5420–5431.

Rope, A. F., Cragun, D. L., Saal, H. M., and Hopkin, R. J. (2009). DiGeorge anomaly in the absence of chromosome 22q11.2 deletion. *J. Pediatr.* **155,** 560–565.

Rubart, M., and Zipes, D. P. (2005). Mechanisms of sudden cardiac death. *J. Clin. Invest.* **115,** 2305–2315.

Ryan, A. K., Goodship, J. A., Wilson, D. I., Philip, N., Levy, A., Seidel, H., Schuffenhauer, S., Oechsler, H., Belohradsky, B., Prieur, M., Aurias, A., Raymond, F. L., *et al.* (1997). Spectrum of clinical features associated with interstitial chromosome 22q11 deletions: A European collaborative study. *J. Med. Genet.* **34,** 798–804.

Sarkozy, A., Conti, E., Neri, C., D'Agostino, R., Digilio, M. C., Esposito, G., Toscano, A., Marino, B., Pizzuti, A., and Dallapiccola, B. (2005). Spectrum of atrial septal defects associated with mutations of NKX2.5 and GATA4 transcription factors. *J. Med. Genet.* **42,** e16.

Sasongko, T. H., Sadewa, A. H., Gunadi, A. H., Lee, M. J., Koterazawa, K., and Nishio, H. (2007). Nonsense mutations of the ZFHX1B gene in two Japanese girls with Mowat-Wilson syndrome. *Kobe J. Med. Sci.* **53,** 157–162.

Satoda, M., Zhao, F., Diaz, G. A., Burn, J., Goodship, J., Davidson, H. R., Pierpont, M. E., and Gelb, B. D. (2000). Mutations in TFAP2B cause Char syndrome, a familial form of patent ductus arteriosus. *Nat. Genet.* **25,** 42–46.

Saunders, C. J., Zhao, W., and Ardinger, H. H. (2009). Comprehensive ZEB2 gene analysis for Mowat-Wilson syndrome in a North American cohort: A suggested approach to molecular diagnostics. *Am. J. Med. Genet. A* **149A,** 2527–2531.

Schluterman, M. K., Krysiak, A. E., Kathiriya, I. S., Abate, N., Chandalia, M., Srivastava, D., and Garg, V. (2007). Screening and biochemical analysis of GATA4 sequence variations identified in patients with congenital heart disease. *Am. J. Med. Genet. A* **143A,** 817–823.

Schott, J. J., Benson, D. W., Basson, C. T., Pease, W., Silberbach, G. M., Moak, J. P., Maron, B. J., Seidman, C. E., and Seidman, J. G. (1998). Congenital heart disease caused by mutations in the transcription factor NKX2-5. *Science* **281,** 108–111.

Searcy, R. D., Vincent, E. B., Liberatore, C. M., and Yutzey, K. E. (1998). A GATA-dependent nkx-2.5 regulatory element activates early cardiac gene expression in transgenic mice. *Development* **125,** 4461–4470.

Sepulveda, J. L., Vlahopoulos, S., Iyer, D., Belaguli, N., and Schwartz, R. J. (2002). Combinatorial expression of GATA4, Nkx2-5, and serum response factor directs early cardiac gene activity. *J. Biol. Chem.* **277,** 25775–25782.

Shen, L., Li, X. F., Shen, A. D., Wang, Q., Liu, C. X., Guo, Y. J., Song, Z. J., and Li, Z. Z. (2010). Transcription factor HAND2 mutations in sporadic Chinese patients with congenital heart disease. *Chin. Med. J. (Engl.)* **123,** 1623–1627.

Sorensen, K. M., Agergaard, P., Olesen, C., Andersen, P. S., Larsen, L. A., Ostergaard, J. R., Schouten, J. P., and Christiansen, M. (2010). Detecting 22q11.2 deletions by use of multiplex ligation-dependent probe amplification on DNA from neonatal dried blood spot samples. *J. Mol. Diagn.* **12,** 147–151.

Sparkes, R., Chernos, J., and Dicke, F. (2005). Duplication of the 22q11.2 region associated with congenital cardiac disease. *Cardiol. Young* **15,** 229–231.

Sperling, S., Grimm, C. H., Dunkel, I., Mebus, S., Sperling, H. P., Ebner, A., Galli, R., Lehrach, H., Fusch, C., Berger, F., and Hammer, S. (2005). Identification and functional analysis of CITED2 mutations in patients with congenital heart defects. *Hum. Mutat.* **26,** 575–582.

Srivastava, D. (2006). Genetic regulation of cardiogenesis and congenital heart disease. *Annu. Rev. Pathol.* **1,** 199–213.

Stallmeyer, B., Fenge, H., Nowak-Gottl, U., and Schulze-Bahr, E. (2010). Mutational spectrum in the cardiac transcription factor gene NKX2.5 (CSX) associated with congenital heart disease. *Clin. Genet.* **78,** 533–540.

Svensson, E. C., Tufts, R. L., Polk, C. E., and Leiden, J. M. (1999). Molecular cloning of FOG-2: A modulator of transcription factor GATA-4 in cardiomyocytes. *Proc. Natl. Acad. Sci. USA* **96,** 956–961.

Svensson, E. C., Huggins, G. S., Lin, H., Clendenin, C., Jiang, F., Tufts, R., Dardik, F. B., and Leiden, J. M. (2000). A syndrome of tricuspid atresia in mice with a targeted mutation of the gene encoding Fog-2. *Nat. Genet.* **25,** 353–356.

Swaby, J. A., Silversides, C. K., Bekeschus, S. C., Piran, S., Oechslin, E. N., Chow, E. W., and Bassett, A. S. (2011). Complex congenital heart disease in unaffected relatives of adults with 22q11.2 deletion syndrome. *Am. J. Cardiol.* **107,** 466–471.

Takeuchi, J. K., Mileikovskaia, M., Koshiba-Takeuchi, K., Heidt, A. B., Mori, A. D., Arruda, E. P., Gertsenstein, M., Georges, R., Davidson, L., Mo, R., Hui, C. C., Henkelman, R. M., *et al.* (2005). Tbx20 dose-dependently regulates transcription factor networks required for mouse heart and motoneuron development. *Development* **132,** 2463–2474.

Tanaka, M., Chen, Z., Bartunkova, S., Yamasaki, N., and Izumo, S. (1999). The cardiac homeobox gene Csx/Nkx2.5 lies genetically upstream of multiple genes essential for heart development. *Development* **126,** 1269–1280.

Tevosian, S. G., Deconinck, A. E., Tanaka, M., Schinke, M., Litovsky, S. H., Izumo, S., Fujiwara, Y., and Orkin, S. H. (2000). FOG-2, a cofactor for GATA transcription factors, is essential for heart morphogenesis and development of coronary vessels from epicardium. *Cell* **101,** 729–739.

Tomita-Mitchell, A., Maslen, C. L., Morris, C. D., Garg, V., and Goldmuntz, E. (2007). GATA4 sequence variants in patients with congenital heart disease. *J. Med. Genet.* **44,** 779–783.

Tomita-Mitchell, A., Mahnke, D. K., Larson, J. M., Ghanta, S., Feng, Y., Simpson, P. M., Broeckel, U., Duffy, K., Tweddell, J. S., Grossman, W. J., Routes, J. M., and Mitchell, M. E. (2010). Multiplexed quantitative real-time PCR to detect 22q11.2 deletion in patients with congenital heart disease. *Physiol. Genomics* **42A,** 52–60.

van Engelen, K., Topf, A., Keavney, B. D., Goodship, J. A., van der Velde, E. T., Baars, M. J., Snijder, S., Moorman, A. F., Postma, A. V., and Mulder, B. J. (2010). 22q11.2 Deletion Syndrome is under-recognised in adult patients with tetralogy of Fallot and pulmonary atresia. *Heart* **96,** 621–624.

Vaughan, C. J., and Basson, C. T. (2000). Molecular determinants of atrial and ventricular septal defects and patent ductus arteriosus. *Am. J. Med. Genet.* **97,** 304–309.

Verhoeven, W., Egger, J., Brunner, H., and de Leeuw, N. (2011). A patient with a de novo distal 22q11.2 microdeletion and anxiety disorder. *Am. J. Med. Genet. A* **155A,** 392–397.

Vincent, S. D., and Buckingham, M. E. (2010). How to make a heart: The origin and regulation of cardiac progenitor cells. *Curr. Top. Dev. Biol.* **90,** 1–41.

Vincentz, J. W., Barnes, R. M., Firulli, B. A., Conway, S. J., and Firulli, A. B. (2008). Cooperative interaction of Nkx2.5 and Mef2c transcription factors during heart development. *Dev. Dyn.* **237,** 3809–3819.

Wang, B., Li, L., Xie, X., Wang, J., Yan, J., Mu, Y., and Ma, X. (2010). Genetic variation of SAL-Like 4 (SALL4) in ventricular septal defect. *Int. J. Cardiol.* **145,** 224–226.

Wang, J., Lu, Y., Chen, H., Yin, M., Yu, T., and Fu, Q. (2011a). Investigation of somatic NKX2-5, GATA4 and HAND1 mutations in patients with tetralogy of Fallot. *Pathology* **43,** 322–326.

Wang, J., Xin, Y. F., Liu, X. Y., Liu, Z. M., Wang, X. Z., and Yang, Y. Q. (2011b). A novel NKX2-5 mutation in familial ventricular septal defect. *Int. J. Mol. Med.* **27,** 369–375.

Ware, S. M., Peng, J., Zhu, L., Fernbach, S., Colicos, S., Casey, B., Towbin, J., and Belmont, J. W. (2004). Identification and functional analysis of ZIC3 mutations in heterotaxy and related congenital heart defects. *Am. J. Hum. Genet.* **74,** 93–105.

Wilson, D. I., Goodship, J. A., Burn, J., Cross, I. E., and Scambler, P. J. (1992). Deletions within chromosome 22q11 in familial congenital heart disease. *Lancet* **340,** 573–575.

Wilson, M., Mowat, D., Dastot-Le Moal, F., Cacheux, V., Kaariainen, H., Cass, D., Donnai, D., Clayton-Smith, J., Townshend, S., Curry, C., Gattas, M., Braddock, S., *et al.* (2003). Further delineation of the phenotype associated with heterozygous mutations in ZFHX1B. *Am. J. Med. Genet. A* **119A,** 257–265.

Worthington, S., Bower, C., Harrop, K., Loh, J., and Walpole, I. (1998). 22q11 deletions in patients with conotruncal heart defects. *J. Paediatr. Child Health* **34,** 438–443.

Wozniak, A., Wolnik-Brzozowska, D., Wisniewska, M., Glazar, R., Materna-Kiryluk, A., Moszura, T., Badura-Stronka, M., Skolozdrzy, J., Krawczynski, M. R., Zeyland, J., Bobkowski, W., Slomski, R., *et al.* (2010). Frequency of 22q11.2 microdeletion in children with congenital heart defects in western poland. *BMC Pediatr.* **10,** 88.

Yakut, T., Kilic, S. S., Cil, E., Yapici, E., and Egeli, U. (2006). FISH investigation of 22q11.2 deletion in patients with immunodeficiency and/or cardiac abnormalities. *Pediatr. Surg. Int.* **22,** 380–383.

Yamagishi, H., Yamagishi, C., Nakagawa, O., Harvey, R. P., Olson, E. N., and Srivastava, D. (2001). The combinatorial activities of Nkx2.5 and dHAND are essential for cardiac ventricle formation. *Dev. Biol.* **239,** 190–203.

Yates, R. W., Raymond, F. L., Cook, A., and Sharland, G. K. (1996). Isomerism of the atrial appendages associated with 22q11 deletion in a fetus. *Heart* **76,** 548–549.

Ye, M., Coldren, C., Liang, X., Mattina, T., Goldmuntz, E., Benson, D. W., Ivy, D., Perryman, M. B., Garrett-Sinha, L. A., and Grossfeld, P. (2010). Deletion of ETS-1, a gene in the Jacobsen syndrome critical region, causes ventricular septal defects and abnormal ventricular morphology in mice. *Hum. Mol. Genet.* **19,** 648–656.

Yehya, A., Souki, R., Bitar, F., and Nemer, G. (2006). Differential duplication of an intronic region in the NFATC1 gene in patients with congenital heart disease. *Genome* **49,** 1092–1098.

Yong, D. E., Booth, P., Baruni, J., Massie, D., Stephen, G., Couzin, D., and Dean, J. C. (1999). Chromosome 22q11 microdeletion and congenital heart disease—A survey in a paediatric population. *Eur. J. Pediatr.* **158,** 566–570.

Yu, S., Shao, L., Kilbride, H., and Zwick, D. L. (2010). Haploinsufficiencies of FOXF1 and FOXC2 genes associated with lethal alveolar capillary dysplasia and congenital heart disease. *Am. J. Med. Genet. A* **152A,** 1257–1262.

Zhang, W., Li, X., Shen, A., Jiao, W., Guan, X., and Li, Z. (2008). GATA4 mutations in 486 Chinese patients with congenital heart disease. *Eur. J. Med. Genet.* **51,** 527–535.

Zhang, W., Li, X., Shen, A., Jiao, W., Guan, X., and Li, Z. (2009a). Screening NKX2.5 mutation in a sample of 230 Han Chinese children with congenital heart diseases. *Genet. Test. Mol. Biomarkers* **13,** 159–162.

Zhang, W. M., Li, X. F., Ma, Z. Y., Zhang, J., Zhou, S. H., Li, T., Shi, L., and Li, Z. Z. (2009b). GATA4 and NKX2.5 gene analysis in Chinese Uygur patients with congenital heart disease. *Chin. Med. J. (Engl.)* **122,** 416–419.

Zhao, F., Weismann, C. G., Satoda, M., Pierpont, M. E., Sweeney, E., Thompson, E. M., and Gelb, B. D. (2001). Novel TFAP2B mutations that cause Char syndrome provide a genotype-phenotype correlation. *Am. J. Hum. Genet.* **69,** 695–703.

Zhu, W., Shiojima, I., Hiroi, Y., Zou, Y., Akazawa, H., Mizukami, M., Toko, H., Yazaki, Y., Nagai, R., and Komuro, I. (2000). Functional analyses of three Csx/Nkx-2.5 mutations that cause human congenital heart disease. *J. Biol. Chem.* **275,** 35291–35296.

Zhu, Y., Gramolini, A. O., Walsh, M. A., Zhou, Y. Q., Slorach, C., Friedberg, M. K., Takeuchi, J. K., Sun, H., Henkelman, R. M., Backx, P. H., Redington, A. N., Maclennan, D. H., *et al.* (2008). Tbx5-dependent pathway regulating diastolic function in congenital heart disease. *Proc. Natl. Acad. Sci. USA* **105,** 5519–5524.

Zweier, C., Thiel, C. T., Dufke, A., Crow, Y. J., Meinecke, P., Suri, M., Ala-Mello, S., Beemer, F., Bernasconi, S., Bianchi, P., Bier, A., Devriendt, K., *et al.* (2005). Clinical and mutational spectrum of Mowat-Wilson syndrome. *Eur. J. Med. Genet.* **48,** 97–111.

CHAPTER TEN

MicroRNAs in Heart Development

Ramón A. Espinoza-Lewis *and* Da-Zhi Wang

Contents

Abstract

MicroRNAs (miRNAs) are a class of small noncoding RNAs of ~22 nt in length which are involved in the regulation of gene expression at the posttranscriptional level by degrading their target mRNAs and/or inhibiting their translation. Expressed ubiquitously or in a tissue-specific manner, miRNAs are involved in the regulation of many biological processes such as cell proliferation, differentiation, apoptosis, and the maintenance of normal cellular physiology. Many miRNAs are expressed in embryonic, postnatal, and adult hearts. Aberrant expression or genetic deletion of miRNAs is associated with abnormal cardiac cell differentiation, disruption of heart development, and cardiac dysfunction. This chapter will summarize the history, biogenesis, and processing of miRNAs as well as their function in heart development, remodeling, and disease.

Cardiovascular Research Division, Department of Cardiology, Children's Hospital Boston, Harvard Medical School, Boston, Massachusetts, USA

Current Topics in Developmental Biology, Volume 100
ISSN 0070-2153, DOI: 10.1016/B978-0-12-387786-4.00009-9

1. Introduction

Gene expression, as a basic biological process, can be characterized as a response to stimuli. However, the level of change in gene expression needs to be carefully controlled to ensure the proper cellular response; similarly, a lack of adjustment may lead to abnormal cellular function. Thus, the regulation of gene expression and its mechanisms of action are fundamental to all organisms and have been thoroughly studied in multiple species and organ systems. Therefore, the recent determination of a new class of small RNAs (microRNAs, miRNAs), as key elements in this regulatory repertoire, has generated enormous interest in this field. It is now known that miRNAs function within the genetic regulatory network at the posttranscriptional level. By imperfect base pairing with mRNAs in a sequence-dependent manner, these miRNAs repress gene expression by degrading target mRNAs and/or inhibiting their translation. Some miRNAs are expressed broadly, while others are restricted to specific tissues or cell types. Roles for miRNAs have been demonstrated in the regulation of a broad range of biological activities and their aberrant expression has been correlated with embryonic malformations and organ dysfunction. Heart function, in particular, is essential for proper vertebrate embryogenesis, and the early morphogenesis and development of this organ are dependent on tightly regulated genetic networks. This chapter is intended to offer a summarized review of the history, biogenesis, and processing of miRNAs, and their function during heart development.

2. An Abridged Overview of Heart Development

Eukaryotic cell functions, including growth, proliferation, differentiation, and survival, rely on several mechanisms including respiration and energy production. Respiration is the diffusion of oxygen into cells, where it will be used as a final acceptor of protons, in the respiratory chain, for the production of energy in the form of ATP (Alberts *et al.*, 1994). In "simple" multicellular organisms, such as the worm (*Caenorhabditis elegans*) oxygen can freely diffuse into the cells; however, in higher invertebrates and vertebrates, oxygen has to be brought directly into contact with the internal tissues and cells by the lymph or blood through a circulatory system. Directional flow within a circulatory system allows the exchange of oxygen and nutrients for waste produced by cells and tissues (Harvey, 1999, 2002). The directionality of the lymph or blood flow is produced by the pumping action of the heart, which beats rhythmically in an organized and regular manner to force fluid movement from the caudal to the cranial end of an organism (Challice and Virágh, 1974; Nishii and Shibata, 2006). This cardiac pump takes multiple forms, from a simple single cardiac tube in invertebrates (i.e., the "fruit fly" *Drosophila melanogaster*) to a two-chambered structure in fish (i.e., the "zebrafish" *Danio rerio*), a three-chambered heart in amphibians (i.e., the "frog" *Xenopus laevis*) and reptiles, and finally to a four-chambered heart in birds (i.e., the "chicken" *Gallus gallus*) and mammals (i.e., the "mouse" *Mus musculus*) (Harvey, 1999, 2002; Kirby, 2007).

Heart function is required early during embryogenesis for survival and the subsequent growth of other tissues and organs. The cells that form the heart originate from cardiac progenitors specified during embryonic gastrulation, a dynamic cellular process which results in the formation of the three germ tissue layers: ectoderm, mesoderm, and endoderm. The concerted and simultaneous activity of endodermal and ectodermal signaling permits the formation of the mesoderm and restricts the location of the cardiogenic cells, as well as their specification. Several molecules with inducing or inhibiting properties are involved in each of the different stages of development (Harvey, 1999; Kirby, 2007; Olson, 2006). For cardiac development, these different stages include (1) specification of cardiac progenitors, (2) initial migration of these cells during gastrulation, (3) further cellular migration to form the cardiac fields, (4) formation of the cardiac crescent, (5) formation of the tubular heart, (6) cellular recruitment at the poles of the heart, and (7) cardiac chamber differentiation (Kirby, 2007).

Tracing analyses utilizing vital dyes and cell/tissue explant experiments have shown that early cardiac progenitor cells are located in the anterior region of the primitive streak in the chick and mouse embryo (Buckingham *et al.*, 2005; Harvey, 2002; Srivastava, 2006). At the onset of gastrulation, initial posterior to anterior migration of these cells is regulated by the activity of the FGF family of signaling factors including FGF2, FGF4, and FGF8 which have been proposed to signal through FGFR1 (Ciruna and Rossant, 2001; Harvey, 2002). Similarly, bone morphogenic protein 2 (BMP2), a member of the TGF-β superfamily of growth factors, is initially expressed in the posterior end of the primitive streak by the onset of gastrulation and implicated in early cell migrations (Andrée *et al.*, 1998).

At embryonic day 6.5 (E6.5), as a result of FGF-8-dependent expression of *Mesp-1* and *Mesp-2*, cardiac progenitor cells migrate bilaterally to the anterior and lateral region of the embryo colonizing the splanchnic layer of the lateral plate mesoderm and form two early lateral cardiac fields (Buckingham *et al.*, 2005; Harvey, 1999; Kitajima *et al.*, 2000; Saga *et al.*, 1999). Canonical *Wnt/β-catenin* signaling within the neural ectoderm induces the *Mesp-1* positive mesodermal lineage during early gastrulation (Bondue *et al.*, 2008; Saga *et al.*, 1999). These data confirm crucial roles for the early action of *Bmp2* and FGF signaling in conferring cardiogenic potential (Lough *et al.*, 1996), guiding the migration of progenitor cells, and restricting the location of the cardiac fields. It is also aided by the activity of a concentration gradient of the Wnt/β-catenin, which will induce the underlying anterior mesoderm into head tissue (Brade *et al.*, 2006; van de Schans *et al.*, 2008).

Approximately a day later (E7.5), these early cardiogenic progenitor cells migrate cranially and converge across the midline region of the embryo to form a horseshoe-shaped epithelial fold, the cardiac crescent. The lateral cardiac fields in mouse and chicken embryos were initially identified by the expression of Nkx2-5, a vertebrate homeodomain transcription factor that is the homologue of the *D. melanogaster tinman* gene (Bodmer *et al.*, 1990; Kasahara *et al.*, 1998; Lyons *et al.*, 1995; Tanaka *et al.*, 1999). In mouse and chick embryos, *Nkx2-5* is the *Nkx* factor most widely expressed in the early lateral cardiac fields and cardiac crescent during cardiogenesis (Lyons *et al.*, 1995; Tanaka *et al.*, 1999). However, in contrast to the function of *tinman* in flies, no *Nkx* gene acts a master regulator of cardiogenesis (Newman and Krieg, 1998). In a dynamic process, the medial region of the cardiac crescent subsequently bulges and moves cranially bringing the lateral regions toward the midline; these two lateral cell populations eventually fuse to form the heart tube (E8.0). It is during the heart tube forming stage that the primitive heart regions can be anatomically and molecularly identified for the first time. From caudal to cranial, these structures are: (1) The inflow tract, where blood is collected to be directed into the heart and which is composed of the right and left horns of the sinus venosus which are the continuation of the common cardinal veins; (2) the primitive atrial chamber; (3) the primitive ventricular chamber; and (4) the outflow tract, which at this stage is known as the conotruncus or bulbus cordis (Buckingham *et al.*, 2005; Kirby, 2007; Srivastava, 2006). During the formation of the heart tube, *Wnt/β-catenin* signaling, especially Wnt-3a and Wnt-8, is necessary for the induction and maintenance of the expression of *Isl-1* (Cohen *et al.*, 2007) and *Fgf10* (Lizhu *et al.*, 2007). These factors are crucial to the correct formation of the second heart field and the expression of *Shh*, *Bmp4*, and *Bmp7* (Tzahor, 2007).

At E9.5, the heart begins a substantial remodeling process with the rightward bending of the heart tube to realign the cardiac chambers (looping); the caudal region of the cardiac tube, including the two sinus venosus horns, is brought cranially to position them dorsally to the outflow tract. By E10.5, the process of cardiac looping is nearly complete as the sinus venosus is shifted to the right side of the heart. The appearance of two outgrowing structures, called the sinus valves, demarcates the junction of the right common cardinal vein and the right atrium. Also, the first signs of septation of the cardiac chambers are evident at E10.5 (Challice and Virágh, 1974; Van Mierop and Gessner, 1970). By E11.5, the caudal region of the heart is finally positioned dorsal to the arterial pole and the subsequent septation within the cardiac chambers, in oxygen breathing vertebrate species, are the final steps in their specification. Initial septation of the outflow tract, by formation of internal cushions, divides it into a pulmonary and arterial arch. The right sinus venosus horn is brought in juxtaposition to the right atrium, and the left horn reduces in size and becomes the coronary artery. The atria and ventricles are separated by the interatrial and interventricular septa, respectively (Buckingham *et al.*, 2005; Harvey, 2002; Kirby, 2007; Van Mierop and Gessner, 1970).

Differentiation of cells within the heart is regulated by a variety of transcription factors which are again dependent on the expression of first or second heart field factors. The bHLH transcription factors, *Hand1* and *Hand2*, whose expression marks the left and right ventricles, respectively, are key examples. *Hand2* is dependent on the expression of *Isl-1*, while *Hand1* is dependent on the expression of *Nkx2-5* (Buckingham *et al.*, 2005; Cai *et al.*, 2003). Nkx2-5 interacts with the T-box family of transcription factors and is involved in the activation or repression of gene expression required for the differentiation of numerous cardiac structures. Several members of the T-box family have been identified to be expressed in cardiac tissue including *Tbx-1*, *-2*, *-3*, *-5*, *-18*, and *Tbx-20* (Plageman and Yutzey, 2005). Interaction of Nkx2-5 and Tbx5 results in activation of the Atrial Natriuretic Factor (*ANF*) gene and the differentiation of the cardiac chambers and cardiac trabeculation in the ventricles (Hiroi *et al.*, 2001). *Nkx2-5* expression is dispensable for the initial formation of the sinoatrial node (SAN) (Blaschke *et al.*, 2007; Espinoza-Lewis *et al.*, 2009, 2011; Kasahara *et al.*, 1998). However, the interaction of Nkx2-5 with Tbx2 or Tbx3 results in the repression of the *ANF* gene and *Connexin-40* (*Cx40*) during differentiation of the cardiac conduction system, including the SAN (Habets *et al.*, 2002; Hoogaars *et al.*, 2004). Later in development, cardiac outgrowth takes place and by E14.5 the final shape of the heart is achieved; the sinus venosus has regressed and slowly integrated into the dorsal wall of the right atrium, the outflow tract has divided into the pulmonary and arterial arches, septation of the cardiac chambers is complete, and the ventricular chambers have become the major volumetric components of the heart (Buckingham *et al.*, 2005; Harvey, 1999, 2002; Kirby, 2007; Srivastava, 2006).

3. The World of the miRNAs

We have briefly summarized some aspects involved in embryonic cardiac morphogenesis and the molecules involved in its regulation. As noted above, cardiac progenitor cells arise from the specification and determination of a special region of the mesoderm. Early determination is driven by the activity of the transcription factors Nkx2-5 and Isl-1 (Cai *et al.*, 2003; Kasahara *et al.*, 1998; Lyons *et al.*, 1995; Tanaka *et al.*, 1999). Also, it is implied that the final cellular response in the form of expression of a genetic profile is a result of not only the activity of transcription factors, but also the concerted activity of extracellular molecules, their signal transduction, and fine-tuning regulatory mechanisms. Gene regulation is achieved by the direct binding of transcription factors (activators or repressors) to *cis*-regulatory elements in order to up- or downregulate gene expression. This process can be enhanced or inhibited by the activity of extracellular signals, the activity of intracellular mediators, the status of chromosomal structure, the activity of transcription cofactors, and posttranslational modification of the protein product. In short, gene expression is regulated at the transcriptional, translational, and/or posttranslational level (Alberts *et al.*, 1994; Srivastava, 2006). In recent years, a novel mechanism has been described which functions at the posttranscriptional level to modulate and fine-tune gene expression by targeting the mRNA. Such mRNA posttranscriptional regulation results in either mRNA degradation or mRNA translation inhibition. This elegant mechanism is carried out by a newly described class of small RNA molecules, the miRNAs.

3.1. miRNAs: A brief history

Embryonic development is a well-organized, highly regulated event with strict spatiotemporal requirements. Stages are categorized by the observation of characteristic anatomical and molecular changes. In the "worm" *C. elegans*, several stages of development can be easily observed by simple visual inspection. From fertilized egg, to embryo, to larva, to adult, the worm's life cycle is completed in about 2 days. The larva stage is the lengthiest and is divided into four stages named L1–L4 (Anderson, 1995). Defects in any of these stages through the disruption of the temporal patterns of cell division and differentiation conducted by regulatory "heterochronic genes" result in larval or adult abnormalities (Chalfie *et al.*, 1981). Mutations in these heterochronic genes result in the induction of cell fate transformations, such as recapitulation of an earlier phenotype at late stages or by adopting late stage phenotypes prematurely (Ambros and Horvitz, 1987; Chalfie *et al.*, 1981).

Initially, heterochronic mutations were related to and identified in four *C. elegans* genes (*lin-4*, *lin-14*, *lin-28*, and *lin-29*) (Ambros and Horvitz, 1984). The *lin-4* mutation resulted in cell lineage reiterations which maintained earlier larval phenotypes. This resulted in supernumerary moults and the continuous production of larval-specific cuticle with the extension of late larval stages (Ambros and Horvitz, 1984; Chalfie *et al.*, 1981). The *lin-14* gain-of-function mutation, which produces a semidominant allele, results in cell lineage retarded development; certain late stage cells adopt fates expressed in cells at earlier stages. In contrast, *lin-14* loss-of-function mutations, which result in null alleles, generate precocious cell lineage development, with certain early stage cells adopting late stage fates prematurely (Ambros and Horvitz, 1987). Interestingly, the *lin-4* and *lin-14* mutant phenotypes were found to occur in the same cell lineages (Ambros and Horvitz, 1987). Additionally, it was found that *lin-14* mutant phenotypes were *lin-4* dependent (Ambros and Horvitz, 1987; Arasu *et al.*, 1991; Ruvkun and Giusto, 1989; Wightman *et al.*, 1993). Those observations suggested a direct regulatory mechanism between these two genes (Ambros and Horvitz, 1987; Arasu *et al.*, 1991; Chalfie *et al.*, 1981). The *lin-14* gene product is a protein that is detected at high levels only in the early larval L1 stage. Surprisingly, however, RNA protection assays revealed that *lin-14* mRNA was expressed and stable in all larval stages (Ruvkun and Giusto, 1989). Extensive screening for *lin-14* mutants facilitated a large analysis of the *lin-14* gene structure and determined a regulatory sequence in the *lin-14* 3′UTR as responsible for the negative regulation of gene expression, specifically in inhibiting mRNA translation (Lee *et al.*, 1993; Wightman *et al.*, 1993). Similarly, gene structure analysis, in addition to a highly laborious gene cloning strategy, suggested that the *lin-4* gene product is not a protein. Indeed, RNA protection analysis demonstrated that this locus produced two small RNA products, a 69-nt-long product (*lin-4L*) and a 21-nt-long product (*lin-4S*) (Lee *et al.*, 1993). Computational analysis indicated that the *lin-4L* was a precursor-like molecule which includes the *lin-4S* product in a hairpin or stem-loop secondary structure and that both were complementary to conserved repeated sequences in the *lin-14* negative regulatory 3′UTR in a nonperfect manner. However, the 5′ 6–8nt of *lin-4S* displayed an exact match with that of the *lin-14* 3′UTR (Lee *et al.*, 1993; Wightman *et al.*, 1993). Simultaneously, it was reported that the *lin-14* 3′UTR could mediate the downregulation of the expression of an unrelated protein (Wightman *et al.*, 1993). A construct comprising the *LacZ-lin-14_3′*UTR driven by the *Collagen-10* promoter (*Col10-LacZ-lin-14_3′*UTR) was engineered and injected into wild-type *C. elegans* embryos. X-gal staining showed the expected fading coloration of cells after the L2 larval stage. In contrast, injection of the construct into a *lin-4* null mutant line (e912) resulted in the maintenance of the X-gal staining in all four larval stages, thus demonstrating the mechanism of the novel predicted regulatory mechanism (Wightman *et al.*, 1993).

Similarly, genetic screening for mutations to suppress the synthetic sterile phenotype in *C. elegans* was performed. Several candidate genes were identified, including *let-7* (Reinhart *et al.*, 2000). *let-7* mutations result in supernumerary moults in the L4-adult stage reiterating earlier larval patterns of cell division. This defect was partially suppressed by mutations in the *lin-41*, *lin-14*, *lin-28*, and *lin-42* genes. Gene structure, Northern blot, and molecular analyses showed that the *let-7* product was not a protein but a 21-nt-long small RNA molecule. Additionally, computational analysis predicted and determined that the *lin-14*, *lin-28*, *lin-41*, *lin-42*, and *daf-12* heterochronic genes contain nonperfect complementary sequences in their 3′UTR. Transgenic *C. elegans* containing a *Col10-LacZ-lin-41_3*′'UTR construct demonstrated that *let-7* acts in similar fashion to *lin-4* (Reinhart *et al.*, 2000). Due to their nature and their specific function in regulating the temporal patterns of cell division in the *C. elegans* larvae, both *lin-4* and *let-7* gene products were termed small temporal RNAs (stRNAs) (Lee *et al.*, 1993; Pasquinelli *et al.*, 2000).

In order to provide a complete overview of this new regulatory mechanism, we believe it is necessary to briefly describe a few experimental landmarks. These discoveries were reported in between the determination of the *lin-4* and the *let-7* mechanism of action and helped shape the idea that small RNAs were indeed important players in the regulation of gene expression.

It was observed that overexpression of a sense or antisense RNA in plants (Hammond *et al.*, 2001b; Jorgensen *et al.*, 1996; Que and Jorgensen, 1998) or the injection of antisense RNA in *C. elegans* resulted in downregulation of gene expression (Fire *et al.*, 1991; Guo and Kemphues, 1995). Based on those empirical results, it was hypothesized that the regulation of gene expression by overexpression of a sense RNA was effective due to the quenching of the necessary factors needed for translation of the targeted mRNA molecule (cosuppression). However, a different mechanism was believed to be responsible for the downregulation of expression by overexpression of an antisense RNA; it was thought to induce the formation of a long double-stranded RNA (dsRNA) molecule, thus inhibiting translation (Fire *et al.*, 1991; Hammond *et al.*, 2001b; Jorgensen *et al.*, 1996; Que and Jorgensen, 1998). It was known that dsRNA could affect gene expression by triggering interferon inducible pathways that inhibit translation, known as a "panic response," through the activation of protein kinases. However, the effects observed when using antisense RNA were gene specific differing from the broad translation inhibition due to kinase activation (Hammond *et al.*, 2001b; Williams, 1999). Strikingly, more dramatic effects in downregulation of gene expression and higher gene specificity were found in *C. elegans* when sense and antisense RNA molecules were introduced at the same time (Fire *et al.*, 1998) leading to the thought that a specific mechanism for RNA silencing must exist. This mechanism was termed RNA interference or RNAi (Fire *et al.*, 1998).

Independently, a similar regulatory system, drawing parallelism to the RNAi mechanism, was previously described in plants, termed posttranscriptional gene silencing (PTGS). This process, which is guided by small ~22–25 nt RNA molecules, was identified after the introduction of a long antisense RNA (Hammond *et al.*, 2001b; Jorgensen *et al.*, 1996; Que and Jorgensen, 1998). Following the methodology described in PTGS studies, similar results were later obtained in *D. melanogaster* S2-cultured cells (Fire *et al.*, 1991) and in the "worm" *C. elegans* (Guo and Kemphues, 1995).

Thus, the description of ~22–25 nt small RNA molecules which guide the PTGS/RNAi mechanism resembled the earlier description of the small temporal genes *lin-4* and *let-7*. stRNAs and RNAi molecules share biochemical and mechanistic characteristics; they are small in size, double stranded, and possess a 5′-phosphate group and a 3′-hydroxyl group in a 2-nt tail (Bartel, 2004; Bartel and Bartel, 2003). It was inferred, from the latter characteristic, that RNase III must be responsible for such end-product. Indeed, *Dicer*, previously identified to be a component of the PTGS complex, was also found to be involved in the maturation of the stRNA (Grishok *et al.*, 2001; Hammond *et al.*, 2000). Ironically, it was by using RNAi against the human *Dicer* gene in mammalian cell cultures that it was demonstrated that *let-7* maturation was DICER dependent (Hutvágner *et al.*, 2001). The analysis of DICER-produced biochemical products, along with an elegant cloning strategy and bioinformatic predictions, led almost simultaneously and fortuitously to the discovery of small RNAs resembling stRNAs in several model systems including *D. melanogaster*, *C. elegans*, and in *HeLa*-cultured cells. Although these new small RNAs were not expressed in a timely fashion as the stRNAs, biochemical similarities proved sufficient to classify them into a new and large class of small RNAs which include *lin-4* and *let-7* as founding members. Due to their short length, they were termed miRNAs (Lagos-Quintana *et al.*, 2001; Lau *et al.*, 2001; Lee and Ambros, 2001).

3.2. miRNA: Biogenesis

The small ~22-nt-long single-strand noncoding miRNA molecules are first expressed as a long transcript known as primary miRNA (pri-RNA) either as a single independent amplicon, as part of a polycistronic RNA molecule, or from intronic sequences within a protein-coding gene host (Fig. 10.1) (Bartel, 2004; Sayed and Abdellatif, 2011). The single or polycistronic molecule is characterized by the adoption of a secondary structure known as the stem-loop or hairpin (Bartel, 2004; Sayed and Abdellatif, 2011). Initial nuclear processing of the long pri-RNA molecule is performed by the microprocessor complex composed of the Class I RNAse type III *Drosha* and the stabilizing protein *Pasha* (Partner of *Drosha*, in *C. elegans*), or *Dgcr8* (*DiGeorge* critical region gene 8, in mice and humans). This process generates a ~70-nt-long RNA molecule in a two-step highly consistent, characteristic, and sequence-independent manner (Denli *et al.*, 2004; Han *et al.*, 2006; Lee *et al.*, 2003; Wu *et al.*, 2000). dsRNA is recognized by Dgcr8 and cleaved by Drosha in a staggered fashion, ~11 nt from the base (Gregory *et al.*, 2004; Han *et al.*, 2006). The result of the staggered cleavage is a hairpin-like molecule, also known as precursor-miRNA (pre-miRNA), with a 5′-phosphate in the forward strand and a 3′-hydroxyl group in the 2 nt-overhang tail on the reverse strand (Han *et al.*, 2006). Intronic miRNAs, known as miRtrons, are expressed along with their host gene and the pre-miRNA is a product of splicing. These miRtrons have been shown to

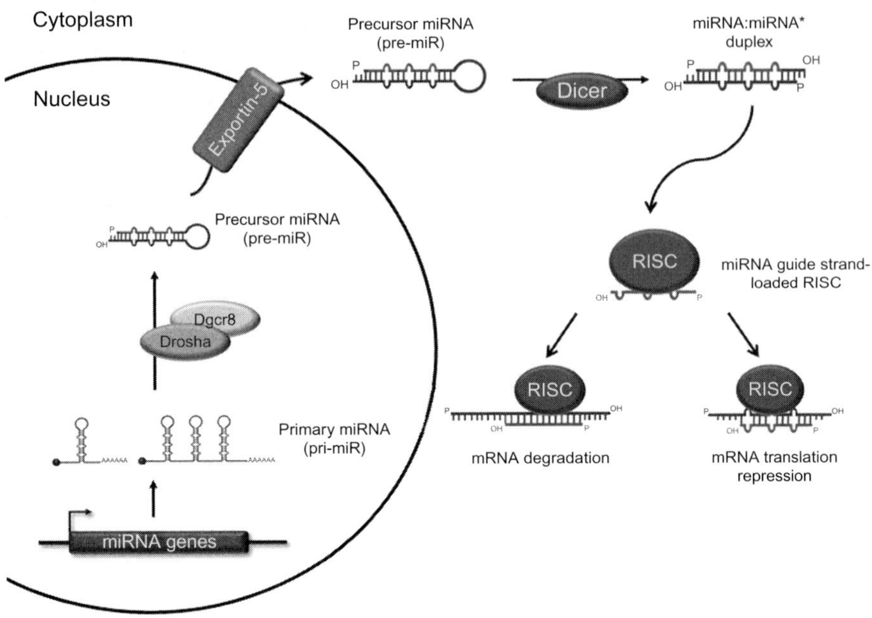

Figure 10.1 *MicroRNA Biogenesis.* In the nucleus, a miRNA gene is transcribed into a long RNA molecule containing a single miRNA sequence or several miRNA sequences (polycistron or cluster), known as primary miRNA (pri-miRNA). Pri-miRNAs are protected by the addition of a 5′-m^{7}G cap (red ball) and a 3′ poly-A tail. Pre-miRNAs are processed by the microprocessor machinery composed of Drosha and the stabilizing protein Dgcr8, thus producing a precursor-miRNA molecule (pre-miRNA). Pre-miRNAs are transported out of the nucleus to the cytoplasm by the active transporter Exportin-5 and further processed by Dicer into a miRNA:miRNA* duplex. The RNA-induced silencing complex (RISC) is assembled and loaded with the miRNA guide strand due to the unwinding of the miRNA:miRNA* duplex. A loaded RISC will facilitate the recognition of mRNA targets leading to the mRNA degradation or the inhibition of translation. (For interpretation of the references to color in this figure legend, the reader is referred to the Web version of this chapter.)

bypass *Drosha/Dgcr8* initial processing and have been reported in *D. melanogaster*, *C. elegans*, and mammals (Ruby *et al.*, 2007).

Importins and Exportins are involved in the active nuclear transport of proteins and small RNAs such as pre-rRNA and tRNA (Bohnsack *et al.*, 2004). Exportin-5 is a Ras–GTP-dependent active transporter involved in the nuclear export of adenovirus VA1 (a small 160-nt small noncoding RNA) and in tRNA transport (Bohnsack *et al.*, 2004; Yi *et al.*, 2003). Exportin-5 binds to the 3′ 2nt-overhang in tRNAs (Bohnsack *et al.*, 2004; Calado *et al.*, 2002), a characteristic also found in Drosha/Dgcr8-processed miRNAs. Indeed, Exportin-5 was identified as the nuclear receptor which mediates miRNA export to the cytoplasm by recognizing and binding to the 3′ 2-nt overhang and dsRNA region of the pre-miRNA in a sequence-independent manner (Bohnsack *et al.*, 2004; Lee *et al.*, 2011; Lund *et al.*, 2004; Yi *et al.*, 2003).

In the cytoplasm, the pre-miRNA is further processed by excision of the loop structure in order to convert the hairpin or stem-loop small RNA into a short dsRNA molecule. Dicer was initially identified as an RNase type III involved in the processing of large dsRNA molecules for the production of siRNA in PTGS in plants (Bartel and Bartel, 2003; Hammond *et al.*, 2001b; Jorgensen *et al.*, 1996; Que and Jorgensen, 1998) and RNAi in *C. elegans* and *D. melanogaster* (Bernstein *et al.*, 2001; Fire *et al.*, 1998; Grishok *et al.*, 2001; Hutvágner *et al.*, 2001). DICER, as an RNase type III, recognizes the double-stranded nature of the pre-miRNA and cleaves in a staggered manner at about two helical turns from the base of the hairpin excising the loop from the hairpin and producing a mature imperfectly matched dsRNA molecule (Grishok *et al.*, 2001; Hutvágner *et al.*, 2001; Ketting and Plasterk, 2000). Dicer homologues have been found in several organisms and are known as *Dicer-like* (*Dcl-1-4*) in *Arabidopsis thaliana*; *Dcr* (*Dcr-1* and *Dcr-2*) in *D. melanogaster*; and *Dicer-1* in mice, rats, and humans (Bartel, 2004).

The mature dsRNA or RNA duplex is composed of the mature miRNA as one of the strands (guide strand), while the other strand is known as miRNA* (passenger strand). Small RNA high-throughput cloning has identified that the majority of the strands screened correspond to miRNAs, while a low number of clones represent miRNA*s, indicating that miRNA:miRNA* duplexes are short lived and that miRNA*s are quickly degraded. However, even though miRNA*s are found in low numbers, a few have been reported to be functional (Bartel, 2004; Sayed and Abdellatif, 2011).

3.3. miRNAs, RISC, and gene silencing mechanism

The two-step RNA processing driven by Drosha and Dicer produces a dsRNA molecule or miRNA:miRNA* duplex which is recognized and bound by a large protein complex known as the RNA-induced silencing complex (RISC). "Loaded" into the RISC complex, the miRNA: miRNA* duplex is unwound and the miRNA guide strand is separated from the miRNA* passenger strand. The passenger strand is readily degraded while the guide strand directs mRNA targeting.

Several genetic screens in *A. thaliana*, *C. elegans*, and *D. melanogaster* suggested the existence of a conserved protein complex implicated in silencing activity induced by small dsRNA. In *A. thaliana*, mutations in the *AGO-1* gene result in aberrations during leaf development and defects in the floral organs. The leaves are smaller and their outgrowth is retarded and their shape resembles a small squid; thus the mutants were known as

Argonautes. Additionally, AGO-1 mutants are ineffective in PTGS (Bohmert *et al.*, 1998). A similar phenotype was previously observed in the suppressor of gene silencing (*SGS-2/SDE-1*) mutant plants (Bartel, 2004; Bartel and Bartel, 2003; Hammond *et al.*, 2001b). Additionally, RNAi deficiencies were found in *Neurospora crassa QDE-2* mutants (Catalanotto *et al.*, 2000), and in *C. elegans rde-1*, *rde-4*, and *mut-7* mutants (Ketting *et al.*, 1999; Tabara *et al.*, 1999) among others (Fagard *et al.*, 2000; Hammond *et al.*, 2001b). Interestingly, these latter genes show shared homology to the *A. thaliana* AGO-1 product (Fagard *et al.*, 2000), indicating the presence of a conserved molecular mechanism involved in gene silencing. In *D. melanogaster*, S2-cultured cells transfected with a mixture of a *Luciferase* expressing construct and dsRNA targeting the Luciferase ORF were subjected to lysate fractionation (Hammond *et al.*, 2000, 2001a). Certain fractions possessed RNAi processing ability, determined by the presence of small RNAs and absence of the *Luciferase* mRNA, and also coprecipitated a low-abundance protein. Protein sequence homology searches again determined that this protein showed a high homology to AGO-1; thus it was termed AGO-2 and classified as a member of the Argonaute superfamily of proteins (Hammond *et al.*, 2000, 2001a). Several other Ago proteins have been identified; *C. elegans* contains 27 members, while only four Ago proteins (AGO-1, AGO-2, AGO-3, AGO-4) are found in mice and humans (Pratt and MacRae, 2009).

The Argonaute superfamily of proteins can be subclassified into three groups: (1) the Ago subfamily, to which siRNA and miRNA bind; (2) the PIWI subfamily, to which piRNAs bind (piRNAs are a subclass of PIWI interacting small RNAs (~27 nt in length) expressed in the germ line regulating gene silencing of retrotransposons particularly during spermatogenesis); and (3) a class of *C. elegans* Ago-like proteins (worm Ago's or WAGO's) (Kapoor *et al.*, 2008). Argonaute proteins are ~100 kDa in average and are at the core of every RISC (Kawamata and Tomari, 2010). Argonaute proteins are also known as PPD proteins due to the conservation of the PAZ and PIWI domains. The PAZ domain is an independently stable domain with a β-barrel core which weakly binds single-stranded RNA as well as dsRNA (Cerutti *et al.*, 2000; Yan *et al.*, 2003). The PIWI domain, also involved in single-stranded RNA binding, recognizes the 5′-phosphate of the guide strand (Cerutti *et al.*, 2000). Human AGO-2 was found to be sufficient and necessary for processing of small RNAs (Hammond *et al.*, 2001a; Meister *et al.*, 2004). Interestingly, although all Ago proteins are capable of binding small dsRNA, it is only AGO-2, in humans and mice, which possesses the catalytic activity known as "slicing" (Kawamata and Tomari, 2010).

In *D. melanogaster*, the Ago-2-containing lysate fraction was determined to be ~500 KDa, much larger than the predicted molecular weight for Ago-2 (Hammond *et al.*, 2001a). This discrepancy indicates that Ago-2, albeit being highly important, is only one component of a larger protein complex. Indeed, several other components of this large protein complex have been discovered and described. TRBP (TAR RNA-binding proteins) has been found and reported to be a Dicer protein partner involved in the "loading" of the RISC (Chendrimada *et al.*, 2005). TRBP functions in a similar manner to *D. melanogaster* R2D2 and Loquacious; however, TRBP, R2D2, and Loquacious are not homologous proteins (Czech *et al.*, 2008; Liu *et al.*, 2003). GW182/Tnrc6 proteins as well as the helicases Rck and MOV10 initially identified to be a component of the RNA processing bodies (P-bodies) recruit "loaded" Ago-2 to P-bodies where final unwinding and RNA processing occurs (Kulkarni *et al.*, 2010). Dcp-1 and Dcp-2 (decapping enzymes) are also present in P-bodies and are involved in RNA processing by excising the 5′-GpC-RNA cap to initiate RNA degradation (Kawamata and Tomari, 2010; Kulkarni *et al.*, 2010).

As noted above, the miRNA:miRNA* duplex is produced in a two-step process directed by the activity of Drosha/Dgcr8 in the nucleus and Dicer in the cytoplasm, respectively. This miRNA:miRNA* duplex is bound by Ago-2 (this complex is also known as a loaded RISC) and further processed. This involves unwinding of the duplex and selection of the functional miRNA guide strand, followed by degradation of the nonfunctional passenger strand and final assembly of the complete RISC (also known as holoenzymatic-RISC or holo-RISC). This process prepares the miRNA for the recognition of and annealing to its target mRNA (Bartel, 2004; Kawamata and Tomari, 2010; Sayed and Abdellatif, 2011).

Loading of the miRNA:miRNA* duplex into Ago-2 is not a random process. First, it is hypothesized that the transfer of the duplex from Dicer to Ago-2 is guided by direct contact. Both Dicer and Ago-2 contain a conserved PAZ domain, involved in weakly binding RNA molecules, and believed to be involved in Dicer:Ago-2 protein–protein interaction. Indeed, Dicer coimmunoprecipitates with Ago-2 in *D. melanogaster* S2-cultured cells (Hammond *et al.*, 2001a). However, Ago proteins are unable to load miRNAs without the involvement of stabilizing factors. In *D. melanogaster*, this Dicer-interacting protein is known as R2D2, while in humans, a factor with a similar function is called the TAR-binding protein 2 (TRBP2); however, these two proteins are not homologous and a true R2D2 homologue has not yet been found in humans (Czech *et al.*, 2008; Kawamata and Tomari, 2010; Liu *et al.*, 2003).

Additionally, loading of the miRNA:miRNA* duplex is directed by the thermodynamic stability of the duplex itself and is sequence dependent, including center mismatches near or in guide positions. Guide positions vary and are specific for the type of Dicer or Ago proteins involved in the loading process; generally, guide positions involve nucleotides at residues 7–11. Each Ago protein has a preference for the 5′end nucleotide present in the miRNA:miRNA* duplex. This 5′nucleotide, along with sequence mismatches, confers the duplex with asymmetry in the thermodynamic stability of the strands directing the loading of the guide strand. The orientation of the duplex is also sensed by the three-dimensional conformation of the recognizing protein. Interestingly, for the *A. thaliana* AGO-1 and AGO-2 proteins, the positional swapping of the MID and PIWI domains switches their nucleotide preference resulting in a preferential loading of the passenger strand. However, in humans, AGO proteins do not have a preference for the 5′nucleotide and it seems that loading is a simpler process. Nevertheless, human AGO proteins appear to have a preference for duplexes with center mismatches and disfavor duplexes with mismatches at the ends (Kawamata and Tomari, 2010; Kawamata *et al.*, 2009).

Two processes have been described for the unwinding and separation of the miRNA:miRNA* duplex. First, Ago-2 (in flies as well as humans) has been shown to cleave and nick the passenger strand (Leuschner *et al.*, 2006; Matranga *et al.*, 2005; Miyoshi *et al.*, 2005; Rand *et al.*, 2005). In an Ago-2 slicing activity-dependent manner, separation by degradation of the passenger strand has been reported as a result of the activity of the Mg^{2+}-dependent endonuclease known as C3PO (component 3 promoter of RISC), in *D. melanogaster*, as well as the activity of the exonuclease QIP (QDE-1 interacting protein), in *N. crassa* (Liu *et al.*, 2009; Maiti *et al.*, 2007).

Second, a slower slicer-independent unwinding and separation of the passenger strand has been reported. This process is accelerated by and is the effect of the lower thermodynamic stability of the duplex compared to that of the miRNA:target duplex in a sequence-dependent manner. Additionally, in an ATP-independent manner, sequence pairing and center sequence mismatches have been shown to be essential for the unwinding of the miRNA:miRNA* duplex. Thus, in this case, it could be noted that the passenger strand is simply replaced by the target mRNA due to a “mirror image” effect in which the passenger strand acts as a first “target” sequence replaced by a more stable mRNA target sequence (Kawamata and Tomari, 2010; Kawamata *et al.*, 2009). This hypothesis is reinforced by the enhanced stability of the miRNA–RISC provided by the target mRNA; in other words, abundance of target mRNA increases accumulation of the cognate miRNA within the RISC. Moreover, enhanced accumulation of passenger strands is observed when a synthetic target is introduced in the system (Chatterjee *et al.*, 2011).

miRNA complementary sequences are most commonly located in the 3′- or 5′UTR of the target mRNA molecules. The miRNA pairing to the mRNA target sequence occurs in a nonperfect manner, commonly presenting central sequence mismatches. However, miRNAs are perfectly complementary to their target sequence at nucleotide ~2–8 in the 5′ of the miRNA, known as the “seed” sequence (Lee *et al.*, 1993; Wightman *et al.*, 1993). The presence of the miRNA in the cellular cytoplasm triggers the aggregation of the RISC. The RISC associates with several proteins involved in mRNA decapping to induce recircularization of the mRNA molecule; these include exonucleases to induce deadenylation and/or polyribosomes in order to block the translation of the mRNA target (Bartel, 2009).

3.4. miRNA expression and regulation

Initially, mutations in the *lin-4* gene were rescued by the introduction of a 683bp genomic DNA fragment, indicating that all or mostly all the regulatory elements for transcription and regulation of the *lin-4* gene were located within this fragment. Also, following gene structure analysis and cloning strategies, it was discovered that the *lin-4* gene was transcribed and expressed as a single amplicon (Ambros and Horvitz, 1987). RNA polymerase III drives the transcription of small RNAs including tRNAs, snRNAs, and U6 RNAs, and it was believed to be the miRNA transcription initiator. However, evidence accumulated to suggest that RNA polymerase II was also involved in the transcription of miRNA genes. Currently, it is believed that RNA polymerase II is the major player responsible for the transcription of miRNAs with few being transcribed by RNA polymerase III (Bartel, 2004; Sayed and Abdellatif, 2011).

Computational prediction and cloning analyses have shown that miRNAs are highly conserved and can be located in the genome embedded within an intron of a host gene (miRtron) or as a single- or polycistronic unit. The location of the miRNA gene is an important determinant of its expression and regulation. As a miRtron, miRNA expression is dictated by regulatory elements present in the promoter region of the host gene. miR-208a and miR-208b are located within intron-31 of the α-*MHC* and intron-29 of the β-*MHC* gene, respectively (Callis *et al.*, 2009; van Rooij *et al.*, 2009). As a single or a polycistronic unit, miRNA genes have their own *cis*-regulatory sequences and transcription is modified in a tissue- and cell-specific manner drawing similarities to transcription regulation of protein-coding genes. miR-1 and miR-133 contain conserved CArG boxes in their promoter regions and are directly regulated by SRF (Liu *et al.*, 2008; Zhao *et al.*, 2005). Indeed, in an elegant model of interregulation, it has been shown that in skeletal muscle, miR-133-modulated SRF expression represses cell proliferation, while miR-1-modulated HDAC4 expression represses MEF2-activated genes and induces cell differentiation. In a feedback regulatory loop, MyoD (a skeletal muscle transcription master regulator) and SRF regulate the expression of the miR-1/miR-133 polycistron (Fig. 10.2) (Chen *et al.*, 2006). Similarly, SRF, Myocardin, and Nkx2-5 directly regulate the expression of miR145/miR143 in cardiac cells as well as in smooth muscle cells (Cordes and Srivastava, 2009; Liu *et al.*, 2007).

Posttranscriptional regulation of miRNA expression has recently been reported in *C. elegans* and involves the activity of the Terminal Urydil

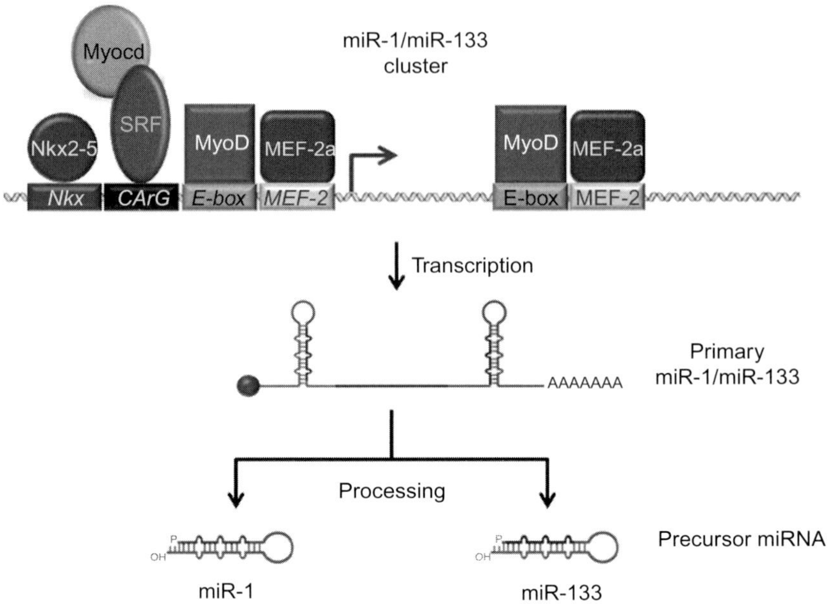

Figure 10.2 *Regulation and processing of the miR-1/miR-133 cluster.* The miR-1/miR-133 clusters are present in different chromosomes (see text). Several transcription factor response elements are shown in the 5′upstream promoter region as well as in the "intergeneic" region. As shown, *Nkx*, *CArG*, *E-box*, and *MEF-2* boxes represent recognition sites for Nkx2-5, SRF, Myogenin/MyoD, and MEF-2a binding, respectively. Further processing of the cluster results in the formation of two hairpin precursor miRNAs, miR-1 (in purple) and miR-133 (in red). Mature miRNA is represented in blue in both hairpins. Primary miRNAs are protected from early degradation by the addition of a 5′-m^{7}G cap (red ball) and a 3′ poly-A tail. (For interpretation of the references to color in this figure legend, the reader is referred to the Web version of this chapter.)

Transferase 4 protein (TUT4). TUT4 is a noncanonical poly (A) polymerase which is recruited to the pre-miRNA by Lin28 protein due to the recognition of a GGAG motif in the terminal loop; it adds a poly (U) tail at the 3′end of the stem-loop, thus inhibiting the processing by the Dicer protein (Heo *et al.*, 2009). Additionally, it has been shown that terminal loop sequences direct the binding of the RNA-binding protein hnRNP A1. Binding of hnRNP A1 to the terminal loop of pri-miRNAs remodels the three-dimensional conformation of the hairpin, thereby modulating miRNA processing. Thus, hnRNP A1 acts as a regulator of Drosha-mediated miRNA processing (Guil and Caceres, 2007; Michlewski and Caceres, 2010; Michlewski *et al.*, 2008, 2010). Similarly, posttranscriptional processing of miRNAs is affected by extracellular signaling. Indeed, pri-miR-21 to pre-miR-21 processing is enhanced by the recruitment of SMAD proteins (the TGF-β and BMP signal transducers) to the Drosha miRNA microprocessor complex by the RNA helicase p68 and the consequent accumulation of pre-miRNA molecules in the complex (Davis *et al.*, 2008).

Additionally, chromatin structure and three-dimensional configuration have also been shown to contribute to the regulation of miRNA expression. Gene compartmentalization in various genomic domains is achieved by the periodic looping of chromatin also known as scaffold/matrix-attachment regions (S/MARs). S/MARs expose or hide specific *cis*-regulatory sequences in a cell- and tissue-specific manner where exposed elements are bound by the transcriptional machinery to up- or downregulate gene expression. Disruption in the expression of S/MAR-binding proteins, such as SATB1 and SMAR1, among others, included within the transcriptional machinery, leads to chromatin modification and gene regulation alteration. Accumulation of S/MAR-binding proteins has been shown in upstream regions of the miR-17-92 cluster, as well as in the individual miRNAs let-7b, miR-17, miR-93, and miR-221 (Chavali *et al.*, 2011).

4. miRNAs in Cardiac Development, Function, and Disease

Retrospectively, the first indication that miRNAs play an important role in embryonic development came to light with the genetic evidence that the small RNAs *lin-4* and *let-7* played a role in the posttranscriptional regulation of the expression of other heterochronic genes (Lee *et al.*, 1993; Reinhart *et al.*, 2000; Wightman *et al.*, 1993). Subsequently, the discovery of miRNAs in other organisms and the description of the molecular machinery responsible for miRNA processing and function consolidated the notion that miRNAs indeed play an important role in several cellular processes. To date, cloning of small RNA molecules, quantitative RT-PCR screening, as well as microarray expression profiling have identified a large number of miRNAs in many tissues and organisms (Glazov *et al.*, 2008; Lagos-Quintana *et al.*, 2001, 2002, 2003; Lau *et al.*, 2001; Lee and Ambros, 2001). Detailed information about miRNAs is accumulated in the miRNA database known as "miRBase." According to its most recent release (Release 18, November 2011), there are 18, 226 entries expressing 21, 643 mature miRNA products in 168 species. In the mouse, there are 741 entries expressing 1283 mature miRNA sequences; in the human database, there are 1527 entries expressing 2108 mature miRNA sequences (miRBase: http://www.mirbase.org).

Expression and functional dissection of the mechanism of miRNA fine-tuning of gene expression has revealed considerable intersection and cross talk with well-described signaling and transcriptional networks for the regulation of development, morphogenesis, cell fate determination, and other cellular processes. In this section, we will discuss the function of miRNAs in heart development and summarize their role during cardiac remodeling and in disease.

4.1. miRNA function during embryonic stem cell differentiation into cardiomyocytes

Processing of mature miRNAs is a multistep biological process driven by the activity of the small RNA microprocessor (*Drosha/Dgcr8*) in the nucleus and Dicer in the cytoplasm. In flies, depletion of the Dicer homologue *Dcr-1* results in the derepression of target genes with noticeable effects during development and patterning (Lee *et al.*, 2004). Similarly, targeted mutation of the mouse *Dicer* gene resulted in early embryonic lethality as a result of the depletion of the embryonic stem cell (ESC) pool (Bernstein *et al.*, 2003). In tissue culture, mouse ES cells carrying a conditional mutation of the *Dicer* gene display severe defects in differentiation, aberrant expression of stem cell markers, reduction in epigenetic silencing, and absence of all small dsRNA molecules (Kanellopoulou *et al.*, 2005). Although these cells maintain a slow rate of proliferation, they do not contribute to the formation of embryonic chimeras when injected into blastocysts (Kanellopoulou *et al.*, 2005). Similarly, a *Dgcr8* null mutation in mouse ES cells results in the accumulation of pri-miRNAs and the absence of precursors and mature miRNA. However, *Dgcr8* null ES cells differ from *Dicer* null ES cells in that the former express stem cell molecular markers normally (Wang *et al.*, 2007).

miRNA and mRNA expression profiles have documented that mouse and human ES-derived cardiomyocytes, compared to fetal and adult heart tissue, exhibit a correlation between miRNA expression and the up- or downregulation of mRNA expression. Although it has been demonstrated that miRNAs target mostly 3′UTR mRNA sequences, such an inverse correlation in miRNA and mRNA expression indicates that miRNAs are surely involved in the cardiac differentiation regulatory network (Synnergren *et al.*, 2011). Indeed, cardiac differentiation of human fetal cardiomyocyte progenitor cells, or mouse ES cells, is accompanied by an upregulation of several miRNAs including miR-1, miR-133, and miR-499. Correspondently, adenoviral overexpression of miR-1 and miR-499 reduced progenitor cell proliferation by 25% and 15%, respectively, while enhancing differentiation (Sluijter *et al.*, 2011). Simultaneously, lower protein levels of several progenitor cell or stem cell molecular markers were observed (Ivey *et al.*, 2008; Sluijter *et al.*, 2011). Downregulation of the Notch signaling pathway has been demonstrated to be necessary for normal muscle differentiation, while its overexpression inhibits such lineage fate. During cardiac differentiation, Delta, a Notch ligand, is targeted by miR-1.

miR-1 lowers the protein levels of Delta as hESC cardiac differentiation progresses (Ivey *et al.*, 2008). Similarly, miR-1 targets and induces the translational repression of HDAC4, a transcriptional repressor of Mef2-dependent activation of muscle-specific genes. Meanwhile, miR-499 has been shown to repress the translation of Sox6 (Sluijter *et al.*, 2011), the sex-determining region-Y box 6 transcription factor which is involved in the maintenance of cardiac development (Cohen-Barak *et al.*, 2003). Moreover, knockdown experiments in ESCs demonstrate that miR-1 is required for normal smooth muscle and cardiac muscle differentiation, at least in part, by modulating the expression of the Kruppel-like factor 4 (Klf4) (Xie *et al.*, 2011). In addition, miR-1 overexpression in ESC attenuates apoptosis by indirectly elevating the levels of p-Akt (which, in turn, diminishes PTEN and Caspase-3 levels) and also by modulating the expression of Hsp60 (Glass and Singla, 2011; Shan *et al.*, 2010). It is predicted that still more miRNAs remain to be identified with significant roles in the regulation of cardiomyocyte cell fate.

4.2. miRNAs during cardiac development

As noted above, miRNAs have been revealed to have an enormous impact during embryonic development, as observed in *Dicer* null mice (Bernstein *et al.*, 2003), and *Dgcr8* null mice (Wang *et al.*, 2007). In the heart, conditional mutation of *Dicer* by employing cardiac tissue-specific Cre lines has revealed that miRNAs are not only necessary for cardiac embryonic development but also essential for postnatal cardiac maintenance and function (Chen *et al.*, 2008; da Costa Martins *et al.*, 2008; Huang *et al.*, 2010b; Saxena and Tabin, 2011). Surprisingly, it has been shown that the developmental program is unaffected when individual miRNAs are mutated. Nevertheless, such miRNAs appear to be powerful mediators in the modulation of gene expression under stress conditions (Cordes and Srivastava, 2009; Huang *et al.*, 2010a; Tatsuguchi *et al.*, 2007; van Rooij *et al.*, 2007).

Myo-miRs or muscle miRNAs is a term used to describe those miRNAs specifically expressed or enriched in smooth, skeletal, and/or cardiac muscle (van Rooij *et al.*, 2009). In particular, miR-1, miR-133, miR-208, and miR-499 have been reported to be highly expressed in cardiac tissue from the early stages of heart development to the adult. These specific myo-miRs have been shown to be involved in cardiac development, regulating cell proliferation, differentiation, and apoptosis, and also playing an important role in several cardiac diseases, such as cardiac hypertrophy, myocardial infarction (MI), cardiac arrhythmia, and heart failure (Sayed and Abdellatif, 2011; van Rooij *et al.*, 2009). In addition, many non-myo-miRs, those expressed in multiple tissues as well as heart muscle, are also involved in the development and differentiation of this organ (Sayed and Abdellatif, 2011).

miR-1 and the closely related miR-206 possess an identical seed sequence and are related in function and expression; however, they are present in different genomic loci and their expression and transcriptional regulation differs. miR-1 is predominantly enriched in cardiac and skeletal muscle tissues, while miR-206 is a skeletal muscle-specific miRNA (Chen *et al.*, 2010). Both miR-1 and miR-206 are expressed along with miR-133 as a bicistronic unit from three different genomic loci: in mice, miR-1-1/miR-133a-2 (chromosome 2), miR-1-2/miR-133a1 (chromosome 18), and miR-206/miR-133b (chromosome 1) (Liu *et al.*, 2007). Cloning screening determined that miR-1 expression accounts for $\sim$45% of the miRNAs found in the adult mouse heart and $\sim$24% in the adult human heart, suggesting its participation in cardiac formation and function (Lagos-Quintana *et al.*, 2002). In chick embryos, *in situ* hybridization shows that miR-1 is expressed in early stages of cardiac development, at the onset of cardiomyocyte differentiation and during the formation of the heart tube, as well as in the developing somites where skeletal muscle progenitors are located (Darnell *et al.*, 2006). Similar results were obtained by X-gal staining in transgenic mice in which the *LacZ* cassette is driven by miR-1 upstream fragments (Zhao *et al.*, 2005). Sequence analysis of this upstream fragment revealed conserved consensus response elements corresponding to the early cardiac determination transcription factor Nkx2-5. Accordingly, the fly transcription factor Tinman and its homologue Nkx2-5 in mice have been shown to regulate the expression of miR-1 by directly binding upstream-specific *cis*-regulatory response elements (Qian *et al.*, 2011). The Rho-GTPase CDC42 interacts with Tinman/Nkx2-5 and cooperates to regulate miR-1 expression; more interestingly, CDC42 itself is a miR-1 target. This feedback regulatory network is involved in cardiac output and the formation of a normal myofibrillar architecture (Qian *et al.*, 2011). Similarly, SRF, Mef2c, MyoD, and Myogenin have been shown to directly bind consensus-conserved response elements in order to regulate miR-1 expression in cardiac and skeletal muscle tissues (Liu *et al.*, 2007; Rao *et al.*, 2006). Interestingly, myocardin, a well-known transcription cofactor of SRF, has been recently reported to regulate the expression of miR-1 in vascular smooth muscle cells (Wang *et al.*, 2001). Inducible overexpression of myocardin results in an upregulation of the expression of miR-1 which, in turn, inhibits cell proliferation due to the downregulation of the serine/threonine kinase Pim-1 (Chen *et al.*, 2011; Jiang *et al.*, 2010).

The early embryonic expression, together with its regulation by early cardiac differentiation determinants, suggests that miR-1 may play an important role in cardiac development and function. Indeed, miR-1 loss-of-function in *D. melanogaster* is embryonic lethal due to a disruption in the patterning of the dorsal vessel as a result of a defect in muscle differentiation. This results in a failure in cardiac progenitor cell determination, leading to an increase in undifferentiated progenitor cells (Kwon *et al.*, 2005; Sokol and Ambros, 2005). This phenotype is strongly related to the miR-1-mediated downregulation of Delta, a Notch ligand involved in cell–cell signaling required for the differentiation and maintenance of skeletal and cardiac muscle (Kwon *et al.*, 2005). Similarly, in mice, null mutation of miR-1-2 results in noticeable embryonic lethality from E15.5 up to 50% lethality at weaning time. miR-1-2 null mutant embryos show pericardial edema and cardiac dysfunction. Juvenile null mice present cardiac hyperplasia due to cell proliferation up to three times higher than normal, ventricular hypertrophy, or dilated cardiomyopathy, in addition to defects in heart function related to defects in the cardiac conduction system. In the surviving mice, higher levels of Irx5 (a miR-1 target which is involved in the regulation of the cardiac repolarization gradient through Kcnd2/Kv4.2) and higher protein levels of Hand2 are observed (Costantini *et al.*, 2005; Zhao *et al.*, 2007). In contrast, miR-1 overexpression driven by the β-MHC promoter in transgenic mice during early cardiac formation results in stunted growth and cardiac development arrest at embryonic day E13.5 (E13.5). The hearts show thin-walled ventricles and heart failure due to lower levels of proliferation compared to wild-type controls (Zhao *et al.*, 2005). miR-1 was shown to target the 3′UTR of Hand2, a bHLH transcription factor expressed early in development beginning at E7.75–E8.0 in the linear heart tube and further regulating differentiation of the right ventricle (Srivastava *et al.*, 1997).

miR-133 is a component of the bicistronic unit along with miR-1, yet miR-133 expression is not as abundant (Lagos-Quintana *et al.*, 2002). However, miR-133 has been shown as an important player during cardiac development. In an effort to uncover miR-133 functions in the heart, miR-133a-1 and miR-133a-2 were targeted for null mutation. Analyses of single miR-133 gene mutation resulted in no differences when compared to wild-type controls. When analyzing double null mutation of miR-133a-1 and miR-133a-2, it was observed that the double mutation resulted in partial embryonic or postnatal lethality. This lethality is likely related to cardiovascular defects including ventricular septal defects (VSDs) and dilation in the atria and ventricles as a result of increased proliferation and, intriguingly, increased apoptosis near the interventricular septum (Liu *et al.*, 2008; Sayed and Abdellatif, 2011). Mutant animals that escape the early lethality develop dilated cardiomyopathy and present thinner ventricular walls and cardiac failure. However, cardiac hypertrophy and VSD are not observed in mutant mice (Liu *et al.*, 2008). These latter results are in contradiction to previous observations where downregulation of miR-133 by specific antagomirs induces cardiac hypertrophy (Carè *et al.*, 2007). These phenotype discrepancies might be attributed to the imperfect downregulation of miR-133 by the antagomir and/or residual expression of miR-133b since it is not targeted in the miR-133a-1/miR-133a-2 null mice. On the other hand, overexpression of miR-133 during mouse embryonic stages driven by the β-MHC promoter results in a phenotype similar to that found in β-MHC-miR-1 overexpressing mice (including embryonic lethality at mid-gestational stages (~E15.5), reduced proliferation inducing thinning of the ventricular walls, VSD, and heart failure) (Liu *et al.*, 2008).

As a member of the bicistronic unit with miR-1, miR-133 expression is also regulated by several transcription factors including SRF. Interestingly, the miR-133a-1/miR-133a-2 double knockout mice show a twofold increase in SRF expression and the activation of a smooth muscle genetic network in cardiac tissue (Liu *et al.*, 2008). SRF overexpression in postnatal hearts driven by the α-MHC promoter induces cardiac dilation, in addition to increased heart weight due to hypertrophy and the reactivation of a fetal genetic program. Moreover, the 3′UTRs of SRF, as well as CyclinD2 and CDC42, are targeted by miR-133 indicating that miR-133 is involved in the control of cell proliferation and cell cycle (Carè *et al.*, 2007; Liu *et al.*, 2008).

Although miR-1 and miR-133 are transcribed as a single molecule, their maturation and function differ. As noted, miR-1 is highly expressed in cardiac and skeletal muscle tissues, while miR-133 appears less abundant. Both miRNAs play important roles in the development of the cardiac tissue, albeit with opposite functions; miR-1 promotes and regulates cell differentiation, while miR-133 induces cell proliferation and the maintenance of the progenitor cell pool (Chen *et al.*, 2006). A similar result is observed when miR-1 is overexpressed in *Xenopus* embryos where myoblast and cardiac precursor proliferation is reduced while differentiation is enhanced. However, injection of miR-133 results in augmented cell proliferation with a reduction in cell differentiation, more so in myoblast than in cardiac progenitor cells (Chen *et al.*, 2006). This apparent contradiction in the effects of miR-1 and miR-133 in cell proliferation and cell differentiation might be explained by the differences in the nature of their targets and their distinct and possible independent transcription regulation, even though both belong to a bicistronic unit. As noted, miR-1 targets HDAC4, which represses myoblast differentiation, while miR-133 has been shown to target SRF, which is involved in transcriptional regulation of proliferation and differentiation of muscle cells (Chen *et al.*, 2006). Interestingly, SRF directly binds to upstream regions of the miR-1/miR-133 bicistronic unit, which contain consensus CArG boxes, to activate its transcription (Zhao *et al.*, 2005). Moreover, MEF2 and MyoD also directly regulate the transcription of miR-1/miR-133 by binding to an intergenic enhancer containing MEF2 response elements and E-box binding sequences. These intergenic enhancers, present in the miR-1-1/miR-133a-2 and miR-1-2/miR-133a-1 bicistrons, are hypothesized to drive miR-1/miR-133 differential expression (Fig. 10.2). Taken together, these results indicate that regulatory feedback networks are in place in order to regulate cell proliferation versus cell differentiation, in part mediated by miR-1 and miR-133 (Liu *et al.*, 2007).

During cardiac development, not only miR-1/miR-133 but also other miRNAs have been shown to exert an effect. Recently, miR-130a was reported to be involved in cardiac development through targeting the transcription factor Friend-of-GATA 2 (Fog-2). miR-130a is broadly expressed in kidney, liver, testis, and brain but highly enriched in lung and heart. Embryonic overexpression of miR-130a, driven by the β-MHC promoter, results in ventricular hypoplasia and embryonic lethality at mid-gestation stages (E13.5–E14.5), a phenotype resembling that described for Fog-2 null mutant mice (Kim *et al.*, 2009). Similarly in mice, miR-195, a member of the miR-15 family, was shown to induce ventricular hypoplasia and VSD when overexpressed in embryonic stages under the β-MHC promoter, in part by affecting the expression of the Check point 1 protein (Check1). Knockdown experiments using locked nucleic acid oligomers against miR-15 derepresses Check1 and induces postnatal myocytes to reenter the cell cycle (Porrello *et al.*, 2011).

In zebrafish, cardiac expression of miR-138 is found to be restricted to the ventricle. Knockdown experiments targeting miR-138 result in pericardial edema at late cardiac stages (48 hpf) reflecting cardiac dysfunction. However, initial cardiac development and formation appears undisturbed. Further, miR-138 knockdown resulted in the expansion of the expression fields of aldh1a2 (retinoic acid dehydrogenase 2-Raldh2), a putative miR-138 target that is involved in retinoic acid synthesis. Derepression of Raldh2 and versican (another putative miR-138 target that encodes a versatile extracellular matrix proteoglycan) in the heart induces a defect in cardiac patterning characterized by the aberrant formation of the atrio-ventricular canal (AVC) (Morton *et al.*, 2008). Knockdown of miR-143, a miRNA highly expressed in the cardiovascular system, derepresses the RA signaling pathway which profoundly influences cardiac patterning (Miyasaka *et al.*, 2011). Similarly, miR-143 regulates ventricular cardiomyocyte F-actin remodeling. Downregulation of miR-143 results in ventricular collapse by targeting *adducin3* (*add3*), an F-capping protein, affecting F-actin dynamics, cell morphology and contractility (Deacon *et al.*, 2010). Further, in an interesting twist, the mechanical input of the heartbeat itself induces the expression of miR-143 although the molecular pathway is not fully known (Miyasaka *et al.*, 2011). Additionally, in a screen performed in *Dicer* mutant zebrafish hearts, *Has2*, a molecular marker of AVC formation, was found to be upregulated. In turn, overexpression of miR-23, a putative targeting miRNA, was found to repress *Has2* expression and to reduce extracellular matrix remodeling and endocardial jelly formation, resulting in impaired cardiac valve formation. Also, miR-23 blocks the EMT (endothelial-to-mesenchymal transformation) induced by the TGF-β signaling pathway possibly due to targeting *Has2*, *icat*, and *tmem2* 3′UTR sequences (Lagendijk *et al.*, 2011). These studies indicate that miRNAs are novel regulators of EMT, cellular migration driven by signaling cues, and are essential for the normal formation of the cardiac tube. The *Slit/Robo* signaling pathway has been shown to be one of the most important migration/repression signals during development (Kidd *et al.*, 1999). miR-218 is expressed from intronic sequences in *SLIT2* and *SLIT3* and targets the *ROBO1* and *ROBO2* mRNA to repress their translation. Overexpression of miR-218 and the consequent repression of the ROBO proteins reduced the activation of the MAPK pathway and disrupted the VEGF-induced migration of endocardial and myocardial cells, leading to an aberrant heart (Fish *et al.*, 2011)

The polycistronic miR-17-92 cluster, also called Oncomir-1, comprises seven miRNAs and related miRNA clusters are ubiquitously expressed. The miR-17-92 cluster has been shown to be related to the enhancement of cell proliferation in several cancers, reduction of apoptosis, and increasing tumor angiogenesis (He *et al.*, 2005). During cardiac development, null mutation of the related miR-106a~363 and miR-106b~25 clusters resulted in no observable phenotype and mice are viable and seemingly normal; however, null mutation of the miR-17-92 cluster has been reported to result in lung hypoplasia and induce the appearance of VSD at late developmental stages leading to immediate postnatal lethality (Ventura *et al.*, 2008).

4.3. miRNAs in cardiac remodeling and pathological cardiac processes

Pathological processes are the cause or the effect of a disruption of the normal physiological status of a cell, tissue, organ, and ultimately an organism. These changes are also seen at the molecular level as a dynamic change in gene expression and manifested by altered genetic interactions. In the heart, cardiac pathologies often refer to stress-related disorders as well as congenital diseases which ultimately lead to cardiac tissue remodeling. These pathological defects include cardiac hypertrophy, cardiac dilation, and MI. Genome-wide analyses have uncovered significant changes in genetic expression signatures when the cardiac tissue is under stress conditions, showing a change in mRNA expression along with up- or down-regulation of miRNAs. In particular, under stress conditions, the 3′UTRs of mRNAs were shortened and the repressive function of miRNAs was reduced. Also, there is an activation of a fetal genetic program, which is indicative of a return to a developmental program in order to mitigate the effects of such a stress (Fig. 10.3) (Park *et al.*, 2011; Thum *et al.*, 2007).

miR-208 is expressed specifically in cardiac tissue and is detected as two isoforms, miR-208a and miR-208b. They are transcribed from intronic sequences in the myosin heavy chain (MHC) genes *Myh6* (α-*MHC*) and *Myh7* (β-*MHC*), respectively. These two miRNAs differ only in 3nt but they share an identical "seed" sequence, indicating that both miRNAs might target the same genes. During cardiac development, the MHC genes are expressed at different developmental stages; the slow contractile

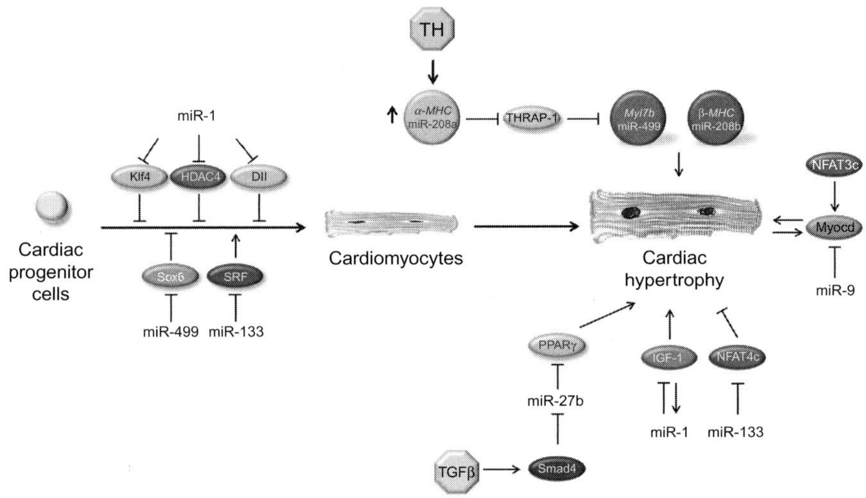

Figure 10.3 *miRNA function and integration in the regulatory networks to regulate cardiomyocyte differentiation and disease.* Graphic representation of several regulatory networks targeted by miRNAs in cardiomyocyte differentiation from cardiac progenitor cells, cardiomyocytes, and their function in the regulation of cardiac hypertrophy. Also, the integration of miRNAs in such regulatory networks by targeting transcription factors, signaling molecules, and/or structural proteins is shown. (See Color Insert.)

β-MHC is predominantly expressed in adult skeletal muscle and in the heart during embryonic stages, whereas the fast contractile α-MHC is expressed at late embryonic stages and in the adult heart, becoming the predominant cardiac MHC shortly after birth. The switch from β-MHC to α-MHC is dependent on the activity of thyroid hormone (TH) signaling at birth which directly regulates *α-MHC* and *β-MHC cis*-response elements, repressing *β-MHC* and activating *α-MHC* expression (Weiss and Leinwand, 1996). Due to their location within introns, miR-208b and miR-208a are coexpressed with their host genes. Therefore, the TH signaling also regulates a miR-208b to miR-208a switch, which was previously unrecognized when investigating the MHC switch, at birth (Callis *et al.*, 2009; van Rooij *et al.*, 2007). In stress models which induce cardiac hypertrophy, such as thoracic aortic banding or transgenic overexpression of calcineurin in the heart (Molkentin *et al.*, 1998), the β-MHC to α-MHC switch at birth is not observed and the β-MHC embryonic levels persist postnatally and in the adult heart (Callis *et al.*, 2009).

In mice, miR-208 null mutations do not show any severe developmental phenotypes. The mice are viable although they exhibit some abnormalities in the cardiac conduction system as adults, including atrial conduction deficiencies and atrial fibrillation. This is consistent with the observation that miR-208a targets GATA4, an early cardiac differentiation marker and a direct regulator of Cx40 expression in the conduction system (Callis *et al.*, 2009). Although miR-208a null mice show normal α-MHC and β-MHC expression levels, under stress conditions, the signature persistence of β-MHC in postnatal stages is not observed (van Rooij *et al.*, 2007). Nevertheless, other cardiac hypertrophy markers such as ANF are upregulated, indicating that the effect of miR-208a mutation is specific in the modulation of β-MHC levels. Indeed, overexpression of miR-208a driven by the α-MHC promoter induces cardiac hypertrophy and the induction of β-MHC but not ANF (Callis *et al.*, 2009). Through bioinformatic screens, it was predicted and later experimentally demonstrated that the thyroid receptor associated protein 1 (THRAP1) was a miR-208a target. THRAP-1 is necessary for TH signaling by recruiting RNA pol II to *cis*-regulatory elements of target genes, in this case, to downregulate β-MHC and upregulate α-MHC at birth when TH levels increase (Canepari *et al.*, 1998). Thus, THRAP-1 expression modulation is determined by miR-208a, providing a direct mechanism for regulation of β-MHC expression. Consequently, the expression of miR-208b is also disturbed in miR-208a null mutant mice under stress conditions. Indeed, miR-208a overexpression represses the expression of miR-208b, and in hypertrophied miR-208a null mice hearts, miR-208b upregulation is stunted (Callis *et al.*, 2009). A third member of the miR-208 family, miR-499, is encoded by intron 19 of the mouse Myh7b gene. Similar to β-MHC, Myh7b and miR-499 are highly enriched in skeletal muscle as well as in the embryonic heart and their expression is absent in miR-208a overexpressing mice. Interestingly, overexpression of miR-499 is sufficient to counteract the effects of loss of miR-208a in mice, including the restoration of β-MHC and miR-208b expression to normal levels, and the normal response to TH inhibition (Fig. 10.3) (van Rooij *et al.*, 2009).

The expression of miR-1 was reduced in hypertrophic hearts induced by transgenic overexpression of calcineurin as well as in the hearts of human patients presenting with cardiac hypertrophy or aortic stenosis (Sayed and Abdellatif, 2011). On the other hand, knockdown of miR-1 in mice proves efficient in induction of cardiac hypertrophy. Intriguingly, therapeutic delivery of miR-1 has been shown to reverse hypertrophy and preserve cardiac function and result in the upregulation of SERCA2a, ANF, and β-MHC (Karakikes *et al.*, 2010). However, results obtained from human patients complicated this issue. It was shown that the expression of miR-1 is up- or downregulated depending on the type of cardiac stress (i.e., cardiac hypertrophy vs. ischemia) indicating that a variety of factors are in play during such events. Indeed, several genes that play an important role in these processes have been shown to be targeted by miR-1, indicating that miR-1 regulates several genetic networks independently in the heart. These genes include Cdk9, involved in the cell cycle; Mef2a, involved in myoblast differentiation; calmodulin, a Ca^{2+}-binding protein involved in several

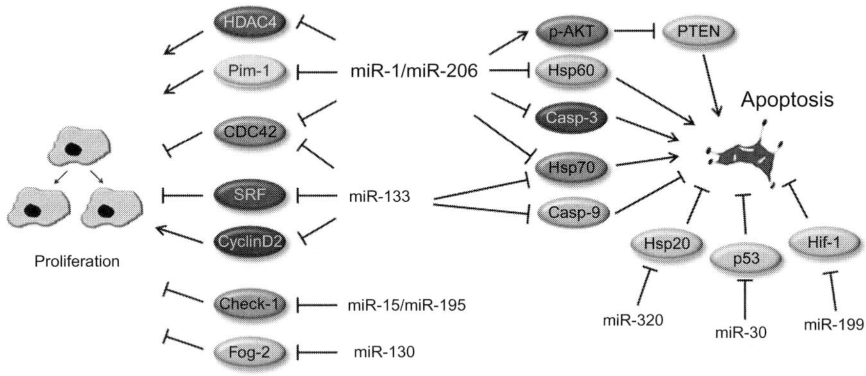

Figure 10.4 *miRNA regulatory effects on cellular processes.* Graphic representation of several miRNAs' regulatory effects on two opposing cellular processes such as proliferation and apoptosis. It is worthy to note that one miRNA might be involved in both processes by targeting different molecules which, in turn, induce or inhibit such process. (For color version of this figure, the reader is referred to the Web version of this chapter.)

cellular processes; insulin-like growth factor-1, involved in cell growth; and the heat shock proteins Hsp60 and Hsp70, which are involved in apoptosis (Fig. 10.4) (Elia *et al.*, 2009; Sayed *et al.*, 2007; Shan *et al.*, 2010.)

In addition to miR-1, miR-133 has also been shown to be downregulated during the onset of cardiac hypertrophy. Indeed, knockdown of miR-133 was sufficient to induce cardiac hypertrophy; however, as described previously, miR-133a-1/miR-133a-2 double ablation results in aberrant myocyte proliferation and apoptosis, VSDs, and partial embryonic lethality without induction of hypertrophy (Liu *et al.*, 2008). The difference in the results might be explained by the residual miR-133b expression in the double knockout model and the difference in the timing of onset of hypertrophy (Abdellatif, 2010; Carè *et al.*, 2007). Alternatively, it is formally possible that the oligonucleotide-based miR-133 knockdown could result in "off-target" effects. Additionally, overexpression of miR-133 under the *α-MHC* promoter shows that miR-133 transgenic mice are protected from stress-induced cardiac fibrosis, although the transgenic mice also present with some cardiac conduction deficiencies (Matkovich *et al.*, 2010). However, transgenic overexpression of miR-133 under the *β-MHC* promoter results in early embryonic lethality. Similar to miR-1, the effects of miR-133 during cardiac development and cardiac hypertrophy are results of the interplay between a wide variety of mRNA targets, including Caspase-9 and Hsp70 in apoptosis, SRF in myocyte differentiation, Cdc42 in the regulation of the cell cycle, among others (Abdellatif, 2010; Carè *et al.*, 2007).

miR-21, another miRNA which is dramatically upregulated during cardiac hypertrophy, has been shown to repress the expression of the sprouty gene 1 (Spry-1), a known inhibitor of the ERK–MAP kinase pathway, thus enhancing cell survival (Thum *et al.*, 2008). Knockdown of endogenous miR-21 in mice under pressure overload conditions results in attenuation of the hypertrophic response measured in terms of collagen fiber deposition and cardiac function (Thum *et al.*, 2008). Conversely, overexpression of miR-21 in mice shows that it exerts a protective effect as determined by the reduction of the myocardial infarct size and reduced apoptosis in the infarct border zone under ischemia conditions. miR-21 was proposed to repress the expression of PDCD24 in this setting (Dong *et al.*, 2009). However, these observations do not correlate with results reported for miR-21 null mutant mice. Deletion of miR-21 does not result in any phenotype and mice do not show any differences compared to wild-type controls, when under stress conditions (Patrick *et al.*, 2010). Certain differences in experimental details might account for the observed contradictory results, highlighting the critical impact of genetic approaches in deciphering the function of individual miRNAs.

A molecular signature of cardiac hypertrophy is the upregulation of myocardin expression. Indeed, myocardin overexpression in cultured cardiomyocytes is sufficient to induce cardiomyocyte hypertrophy (Xing *et al.*, 2006). During hypertrophy, the nuclear factor activator of T cells (NFAT3c and NFATc4) directly regulates myocardin expression by binding to the upstream regions of the myocardin promoter. In turn, miR-9 suppresses myocardin expression by targeting the myocardin 3′UTR, attenuating the effects of hypertrophy. Similar effects are observed upon miR-133a repression of NFAT4c and lowering of its protein levels (Li *et al.*, 2010; Wang *et al.*, 2010).

During cardiac remodeling, fibrosis or fibrotic deposition is enhanced, resulting in an increase of muscular thickening and the loss of flexibility. Fibrotic deposition involves the enhanced proliferation of cardiac fibroblasts and the enhanced production of extracellular matrix components. Several molecular pathways, including the TGF-β signaling pathway, are involved in these processes. Generation of mice with a null mutation in *Smad-4* results in animals with hypertrophic hearts and is correlated with an upregulation of miR-27b. Silencing of miR-27b is sufficient to mitigate the effects of *Smad-4* null mutation, or the hypertrophy induced in a pressure overload model (Wang *et al.*, 2011). Conversely, TGF-β1, an inhibitor of myoblast differentiation, downregulates miR-24 expression by direct binding of Smad-3 and Smad-4 to upstream promoter sequences. In *Smad-4* null hearts, miR-24 is upregulated which then induces endothelial cell apoptosis under cardiac ischemic conditions by targeting the endothelium-rich transcription factor GATA2 and the p21-activated kinase PAK4 (Sun *et al.*, 2008). Overexpression of miR-24 in zebrafish impairs normal cardiac angiogenesis, and knockdown experiments result in a reduced infarct size by vascularization enhancement (Fiedler *et al.*, 2011).

Higher levels of cellular apoptosis have been observed following MI and ischemia/reperfusion (I/R) injury and several miRNAs have been shown to regulate such cellular process. In mice, miR-320 expression is decreased following I/R injury. miR-320 overexpression increases apoptosis and increases the extent of the infarct size after I/R, while miR-320 knockdown shows cell-protective effects through the derepression of Hsp20 (Dong *et al.*, 2009; Ren *et al.*, 2009). Similarly, miR-1/206 has been shown to function as a cardioprotective factor through the repression of Hsp60 and Hsp70 (Fig. 10.4). Additionally, miR-30, miR-133, and miR-199a were shown to function as apoptotic factors through targeting p53, Caspase-9, and Hif-1a and Sirt-1, respectively (Li, 2010).

5. Final Conclusions

In the past decade, miRNAs have clearly established themselves as important manipulators of gene expression and key regulators of cell fate determination, proliferation, differentiation, and cell death during embryogenesis as well as in postnatal life. Yet, only a fraction of the total number of miRNAs have been studied in detail and the true magnitude of their regulatory ability remains to be determined. Indeed, the miRNA's characteristic "seed" sequence along with its imperfect base pairing allows for one miRNA to target multiple mRNAs; as a corollary, a single mRNA could be targeted by many miRNAs, increasing the level of complexity in the regulation of these intertwined genetic networks. We are confident that we will see more and more reports on the roles of miRNAs in the regulation of a variety of essential biological processes.

Human disease, in general, and cardiac diseases, specifically, are commonly associated with genetic haploinsufficiencies and/or point mutations. Such mutations often lead to the reduction of gene expression and/or function. Therefore, miRNAs, which function as reagents for fine-tuning levels of mRNAs and their protein products, are potential pharmacological targets that can be exploited for therapeutic purposes. With further characterization and the elucidation of context-dependent functions, miRNAs will provide new avenues for the diagnosis, prognosis, and treatment of cardiovascular diseases.

Acknowledgments

We thank members of the Wang laboratory for discussion and support. We thank Dr. John Mably for careful reading of the chapter and stimulating discussion. Research in the Wang lab was supported by the March of Dimes Birth Defect Foundation and the National Institutes of Health. R. A. E. L. is a postdoctoral fellow grantee of the American Heart Association. D. Z. W. is an established investigator of the American Heart Association.

REFERENCES

Abdellatif, M. (2010). The role of microRNA-133 in cardiac hypertrophy uncovered. *Circ. Res.* **106,** 16–18.

Alberts, B., Bray, D., Lewis, J., Raff, M., Roberts, K., and Watson, J. (eds.), (1994). Molecular Biology of the Cell, 3rd edn., pp. 72–73 & 292–334. Garland publishing, New York & London.

Ambros, V., and Horvitz, H. R. (1984). Heterochronic mutants of the Nematode Caenorhabditis elegans. *Science* **226,** 409–416.

Ambros, V., and Horvitz, H. R. (1987). The lin-14 locus of Caenorhabditis elegans controls the time of expression of specific postembryonic developmental events. *Genes Dev.* **1,** 398–414.

Anderson, P. (1995). Mutagenesis. *In* "Caenorhabditis elegans: Modern Biological Analysis of an Organism," (F. Epstein and C. Shakes, eds.), pp. 31–54. Academic press, California.

Andrée, B., Duprez, D., Vorbusch, B., Arnold, H. H., and Brand, T. (1998). BMP-2 induces ectopic expression of cardiac lineage markers and interferes with somite formation in chicken embryos. *Mech. Dev.* **70,** 119–131.

Arasu, P., Wightman, B., and Ruvkun, G. (1991). Temporal regulation of lin-14 by the antagonistic action of two other heterochronic genes, lin-4 and lin-28. *Genes Dev.* **5,** 1825–1833.

Bartel, D. P. (2004). MicroRNAs: Genomics, biogenesis, mechanism, and function. *Cell* **116,** 281–297.

Bartel, D. P. (2009). MicroRNAs: Target recognition and regulatory functions. *Cell* **136,** 215–233.

Bartel, B., and Bartel, D. P. (2003). MicroRNAs: At the root of plant development? *Plant Physiol.* **132,** 709–717.

Bernstein, E., Caudy, A. A., Hammond, S. M., and Hannon, G. J. (2001). Role for a bidentate ribonuclease in the initiation step of RNA interference. *Nature* **409,** 363–366.

Bernstein, E., Kim, S. Y., Carmell, M. A., Murchison, E. P., Alcorn, H., Li, M. Z., Mills, A. A., Elledge, S. J., Anderson, K. V., and Hannon, G. J. (2003). Dicer is essential for mouse development. *Nat. Genet.* **35,** 215–217.

Blaschke, R. J., Hahurij, N. D., Kuijper, S., Just, S., Wisse, L. J., Deissler, K., Maxelon, T., Anastassiadis, K., Spitzer, J., Hardt, S. E., Scholer, H., Feitsma, H., *et al.* (2007). Targeted mutation reveals essential functions of the homeodomain transcription factor Shox2 in sinoatrial and pacemaking development. *Circulation* **115,** 1830–1838.

Bodmer, R., Jan, L. Y., and Jan, Y. N. (1990). A new homeobox-containing gene, msh-2, is transiently expressed early during mesoderm formation of Drosophila. *Development* **110,** 661–669.

Bohmert, K., Camus, I., Bellini, C., Bouchez, D., Caboche, M., and Benning, C. (1998). AGO1 defines a novel locus of Arabidopsis controlling leaf development. *EMBO J.* **17,** 170–180.

Bohnsack, M. T., Czaplinski, K., and Görlich, D. (2004). Exportin 5 is a RanGTP-dependent dsRNA-binding protein that mediates nuclear export of pre-miRNAs. *RNA* **10,** 185–191.

Bondue, A., Lapouge, G., Paulissen, C., Semeraro, C., Iacovino, M., Kyba, M., and Blanpain, C. (2008). Mesp1 acts as a master regulator of multipotent cardiovascular progenitor specification. *Cell Stem Cell* **3,** 69–84.

Brade, T., Manner, J., and Kuhl, M. (2006). The role of Wnt signalling in cardiac development and tissue remodelling in the mature heart. *Cardiovasc. Res.* **72,** 198–209.

Buckingham, M., Meilhac, S., and Zaffran, S. (2005). Building the mammalian heart from two sources of myocardial cells. *Nat. Rev. Genet.* **6,** 826–835.

Cai, C.-L., Liang, X., Shi, Y., Chu, P.-H., Pfaff, S. L., Chen, J., and Evans, S. (2003). Isl1 identifies a cardiac progenitor population that proliferates prior to differentiation and contributes a majority of cells to the heart. *Dev. Cell* **5,** 877–889.

Calado, A., Treichel, N., Muller, E.-C., Otto, A., and Kutay, U. (2002). Exportin-5-mediated nuclear export of eukaryotic elongation factor 1A and tRNA. *EMBO J.* **21,** 6216–6224.

Callis, T. E., Pandya, K., Seok, H. Y., Tang, R. H., Tatsuguchi, M., Huang, Z. P., Chen, J. F., Deng, Z., Gunn, B., and Shumate, J. (2009). MicroRNA-208a is a regulator of cardiac hypertrophy and conduction in mice. *J. Clin. Invest.* **119,** 2772–2786.

Canepari, M., Cappelli, V., Pellegrino, M. A., Zanardi, M. C., and Reggiani, C. (1998). Thyroid hormone regulation of MHC isoform composition and myofibrillar ATPase activity in rat skeletal muscles. *Arch. Physiol. Biochem.* **106,** 308–315.

Carè, A., Catalucci, D., Felicetti, F., Bonci, D., Addario, A., Gallo, P., Bang, M.-L., Segnalini, P., Gu, Y., Dalton, N. D., Elia, L., Latronico, M. V. G., *et al.* (2007). MicroRNA-133 controls cardiac hypertrophy. *Nat. Med.* **13,** 613–618.

Catalanotto, C., Azzalin, G., Macino, G., and Cogoni, C. (2000). Transcription: Gene silencing in worms and fungi. *Nature* **404,** 245.

Cerutti, L., Mian, N., and Bateman, A. (2000). Domains in gene silencing and cell differentiation proteins: The novel PAZ domain and redefinition of the Piwi domain. *Trends Biochem. Sci.* **25,** 481–482.

Chalfie, M., Horvitz, H. R., and Sulston, J. E. (1981). Mutations that lead to reiterations in the cell lineages of C. elegans. *Cell* **24,** 59–69.

Challice, C. E., and Virágh, S. (1974). The architectural development of the early mammalian heart. *Tissue Cell* **6,** 447–462.

Chatterjee, S., Fasler, M., Büssing, I., and Großhans, H. (2011). Target-mediated protection of endogenous microRNAs in C. elegans. *Dev. Cell* **20,** 388–396.

Chavali, P. L., Funa, K., and Chavali, S. (2011). Cis-regulation of microRNA expression by scaffold/matrix-attachment regions. *Nucleic Acids Res.* **39,** 6908–6918.

Chen, J. F., Mandel, E. M., Thomson, J. M., Wu, Q., Callis, T. E., Hammond, S. M., Conlon, F. L., and Wang, D. Z. (2006). The role of microRNA-1 and microRNA-133 in skeletal muscle proliferation and differentiation. *Nat. Genet.* **38,** 228–233.

Chen, J.-F., Murchison, E. P., Tang, R., Callis, T. E., Tatsuguchi, M., Deng, Z., Rojas, M., Hammond, S. M., Schneider, M. D., Selzman, C. H., Meissner, G., Patterson, C., *et al.* (2008). Targeted deletion of Dicer in the heart leads to dilated cardiomyopathy and heart failure. *Proc. Natl. Acad. Sci. USA* **105,** 2111–2116.

Chen, J. F., Tao, Y., Li, J., Deng, Z., Yan, Z., Xiao, X., and Wang, D. Z. (2010). MicroRNA-1 and microRNA-206 regulate skeletal muscle satellite cell proliferation and differentiation by repressing Pax7. *J. Cell Biol.* **190,** 867–879.

Chen, J., Yin, H., Jiang, Y., Radhakrishnan, S. K., Huang, Z.-P., Li, J., Shi, Z., Kilsdonk, E. P. C., Gui, Y., Wang, D.-Z., and Zheng, X.-L. (2011). Induction of microRNA-1 by myocardin in smooth muscle cells inhibits cell proliferation. *Arterioscler. Thromb. Vasc. Biol.* **31,** 368–375.

Chendrimada, T. P., Gregory, R. I., Kumaraswamy, E., Norman, J., Cooch, N., Nishikura, K., and Shiekhattar, R. (2005). TRBP recruits the Dicer complex to Ago2 for microRNA processing and gene silencing. *Nature* **436,** 740–744.

Ciruna, B., and Rossant, J. (2001). FGF signaling regulates mesoderm cell fate specification and morphogenetic movement at the primitive streak. *Dev. Cell* **1,** 37–49.

Cohen, E. D., Wang, Z., Lepore, J. J., Lu, M. M., Taketo, M. M., Epstein, D. J., and Morrisey, E. E. (2007). Wnt/beta-catenin signaling promotes expansion of Isl-1-positive cardiac progenitor cells through regulation of FGF signaling. *J. Clin. Invest.* **117,** 1794–1804.

Cohen-Barak, O., Yi, Z., Hagiwara, N., Monzen, K., Komuro, I., and Brilliant, M. H. (2003). Sox6 regulation of cardiac myocyte development. *Nucleic Acids Res.* **31,** 5941–5948.

Cordes, K. R., and Srivastava, D. (2009). MicroRNA regulation of cardiovascular development. *Circ. Res.* **104,** 724–732.

Costantini, D. L., Arruda, E. P., Agarwal, P., Kim, K.-H., Zhu, Y., Zhu, W., Lebel, M., Cheng, C. W., Park, C. Y., Pierce, S. A., Guerchicoff, A., Pollevick, G. D., *et al.* (2005). The homeodomain transcription factor Irx5 establishes the mouse cardiac ventricular repolarization gradient. *Cell* **123,** 347–358.

Czech, B., Malone, C. D., Zhou, R., Stark, A., Schlingeheyde, C., Dus, M., Perrimon, N., Kellis, M., Wohlschlegel, J. A., Sachidanandam, R., Hannon, G. J., and Brennecke, J. (2008). An endogenous small interfering RNA pathway in Drosophila. *Nature* **453,** 798–802.

da Costa Martins, P. A., Bourajjaj, M., Gladka, M., Kortland, M., van Oort, R. J., Pinto, Y. M., Molkentin, J. D., and De Windt, L. J. (2008). Conditional Dicer gene deletion in the postnatal myocardium provokes spontaneous cardiac remodeling. *Circulation* **118,** 1567–1576.

Darnell, D. K., Kaur, S., Stanislaw, S., Konieczka, J. K., Yatskievych, T. A., and Antin, P. B. (2006). MicroRNA expression during chick embryo development. *Dev. Dyn.* **235,** 3156–3165.

Davis, B. N., Hilyard, A. C., Lagna, G., and Hata, A. (2008). SMAD proteins control DROSHA-mediated microRNA maturation. *Nature* **454,** 56–61.

Deacon, D. C., Nevis, K. R., Cashman, T. J., Zhou, Y., Zhao, L., Washko, D., Guner-Ataman, B., Burns, C. G., and Burns, C. E. (2010). The miR-143-adducin3 pathway is essential for cardiac chamber morphogenesis. *Development* **137,** 1887–1896.

Denli, A. M., Tops, B. B. J., Plasterk, R. H. A., Ketting, R. F., and Hannon, G. J. (2004). Processing of primary microRNAs by the Microprocessor complex. *Nature* **432,** 231–235.

Dong, S., Cheng, Y., Yang, J., Li, J., Liu, X., Wang, X., Wang, D., Krall, T. J., Delphin, E. S., and Zhang, C. (2009). MicroRNA expression signature and the role of microRNA-21 in the early phase of acute myocardial infarction. *J. Biol. Chem.* **284,** 29514–29525.

Elia, L., Contu, R., Quintavalle, M., Varrone, F., Chimenti, C., Russo, M. A., Cimino, V., De Marinis, L., Frustaci, A., Catalucci, D., and Condorelli, G. (2009). Reciprocal regulation of microRNA-1 and insulin-like growth factor-1 signal transduction cascade in cardiac and skeletal muscle in physiological and pathological conditions. *Circulation* **120,** 2377–2385.

Espinoza-Lewis, R. A., Yu, L., He, F., Liu, H., Tang, R., Shi, J., Sun, X., Martin, J. F., Wang, D., Yang, J., and Chen, Y. (2009). Shox2 is essential for the differentiation of cardiac pacemaker cells by repressing Nkx2-5. *Dev. Biol.* **327,** 376–385.

Espinoza-Lewis, R. N. A., Liu, H., Sun, C., Chen, C., Jiao, K., and Chen, Y. (2011). Ectopic expression of Nkx2.5 suppresses the formation of the sinoatrial node in mice. *Dev. Biol.* **356,** 359–369.

Fagard, M., Boutet, S. P., Morel, J.-B., Bellini, C., and Vaucheret, H. (2000). AGO1, QDE-2, and RDE-1 are related proteins required for post-transcriptional gene silencing in plants, quelling in fungi, and RNA interference in animals. *Proc. Natl. Acad. Sci. USA* **97,** 11650–11654.

Fiedler, J., Jazbutyte, V., Kirchmaier, B. C., Gupta, S. K., Lorenzen, J., Hartmann, D., Galuppo, P., Kneitz, S., Pena, J. T. G., Sohn-Lee, C., Loyer, X., Soutschek, J., *et al.* (2011). MicroRNA-24 regulates vascularity after myocardial infarction/clinical perspective. *Circulation* **124,** 720–730.

Fire, A., Albertson, D., Harrison, S. W., and Moerman, D. G. (1991). Production of antisense RNA leads to effective and specific inhibition of gene expression in C. elegans muscle. *Development* **113,** 503–514.

Fire, A., Xu, S., Montgomery, M. K., Kostas, S. A., Driver, S. E., and Mello, C. C. (1998). Potent and specific genetic interference by double-stranded RNA in Caenorhabditis elegans. *Nature* **391,** 806–811.

Fish, J. E., Wythe, J. D., Xiao, T., Bruneau, B. G., Stainier, D. Y. R., Srivastava, D., and Woo, S. (2011). A Slit/miR-218/Robo regulatory loop is required during heart tube formation in zebrafish. *Development* **138,** 1409–1419.

Glass, C., and Singla, D. K. (2011). MicroRNA-1 transfected embryonic stem cells enhance cardiac myocyte differentiation and inhibit apoptosis by modulating PTEN/Akt pathway in the infarcted heart. *Am. J. Physiol. Heart Circ. Physiol.* **301,** H2038–H2049.

Glazov, E. A., Cottee, P. A., Barris, W. C., Moore, R. J., Dalrymple, B. P., and Tizard, M. L. (2008). A microRNA catalog of the developing chicken embryo identified by a deep sequencing approach. *Genome Res.* **18,** 957–964.

Gregory, R. I., Yan, K.-P., Amuthan, G., Chendrimada, T., Doratotaj, B., Cooch, N., and Shiekhattar, R. (2004). The Microprocessor complex mediates the genesis of microRNAs. *Nature* **432,** 235–240.

Grishok, A., Pasquinelli, A. E., Conte, D., Li, N., Parrish, S., Ha, I., Baillie, D. L., Fire, A., Ruvkun, G., and Mello, C. C. (2001). Genes and mechanisms related to RNA interference regulate expression of the small temporal RNAs that control C. elegans developmental timing. *Cell* **106,** 23–34.

Guil, S., and Caceres, J. F. (2007). The multifunctional RNA-binding protein hnRNP A1 is required for processing of miR-18a. *Nat. Struct. Mol. Biol.* **14,** 591–596.

Guo, S., and Kemphues, K. J. (1995). par-1, a gene required for establishing polarity in C. elegans embryos, encodes a putative Ser/Thr kinase that is asymmetrically distributed. *Cell* **81,** 611–620.

Habets, P. E. M. H., Moorman, A. F. M., Clout, D. E. W., van Roon, M. A., Lingbeek, M., van Lohuizen, M., Campione, M., and Christoffels, V. M. (2002). Cooperative action of Tbx2 and Nkx2.5 inhibits ANF expression in the atrioventricular canal: Implications for cardiac chamber formation. *Genes Dev.* **16,** 1234–1246.

Hammond, S. M., Bernstein, E., Beach, D., and Hannon, G. J. (2000). An RNA-directed nuclease mediates post-transcriptional gene silencing in Drosophila cells. *Nature* **404,** 293–296.

Hammond, S. M., Boettcher, S., Caudy, A. A., Kobayashi, R., and Hannon, G. J. (2001a). Argonaute2, a link between genetic and biochemical analyses of RNAi. *Science* **293,** 1146–1150.

Hammond, S. M., Caudy, A. A., and Hannon, G. J. (2001b). Post-transcriptional gene silencing by double-stranded RNA. *Nat. Rev. Genet.* **2,** 110–119.

Han, J., Lee, Y., Yeom, K.-H., Nam, J.-W., Heo, I., Rhee, J.-K., Sohn, S. Y., Cho, Y., Zhang, B.-T., and Kim, V. N. (2006). Molecular basis for the recognition of primary microRNAs by the Drosha-DGCR8 complex. *Cell* **125,** 887–901.

Harvey, R. P. (1999). Heart Development. Academic Press, London.

Harvey, R. P. (2002). Patterning the vertebrate heart. *Nat. Rev. Genet.* **3,** 544–556.

He, L., Thomson, J. M., Hemann, M. T., Hernando-Monge, E., Mu, D., Goodson, S., Powers, S., Cordon-Cardo, C., Lowe, S. W., Hannon, G. J., and Hammond, S. M. (2005). A microRNA polycistron as a potential human oncogene. *Nature* **435,** 828–833.

Heo, I., Joo, C., Kim, Y.-K., Ha, M., Yoon, M.-J., Cho, J., Yeom, K.-H., Han, J., and Kim, V. N. (2009). TUT4 in concert with Lin28 suppresses microRNA biogenesis through pre-microRNA uridylation. *Cell* **138,** 696–708.

Hiroi, Y., Kudoh, S., Monzen, K., Ikeda, Y., Yazaki, Y., Nagai, R., and Komuro, I. (2001). Tbx5 associates with Nkx2-5 and synergistically promotes cardiomyocyte differentiation. *Nat. Genet.* **28,** 276–280.

Hoogaars, W. M. H., Tessari, A., Moorman, A. F. M., de Boer, P. A. J., Hagoort, J., Soufan, A. T., Campione, M., and Christoffels, V. M. (2004). The transcriptional repressor Tbx3 delineates the developing central conduction system of the heart. *Cardiovasc. Res.* **62,** 489–499.

Huang, Z.-P., Neppl, R., and Wang, D.-Z. (2010a). MicroRNAs in cardiac remodeling and disease. *J. Cardiovasc. Transl. Res.* **3,** 212–218.

Huang, Z. P., Chen, J. F., Regan, J. N., Maguire, C. T., Tang, R. H., Dong, X. R., Majesky, M. W., and Wang, D. Z. (2010b). Loss of microRNAs in neural crest leads to cardiovascular syndromes resembling human congenital heart defects. *Arterioscler. Thromb. Vasc. Biol.* **30,** 2575–2586.

Hutvágner, G., McLachlan, J., Pasquinelli, A. E., Bálint, E., Tuschl, T., and Zamore, P. D. (2001). A cellular function for the RNA-interference enzyme Dicer in the maturation of the let-7 small temporal RNA. *Science* **293,** 834–838.

Ivey, K. N., Muth, A., Arnold, J., King, F. W., Yeh, R.-F., Fish, J. E., Hsiao, E. C., Schwartz, R. J., Conklin, B. R., Bernstein, H. S., and Srivastava, D. (2008). MicroRNA regulation of cell lineages in mouse and human embryonic stem cells. *Cell Stem Cell* **2,** 219–229.

Jiang, Y., Yin, H., and Zheng, X.-L. (2010). MicroRNA-1 inhibits myocardin-induced contractility of human vascular smooth muscle cells. *J. Cell. Physiol.* **225,** 506–511.

Jorgensen, R. A., Cluster, P. D., English, J., Que, Q., and Napoli, C. A. (1996). Chalcone synthase cosuppression phenotypes in petunia flowers: Comparison of sense vs. antisense constructs and single-copy vs. complex T-DNA sequences. *Plant Mol. Biol.* **31,** 957–973.

Kanellopoulou, C., Muljo, S. A., Kung, A. L., Ganesan, S., Drapkin, R., Jenuwein, T., Livingston, D. M., and Rajewsky, K. (2005). Dicer-deficient mouse embryonic stem cells are defective in differentiation and centromeric silencing. *Genes Dev.* **19,** 489–501.

Kapoor, M., Arora, R., Lama, T., Nijhawan, A., Khurana, J., Tyagi, A., and Kapoor, S. (2008). Genome-wide identification, organization and phylogenetic analysis of Dicer-like, Argonaute and RNA-dependent RNA polymerase gene families and their expression analysis during reproductive development and stress in rice. *BMC Genomics* **9,** 451.

Karakikes, I., Chaanine, A., Kim, J., Lebeche, D., and Hajjar, R. (2010). Therapeutic cardiac-targeted delivery of mir-1 reverses hypertrophy and preserves cardiac function in a pressure overload animal model. *Circulation* A20916.

Kasahara, H., Bartunkova, S., Schinke, M., Tanaka, M., and Izumo, S. (1998). Cardiac and extracardiac expression of Csx/Nkx2.5 homeodomain protein. *Circ. Res.* **82,** 936–946.

Kawamata, T., and Tomari, Y. (2010). Making RISC. *Trends Biochem. Sci.* **35,** 368–376.

Kawamata, T., Seitz, H., and Tomari, Y. (2009). Structural determinants of miRNAs for RISC loading and slicer-independent unwinding. *Nat. Struct. Mol. Biol.* **16,** 953–960.

Ketting, R. F., and Plasterk, R. H. A. (2000). A genetic link between co-suppression and RNA interference in C. elegans. *Nature* **404,** 296–298.

Ketting, R. F., Haverkamp, T. H. A., van Luenen, H. G. A. M., and Plasterk, R. H. A. (1999). Mut-7 of C. elegans, required for transposon silencing and RNA interference, is a homolog of Werner syndrome helicase and RNaseD. *Cell* **99,** 133–141.

Kidd, T., Bland, K. S., and Goodman, C. S. (1999). Slit is the midline repellent for the robo receptor in Drosophila. *Cell* **96,** 785–794.

Kim, G. H., Samant, S. A., Earley, J. U., and Svensson, E. C. (2009). Translational control of FOG-2 expression in cardiomyocytes by microRNA-130a. *PLoS One* **4,** e6161.

Kirby, M. L. (2007). Cardiac Development. 1st edn. Oxford University Press, New York.

Kitajima, S., Takagi, A., Inoue, T., and Saga, Y. (2000). MesP1 and MesP2 are essential for the development of cardiac mesoderm. *Development* **127,** 3215–3226.

Kulkarni, M., Ozgur, S., and Stoecklin, G. (2010). On track with P-bodies. *Biochem. Soc. Trans.* **38,** 242–251.

Kwon, C., Han, Z., Olson, E. N., and Srivastava, D. (2005). MicroRNA1 influences cardiac differentiation in Drosophila and regulates Notch signaling. *Proc. Natl. Acad. Sci. USA* **102,** 18986–18991.

Lagendijk, A. K., Goumans, M. J., Burkhard, S. B., and Bakkers, J. (2011). MicroRNA-23 restricts cardiac valve formation by inhibiting Has2 and extracellular hyaluronic acid production/novelty and significance. *Circ. Res.* **109,** 649–657.

Lagos-Quintana, M., Rauhut, R., Lendeckel, W., and Tuschl, T. (2001). Identification of novel genes coding for small expressed RNAs. *Science* **294,** 853–858.

Lagos-Quintana, M., Rauhut, R., Yalcin, A., Meyer, J., Lendeckel, W., and Tuschl, T. (2002). Identification of tissue-specific microRNAs from mouse. *Curr. Biol.* **12,** 735–739.

Lagos-Quintana, M., Rauhut, R., Meyer, J., Borkhardt, A., and Tuschl, T. (2003). New microRNAs from mouse and human. *RNA* **9,** 175–179.

Lau, N. C., Lim, L. P., Weinstein, E. G., and Bartel, D. P. (2001). An abundant class of tiny RNAs with probable regulatory roles in Caenorhabditis elegans. *Science* **294,** 858–862.

Lee, R. C., and Ambros, V. (2001). An extensive class of small RNAs in Caenorhabditis elegans. *Science* **294,** 862–864.

Lee, R. C., Feinbaum, R. L., and Ambros, V. (1993). The C. elegans heterochronic gene lin-4 encodes small RNAs with antisense complementarity to lin-14. *Cell* **75,** 843–854.

Lee, Y., Ahn, C., Han, J., Choi, H., Kim, J., Yim, J., Lee, J., Provost, P., Radmark, O., Kim, S., and Kim, V. N. (2003). The nuclear RNase III Drosha initiates microRNA processing. *Nature* **425,** 415–419.

Lee, Y. S., Nakahara, K., Pham, J. W., Kim, K., He, Z., Sontheimer, E. J., and Carthew, R. W. (2004). Distinct roles for Drosophila Dicer-1 and Dicer-2 in the siRNA/miRNA silencing pathways. *Cell* **117,** 69–81.

Lee, S. J., Jiko, C., Yamashita, E., and Tsukihara, T. (2011). Selective nuclear export mechanism of small RNAs. *Curr. Opin. Struct. Biol.* **21,** 101–108.

Leuschner, P. J. F., Ameres, S. L., Kueng, S., and Martinez, J. (2006). Cleavage of the siRNA passenger strand during RISC assembly in human cells. *EMBO Rep.* **7,** 314–320.

Li, P. (2010). MicroRNAs in cardiac apoptosis. *J. Cardiovasc. Transl. Res.* **3,** 219–224.

Li, Q., Lin, X., Yang, X., and Chang, J. (2010). NFATc4 is negatively regulated in miR-133a-mediated cardiomyocyte hypertrophic repression. *Am. J. Physiol. Heart Circ. Physiol.* **298,** H1340–H1347.

Liu, Q., Rand, T. A., Kalidas, S., Du, F., Kim, H.-E., Smith, D. P., and Wang, X. (2003). R2D2, a bridge between the initiation and effector steps of the Drosophila RNAi pathway. *Science* **301,** 1921–1925.

Liu, N., Williams, A. H., Kim, Y., McAnally, J., Bezprozvannaya, S., Sutherland, L. B., Richardson, J. A., Bassel-Duby, R., and Olson, E. N. (2007). An intragenic MEF2-dependent enhancer directs muscle-specific expression of microRNAs 1 and 133. *Proc. Natl. Acad. Sci. USA* **104,** 20844–20849.

Liu, N., Bezprozvannaya, S., Williams, A. H., Qi, X., Richardson, J. A., Bassel-Duby, R., and Olson, E. N. (2008). MicroRNA-133a regulates cardiomyocyte proliferation and suppresses smooth muscle gene expression in the heart. *Genes Dev.* **22,** 3242–3254.

Liu, Y., Ye, X., Jiang, F., Liang, C., Chen, D., Peng, J., Kinch, L. N., Grishin, N. V., and Liu, Q. (2009). C3PO, an endoribonuclease that promotes RNAi by facilitating RISC activation. *Science* **325,** 750–753.

Lizhu, L., Li, C., Wenlai, Z., Daniel, D., Xiaoxue, Z., Chen-Leng, C., Lei, B., Lei, Y., Jody, M., Rolf, K., Michael, G. R., Ju, C., *et al.* (2007). β-Catenin directly regulates Islet1 expression in cardiovascular progenitors and is required for multiple aspects of cardiogenesis. *Proc. Natl. Acad. Sci. USA* **104,** 9313–9318.

Lough, J., Barron, M., Brogley, M., Sugi, Y., Bolender, D. L., and Zhu, X. (1996). Combined BMP-2 and FGF-4, but neither factor alone, induces cardiogenesis in non-precardiac embryonic mesoderm. *Dev. Biol.* **178,** 198–202.

Lund, E., Güttinger, S., Calado, A., Dahlberg, J. E., and Kutay, U. (2004). Nuclear export of microRNA precursors. *Science* **303,** 95–98.

Lyons, I., Parsons, L. M., Hartley, L., Li, R., Andrews, J. E., Robb, L., and Harvey, R. P. (1995). Myogenic and morphogenetic defects in the heart tubes of murine embryos lacking the homeo box gene Nkx2-5. *Genes Dev.* **9,** 1654–1666.

Maiti, M., Lee, H.-C., and Liu, Y. (2007). QIP, a putative exonuclease, interacts with the Neurospora Argonaute protein and facilitates conversion of duplex siRNA into single strands. *Genes Dev.* **21,** 590–600.

Matkovich, S. J., Wang, W., Tu, Y., Eschenbacher, W. H., Dorn, L. E., Condorelli, G., Diwan, A., Nerbonne, J. M., and Dorn, G. W. (2010). MicroRNA-133a protects against myocardial fibrosis and modulates electrical repolarization without affecting hypertrophy in pressure-overloaded adult hearts. *Circ. Res.* **106,** 166–175.

Matranga, C., Tomari, Y., Shin, C., Bartel, D. P., and Zamore, P. D. (2005). Passenger-strand cleavage facilitates assembly of siRNA into Ago2-containing RNAi enzyme complexes. *Cell* **123,** 607–620.

Meister, G., Landthaler, M., Patkaniowska, A., Dorsett, Y., Teng, G., and Tuschl, T. (2004). Human Argonaute2 mediates RNA cleavage targeted by miRNAs and siRNAs. *Mol. Cell* **15,** 185–197.

Michlewski, G., and Caceres, J. F. (2010). Antagonistic role of hnRNP A1 and KSRP in the regulation of let-7a biogenesis. *Nat. Struct. Mol. Biol.* **17,** 1011–1018.

Michlewski, G., Guil, S., Semple, C. A., and Cãceres, J. F. (2008). Posttranscriptional regulation of miRNAs harboring conserved terminal loops. *Mol. Cell* **32,** 383–393.

Michlewski, G., Guil, S., CÃ¡ceres, J. F., and GroÃŸhans, H. (2010). *Stimulation of pri-miR-18a processing by hnRNP A1. Regulation of MicroRNAs, Regulation of MicroRNAs.* Springer, New York. pp. 28–35.

Miyasaka, K. Y., Kida, Y. S., Banjo, T., Ueki, Y., Nagayama, K., Matsumoto, T., Sato, M., and Ogura, T. (2011). Heartbeat regulates cardiogenesis by suppressing retinoic acid signaling via expression of miR-143. *Mech. Dev.* **128,** 18–28.

Miyoshi, K., Tsukumo, H., Nagami, T., Siomi, H., and Siomi, M. C. (2005). Slicer function of Drosophila Argonautes and its involvement in RISC formation. *Genes Dev.* **19,** 2837–2848.

Molkentin, J. D., Lu, J. R., Antos, C. L., Markham, B., Richardson, J., Robbins, J., Grant, S. R., and Olson, E. N. (1998). A calcineurin-dependent transcriptional pathway for cardiac hypertrophy. *Cell* **93,** 215–228.

Morton, S. U., Scherz, P. J., Cordes, K. R., Ivey, K. N., Stainier, D. Y. R., and Srivastava, D. (2008). MicroRNA-138 modulates cardiac patterning during embryonic development. *Proc. Natl. Acad. Sci. USA* **105,** 17830–17835.

Newman, C. S., and Krieg, P. A. (1998). *Tinman*-related genes expressed during heart development in *Xenopus*. *Dev. Genet.* **22,** 230–238.

Nishii, K., and Shibata, Y. (2006). Mode and determination of the initial contraction stage in the mouse embryo heart. *Anat. Embryol.* **211,** 95–100.

Olson, E. N. (2006). Gene regulatory networks in the evolution and development of the heart. *Science* **313,** 1922–1927.

Park, J. Y., Li, W., Zheng, D., Zhai, P., Zhao, Y., Matsuda, T., Vatner, S. F., Sadoshima, J., and Tian, B. (2011). Comparative analysis of mRNA isoform expression in cardiac hypertrophy and development reveals multiple post-transcriptional regulatory modules. *PLoS One* **6,** e22391.

Pasquinelli, A. E., Reinhart, B. J., Slack, F., Martindale, M. Q., Kuroda, M. I., Maller, B., Hayward, D. C., Ball, E. E., Degnan, B., Muller, P., Spring, J., Srinivasan, A., *et al.* (2000). Conservation of the sequence and temporal expression of let-7 heterochronic regulatory RNA. *Nature* **408,** 86–89.

Patrick, D. M., Montgomery, R. L., Qi, X., Obad, S., Kauppinen, S., Hill, J. A., van Rooij, E., and Olson, E. N. (2010). Stress-dependent cardiac remodeling occurs in the absence of microRNA-21 in mice. *J. Clin. Invest.* **120,** 3912–3916.

Plageman, T. F., Jr., and Yutzey, K. E. (2005). T-box genes and heart development: Putting the "T" in heart. *Dev. Dyn.* **232,** 11–20.

Porrello, E. R., Johnson, B. A., Aurora, A. B., Simpson, E., Nam, Y.-J., Matkovich, S. J., Dorn, G. W., Van Rooij, E., and Olson, E. N. (2011). MiR-15 family regulates postnatal mitotic arrest of cardiomyocytes. *Circ. Res.* **109,** 670–679.

Pratt, A. J., and MacRae, I. J. (2009). The RNA-induced silencing complex: A versatile gene-silencing machine. *J. Biol. Chem.* **284,** 17897–17901.

Qian, L., Wythe, J. D., Liu, J., Cartry, J., Vogler, G., Mohapatra, B., Otway, R. T., Huang, Y., King, I. N., Maillet, M., Zheng, Y., Crawley, T., *et al.* (2011). Tinman/Nkx2-5 acts via miR-1 and upstream of Cdc42 to regulate heart function across species. *J. Cell Biol.* **193,** 1181–1196.

Que, Q., and Jorgensen, R. A. (1998). Homology-based control of gene expression patterns in transgenic petunia flowers. *Dev. Genet.* **22,** 100–109.

Rand, T. A., Petersen, S., Du, F., and Wang, X. (2005). Argonaute2 cleaves the anti-guide strand of siRNA during RISC activation. *Cell* **123,** 621–629.

Rao, P. K., Kumar, R. M., Farkhondeh, M., Baskerville, S., and Lodish, H. F. (2006). Myogenic factors that regulate expression of muscle-specific microRNAs. *Proc. Natl. Acad. Sci. USA* **103,** 8721–8726.

Reinhart, B. J., Slack, F. J., Basson, M., Pasquinelli, A. E., Bettinger, J. C., Rougvie, A. E., Horvitz, H. R., and Ruvkun, G. (2000). The 21-nucleotide let-7 RNA regulates developmental timing in Caenorhabditis elegans. *Nature* **403,** 901–906.

Ren, X.-P., Wu, J., Wang, X., Sartor, M. A., Qian, J., Jones, K., Nicolaou, P., Pritchard, T. J., and Fan, G.-C. (2009). MicroRNA-320 is involved in the regulation of cardiac ischemia/reperfusion injury by targeting heat-shock protein 20. *Circulation* **119,** 2357–2366.

Ruby, J. G., Jan, C. H., and Bartel, D. P. (2007). Intronic microRNA precursors that bypass Drosha processing. *Nature* **448,** 83–86.

Ruvkun, G., and Giusto, J. (1989). The Caenorhabditis elegans heterochronic gene lin-14 encodes a nuclear protein that forms a temporal developmental switch. *Nature* **338,** 313–319.

Saga, Y., Miyagawa-Tomita, S., Takagi, A., Kitajima, S., Miyazaki, J., and Inoue, T. (1999). MesP1 is expressed in the heart precursor cells and required for the formation of a single heart tube. *Development* **126,** 3437–3447.

Saxena, A., and Tabin, C. J. (2011). miRNA-processing enzyme Dicer is necessary for cardiac outflow tract alignment and chamber septation. *Proc. Natl. Acad. Sci. USA* **107,** 87–91.

Sayed, D., and Abdellatif, M. (2011). MicroRNAs in development and disease. *Physiol. Rev.* **91,** 827–887.

Sayed, D., Hong, C., Chen, I.-Y., Lypowy, J., and Abdellatif, M. (2007). MicroRNAs play an essential role in the development of cardiac hypertrophy. *Circ. Res.* **100,** 416–424.

Shan, Z.-X., Lin, Q.-X., Deng, C.-Y., Zhu, J.-N., Mai, L.-P., Liu, J.-L., Fu, Y.-H., Liu, X.-Y., Li, Y.-X., Zhang, Y.-Y., Lin, S.-G., and Yu, X.-Y. (2010). miR-1/miR-206 regulate Hsp60 expression contributing to glucose-mediated apoptosis in cardiomyocytes. *FEBS Lett.* **584,** 3592–3600.

Sluijter, J. P. G., van Mil, A., van Vliet, P., Metz, C. H. G., Liu, J., Doevendans, P. A., and Goumans, M.-J. (2011). MicroRNA-1 and -499 regulate differentiation and proliferation in human-derived cardiomyocyte progenitor cells. *Arterioscler. Thromb. Vasc. Biol.* **30,** 859–868.

Sokol, N. S., and Ambros, V. (2005). Mesodermally expressed Drosophila microRNA-1 is regulated by Twist and is required in muscles during larval growth. *Genes Dev.* **19,** 2343–2354.

Srivastava, D. (2006). Making or breaking the heart: From lineage determination to morphogenesis. *Cell* **126,** 1037–1048.

Srivastava, D., Thomas, T., Lin, Q., Kirby, M. L., Brown, D., and Olson, E. N. (1997). Regulation of cardiac mesodermal and neural crest development by the bHLH transcription factor, dHAND. *Nat. Genet.* **16,** 154–160.

Sun, Q., Zhang, Y., Yang, G., Chen, X., Zhang, Y., Cao, G., Wang, J., Sun, Y., Zhang, P., Fan, M., Shao, N., and Yang, X. (2008). Transforming growth factor-β-regulated miR-24 promotes skeletal muscle differentiation. *Nucleic Acids Res.* **36,** 2690–2699.

Synnergren, J., Améen, C., Lindahl, A., Olsson, B., and Sartipy, P. (2011). Expression of microRNAs and their target mRNAs in human stem cell-derived cardiomyocyte clusters and in heart tissue. *Physiol. Genomics* **43,** 581–594.

Tabara, H., Sarkissian, M., Kelly, W. G., Fleenor, J., Grishok, A., Timmons, L., Fire, A., and Mello, C. C. (1999). The rde-1 gene, RNA interference, and transposon silencing in C. elegans. *Cell* **99,** 123–132.

Tanaka, M., Chen, Z., Bartunkova, S., Yamasaki, N., and Izumo, S. (1999). The cardiac homeobox gene Csx/Nkx2.5 lies genetically upstream of multiple genes essential for heart development. *Development* **126,** 1269–1280.

Tatsuguchi, M., Seok, H. Y., Callis, T. E., Thomson, J. M., Chen, J. F., Newman, M., Rojas, M., Hammond, S. M., and Wang, D. Z. (2007). Expression of microRNAs is dynamically regulated during cardiomyocyte hypertrophy. *J. Mol. Cell. Cardiol.* **42,** 1137–1141.

Thum, T., Galuppo, P., Wolf, C., Fiedler, J., Kneitz, S., van Laake, L. W., Doevendans, P. A., Mummery, C. L., Borlak, J., and Haverich, A. (2007). MicroRNAs in the human heart: A clue to fetal gene reprogramming in heart failure. *Circulation* **116,** 258–267.

Thum, T., Gross, C., Fiedler, J., Fischer, T., Kissler, S., Bussen, M., Galuppo, P., Just, S., Rottbauer, W., Frantz, S., Castoldi, M., Soutschek, J., *et al.* (2008). MicroRNA-21 contributes to myocardial disease by stimulating MAP kinase signalling in fibroblasts. *Nature* **456,** 980–984.

Tzahor, E. (2007). Wnt/beta-catenin signaling and cardiogenesis: Timing does matter. *Dev. Cell* **13,** 10–13.

van de Schans, V. A. M., Smits, J. F. M., and Blankesteijn, W. M. (2008). The Wnt/frizzled pathway in cardiovascular development and disease: Friend or foe? *Eur. J. Pharmacol.* **585,** 338–345.

Van Mierop, L. H., and Gessner, I. H. (1970). The morphologic development of the sinoatrial node in the mouse. *Am. J. Cardiol.* **25,** 204–212.

van Rooij, E., Sutherland, L. B., Qi, X., Richardson, J. A., Hill, J., and Olson, E. N. (2007). Control of stress-dependent cardiac growth and gene expression by a microRNA. *Science* **316,** 575–579.

van Rooij, E., Quiat, D., Johnson, B. A., Sutherland, L. B., Qi, X., Richardson, J. A., Kelm, R. J., and Olson, E. N. (2009). A family of microRNAs encoded by myosin genes governs myosin expression and muscle performance. *Dev. Cell* **17,** 662–673.

Ventura, A., Young, A. G., Winslow, M. M., Lintault, L., Meissner, A., Erkeland, S. J., Newman, J., Bronson, R. T., Crowley, D., Stone, J. R., Jaenisch, R., Sharp, P. A., *et al.* (2008). Targeted deletion reveals essential and overlapping functions of the miR-17∼92 family of miRNA clusters. *Cell* **132,** 875–886.

Wang, D., Chang, P. S., Wang, Z., Sutherland, L., Richardson, J. A., Small, E., Krieg, P. A., and Olson, E. N. (2001). Activation of cardiac gene expression by myocardin, a transcriptional cofactor for serum response factor. *Cell* **105,** 851–862.

Wang, Y., Medvid, R., Melton, C., Jaenisch, R., and Blelloch, R. (2007). DGCR8 is essential for microRNA biogenesis and silencing of embryonic stem cell self-renewal. *Nat. Genet.* **39,** 380–385.

Wang, K., Long, B., Zhou, J., and Li, P.-F. (2010). miR-9 and NFATc3 regulate myocardin in cardiac hypertrophy. *J. Biol. Chem.* **285,** 11903–11912.

Wang, J., Song, Y., Zhang, Y., Xiao, H., Sun, Q., Hou, N., Guo, S., Wang, Y., Fan, K., Zhan, D., Zha, L., Cao, Y., *et al.* (2011). Cardiomyocyte overexpression of miR-27b induces cardiac hypertrophy and dysfunction in mice. *Cell Res.* Aug 16, 2011. doi:10.1038/cr.2011.

Weiss, A., and Leinwand, L. A. (1996). The mammalian myosin heavy chain gene family. *Annu. Rev. Cell Dev. Biol.* **12,** 417–439.

Wightman, B., Ha, I., and Ruvkun, G. (1993). Posttranscriptional regulation of the heterochronic gene lin-14 by lin-4 mediates temporal pattern formation in C. elegans. *Cell* **75,** 855–862.

Williams, B. R. (1999). PKR: A sentinel kinase for cellular stress. *Oncogene* **18,** 6112–6120.

Wu, H., Xu, H., Miraglia, L. J., and Crooke, S. T. (2000). Human RNase III Is a 160-kDa protein involved in preribosomal RNA processing. *J. Biol. Chem.* **275,** 36957–36965.

Xie, C., Huang, H., Sun, X., Guo, Y., Hamblin, M., Ritchie, R. P., Garcia-Barrio, M. T., Zhang, J., and Chen, Y. E. (2011). MicroRNA-1 regulates smooth muscle cell differentiation by repressing Kruppel-like factor 4. *Stem Cells Dev.* **20,** 205–210.

Xing, W., Zhang, T. C., Cao, D., Wang, Z., Antos, C. L., Li, S., Wang, Y., Olson, E. N., and Wang, D. Z. (2006). Myocardin induces cardiomyocyte hypertrophy. *Circ. Res.* **98,** 1089–1097.

Yan, K. S., Yan, S., Farooq, A., Han, A., Zeng, L., and Zhou, M.-M. (2003). Structure and conserved RNA binding of the PAZ domain. *Nature* **426,** 469–474.

Yi, R., Qin, Y., Macara, I. G., and Cullen, B. R. (2003). Exportin-5 mediates the nuclear export of pre-microRNAs and short hairpin RNAs. *Genes Dev.* **17,** 3011–3016.

Zhao, Y., Samal, E., and Srivastava, D. (2005). Serum response factor regulates a muscle-specific microRNA that targets Hand2 during cardiogenesis. *Nature* **436,** 214–220.

Zhao, Y., Ransom, J. F., Li, A., Vedantham, V., Von, M., Muth, A. N., Tsuchihashi, T., McManus, M. T., Schwartz, R. J., and Srivastava, D. (2007). Dysregulation of cardiogenesis, cardiac conduction, and cell cycle in mice lacking miRNA-1-2. *Cell* **129,** 303–317.

CHAPTER ELEVEN

Cardiac Regeneration

Wen-Yee Choi *and* Kenneth D. Poss

Contents

Abstract

The heart is a pump that is comprised of cardiac myocytes and other cell types and whose proper function is critical to quality of life. The ability to trigger regeneration of heart muscle following injury eludes adult mammals, a deficiency of great clinical impact. Major research efforts are attempting to change this through advances in cell therapy or activating endogenous regenerative mechanisms that exist only early in life. In contrast with mammals, lower vertebrates like zebrafish demonstrate an impressive natural capacity for cardiac regeneration throughout life. This review will cover recent progress in the field of heart regeneration with a focus on endogenous regenerative capacity and its potential manipulation.

Department of Cell Biology, Howard Hughes Medical Institute, Duke University Medical Center, Durham, North Carolina, USA

Current Topics in Developmental Biology, Volume 100
ISSN 0070-2153, DOI: 10.1016/B978-0-12-387786-4.00010-5

1. Introduction

As more nations have become developed, heart failure has come to be a leading cause of death worldwide. This condition, in which the heart is unable to supply sufficient blood flow to other vital organs, is most often triggered by an ischemic event like myocardial infarction (MI) or nonischemic events also causing myocardial loss. After cardiomyocyte death, there is no significant replacement of old muscle with new, and a collagen-rich scar is generated by fibroblasts concomitant with removal of cell debris. While scarring is essential as a quick fix, a noncontractile scar cannot restore cardiac output and can cause arrhythmia. In many cases, heart failure signs can be medically managed, while other cases require cardiac transplantation. Undoubtedly, regenerative strategies that improve cardiac function by restoring electrically coupled cardiomyocytes to injured or diseased hearts would be manna from heaven to cardiologists and their patients. In this review, we will focus on the efforts that have been made to uncover, understand, and enhance regenerative capacity after cardiomyocyte loss.

One option for replenishing a depleted cardiomyocyte population is transplantation of exogenous cells (Fig. 11.1A). Recent studies of embryogenesis and stem cell biology have improved our understanding of cardiomyocyte differentiation from progenitor cells. Drawing from these advances, various cell types from embryonic stem (ES) cells to differentiated cardiomyocytes have been injected into injured mammalian hearts and examined for their potential to engraft and generate new cardiomyocytes. While progress has been made, there remain technical and biological hurdles that limit current success of these approaches. Most notably, while functional improvements have been reported, there exists no consensus agreement on a cell type and delivery strategy that provides faithful incorporation and electrical coupling of the transplanted cells and/or their progeny.

Complementary regenerative strategies to cell transplantation are simultaneously being pursued by the field. A seemingly less invasive option to cell therapy is to stimulate growth of new cardiomyocytes from within the milieu of the cardiac muscle spared by injury, regenerating new heart tissue from endogenous players (Fig. 11.1B). While there are currently few clues on how to unlock the regenerative capacity of the mammalian heart, this notion is not far-fetched. Regeneration is an ancestral trait of animals that is selectively distributed among species and their organ systems (Poss, 2010). While certain organisms are able to regrow missing body parts like heads and appendages, the regenerative capabilities of mammals are relatively modest. In humans, skin, blood, skeletal muscle, and liver regenerate efficiently following injury. In fact, evidence from studies in zebrafish and

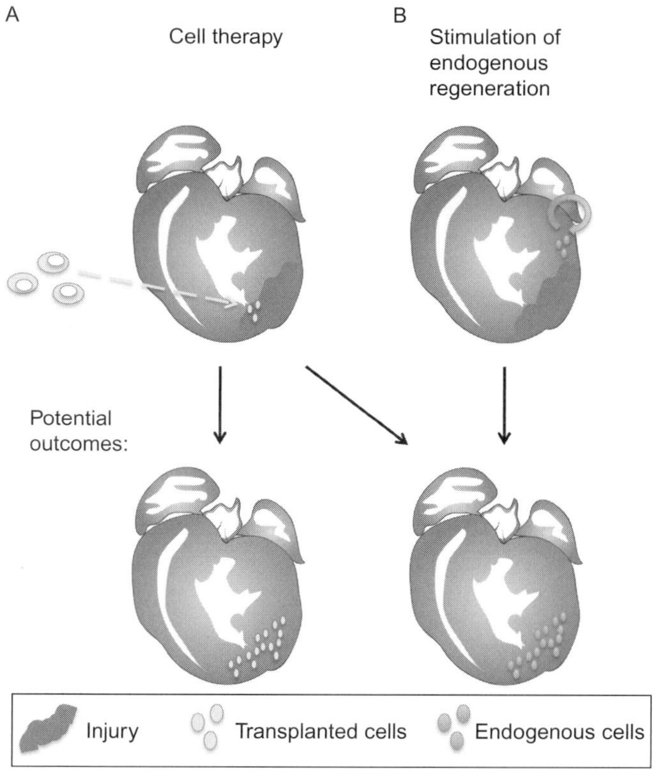

Figure 11.1 Approaches to stimulating heart regeneration. (A) One strategy for heart regeneration involves the injection of exogenous cells into the infarcted area of the heart. These transplanted cells could proliferate and repopulate the injured area with myocardium or may signal to endogenous cardiac cells and promote their proliferation. (B) A second fundamental approach relies on the regenerative capacity of uninjured progenitor cells or cardiomyocytes adjacent to the infarcted area. These cells proliferate and/or differentiate *in situ* to build new cardiac muscle. (For color version of this figure, the reader is referred to the Web version of this chapter.)

neonatal mammals has stoked the idea that the regenerative capacity can be provoked after an injury to the adult mammalian heart.

2. Cell Therapy and Heart Regeneration

Cardiomyocyte proliferation is the dominant mechanism for cardiac growth throughout the first several days of life in rodent model systems, while cardiomyocyte hypertrophy and limited cell division is the hallmark of juvenile and adult hearts (discussed later; Adler and Costabel, 1975;

Li *et al.*, 1996). Given the relative scarcity of natural cardiomyocyte hyperplasia in mature animals, significant effort has been invested in identifying cells that might differentiate into cardiomyocytes upon transplantation into an injured heart. The burgeoning field of stem cell biology has driven this quest, providing opportunity to investigate and direct cardiomyocyte differentiation from undifferentiated cells isolated from the heart or other tissues. One goal of these studies has been to identify markers to allow prospective isolation of cardiac stem and progenitor cell populations from embryonic and adult tissue. There are multiple sources, techniques, and markers used by different groups for isolating progenitor cells, as is the case with any tissue stem cell(s). In addition, there are new paradigms for generating cardiomyocytes *in vitro* from pluripotent stem cells, advances that are relevant to potential cellular therapies. Here, we will provide an overview with descriptions of a few of these promising cell populations, and we refer readers to other reviews with more extensive coverage (Hansson *et al.*, 2009; Laflamme and Murry, 2011; Laflamme *et al.*, 2007b; Seidel *et al.*, 2009; Wu *et al.*, 2008).

2.1. Adult cardiac progenitor cells

Studies of heart development in embryos have yielded a wealth of information about embryonic progenitors of cardiomyocytes, as described earlier in this issue. Logically, there is value in determining the presence and nature of analogous cardiac progenitor cells that might exist in the adult heart, as they would represent a natural source of regenerating cardiomyocytes and perhaps other important cell types. These investigations have revealed multiple distinct cell populations that can be isolated from the adult murine heart; for example, defined by surface expression of the markers c-kit, Sca1, or the ATP-binding cassette transporter (Beltrami *et al.*, 2003; Martin *et al.*, 2004; Oh *et al.*, 2003). Such cells comprise less than 1% of the total cell population in the mouse heart and, when cultured *in vitro*, are able to differentiate into cells expressing contractile genes.

While these adult cell populations may be related to each other, their origins and potential relationship to embryonic cardiac progenitor cells remain unclear. Also, the success of these cardiac progenitor cells, particularly the population expressing the surface marker c-kit, in generating mature cardiomyocytes *in vivo* has varied in the literature. A population of cells expressing c-kit was isolated from the human heart and showed similar abilities to give rise to cardiomyocytes *in vivo* (Bearzi *et al.*, 2007), prompting cell therapy studies. While some work indicated significant formation of new myocardium and blood vessels after addition of c-kit+ cells to a rodent MI (Bearzi *et al.*, 2007), other results suggested modest gains (Tang *et al.*, 2010). Some reports have been unable to detect adult myogenesis from transplanted c-kit+ cells or derivation of cardiomyocytes from adult c-kit+ cells in culture (Tallini *et al.*, 2009; Zaruba *et al.*, 2010). Further characterization of these cardiac cell populations is necessary both to understand their endogenous roles in healthy and injured hearts and to optimize their use in cell therapy.

2.2. Pluripotent stem cells

While cardiac progenitor cells have relatively limited lineage potential, ES cells and induced pluripotent stem (iPS) cells have the capacity to develop into almost any cell type. The key to harnessing these potentially unlimited sources of cardiomyocytes lies in the ability to consistently initiate differentiation into functional, mature cells. Using traditional differentiation protocols, ES cells give rise to approximately 1% cardiomyocytes in embryoid body cultures (Kehat *et al.*, 2001). Guided by principles of developmental biology, multiple groups have optimized these protocols for cardiogenic purposes with reported yields of ~30–60% human ES cell-derived cardiomyocytes (Laflamme *et al.*, 2007a; Zhu *et al.*, 2010). These derived cells express cardiac transcription factors such as Nkx2.5 and Islet1 and structural proteins characteristic of developing cardiomyocytes. However, these cells are still distinct from adult cardiomyocytes in morphology, function, and expression profiles. There is also interest in differentiating pluripotent cells to a mesodermal or cardiovascular progenitor cell state, as they may be able to self-renew and proliferate upon transplantation (Bu *et al.*, 2009; Kattman *et al.*, 2011).

Progress has been made in cardiomyocyte derivation protocols, but challenges remain before ES cell- or iPS cell-derived cardiomyocytes can be used for therapeutic purposes. Human ES cell use remains controversial and any cells derived from these lineages will require immunosuppression upon transplantation. iPS cells eliminate the need for immunosuppression, but the methods used to deliver reprogramming factors have occasionally been tumorigenic, raising questions about their subsequent use in cell therapy. While these issues are being addressed, human ES cell-derived cardiomyocytes have been injected into various infarcted animal hearts. These cardiomyocytes have been shown to engraft and proliferate within the scar zone with reported short-term improvements in overall heart function (Caspi *et al.*, 2007; Fernandes *et al.*, 2010; Laflamme *et al.*, 2005). However, cardiomyocytes of different species have important differences (e.g., they contract at different rates), that in combination with noncontractile scar tissue obstructs successful coupling of transplanted and host cardiomyocytes.

2.3. Noncardiac tissue stem cells

Several cell types of noncardiac lineages have also been tested for therapeutic effects, some based on reports that they might transdifferentiate into cardiomyocytes in the heart. For instance, both hematopoietic stem cells and stromal cells derived from the bone marrow were initially reported to transdifferentiate into cardiomyocytes (Orlic *et al.*, 2001; Toma *et al.*, 2002), although subsequent studies have argued against these interpretations (Balsam *et al.*, 2004; Murry *et al.*, 2004). Interestingly, infarcted hearts that were injected with cells of either lineage have shown functional improvement in multiple studies, suggesting that they provide beneficial paracrine effects (Hatzistergos *et al.*, 2010; Mirotsou *et al.*, 2007). In marrow stromal cells, the release of antagonists of the Wnt pathway, among other secreted proteins, are likely contributors to these effects (Mirotsou *et al.*, 2007). Such paracrine effects continue past the death of the injected marrow stromal cells, which only survive in the infarcted hearts for a few days or weeks (Murry *et al.*, 2004). Skeletal muscle cells have also been examined as a potential source of cardiomyocytes for the injured heart. Similarly to bone marrow-derived cells, skeletal myoblasts engraft at least temporarily and can lead to improvements in heart function. Close examination of the transplanted cells in these studies have confirmed that transdifferentiation does not occur, as the myoblasts do not express typical cardiac markers or the cell junction proteins required for coupling to endogenous cardiomyocytes (Reinecke *et al.*, 2002). As with bone marrow-derived cells, the majority of engrafted myoblasts that survive transplantation die within weeks, and their mechanism of action remains unclear (Seidel *et al.*, 2009).

3. Natural Cardiac Regeneration

While the various approaches toward cell therapy provide hope for cardiac repair in mammalian models, significant steps are required to optimize their survival and functional incorporation into an injured heart. Continued fate-mapping experiments are necessary to definitively determine the outcome of the transplanted cells, as is explanation and extension of short-term beneficial effects toward possible long-term healing. Notably, any improvement observed from paracrine effects indicates that there might be value in examining and enhancing the regenerative potential of endogenous heart cells, a line of investigation relevant to the study of naturally occurring heart regeneration (Fig. 11.1B).

In fact, natural cardiac regeneration is now well documented in multiple nonmammalian species. Initial experiments performed in the 1960s and 1970s probed amphibians like frogs and salamanders for the ability to heal mechanical injuries to the cardiac ventricle (Rumyantsev, 1973, 1977). These studies were inspired by the discoveries that salamanders like newts regenerate many tissues that adult mammals cannot, including major appendages, tail, spinal cord, lens, jaws, portions of intestine, and retina (Tsonis, 1996). The extent of regeneration after cardiac injury varied among the reports on the topic, but a synthesis of these and more recent findings highlights a sharp contrast with responses by mammals. Markers of cell proliferation such as ^{3}H-thymidine and BrdU incorporation were detected in cardiomyocytes, as were presence of mitotic figures (Flink, 2002; Oberpriller and Oberpriller, 1974). Removal of ~25% of the ventricle at the apex typically resulted in cardiac scarring, although minced ventricular muscle transplanted to the injury site displayed a more vigorous response of cardiomyocyte proliferation and reorganization (Bader and Oberpriller, 1978, 1979; Oberpriller and Oberpriller, 1974). A recent report indicated that newts show more robust regeneration when the injury is inflicted at the base of the ventricle (Witman *et al.*, 2011). Bolstering these *in vivo* results is demonstrable proliferation by adult newt cardiomyocytes cultured *in vitro* (Oberpriller *et al.*, 1995), events that are rare in cultured adult mammalian cardiomyocytes (Engel *et al.*, 2005). Thus, early studies in leopard frogs, newts, and axolotls laid the foundation for developing and understanding additional models of natural heart regeneration.

4. Heart Regeneration in Zebrafish

4.1. Injury models

The examination of heart regeneration in lower vertebrates has continued in zebrafish, a popular vertebrate model organism with embryologists for the past 20 years. Zebrafish have been a particularly tractable system with low maintenance costs, relatively short generation times, easy breeding and large clutch sizes, and transparency during early development. A large collection of mutant and transgenic strains exists, as does a growing body of molecular genetic tools. An increasing number of groups have become interested in the regenerative potential of adult zebrafish, which compare to salamanders in that transected spinal cords, damaged retinae, and amputated fins regenerate (Becker *et al.*, 1997; Otteson and Hitchcock, 2003; Poss *et al.*, 2003; Vihtelic and Hyde, 2000). Moreover, almost 10 years ago, it was discovered that zebrafish have a robust capacity for cardiac muscle regeneration.

Initial studies of heart regeneration in zebrafish employed a resection model as had been performed in amphibians (Poss *et al.*, 2002b). In these studies, approximately 20% of their single ventricle was removed with iridectomy scissors—a large injury that penetrates the lumen and causes massive bleeding. Wounds clotted quickly, become filled with mature fibrin within 1 week, and then with cardiac muscle over the next several days to weeks. Adult zebrafish cardiomyocytes showed indices of elevated proliferation, beginning at the end of the first week after injury. Cardiomyocyte proliferation was highest within the first 2 weeks of resection but could be observed for weeks beyond this. By 60 days post-amputation (dpa), the ventricular wall had recovered approximately the amount of tissue that was removed (Fig. 11.2A). The zebrafish ventricle is comprised of two main types of cardiac muscle: (1) inner myofibers organized into long, elaborate trabeculae separated from the lumen by the endocardial layer and (2) a thin wall of compact muscle enveloped by the epicardial layer. The regenerated muscle returned largely in the form of a thickened compact myocardial wall (Poss *et al.*, 2002b).

Resection provides a major, complex injury and stimulates a dramatic regenerative response akin to that which regenerates an amputated limb. Because an MI is characterized by local death of cardiac muscle rather than physical removal, there was considerable interest in probing whether

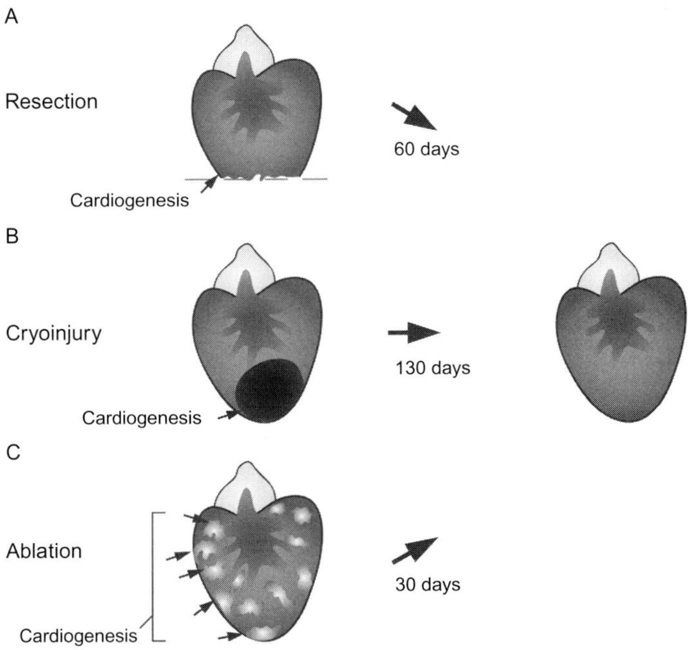

Figure 11.2 Injury and regeneration in zebrafish. (A) In the resection model, 20% of the ventricular apex is surgically removed. Within 2 months, the resected tissue is replaced through local cardiomyocyte proliferation. (B) In the cryoablation model, a cooled probe is applied directly to the heart to induce localized cell death affecting up to 25% of the ventricle. The lesions can remain detectable past 60 days postinjury, but the majority of collagen is eventually cleared coincident with local cardiomyocyte proliferation. (C) Using a conditional genetic approach, over 60% of the ventricular (and atrial) myocardium is destroyed throughout the heart. Regeneration is particularly rapid and robust, with full recovery within 1 month after ablation. (See Color Insert.)

highregenerative capacity as in the case of zebrafish can function in the setting of an infarct. In mouse and rat, this is achieved by occluding a coronary artery. However, the extent to which one could create an ischemic MI in zebrafish is unclear, due to the small ventricular size (~1 μl) and the fact that the coronary vasculature perfuses only a relatively small proportion of the ventricular muscle. Therefore, multiple groups recently used a cryoinjury model, creating a large region of necrotic myocardium through application of dry ice or a liquid nitrogen-cooled probe directly to the ventricle (Chablais *et al.*, 2011; Gonzalez-Rosa *et al.*, 2011; Schnabel *et al.*, 2011). Like the resection model, cryoinjury killed a target region of ~25% of the ventricular muscle without impacting animal survival (Fig. 11.2B). Moreover, while taking slightly longer to complete, the dead tissue was cleared and replaced by a contiguous myocardial wall. Vigorous cardiomyocyte proliferation accompanied these events, clear evidence for heart regeneration.

Finally, another recent study employed an inducible, genetic ablation technique for producing major cardiac injuries in zebrafish (Wang *et al.*, 2011). Here, a double transgenic system was generated consisting of a tamoxifen-inducible CreER recombinase driven specifically in cardiac muscle cells via the *cardiac myosin light chain 2* (*cmlc2*) promoter, combined with a transgene enabling the release of expression of cytotoxic diptheria toxin A chain (DTA) by Cre-mediated recombination. After tamoxifen application, cardiomyocyte cell death was triggered diffusely throughout both cardiac chambers over the course of the next 5–7 days. The amount of ablation was tunable by tamoxifen administration, with histological analysis indicating that more than 60% of the ventricular myocardium can be ablated without effects on animal survival (Fig. 11.2C). Robust cardiomyocyte proliferation and muscle recovery followed, with almost half of the ventricular cardiomyocytes replaced within a week after initial assessment (14 days post-tamoxifen treatment), and full muscular recovery by 30 days posttreatment. This injury model presents certain advantages. First, because of the extreme severity of the injuries versus other models, zebrafish showed clear, outward signs of heart failure after injury, including lethargy, reduced exercise tolerance, and stress sensitivity, that were ameliorated during regeneration. This reversal of heart failure is a striking example of the impact of highcardiac regenerative capacity. Second, as a nonsurgical model, the procedure for injury is simple and offers the potential for more consistent cardiomyocyte loss and recovery among different animals. Third, genetic ablation directly injures only one of the major cardiac cell types, providing specificity that could also be employed with similar tools to ablate other cardiac cell types, and suggesting that cardiomyocyte cell death is sufficient to stimulate regeneration.

Overall, the similarities between the regenerative events after the three cardiac injury models are greater than their differences, and several themes of heart regeneration can be highlighted from this work.

4.2. Cellular hallmarks of heart regeneration

4.2.1. Lack of scarring

Scarring is a major obstacle to functional recovery in the infarcted mammalian heart. By contrast, scarring is a minimal component of repair in the injured zebrafish heart. The initial fibrin clot that fills a resection injury accumulates low levels of collagen, but there is typically no significant accumulation after an injury (Poss *et al.*, 2002b). Cardiomyocyte ablation injuries also show very little collagen deposition (Wang *et al.*, 2011). Importantly, this is not to say that zebrafish hearts cannot repair through scarring. This was first deduced from experiments using mutants in the mitotic checkpoint gene, *mps1*, which fail to regenerate heart muscle after ventricular resection. Instead of filling the wound with cardiac muscle, injured mutants healed injuries with large, collagen-rich scars (Poss *et al.*, 2002b). Similar results were obtained when cardiac regeneration was inhibited by transgenic blockade of Fgf receptors (see below) (Lepilina *et al.*, 2006). Thus, zebrafish have the capacity to repair cardiac injuries by scarring, but this mechanism is typically overridden by a robust regenerative process. Recovery from cryoablation injuries includes collagen deposition to greater extents, yet the majority of this collagen is cleared during the regenerative process (Gonzalez-Rosa *et al.*, 2011). Therefore, while some collagen deposition occurs after resection injury, and more so after cryoinjury, a mature scar forms ostensibly only when muscle regenerative mechanisms are disabled.

Cardiac regenerative capacity would potentially have its greatest impact in humans if it was effective in MI survivors with existing ventricular scars. To test whether the zebrafish heart mounts a regenerative response in the presence of mature scar tissue, Kikuchi *et al.* performed resection surgery and arrested regeneration by Fgfr inhibition to create a scar (Kikuchi *et al.*, 2010). After restoring Fgf signaling, many animals built a new, contiguous wall of muscle around the scar. These findings indicated that even in the presence of established scars, zebrafish cardiomyocytes retain regenerative capacity.

4.2.2. A cardiomyocyte source

Arguably the most important initial goal in investigating a regenerative process is to identify the cellular source(s) of new tissue (Tanaka and Reddien, 2011), in this case, new cardiomyocytes. In zebrafish, as in amphibians, it is clear that adult cardiomyocytes proliferate to replace new muscle, but there are several mechanisms by which these cardiomyocytes might initially arise. These include: (1) a stem or progenitor cell as during embryonic development; (2) direct proliferation by cardiomyocytes; and/or (3) reduction of the contractile program in mature myocytes, or dedifferentiation, to facilitate cell proliferation. Early studies suggested that undifferentiated progenitor cells might give rise to the regenerating cardiomyocytes, based on the induction of embryonic cardiogenic markers and the indication of cells activating the contractile gene *cmlc2 de novo* (Lepilina *et al.*, 2006).

More recently, genetic fate mapping was applied in attempts to conclusively assess existing cardiomyocytes as a source tissue (Jopling *et al.*, 2010; Kikuchi *et al.*, 2010). Using inducible fate-mapping techniques that employ Cre recombinase (Buckingham and Meilhac, 2011), two groups independently labeled *cmlc2*-expressing cardiomyocytes prior to injury. Both found that most, if not all, newly regenerated cardiomyocytes were derived from existing cardiomyocytes (Fig. 11.3A). Similar results were found when

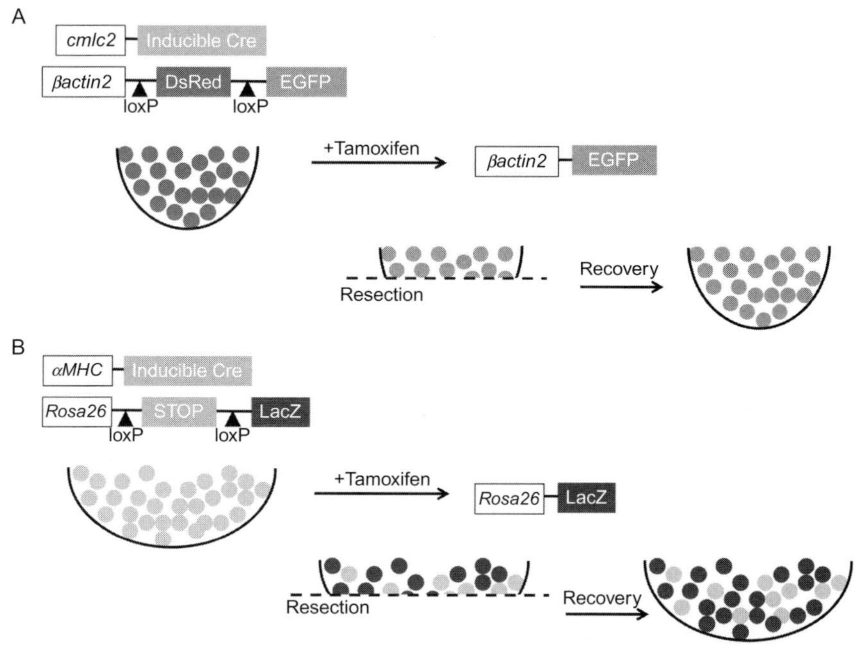

Figure 11.3 Cellular source of regenerating cardiomyocytes. (A) Genetic fate mapping to determine the source of new cardiomyocytes during zebrafish heart regeneration. Treatment with tamoxifen leads to fluorescent EGFP marking of nearly all cardiomyocytes. Following resection, new cardiomyocytes expressed EGFP, indicating derivation from existing cardiomyocytes rather than a different unmarked cell source. (B) A similar fate-mapping experiment was performed in the neonatal mouse heart, in this case with an inducible LacZ reporter. Although the labeling of cardiomyocytes with this system was ~60%, the ratio of labeled cells was maintained between the regenerate and uninjured myocardium. This result indicates that the new cardiomyocytes are derived from existing cardiomyocytes rather than a progenitor cell. (For color version of this figure, the reader is referred to the Web version of this chapter.)

cardiac muscle regeneration after genetic ablation injury was assessed by this method (Wang *et al.*, 2011), suggesting a common source mechanism that is likely to also apply to the cryoinjury model.

Kikuchi *et al.* additionally found using a transgenic reporter line that regulatory sequences of the cardiogenic transcription factor *gata4* are activated throughout the peripheral compact muscle by 7 dpa (Fig. 11.4A) and appears to label most of the regenerated ventricular wall by 14 and 30 dpa (Kikuchi *et al.*, 2010). At 7 and 14 dpa, many of these cells colabel with BrdU, suggesting that they have recently undergone DNA synthesis. While it is known that *gata4* is necessary for embryonic heart development and vascularization (Heicklen-Klein and Evans, 2004; Holtzinger and Evans,

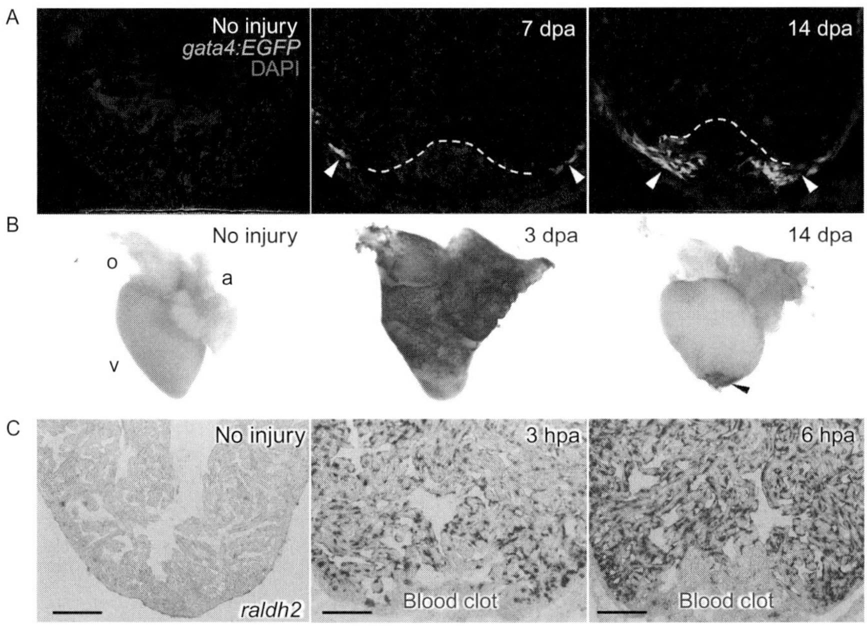

Figure 11.4 Three major cardiac cell types activate developmental gene expression in a chamber-wide manner after resection injury. (A) Myocardial activation, represented by the activation of a *gata4*:EGFP reporter, is first detected throughout the compact layer of the myocardium by 7 days post-amputation (dpa) and becomes localized to the regenerating myocardium by 14–30 dpa (arrowheads). Dotted lines indicate amputation planes. (B) Ostensibly the entire epicardium activates developmental genes such as *raldh2* (violet) within several days following injury. Enhanced gene expression localizes to the wound area by 2 weeks postinjury (arrowheads). (C) Endocardial cells undergo structural remodeling following injury and upregulate *raldh2* (violet) within hours of amputation. Modified from Lepilina *et al.* (2006) and Kikuchi *et al.* (2010, 2011b). (See Color Insert.)

2005), its role in heart regeneration is not yet determined. That *gata4* induction marks regenerating cardiac muscle suggests that the tissue has activated an embryonic program. This is consistent with results of transmission electron microscopy and sarcomere stains, which indicated that regenerating cardiomyocytes acquire a less-organized sarcomeric structure during regeneration (Jopling *et al.*, 2010; Kikuchi *et al.*, 2010; Wang *et al.*, 2011). Thus, the current thinking in the field is that dedifferentiation of existing cardiomyocytes is the dominant source mechanism for heart regeneration in zebrafish.

The reentry of resident cardiomyocytes into the cell cycle is a crucial aspect of replacing the injured cardiomyocyte population. Equally important is that these new cardiomyocytes incorporate functionally with existing cardiac muscle and do not generate arrhythmias. This was investigated using optical voltage mapping of surface myocardium at various stages of regeneration following resection surgery (Kikuchi *et al.*, 2010). At 7 dpa, when cardiomyocytes begin to proliferate, muscle at the regenerating apex was uncoupled. A week later, coupling was evident, and by 30 dpa, electrical conduction through the apex occurred at normal velocities. Loss, and recovery, of conduction velocities was also evident after diffuse genetic ablation, and then regeneration of cardiomyocytes (Wang *et al.*, 2011). Thus, in addition to replacing lost tissue, the newly regenerated cardiomyocytes show evidence of functional integration.

4.2.3. Epicardial and endocardial activation

Another characteristic of the regenerative process in zebrafish is the activation and participation of two other major cardiac cell types, the epicardium and the endocardium, mentioned briefly earlier. These cell layers that line the cardiac muscle appear to play important structural and signaling roles during heart regeneration.

The epicardium covers the periphery of the chambers the heart and interacts closely with the myocardium during development, both as a source of fibroblasts and vascular support cells as well as cardiomyocyte mitogens. In all three injury models, trauma stimulates the epicardium to proliferate and upregulate several genes that mark the embryonic epicardium. These include *wt1b* and *tbx18* genes that are markers of epicardium in the developing embryo (Chablais *et al.*, 2011; Gonzalez-Rosa *et al.*, 2011; Kikuchi *et al.*, 2011a; Lepilina *et al.*, 2006; Schnabel *et al.*, 2011; Wang *et al.*, 2011). Additionally, expression of *raldh2*, an enzyme involved in the synthesis of retinoic acid (RA), is strongly induced in epicardial cells throughout the chambers within the first 1–3 days of injury (Lepilina *et al.*, 2006; Fig. 11.4B). These markers localize to cells within the injury site within the first week, where they subsequently incorporate into the regenerate. Various reports have suggested that epicardial cells might have the capacity to transdifferentiate and give rise to cardiomyocytes in mice (Cai *et al.*, 2008; Smart *et al.*, 2011; Zhou *et al.*, 2008a). However, lineage-tracing experiments in zebrafish using a tamoxifen-inducible Cre driven by the epicardial specific *tcf21* promoter indicated that this is not the case during heart development or regeneration in zebrafish. Rather, epicardial cells play a supporting role for neovascularization that occurs during the process (Kikuchi *et al.*, 2011a). Additionally, these cells appear to promote cardiomyocyte proliferation, at least partially through the upregulation of the RA signaling pathway (described further below; Kikuchi *et al.*, 2011b).

The endocardial cell layer lining the inside of the cardiac chambers also plays an important role in development of the myocardium (Smith and Bader, 2007), suggesting that it may also facilitate adult heart regeneration. Following ventricular resection injury, endocardial cells not only adjacent to the injury site, but throughout the ventricle, undergo rapid morphological changes (Kikuchi *et al.*, 2011b). Additionally, within hours, endocardial cells throughout the ventricle upregulate developmental genes including *raldh2* and *heart of glass* (Fig. 11.4C). This developmental activation of the endocardium completes a trifecta of chamber-wide cellular responses: the compact myocardium (*gata4*), the epicardium (*raldh2* and others), and the endocardium each respond to a local injury in a rapid, global manner (Fig. 11.4). How the appropriate signals are transmitted throughout the chamber, and just how much the response underlies regenerative capacity, are key issues being pursued.

This early endocardial response is transient, as by 1 dpa, endocardial cells away from the wound recover their uninjured appearance. By contrast, the morphological changes and *raldh2* expression persist near the injury site at least through 7 dpa. This suggests that paracrine RA signaling from the endocardium, and presumably also from the epicardium, plays a role in localizing cardiomyocyte proliferation to the injury site. Indeed, the use of transgenic lines that inhibited RA signaling had very strong inhibitory effects on proliferation during regeneration (Kikuchi *et al.*, 2011b). Thus, both the epicardium and endocardium appear to be important cell layers for regenerating cardiac muscle, as they are for building embryonic myocardium.

4.3. Molecular factors controlling heart regeneration

The primary goal of most heart regeneration research is to identify molecules and manipulations that block or enhance the process. These discoveries in turn can be considered in the context of the poorly regenerative mammalian heart. RA signaling mentioned above is an intriguing pathway whose manipulation impacts heart regeneration. While expression analyses have identified many possible candidates (Lien *et al.*, 2006; Raya *et al.*, 2003), only a small list of additional developmental signaling pathways has also been functionally implicated in zebrafish heart regeneration. For example, members of the Fgf signaling pathway are upregulated following resection injury (Lepilina *et al.*, 2006). Expression of the ligand *fgf17b* is activated in injured myocardium, corresponding with the upregulation of the receptors *fgfr2* and *fgfr4* in epicardial cells within the regenerate. Pathway inhibition using a dominant-negative Fgfr transgene prevented completion of regeneration and led to the formation of a scar, as mentioned earlier. During Fgfr inhibition, epicardial cell integration and neovascularization of regenerating myocardium was decreased, suggesting that Fgfs recruit epicardial cells into the regenerate to enable its vascularization during cardiomyocyte proliferation. Lien and colleagues performed a microarray to compare gene expression at 3, 7, and 17 dpa and identified two members of the platelet-derived growth factor (Pdgf) family as exhibiting increased expression in the regenerating heart. Pdgf was demonstrated to promote DNA synthesis in cultured cardiomyocytes, suggesting a possible role in activating proliferation (Lien *et al.*, 2006). Follow-up studies determined that Pdgf signaling involved the epicardium and revealed that the likely primary role of Pdgf signaling during heart regeneration involves reactivating vascular development within the injury site (Kim *et al.*, 2010).

As heart regeneration is driven by cardiomyocyte proliferation, manipulation of genes associated with the cell cycle can impact regeneration. In *mps1* mutants, mentioned earlier, the injured heart retains collagen and fibrin in the place of a new, regenerated myocardial wall. On an interesting note, the *mps1* mutation was identified through a forward genetic screen (in this case designed toward identifying genes involved in caudal fin regeneration), highlighting one of the advantages of studying regeneration in a genetic model organism where screens are common practice (Poss *et al.*, 2002a). Pharmacologic inhibition of Polo-like kinase 1, a regulator of cell-cycle progression, reduced cardiomyocyte proliferation and slowed regeneration (Jopling *et al.*, 2010). With a growing number of highly specific pharmacologic agents, this approach is quick and might translate more easily for comparing responses and requirements in mammalian tissues.

5. Endogenous Regenerative Capacity of Mice

The existence of high-regenerative capacity in lower vertebrates is a unique vantage point, but key differences in cardiac and cardiomyocyte biology from mammalians model system must be taken into consideration. The four-chambered human heart is several thousand times larger and functions at much higher pressure than the two-chambered zebrafish heart. Notwithstanding, the heart rate of adult zebrafish (~120 beats/min) is much closer than mouse (300–600) to the human resting heart rate (60–90). The majority of fish cardiomyocytes remain diploid and mononucleated into adulthood and lack the T-tubule system characteristic of mammalian cardiomyocytes (Poss, 2007). By contrast, a high percentage of mammalian cardiomyocytes increase their DNA in a stage-dependent manner. In mouse, this occurs as most cardiomyocytes withdraw from the cell cycle during the first week after birth, with the majority of these cells undergoing a final round of DNA replication followed by nuclear division to generate binucleated cardiomyocytes (Li *et al.*, 1996). In humans, cardiomyocyte proliferation continues for several months after birth before withdrawal from the cell cycle, when cardiomyocytes undergo DNA replication without nuclear or cell division resulting in mononucleated cells with tetraploid (or higher) DNA content (Adler and Costabel, 1975; Bergmann *et al.*, 2009). Because of available genetic tools, mice are especially attractive mammalian models for interpreting and potentially influencing human cardiac regenerative capacity.

5.1. Fetal and neonatal capacity

The paucity of cardiomyocyte proliferation in adult mammalian hearts has been well documented (detailed below), but, as mentioned earlier, cardiomyocytes in developing embryos proliferate to build muscle. To investigate whether this proliferation might be able to compensate for injury in the fetal heart, Drenckhahn *et al.* used a conditional X-linked knockout mouse to disrupt half of the cardiomyocytes in a fetal female mouse heart (Drenckhahn *et al.*, 2008). The deleted gene, *Holochrome c synthase* (*Hccs*), is required for energy generation in cardiomyocytes, and cells lacking the gene can no longer function appropriately. Interestingly, in spite of the loss of roughly half the cardiomyocytes at 12.5 dpc, the heterozygous knockout females survived until birth. At this stage, lineage-tracing studies demonstrated that the hearts are composed of only ~10% *Hccs*-deficient cells, with the change in proportion being driven by an increase in the number of wild-type cells. While the precise source of the new cells was not determined by this study, these results suggested that the fetal mouse heart possesses high regenerative capacity following cardiomyocyte loss.

Intriguingly, it was recently reported that mice into the neonatal period are able to regenerate cardiac muscle following injury (Porrello *et al.*, 2011). When the ventricular apex of a 1-day-old mouse was resected, an initial blood clot was replaced by regenerating cardiomyocytes over the following weeks. Most new cardiomyocytes were derived from existing cardiomyocytes, as determined by a *Myh6*-Cre lineage tracing experiment (Fig. 11.3B; see below), and proliferation was accompanied by sarcomere disassembly. This regenerative capacity was lost by 7 days after birth, when resection surgery led to the formation of a fibrotic scar (Fig. 11.5). It will be critical to explore the differences in gene expression between cardiac tissues at these two developmental stages to perhaps uncover cues that enable

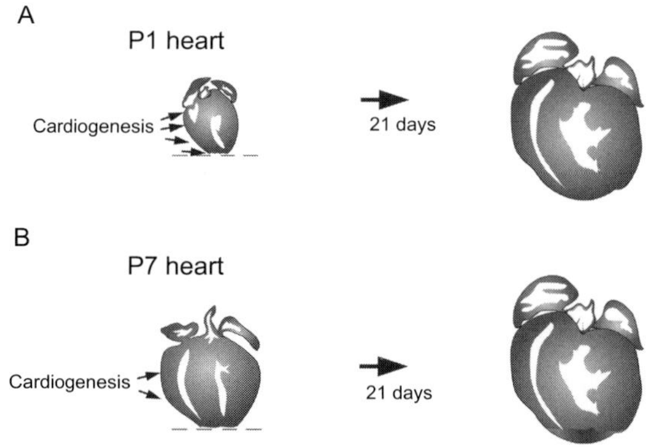

Figure 11.5 Transient regenerative capacity of the neonatal mouse heart after ventricular resection. (A) Rather than inducing fibrosis, resection injuries to the 1-day neonate stimulate replacement with new cardiomyocytes, as assessed 21 days later. Cardiomyocyte proliferation stimulated away from the injury is also likely to be a key component of this regenerative potential. (B) By contrast, fibrosis is the dominant response after resection of ventricular muscle in the 7-day neonate.

cardiomyocyte regeneration. Neonatal mouse heart regeneration appears similar to adult zebrafish heart regeneration in many mechanistic ways, like muscle source and epicardial activation (Porrello *et al.*, 2011). Different from the phenomenon in zebrafish is that the mouse heart is rapidly increasing its size during its window of regenerative potential, and that baseline cardiomyocyte proliferation is quite high both near to and away from the resection injury. Thus, the regenerated myocardium is likely to be a product of an interesting compensatory proliferation response away from the injury in addition to local proliferation. It is exciting to have a new model for natural cardiac regeneration in mammals, and its commonalities and differences with respect to events in zebrafish can inform the biology of required and flexible mechanisms for regenerative success.

5.2. Adult capacity

Cardiomyocyte cell cycle entry is extremely rare in adult mammals, with one study identifying just one murine ventricular cardiomyocyte in 180,000 undergoing DNA synthesis (0.0006%) over the course of 36 h (Soonpaa and Field, 1997). Cardiomyocyte DNA synthesis was measured after focal cauterization of the myocardium in the same study, and a slightly higher proliferation index of 0.0083% was observed. Even after cardiomyocyte cell cycle entry, an additional event of cytokinesis would be necessary to generate new contractile cells. This is a hurdle indeed, given that unlike zebrafish cardiomyocytes, a high percentage (~75%) of adult mouse cardiomyocytes are binucleate (Li *et al.*, 1996). However, even that rare cell cycle entry events occur suggests that there is opportunity for enhancing a naturally occurring process.

Many studies have attempted to override the roadblocks to adult mammalian cardiomyocyte proliferation *in vitro* or *in vivo* by experimental introduction of potential cell-autonomous or paracrine effectors. A recent study examined the effects of continuous transgenic expression of the cell cycle activator cyclin D2 specifically in cardiomyocytes (Hassink *et al.*, 2008). This manipulation contributed to improved cardiac function after MI, an effect attributed to more myocardium and less scar tissue in the transgenic hearts. The result suggests, as in zebrafish, that cardiomyocyte proliferation may be sufficient to outcompete the scarring process in mammals. In a second study, Bersell and colleagues found that neuregulin1 (NRG1) promotes proliferation of differentiated mononucleated cardiomyocytes in cell culture (Bersell *et al.*, 2009). Following MI injury in mouse, injection of NRG1 for 12 weeks improved cardiac function as measured by ejection fraction and decreased scar size by 46%. This improved function was correlated with a 4.4-fold increase in cardiomyocyte-proliferation indices, expected to result in approximately 700,000 new cardiomyocytes after 15 weeks of treatment. This group also visualized proliferating cardiomyocytes and found evidence that they disassemble their sarcomeres to facilitate cytokinesis. That a secreted factor could be added systemically to bolster a cardiac regenerative response is an attractive idea. It will be important to understand which cardiomyocytes respond favorably to NRG1 and to reconcile its apparent hyperplastic effects with its reported hypertrophic effects (Baliga *et al.*, 1999).

5.3. Possible nonmyocyte sources

There is great interest in determining the cell types with the potential to give rise to new cardiomyocytes in adult mice. To this end, a fate-mapping experiment similar to those described above was performed in injured and uninjured animals (Hsieh *et al.*, 2007). In this study, transgenic mice carrying a tamoxifen-inducible Cre under the control of the cardiomyocyte-specific α-myosin heavy chain (*Myh6*) promoter were used along with a lineage labeling strain. The researchers found that during normal aging, the percentage of labeled cardiomyocytes did not change, suggesting that any new cardiomyocytes that arise during homeostasis are not contributed by nonmyocytes. Interestingly, following injury, the percentage of labeled cardiomyocytes decreased, a result that could be interpreted as dilution with cardiomyocytes derived from unlabeled cardiac progenitors. Because the *Myh6*-Cre initially labeled only 80% of the cardiomyocyte population, it could not be ruled out that the new cardiomyocytes might also arise from existing, unlabeled cardiomyocytes.

In a follow-up study, this system was used to examine the effects of injecting bone marrow-derived c-kit+ cells into the infarcted heart (Loffredo *et al.*, 2011). Lineage tracing revealed that upon cell injection, the percentage of labeled cardiomyocytes was further decreased compared to sham or injections with bone marrow-derived mesenchymal stem cells. These results suggest that the addition of c-kit+ cells to the injured heart stimulates greater amounts of cardiomyocyte regeneration from an endogenous progenitor or unlabeled cardiomyocyte source than in hearts without added c-kit+ cells. Additionally, by reversing the system and transplanting c-kit+ cells of the lineage-tracing line into sex-mismatched wild-type recipients, Loffredo and colleagues could exclude transdifferentiation with a high degree of certainty. The sex-mismatched aspect of the experiment also allowed the group to examine and argue against cell fusion of the c-kit+ cells with host cells as a possible source of new cardiomyocytes. Taken together, the results point to paracrine effects from a yet-uncharacterized effector signal(s) from the transplanted cells that stimulate *de novo* cardiomyocyte creation.

Because of their abundance, epicardial cells and fibroblasts represent potential sources of cardiomyocytes for the injured mammalian heart. As mentioned earlier, fate-mapping experiments performed a few years ago with regulatory sequences of the epicardial markers *Wt1* and *Tbx18* suggested that embryonic cells of this lineage contribute significantly to ventricular myocardium (Cai *et al.*, 2008; Zhou *et al.*, 2008a). Conclusive interpretation of these data requires that *Wt1* and *Tbx18* regulatory sequences are inactive in embryonic cardiomyocytes, yet subsequent findings revealed *Tbx18* expression in these cells (Christoffels *et al.*, 2009). Although zebrafish *wt1* sequences drive expression in embryonic zebrafish cardiomyocytes in addition to epicardium (Kikuchi *et al.*, 2011a), the parallel in mice has not yet been reported. Interestingly, Wt1 is upregulated in epicardial cells in the adult mouse heart following MI, where lineage-tracing evidence indicates Wt1-expressing cells normally contribute to the canonical epicardial lineage (epicardium, fibroblasts, perivascular cells; Zhou *et al.*, 2011). Yet, a recent study indicates that these cells might at low frequencies differentiate into mature cardiomyocytes in the presence of injury and exogenously added Thymosin beta-4 (Smart *et al.*, 2011). These are intriguing results with major implications, and advancing these observations with additional markers and mechanistic exploration is anticipated. This finding might explain in part the reported effects of Thymosin beta-4 on improving cardiac function in rodents after MI (Bock-Marquette *et al.*, 2004).

While the extent to which transdifferentiation between cardiac lineages occurs naturally is debated, recent studies indicate that overexpression of certain cardiogenic genes can facilitate this process. This line of research is based on successes in fibroblast reprogramming to iPS cells (Takahashi and Yamanaka, 2006) or neurons (Vierbuchen *et al.*, 2010), and direct reprogramming of pancreatic exocrine cells to beta cells (Zhou *et al.*, 2008b). An initial study along these lines uncovered that a minimal combination of factors (Gata4, Tbx5, and BAF60C) is necessary to differentiate mouse mesoderm into beating cardiomyocytes (Takeuchi and Bruneau, 2009). Building on this research, a cocktail of three overexpressed transcription factors (Mef2C, Gata4, and Tbx5) was found to transform cardiac or skin fibroblasts into beating cardiomyocytes at low frequency (Ieda *et al.*, 2010). A similar study reprogrammed mouse embryonic fibroblasts to cardiomyocytes, but in this case the cells were initially directed toward an IPS cell state with the transcription factors Oct4, Sox2, Klf4, and c-Myc before provocation to a cardiogenic fate by Bmp4 addition (Efe *et al.*, 2011). Thus far, these studies have been performed *in vitro*, but the idea of reprogramming scar-forming fibroblasts into cardiomyocytes within the infarcted heart is undeniably attractive.

6. The Regenerative Capacity of Human Hearts

The consensus of the majority in the field is that adult human hearts have little or no significant regenerative capacity after injury. The detection of cardiomyocyte proliferation itself has historically been challenging, given reliance on pathology samples. Several groups have attempted to count cardiomyocyte numbers at different life stages with varying results (Adler, 1975; Adler and Costabel, 1975; Herget *et al.*, 1997). To address this problem in a lineage-tracing paradigm, a recent study took advantage of the residual ^{14}C left in the atmosphere from the testing of nuclear weapons in the 1950s (Bergmann *et al.*, 2009). Following the agreement of the Limited Nuclear Test Ban Treaty in 1963, the levels of atmospheric ^{14}C dropped sharply. These conditions enabled an elegant pulse-chase experiment where people alive during that time could be evaluated posthumously for the age of their cardiomyocytes versus actual birthdate. This was achieved by sorting cardiomyocyte nuclei from noncardiomyocytes (using a specific marker, cardiac troponin T) and dating them according to ^{14}C concentration. These experiments indicated that cardiomyocyte DNA was synthesized many years after birth in this population, an indication of cell turnover. Interestingly, the rate of turnover appeared to decrease with age, as $\sim$1% of cardiomyocytes is estimated to turn over annually at age 25, and only 0.45% at age 75. Thus, over the course of an average lifetime, almost half of the cardiomyocyte population is replaced, suggesting existence of an endogenous mechanism of low-frequency, homeostatic regeneration. Importantly, it is unclear whether the cellular origin of new muscle is a progenitor cell or differentiated cardiomyocyte, or both. As there are large communities attempting to define mechanisms of regeneration and identify how to stimulate regeneration from both types of source, there is reason to be optimistic that either source mechanism could be accommodated and activated.

7. Perspectives

Cardiac regeneration is a fascinating biological phenomenon that is distributed unequally among species. As a potential panacea for common human heart diseases, there is urgency to secure the blueprint. While various strategies may appeal differently to developmental biologists, stem cell biologists, cardiologists, or pathologists in the field, stimulating human heart regeneration is the shared goal. This will ultimately occur via success in cell therapy and/or stimulating endogenous regenerative mechanisms. For the former, continued searches for sources, and systematic tracking of transplanted cells, are needed to improve our understanding of beneficial effects and guide the development of consistently successful therapies. For the latter, it is key to interpret examples of highregenerative capacity and to use these systems to identify the cells and molecules that drive successful heart regeneration.

From zebrafish and early studies in neonatal mice, we know that regenerative capacity is a product of at least two major components. These are (1) a cardiomyocyte source that can be activated for regeneration and (2) a dynamic cardiac environment comprised of nonmyocardial tissue that facilitates regeneration from this source. Future studies anticipate procurement of molecular strategies for stimulating regeneration and may inform cell therapy approaches by suggesting optimal combinations of cardiac cell types and/or cofactors, or inhibitory responses to be suppressed after cell delivery. The heart regeneration research community has gained momentum and become multifaceted in the past 10–15years, and it is exciting to consider what interventions will arise in the near future from a synthesis of findings.

Acknowledgments

We thank Amy Dickson for artwork. The authors were funded by postdoctoral fellowships from National Heart, Lung, and Blood Institute and American Heart Association to W. Y. C., and grants from National Heart, Lung, and Blood Institute, American Heart Association, and American Federation for Aging Research to K. D. P. K. D. P. is an Early Career Scientist of the Howard Hughes Medical Institute.

References

Adler, C. P. (1975). Relationship between deoxyribonucleic acid content and nucleoli in human heart muscle cells and estimation of cell number during cardiac growth and hyperfunction. *Recent Adv. Stud. Cardiac Struct. Metab.* **8,** 373–386.

Adler, C. P., and Costabel, U. (1975). Cell number in human heart in atrophy, hypertrophy, and under the influence of cytostatics. *Recent Adv. Stud. Cardiac Struct. Metab.* **6,** 343–355.

Bader, D., and Oberpriller, J. O. (1978). Repair and reorganization of minced cardiac muscle in the adult newt (Notophthalmus viridescens). *J. Morphol.* **155,** 349–357.

Bader, D., and Oberpriller, J. (1979). Autoradiographic and electron microscopic studies of minced cardiac muscle regeneration in the adult newt, notophthalmus viridescens. *J. Exp. Zool.* **208,** 177–193.

Baliga, R. R., Pimental, D. R., Zhao, Y. Y., Simmons, W. W., Marchionni, M. A., Sawyer, D. B., and Kelly, R. A. (1999). NRG-1-induced cardiomyocyte hypertrophy. Role of PI-3-kinase, p70(S6K), and MEK-MAPK-RSK. *Am. J. Physiol.* **277,** H2026–H2037.

Balsam, L. B., Wagers, A. J., Christensen, J. L., Kofidis, T., Weissman, I. L., and Robbins, R. C. (2004). Haematopoietic stem cells adopt mature haematopoietic fates in ischaemic myocardium. *Nature* **428,** 668–673.

Bearzi, C., Rota, M., Hosoda, T., Tillmanns, J., Nascimbene, A., De Angelis, A., Yasuzawa-Amano, S., Trofimova, I., Siggins, R. W., Lecapitaine, N., Cascapera, S., Beltrami, A. P., *et al.* (2007). Human cardiac stem cells. *Proc. Natl. Acad. Sci. USA* **104,** 14068–14073.

Becker, T., Wullimann, M. F., Becker, C. G., Bernhardt, R. R., and Schachner, M. (1997). Axonal regrowth after spinal cord transection in adult zebrafish. *J. Comp. Neurol.* **377,** 577–595.

Beltrami, A. P., Barlucchi, L., Torella, D., Baker, M., Limana, F., Chimenti, S., Kasahara, H., Rota, M., Musso, E., Urbanek, K., Leri, A., Kajstura, J., *et al.* (2003). Adult cardiac stem cells are multipotent and support myocardial regeneration. *Cell* **114,** 763–776.

Bergmann, O., Bhardwaj, R. D., Bernard, S., Zdunek, S., Barnabe-Heider, F., Walsh, S., Zupicich, J., Alkass, K., Buchholz, B. A., Druid, H., Jovinge, S., and Frisen, J. (2009). Evidence for cardiomyocyte renewal in humans. *Science* **324,** 98–102.

Bersell, K., Arab, S., Haring, B., and Kuhn, B. (2009). Neuregulin1/ErbB4 signaling induces cardiomyocyte proliferation and repair of heart injury. *Cell* **138,** 257–270.

Bock-Marquette, I., Saxena, A., White, M. D., Dimaio, J. M., and Srivastava, D. (2004). Thymosin beta4 activates integrin-linked kinase and promotes cardiac cell migration, survival and cardiac repair. *Nature* **432,** 466–472.

Bu, L., Jiang, X., Martin-Puig, S., Caron, L., Zhu, S., Shao, Y., Roberts, D. J., Huang, P. L., Domian, I. J., and Chien, K. R. (2009). Human ISL1 heart progenitors generate diverse multipotent cardiovascular cell lineages. *Nature* **460,** 113–117.

Buckingham, M. E., and Meilhac, S. M. (2011). Tracing cells for tracking cell lineage and clonal behavior. *Dev. Cell* **21,** 394–409.

Cai, C. L., Martin, J. C., Sun, Y., Cui, L., Wang, L., Ouyang, K., Yang, L., Bu, L., Liang, X., Zhang, X., Stallcup, W. B., Denton, C. P., *et al.* (2008). A myocardial lineage derives from Tbx18 epicardial cells. *Nature* **454,** 104–108.

Caspi, O., Huber, I., Kehat, I., Habib, M., Arbel, G., Gepstein, A., Yankelson, L., Aronson, D., Beyar, R., and Gepstein, L. (2007). Transplantation of human embryonic stem cell-derived cardiomyocytes improves myocardial performance in infarcted rat hearts. *J. Am. Coll. Cardiol.* **50,** 1884–1893.

Chablais, F., Veit, J., Rainer, G., and Jazwinska, A. (2011). The zebrafish heart regenerates after cryoinjury-induced myocardial infarction. *BMC Dev. Biol.* **11,** 21.

Christoffels, V. M., Grieskamp, T., Norden, J., Mommersteeg, M. T., Rudat, C., and Kispert, A. (2009). Tbx18 and the fate of epicardial progenitors. *Nature* **458,** E8–E9; discussion E9-10.

Drenckhahn, J. D., Schwarz, Q. P., Gray, S., Laskowski, A., Kiriazis, H., Ming, Z., Harvey, R. P., Du, X. J., Thorburn, D. R., and Cox, T. C. (2008). Compensatory growth of healthy cardiac cells in the presence of diseased cells restores tissue homeostasis during heart development. *Dev. Cell* **15,** 521–533.

Efe, J. A., Hilcove, S., Kim, J., Zhou, H., Ouyang, K., Wang, G., Chen, J., and Ding, S. (2011). Conversion of mouse fibroblasts into cardiomyocytes using a direct reprogramming strategy. *Nat. Cell Biol.* **13,** 215–222.

Engel, F. B., Schebesta, M., Duong, M. T., Lu, G., Ren, S., Madwed, J. B., Jiang, H., Wang, Y., and Keating, M. T. (2005). p38 MAP kinase inhibition enables proliferation of adult mammalian cardiomyocytes. *Genes Dev.* **19,** 1175–1187.

Fernandes, S., Naumova, A. V., Zhu, W. Z., Laflamme, M. A., Gold, J., and Murry, C. E. (2010). Human embryonic stem cell-derived cardiomyocytes engraft but do not alter cardiac remodeling after chronic infarction in rats. *J. Mol. Cell. Cardiol.* **49,** 941–949.

Flink, I. L. (2002). Cell cycle reentry of ventricular and atrial cardiomyocytes and cells within the epicardium following amputation of the ventricular apex in the axolotl, Amblystoma mexicanum: Confocal microscopic immunofluorescent image analysis of bromodeoxyuridine-labeled nuclei. *Anat. Embryol. (Berl).* **205,** 235–244.

Gonzalez-Rosa, J. M., Martin, V., Peralta, M., Torres, M., and Mercader, N. (2011). Extensive scar formation and regression during heart regeneration after cryoinjury in zebrafish. *Development* **138,** 1663–1674.

Hansson, E. M., Lindsay, M. E., and Chien, K. R. (2009). Regeneration next: Toward heart stem cell therapeutics. *Cell Stem Cell* **5,** 364–377.

Hassink, R. J., Pasumarthi, K. B., Nakajima, H., Rubart, M., Soonpaa, M. H., de la Riviere, A. B., Doevendans, P. A., and Field, L. J. (2008). Cardiomyocyte cell cycle activation improves cardiac function after myocardial infarction. *Cardiovasc. Res.* **78,** 18–25.

Hatzistergos, K. E., Quevedo, H., Oskouei, B. N., Hu, Q., Feigenbaum, G. S., Margitich, I. S., Mazhari, R., Boyle, A. J., Zambrano, J. P., Rodriguez, J. E., Dulce, R., Pattany, P. M., *et al.* (2010). Bone marrow mesenchymal stem cells stimulate cardiac stem cell proliferation and differentiation. *Circ. Res.* **107,** 913–922.

Heicklen-Klein, A., and Evans, T. (2004). T-box binding sites are required for activity of a cardiac GATA-4 enhancer. *Dev. Biol.* **267,** 490–504.

Herget, G. W., Neuburger, M., Plagwitz, R., and Adler, C. P. (1997). DNA content, ploidy level and number of nuclei in the human heart after myocardial infarction. *Cardiovasc. Res.* **36,** 45–51.

Holtzinger, A., and Evans, T. (2005). Gata4 regulates the formation of multiple organs. *Development* **132,** 4005–4014.

Hsieh, P. C., Segers, V. F., Davis, M. E., Macgillivray, C., Gannon, J., Molkentin, J. D., Robbins, J., and Lee, R. T. (2007). Evidence from a genetic fate-mapping study that stem cells refresh adult mammalian cardiomyocytes after injury. *Nat. Med.* **13,** 970–974.

Ieda, M., Fu, J. D., Delgado-Olguin, P., Vedantham, V., Hayashi, Y., Bruneau, B. G., and Srivastava, D. (2010). Direct reprogramming of fibroblasts into functional cardiomyocytes by defined factors. *Cell* **142,** 375–386.

Jopling, C., Sleep, E., Raya, M., Marti, M., Raya, A., and Belmonte, J. C. (2010). Zebrafish heart regeneration occurs by cardiomyocyte dedifferentiation and proliferation. *Nature* **464,** 606–609.

Kattman, S. J., Witty, A. D., Gagliardi, M., Dubois, N. C., Niapour, M., Hotta, A., Ellis, J., and Keller, G. (2011). Stage-specific optimization of activin/nodal and BMP signaling promotes cardiac differentiation of mouse and human pluripotent stem cell lines. *Cell Stem Cell* **8,** 228–240.

Kehat, I., Kenyagin-Karsenti, D., Snir, M., Segev, H., Amit, M., Gepstein, A., Livne, E., Binah, O., Itskovitz-Eldor, J., and Gepstein, L. (2001). Human embryonic stem cells can differentiate into myocytes with structural and functional properties of cardiomyocytes. *J. Clin. Invest.* **108,** 407–414.

Kikuchi, K., Holdway, J. E., Werdich, A. A., Anderson, R. M., Fang, Y., Egnaczyk, G. F., Evans, T., Macrae, C. A., Stainier, D. Y., and Poss, K. D. (2010). Primary contribution to zebrafish heart regeneration by gata4(+) cardiomyocytes. *Nature* **464,** 601–605.

Kikuchi, K., Gupta, V., Wang, J., Holdway, J. E., Wills, A. A., Fang, Y., and Poss, K. D. (2011a). tcf21+ epicardial cells adopt non-myocardial fates during zebrafish heart development and regeneration. *Development* **138,** 2895–2902.

Kikuchi, K., Holdway, J. E., Major, R. J., Blum, N., Dahn, R. D., Begemann, G., and Poss, K. D. (2011b). Retinoic acid production by endocardium and epicardium is an injury response essential for zebrafish heart regeneration. *Dev. Cell* **20,** 397–404.

Kim, J., Wu, Q., Zhang, Y., Wiens, K. M., Huang, Y., Rubin, N., Shimada, H., Handin, R. I., Chao, M. Y., Tuan, T. L., Starnes, V. A., and Lien, C. L. (2010). PDGF signaling is required for epicardial function and blood vessel formation in regenerating zebrafish hearts. *Proc. Natl. Acad. Sci. USA* **107,** 17206–17210.

Laflamme, M. A., and Murry, C. E. (2011). Heart regeneration. *Nature* **473,** 326–335.

Laflamme, M. A., Gold, J., Xu, C., Hassanipour, M., Rosler, E., Police, S., Muskheli, V., and Murry, C. E. (2005). Formation of human myocardium in the rat heart from human embryonic stem cells. *Am. J. Pathol.* **167,** 663–671.

Laflamme, M. A., Chen, K. Y., Naumova, A. V., Muskheli, V., Fugate, J. A., Dupras, S. K., Reinecke, H., Xu, C., Hassanipour, M., Police, S., O'Sullivan, C., Collins, L., *et al.* (2007a). Cardiomyocytes derived from human embryonic stem cells in pro-survival factors enhance function of infarcted rat hearts. *Nat. Biotechnol.* **25,** 1015–1024.

Laflamme, M. A., Zbinden, S., Epstein, S. E., and Murry, C. E. (2007b). Cell-based therapy for myocardial ischemia and infarction: Pathophysiological mechanisms. *Annu. Rev. Pathol.* **2,** 307–339.

Lepilina, A., Coon, A. N., Kikuchi, K., Holdway, J. E., Roberts, R. W., Burns, C. G., and Poss, K. D. (2006). A dynamic epicardial injury response supports progenitor cell activity during zebrafish heart regeneration. *Cell* **127,** 607–619.

Li, F., Wang, X., Capasso, J. M., and Gerdes, A. M. (1996). Rapid transition of cardiac myocytes from hyperplasia to hypertrophy during postnatal development. *J. Mol. Cell. Cardiol.* **28,** 1737–1746.

Lien, C. L., Schebesta, M., Makino, S., Weber, G. J., and Keating, M. T. (2006). Gene expression analysis of zebrafish heart regeneration. *PLoS Biol.* **4,** e260.

Loffredo, F. S., Steinhauser, M. L., Gannon, J., and Lee, R. T. (2011). Bone marrow-derived cell therapy stimulates endogenous cardiomyocyte progenitors and promotes cardiac repair. *Cell Stem Cell* **8,** 389–398.

Martin, C. M., Meeson, A. P., Robertson, S. M., Hawke, T. J., Richardson, J. A., Bates, S., Goetsch, S. C., Gallardo, T. D., and Garry, D. J. (2004). Persistent expression of the ATP-binding cassette transporter, Abcg2, identifies cardiac SP cells in the developing and adult heart. *Dev. Biol.* **265,** 262–275.

Mirotsou, M., Zhang, Z., Deb, A., Zhang, L., Gnecchi, M., Noiseux, N., Mu, H., Pachori, A., and Dzau, V. (2007). Secreted frizzled related protein 2 (Sfrp2) is the key Akt-mesenchymal stem cell-released paracrine factor mediating myocardial survival and repair. *Proc. Natl. Acad. Sci. USA* **104,** 1643–1648.

Murry, C. E., Soonpaa, M. H., Reinecke, H., Nakajima, H., Nakajima, H. O., Rubart, M., Pasumarthi, K. B., Virag, J. I., Bartelmez, S. H., Poppa, V., Bradford, G., Dowell, J. D., *et al.* (2004). Haematopoietic stem cells do not transdifferentiate into cardiac myocytes in myocardial infarcts. *Nature* **428,** 664–668.

Oberpriller, J. O., and Oberpriller, J. C. (1974). Response of the adult newt ventricle to injury. *J. Exp. Zool.* **187,** 249–253.

Oberpriller, J. O., Oberpriller, J. C., Matz, D. G., and Soonpaa, M. H. (1995). Stimulation of proliferative events in the adult amphibian cardiac myocyte. *Ann. NY Acad. Sci.* **752,** 30–46.

Oh, H., Bradfute, S. B., Gallardo, T. D., Nakamura, T., Gaussin, V., Mishina, Y., Pocius, J., Michael, L. H., Behringer, R. R., Garry, D. J., Entman, M. L., and Schneider, M. D. (2003). Cardiac progenitor cells from adult myocardium: Homing, differentiation, and fusion after infarction. *Proc. Natl. Acad. Sci. USA* **100,** 12313–12318.

Orlic, D., Kajstura, J., Chimenti, S., Jakoniuk, I., Anderson, S. M., Li, B., Pickel, J., McKay, R., Nadal-Ginard, B., Bodine, D. M., Leri, A., and Anversa, P. (2001). Bone marrow cells regenerate infarcted myocardium. *Nature* **410,** 701–705.

Otteson, D. C., and Hitchcock, P. F. (2003). Stem cells in the teleost retina: Persistent neurogenesis and injury-induced regeneration. *Vision Res.* **43,** 927–936.

Porrello, E. R., Mahmoud, A. I., Simpson, E., Hill, J. A., Richardson, J. A., Olson, E. N., and Sadek, H. A. (2011). Transient regenerative potential of the neonatal mouse heart. *Science* **331,** 1078–1080.

Poss, K. D. (2007). Getting to the heart of regeneration in zebrafish. *Semin. Cell Dev. Biol.* **18,** 36–45.

Poss, K. D. (2010). Advances in understanding tissue regenerative capacity and mechanisms in animals. *Nat. Rev. Genet.* **11,** 710–722.

Poss, K. D., Nechiporuk, A., Hillam, A. M., Johnson, S. L., and Keating, M. T. (2002a). Mps1 defines a proximal blastemal proliferative compartment essential for zebrafish fin regeneration. *Development* **129,** 5141–5149.

Poss, K. D., Wilson, L. G., and Keating, M. T. (2002b). Heart regeneration in zebrafish. *Science* **298,** 2188–2190.

Poss, K. D., Keating, M. T., and Nechiporuk, A. (2003). Tales of regeneration in zebrafish. *Dev. Dyn.* **226,** 202–210.

Raya, A., Koth, C. M., Buscher, D., Kawakami, Y., Itoh, T., Raya, R. M., Sternik, G., Tsai, H. J., Rodriguez-Esteban, C., and Izpisua-Belmonte, J. C. (2003). Activation of Notch signaling pathway precedes heart regeneration in zebrafish. *Proc. Natl. Acad. Sci. USA* **100**(Suppl 1), 11889–11895.

Reinecke, H., Poppa, V., and Murry, C. E. (2002). Skeletal muscle stem cells do not transdifferentiate into cardiomyocytes after cardiac grafting. *J. Mol. Cell. Cardiol.* **34,** 241–249.

Rumyantsev, P. P. (1973). Post-injury DNA synthesis, mitosis and ultrastructural reorganization of adult frog cardiac myocytes. An electron microscopic-autoradiographic study. *Z. Zellforsch. Mikrosk. Anat.* **139,** 431–450.

Rumyantsev, P. P. (1977). Interrelations of the proliferation and differentiation processes during cardiac myogenesis and regeneration. *Int. Rev. Cytol.* **51,** 187–273.

Schnabel, K., Wu, C. C., Kurth, T., and Weidinger, G. (2011). Regeneration of cryoinjury induced necrotic heart lesions in zebrafish is associated with epicardial activation and cardiomyocyte proliferation. *PLoS One* **6,** e18503.

Seidel, M., Borczynska, A., Rozwadowska, N., and Kurpisz, M. (2009). Cell-based therapy for heart failure: Skeletal myoblasts. *Cell Transplant.* **18,** 695–707.

Smart, N., Bollini, S., Dube, K. N., Vieira, J. M., Zhou, B., Davidson, S., Yellon, D., Riegler, J., Price, A. N., Lythgoe, M. F., Pu, W. T., and Riley, P. R. (2011). De novo cardiomyocytes from within the activated adult heart after injury. *Nature* **474,** 640–644.

Smith, T. K., and Bader, D. M. (2007). Signals from both sides: Control of cardiac development by the endocardium and epicardium. *Semin. Cell Dev. Biol.* **18,** 84–89.

Soonpaa, M. H., and Field, L. J. (1997). Assessment of cardiomyocyte DNA synthesis in normal and injured adult mouse hearts. *Am. J. Physiol.* **272,** H220–H226.

Takahashi, K., and Yamanaka, S. (2006). Induction of pluripotent stem cells from mouse embryonic and adult fibroblast cultures by defined factors. *Cell* **126,** 663–676.

Takeuchi, J. K., and Bruneau, B. G. (2009). Directed transdifferentiation of mouse mesoderm to heart tissue by defined factors. *Nature* **459,** 708–711.

Tallini, Y. N., Greene, K. S., Craven, M., Spealman, A., Breitbach, M., Smith, J., Fisher, P. J., Steffey, M., Hesse, M., Doran, R. M., Woods, A., Singh, B., *et al.* (2009). c-kit expression identifies cardiovascular precursors in the neonatal heart. *Proc. Natl. Acad. Sci. USA* **106,** 1808–1813.

Tanaka, E. M., and Reddien, P. W. (2011). The cellular basis for animal regeneration. *Dev. Cell* **21,** 172–185.

Tang, X. L., Rokosh, G., Sanganalmath, S. K., Yuan, F., Sato, H., Mu, J., Dai, S., Li, C., Chen, N., Peng, Y., Dawn, B., Hunt, G., *et al.* (2010). Intracoronary administration of cardiac progenitor cells alleviates left ventricular dysfunction in rats with a 30-day-old infarction. *Circulation* **121,** 293–305.

Toma, C., Pittenger, M. F., Cahill, K. S., Byrne, B. J., and Kessler, P. D. (2002). Human mesenchymal stem cells differentiate to a cardiomyocyte phenotype in the adult murine heart. *Circulation* **105,** 93–98.

Tsonis, P. A. (1996). Limb Regeneration. Cambridge University Press, Cambridge.

Vierbuchen, T., Ostermeier, A., Pang, Z. P., Kokubu, Y., Sudhof, T. C., and Wernig, M. (2010). Direct conversion of fibroblasts to functional neurons by defined factors. *Nature* **463,** 1035–1041.

Vihtelic, T. S., and Hyde, D. R. (2000). Light-induced rod and cone cell death and regeneration in the adult albino zebrafish (Danio rerio) retina. *J. Neurobiol.* **44,** 289–307.

Wang, J., Panakova, D., Kikuchi, K., Holdway, J. E., Gemberling, M., Burris, J. S., Singh, S. P., Dickson, A. L., Lin, Y. F., Sabeh, M. K., Werdich, A. A., Yelon, D., *et al.* (2011). The regenerative capacity of zebrafish reverses cardiac failure caused by genetic cardiomyocyte depletion. *Development* **138,** 3421–3430.

Witman, N., Murtuza, B., Davis, B., Arner, A., and Morrison, J. I. (2011). Recapitulation of developmental cardiogenesis governs the morphological and functional regeneration of adult newt hearts following injury. *Dev. Biol.* **354,** 67–76.

Wu, S. M., Chien, K. R., and Mummery, C. (2008). Origins and fates of cardiovascular progenitor cells. *Cell* **132,** 537–543.

Zaruba, M. M., Soonpaa, M., Reuter, S., and Field, L. J. (2010). Cardiomyogenic potential of C-kit(+)-expressing cells derived from neonatal and adult mouse hearts. *Circulation* **121,** 1992–2000.

Zhou, B., Ma, Q., Rajagopal, S., Wu, S. M., Domian, I., Rivera-Feliciano, J., Jiang, D., von Gise, A., Ikeda, S., Chien, K. R., and Pu, W. T. (2008a). Epicardial progenitors contribute to the cardiomyocyte lineage in the developing heart. *Nature* **454,** 109–113.

Zhou, Q., Brown, J., Kanarek, A., Rajagopal, J., and Melton, D. A. (2008b). In vivo reprogramming of adult pancreatic exocrine cells to beta-cells. *Nature* **455,** 627–632.

Zhou, B., Honor, L. B., He, H., Ma, Q., Oh, J. H., Butterfield, C., Lin, R. Z., Melero-Martin, J. M., Dolmatova, E., Duffy, H. S., Gise, A., Zhou, P., *et al.* (2011). Adult mouse epicardium modulates myocardial injury by secreting paracrine factors. *J. Clin. Invest.* **121,** 1894–1904.

Zhu, W. Z., Xie, Y., Moyes, K. W., Gold, J. D., Askari, B., and Laflamme, M. A. (2010). Neuregulin/ErbB signaling regulates cardiac subtype specification in differentiating human embryonic stem cells. *Circ. Res.* **107,** 776–786.

Index

Note: Page numbers followed by "*f*" indicate figures, and "*t*" indicate tables.

C

D

E

Contents of Previous Volumes

Volume 47

Volume 48

Volume 49

Volume 52

Volume 53

Volume 54

Volume 55

Volume 56

Volume 57

Volume 58

Volume 59

Volume 62

Volume 63

Volume 64

Volume 65

Volume 66

Volume 67

Volume 68

Volume 69

Volume 70

Volume 71

Volume 72

Volume 73

Volume 74

Volume 75

Volume 76

Volume 77

Volume 78

Volume 79

Volume 80

Volume 81

Volume 82

Volume 83

Volume 84

Volume 85

Volume 86

Volume 87

Volume 88

Volume 89

Volume 90

Volume 91

Volume 92

Volume 95

Volume 98

Volume 99

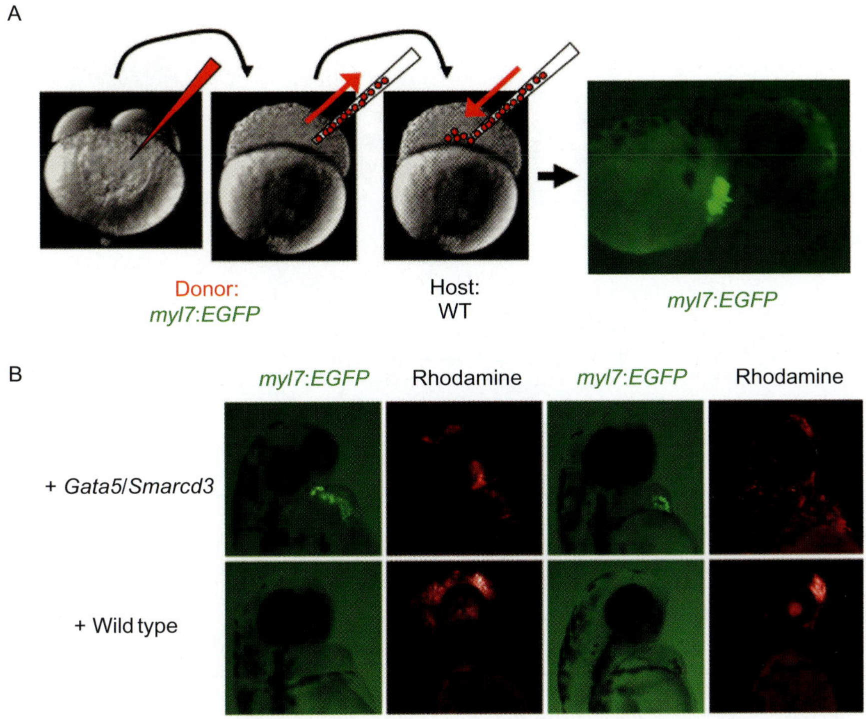

Ian C. Scott, Figure 1.2 Gata5/Smarcd3 promotes a CPC-like state in zebrafish. (A) Schematics of transplantation experiments in zebrafish to assay cardiac potential. Cells from a transgenic 4-hpf *myl7:EGFP* donor embryo (injected with Rhodamine tracer at the one-cell stage) are transplanted to a host embryo at 4 hpf. EGFP signal in the heart at 24–48 hpf indicates donor cells that have differentiated as cardiomyocytes in the host heart. By manipulating host or donor cells, gene function in cardiac fate can be examined. (B) Transplants where donor cells are placed in the animal cap of host embryos. Wild-type donor cells do not take on a cardiac fate (no EGFP), with Rhodamine tracer showing cells remain in the brain/retina. Overexpression of *Gata5/Smarcd3* in donor cells promotes myocardial fate in the heart (EGFP and Rhodamine signal now evident in heart).

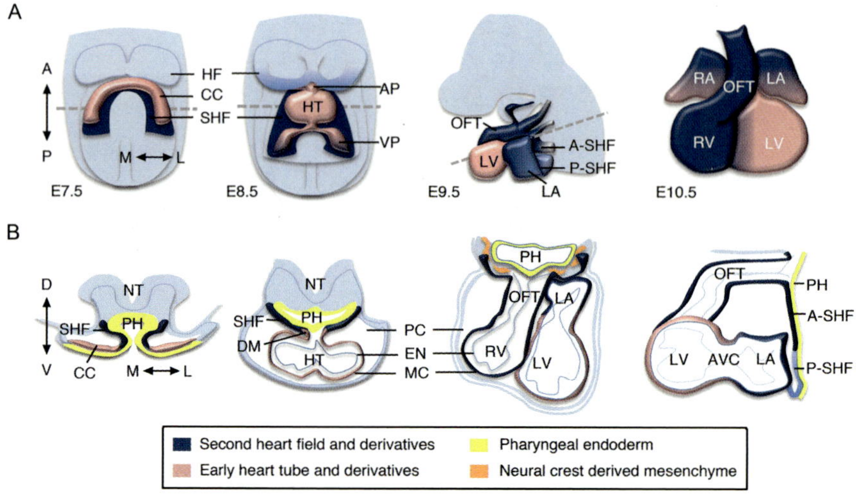

Robert G. Kelly, Figure 2.1 Topography of the second heart field. (A) Cartoons showing the stages of mouse heart development during which second heart field progenitors are added to the heart. From left to right: embryonic day (E) 7.5 (cardiac crescent, ventral view), E8.5 (early heart tube, ventral view), E9.5 (looped heart, left lateral view), and E10.5 (ventral view). Second heart heart field progenitor cells and their contribution to the heart are shown in dark blue and the early heart tube giving rise to the left ventricle in pink. Anterior (A) posterior (P) and medial (M) lateral (L) embryonic axes are indicated. (B) Schematized histological sections showing the location of second heart field cells and their myocardial derivatives (dark blue) at the stages shown in (A) relative to pharyngeal endoderm (yellow), myoctes derived from the early heart tube (pink), and neural crest-derived mesenchyme (orange). The first three panels show transverse sections at the levels indicated by the dotted lines in (A). The right hand panel shows a mid-saggital E9.5 section. Dorsal (D) ventral (V) and medial (M) lateral (L) embryonic axes are indicated. CC, cardiac crescent; SHF, second heart field; HF, head fold; HT, heart tube, AP, arterial pole; VP, venous pole; OFT, outflow tract; LV, left ventricle; LA, left atrium; RV, right ventricle; RA, right atrium; A-SHF, anterior component of the second heart field; P-SHF, posterior component of the second heart field; NT, neural tube; PH, pharynx; PC, pericardial cavity; DM, dorsal mesocardium; EN, enodcardium; MC, myocardium.

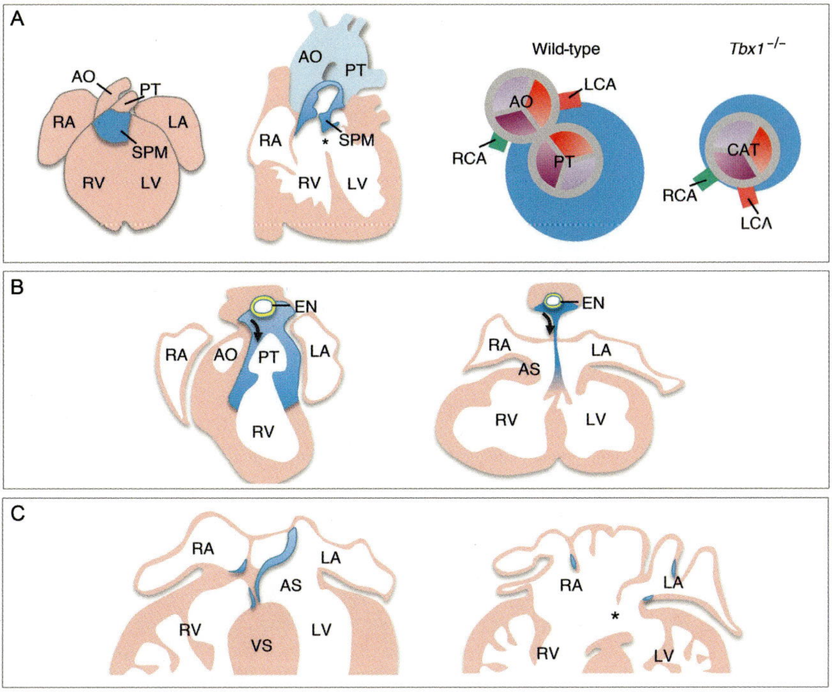

Robert G. Kelly, Figure 2.3 Clinically relevant subdomains of the second heart field. (A) Subpulmonary myocardium at the base of the outlet of the right ventricle (blue) in a mouse heart at E14.5 and a human heart with tetralogy of Fallot; note the hypoplastic right ventricular outlet and ventricular septal defect (asterisk). Subpulmonary myocardium is severely reduced in the common arterial trunk of *Tbx1* mutant hearts; note the abnormal ventrally positioned proximal left coronary artery (LCA) compared to the wild-type situation. (B) Cartoons showing the migration (arrows) of Hedgehog responsive cells (blue) from mesoderm adjacent to pharyngeal endoderm (yellow) into the base of the pulmonary trunk (left) and interatrial septum (right). (C) Schematized histological sections showing the contribution of progenitor cells expressing the hedgehog target gene *Gli1* to the interatrial septum (left) and the failure of atrial and atrioventricular septation (asterisk) when the hedgehog receptor smoothened is deleted from *Gli* expressing cells at early developmental stages (right). (B) and (C) adapted from Hoffmann *et al.* (2009) with permission. AO, aorta; PT, pulmonary trunk; RA, right atrium; RV, right ventricle; LA, left atrium; LV, left ventricle; SPM, subpulmonary myocardium; RCA, right coronary artery; CAT, common arterial trunk; EN, pharyngeal endoderm; AS, atrial septum; VS, ventricular septum.

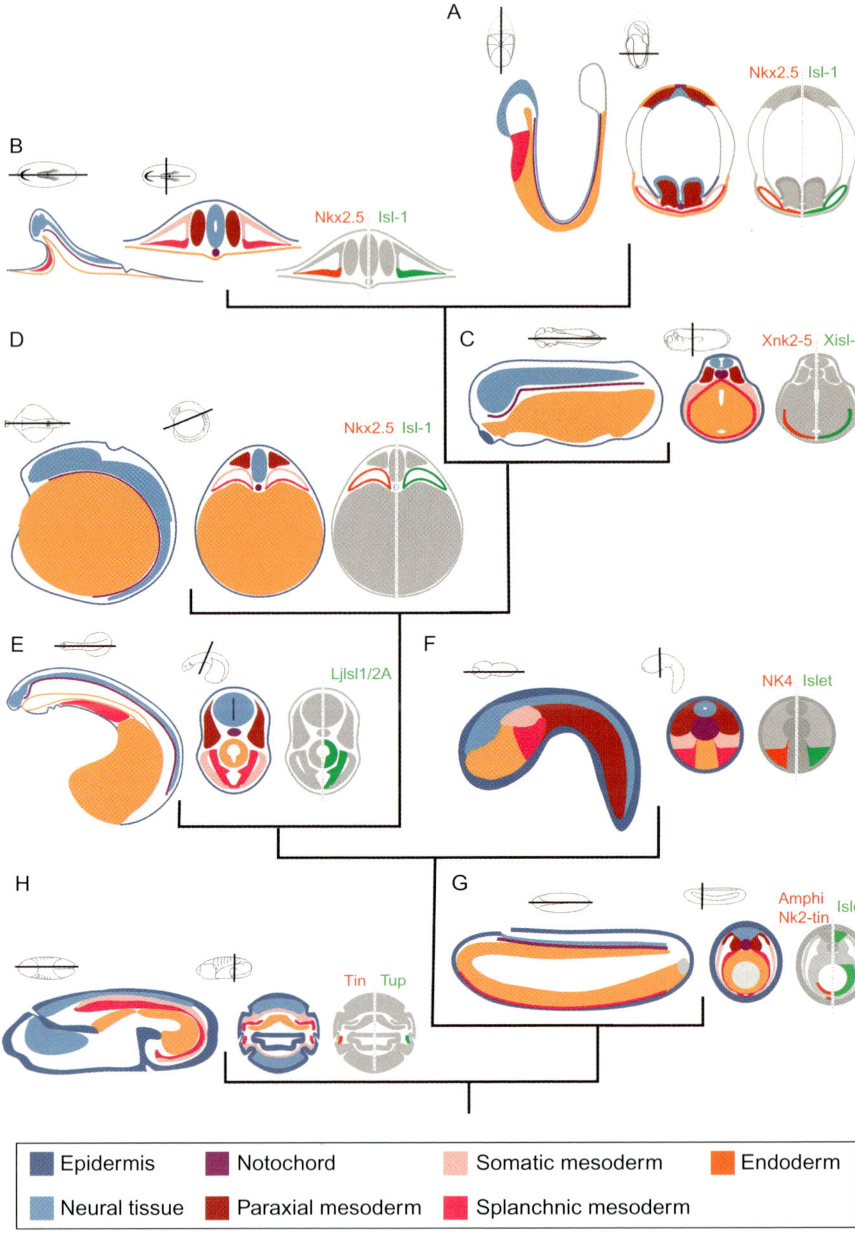

José Xavier-Neto *et al.*, Figure 3.3 Isl and NK4 ortholog expression during circulatory pump ontogenesis in bilaterian model organisms. Schemes of longitudinal and transverse sections of embryos at post-segmentation stages. Section level is indicated above each drawing. Embryonic territories are color-coded: epidermis (dark blue), neural tissue (light blue), notochord (violet), paraxial mesoderm (red), somatic mesoderm (light pink), splanchnic mesoderm (dark pink) and endoderm (orange). Right transversal sections show the endogenous expression of NK4 orthologs on the left side of the embryo and of Isl orthologs on the right side to highlight the substantial overlap between these expression patterns. (A) Schematic sections of mouse embryo at 7.5 dpc, showing Isl-1 and Nkx2.5 coexpression in somatic and splanchnic mesoderm. (B) Sections of chicken embryo at HH8 showing Isl-1 and Nkx2.5 in the splanchnic mesoderm. (C) Illustrations of a *Xenopus* embryo at NF 25 with Isl-1 and Nkx2.5 expression domains overlapping in the cardiogenic-splanchnic mesoderm. (D) Schematic sections of zebrafish embryo at 16 hpf displaying Isl-1 and Nkx2.5 coexpression in splanchnic and somatic mesoderm. (E) Scheme of a lamprey embryo at stage 25 this is the initial stage of Isl-1/2 expression restricted to the myocardium of the linear heart tube and to the subpharyngeal mesoderm. Lamprey Nkx2.5 ortholog expression is as yet unavailable. (F) Illustration of *Ciona intestinalis* mid-tailbud stage with *Ciona* Isl-1/2 (islet) weakly expressed in some pump precursors (medial trunk ventral cells, TVCs), while the *Ciona* Nkx2.5 ortholog (NK4) is expressed in all TVCs, indicating that *Islet* and *NK4* are coexpressed in pump precursors. (G) Schematic representations of an amphioxus embryo at late neurula stage (18 h) show AmphiNk2-tin expression in splanchnic mesoderm and ventral endoderm. Isl-1/2 ortholog is expressed in neural plate and endoderm, suggesting that Nk2-tin and Isl-1/2 are coexpressed in endoderm-derived tissues. (H) Schematic representation of *Drosophila* embryo (ventral side up) at late stage 11, when Tailup (Tup), the fly Isl-1/2 ortholog, is expressed with Tinman (Tin), the fly NK4 gene, which is expressed in all myocardial and pericardial cells of cardiac mesoderm.

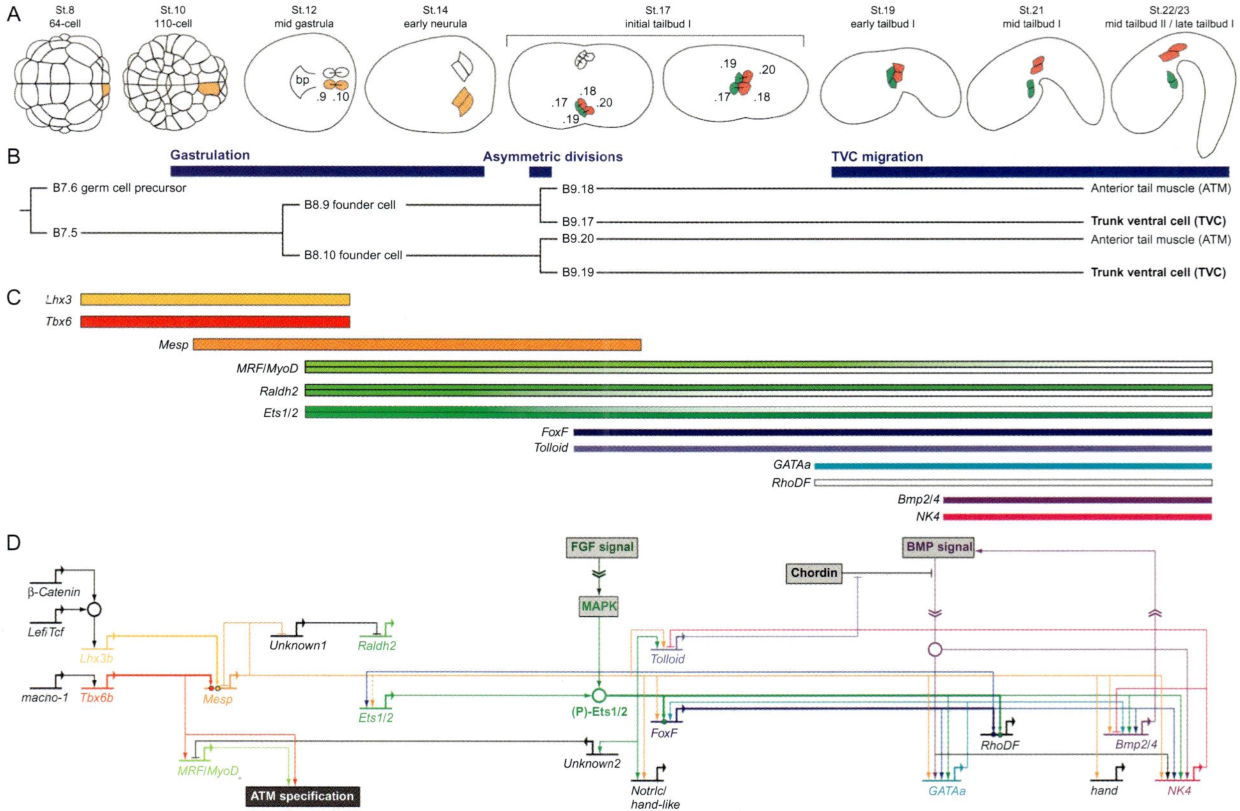

Theadora Tolkin and Lionel Christiaen, Figure 4.1 Overview of early cardiac progenitors specification in *Ciona intestinalis*. (A) Schematic representations of early embryo showing colored B7.5 lineage cells. Orange: B7.5 and daughter cells; green: trunk ventral cells (TVCs); red: anterior tail muscles (ATMs). Stage numbers are according to Hotta *et al*. (2007). (B) B7.5 lineage during the stages represented above. The main events are indicated in blue. (C) Temporal gene expression profiles in the B7.5 lineage cells. For *MRF/MyoD*, *Raldh2*, and *Ets1/2*, expression is split between early pan-B7.5 and late ATM (upper stripe) or TVC (lower stripe) specific expression. (D) Provisional Gene Regulatory Network (see the text for details). Unknown regulators are invoked to account for possible indirect regulatory inputs of Mesp on *Raldh2* and FGF-MAPK-Ets1/2 on MRF/MyoD. Other regulatory connections are based on experimental evidence. Direct connections supported by *cis*-regulatory analyses are marked by a dot on the target gene.

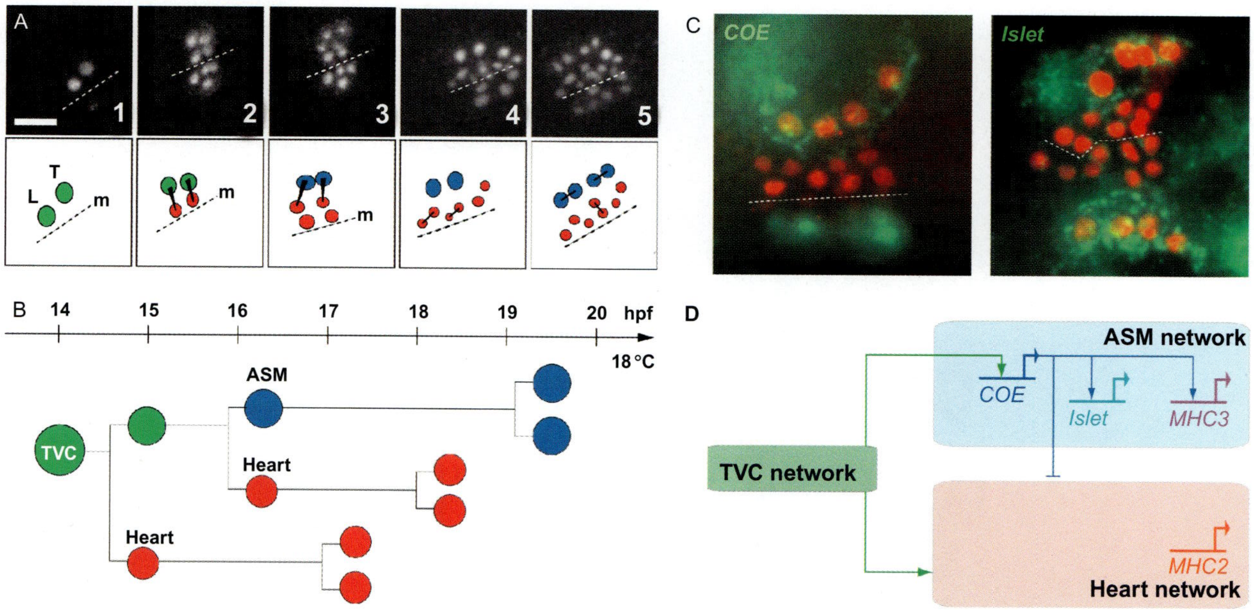

Theadora Tolkin and Lionel Christiaen, Figure 4.5 Heart versus ATM fate specification. (A) After they stop migrating, TVCs (green dots) divide asymmetrically into heart (red dots) and ASM precursors (blue dots). "1" to "5" indicate successive time points extracted from time-lapse movies in which TVCs express an Histone2B–GFP fusion (Stolfi *et al.*, 2010). Black segments in lower panels connect sister cells. m, midline (dotted line); L, Leader; T, Trailer. Scale bar: 20 μm. (B) Cell lineage showing the TVCs, ASMs, and heart precursors. hpf, hours postfertilization. (C) Double fluorescent *in situ* hybridization and immunostaining on 21 hpf larvae expressing the Mesp>lacZ transgene (red dots mark nuclear β-galactosidase in TVCs) with *COE* and *Islet* specific digoxigenin-labeled antisense RNA probes, respectively. Note the ASM-specific expression of *COE* and *Islet* (*Islet* may also be expressed at a weaker level in the median heart precursors). On the left panel, the *COE*-positive cells initiate migration toward the atrial siphon placode. White dotted line: midline. At this stage, each side displays eight heart and four ASM precursors. (D) Simplified network showing regulatory connections between the heart TVC network and the later ASM and heart-specific networks (Stolfi *et al.*, 2010).

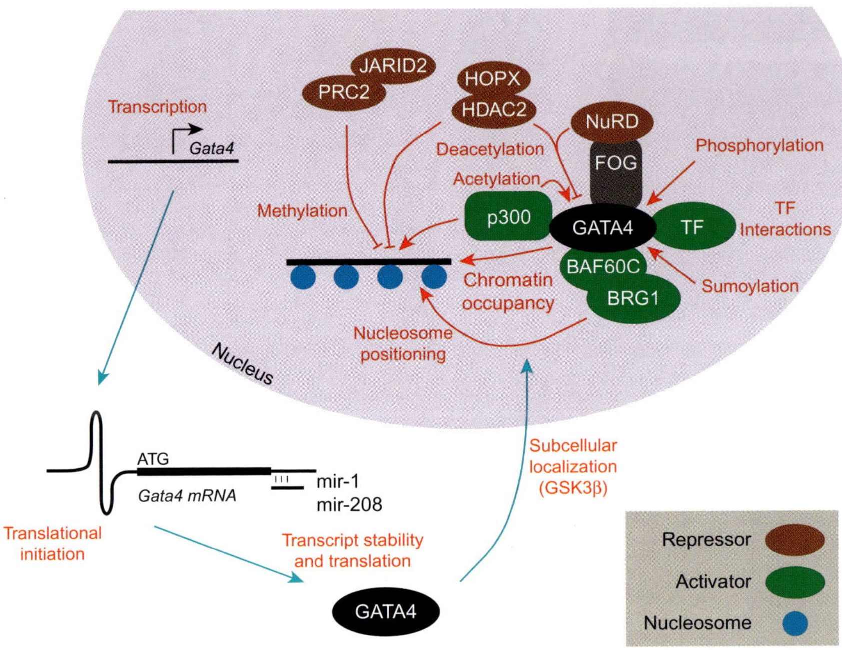

Pingzhu Zhou *et al.*, Figure 5.3 Regulation of GATA4 expression and activity. GATA4 is regulated by transcriptional, translational, and posttranslational mechanisms. Locus-specific transcriptional activity is regulated by interactions with other TFs, coactivators, corepressors, covalent modifications, and interactions with chromatin remodeling and modifying enzymes.

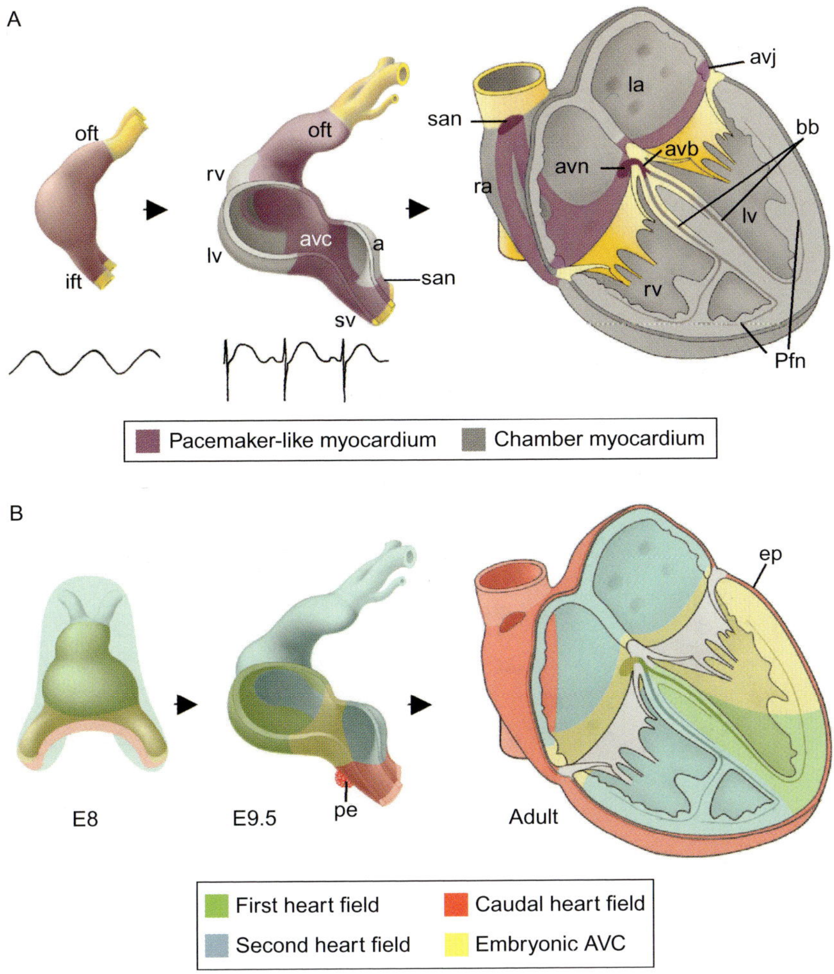

Phil Barnett *et al.*, Figure 6.1 Development of the heart. (A) Formation of the atrial and ventricular chambers, and the development of the pacemaker and conduction system. A mature type of ECG can be observed already during chamber development. (B) Contribution of different cell lineages to the heart. At E8.0 the primary heart tube primarily consists of first heart field (FHF) derived cells, shown in green. This region will become restricted to the left part of the interventricular septum and part of the left ventricle in the adult heart. The second heart field (blue) enters at the inflow tract, dorsal mesocardium, and outflow tract and will eventually derive a large part of the atria in the right ventricle. Part of the inflow tract (yellow) will form the embryonic atrioventricular canal (AVC) and thereafter contribute to the ventricle, particularly the left ventricular wall. The caudal heart field will contribute to the formation of the (pro) epicardium and sinus venosus. l/ra, left/right atrium; l/rv, left/right ventricle; avc/n, atrioventricular canal/node; san, sinus node; pe, proepicardium; ep, epicardium. Modified from Moorman and Christoffels (2003).

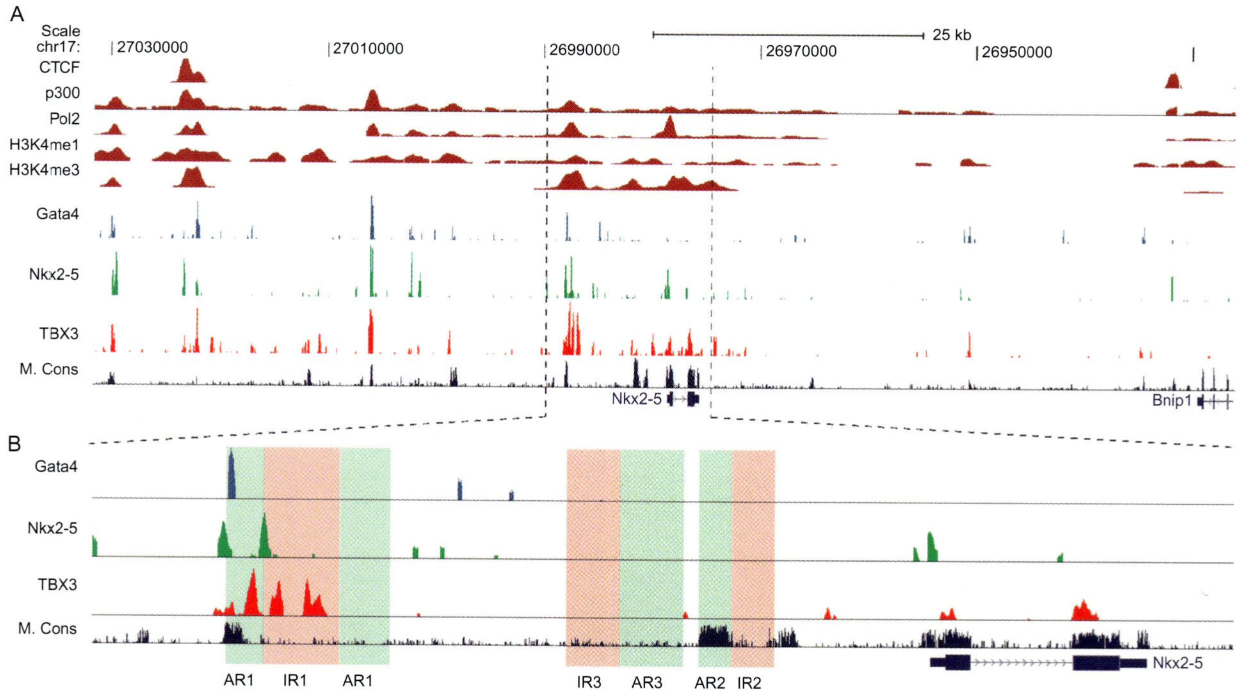

Phil Barnett *et al.*, Figure 6.3 ChIP-sequence analysis of the *Nkx2-5* gene region. (A, B) ChIP-seq traces for various transcription factor analyses carried out *in vivo* using mouse hearts. (A) specifically shows ENCODE/UCSC (Rosenbloom *et al.*, 2010) adult mouse heart ChIP-seq traces for insulator associated factor CTCF, polymerase II (Pol2), histone monomethylation and trimethylation of lysine residue 4 (H3K4me1, H3K4me3), and the enhancer associated protein P300 in and around the *Nkx2-5* gene locus. (B) Closeup of the region directly surrounding *Nkx2-5* showing adult mouse heart ChIP-seq data traces (our unpublished results) for Gata4, Nkx2-5, and Tbx3. Both figures also highlight (dotted lines) the relative positions of previously discovered enhancer elements that have been identified and verified *in vivo* (see text). Significant overlap can be seen between the various datasets and the verified enhancer positions, particularly at the most distally located elements, shown to mark expression specifically in the heart.

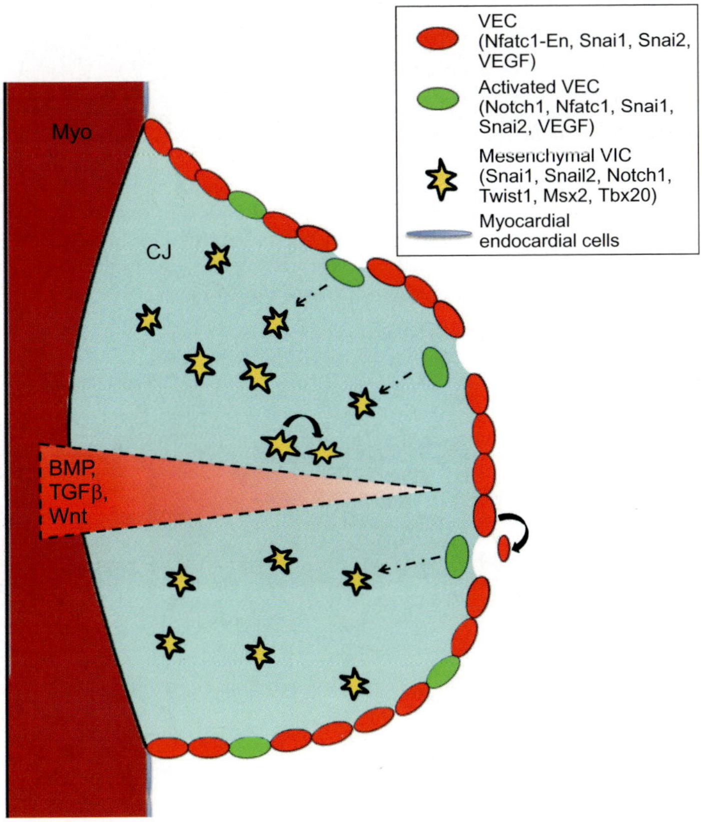

Ge Tao *et al.*, Figure 7.2 EMT and endocardial cushion formation. Schematic diagram to indicate molecular and cellular processes involved in endocardial cushion endothelial-to-mesenchymal transformation (EMT). Growth factor signals emanating from the myocardium (myo) diffuse through the cardiac jelly (CJ) to activate VECs and initiate EMT. During this process, VECs lose cell–cell contacts with neighboring cells, transform, and migrate into the underlying CJ where they proliferate to expand the population size (curved arrow).

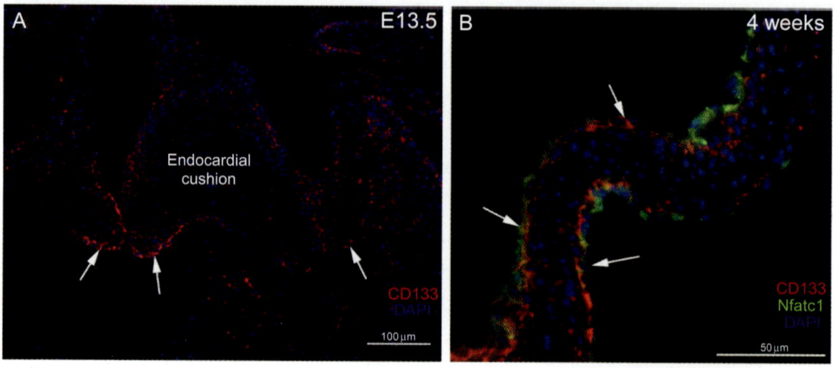

Ge Tao *et al.*, Figure 7.3 CD133 expression in developing and mature murine heart valve structures. Immunohistochemistry to show expression of the endothelial progenitor cell marker CD133 in VECs of the developing endocardial cushion at E13.5 (A) and mature mitral valve leaflets at 4 weeks of age (B). VECs are highlighted by Nfatc1 immunoreactivity in (B) and DAPI indicates nuclei in (A) and (B).

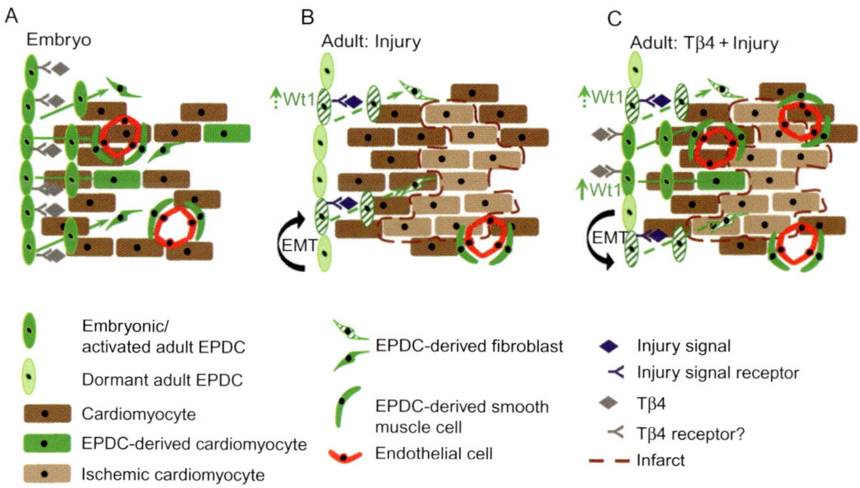

Paul R. Riley, Figure 8.2 Model for how Thymosin β4 and injury might restore embryonic epicardial potential in adult heart repair. (A) During heart development, a myocardial source of thymosin β4 (Tβ4) acts in a paracrine fashion to stimulate migration of embryonic epicardium-derived cells (EPDCs) into the underlying myocardium where they can contribute vSMCs, fibroblasts, and cardiomyocytes. (B) Following ischemic injury in the adult heart, the epicardium can respond to, as-yet-unknown (?), injury-induced signals to partially upregulate the embryonic epicardial gene Wt1, proliferate, and undergo an epithelial-to-mesenchmye transformation (EMT) followed by restricted migration and contribution of fibroblasts to partially recapturing embryonic potential. (C) Pretreatment (priming) with ectopic Tβ4 prior to injury is proposed to recruit additional epicardial progenitor subpopulation(s), via significant and precocious upregulation of Wt1 (relative to injury alone). Tβ4-dependent subpopulations of activated epicardial progenitors can contribute vSMCs and cardiomyocytes. Tβ4 priming plus injury thus restores complete embryonic potential to the adult epicardium to establish the basis for neovascular and myocardial repair.

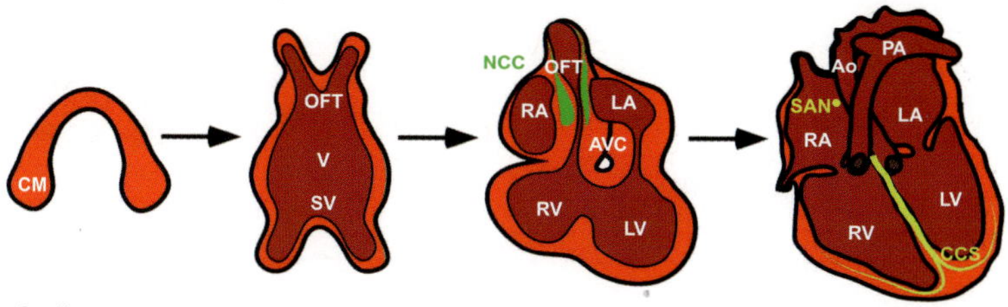

Stage	Cardiac crescent	Linear heart tube	Looped heart tube	Mature heart
Human development	Day 15	Day 20	Day 30	Day 40-birth
Mouse development	E7.5	E8.0	E10–12	E14–18
Key events	Myocyte differentiation Migration to the midline	Bidirection contractions	Looping morphogenesis Addition of SHF progenitors Cushion formation Neural crest addition Ventricular septation begins	Conduction system OFT septation Cushion remodeling

David J. McCulley and Brian L. Black, Figure 9.1 Schematic representations of several stages of mammalian heart development with key events in mouse and human heart development denoted. Mouse development is denoted in embryonic days (E). Ao, aorta; AVC, atrioventricular canal; CCS, cardiac conduction system; CM, cardiac mesoderm; LA, left atrium; LV, left ventricle; NCC, neural crest cells; OFT, outflow tract; PA, pulmonary artery; RA, right atrium; RV, right ventricle; SAN, sinoatrial node; SV, sinus venosus; V, ventricle.

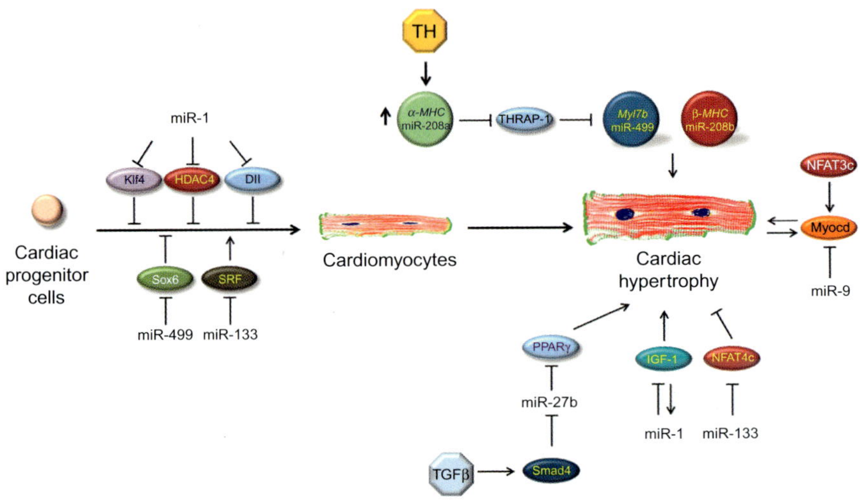

Ramón A. Espinoza-Lewis and Da-Zhi Wang, Figure 10.3 *miRNA function and integration in the regulatory networks to regulate cardiomyocyte differentiation and disease.* Graphic representation of several regulatory networks targeted by miRNAs in cardiomyocyte differentiation from cardiac progenitor cells, cardiomyocytes, and their function in the regulation of cardiac hypertrophy. Also, the integration of miRNAs in such regulatory networks by targeting transcription factors, signaling molecules, and/or structural proteins is shown.

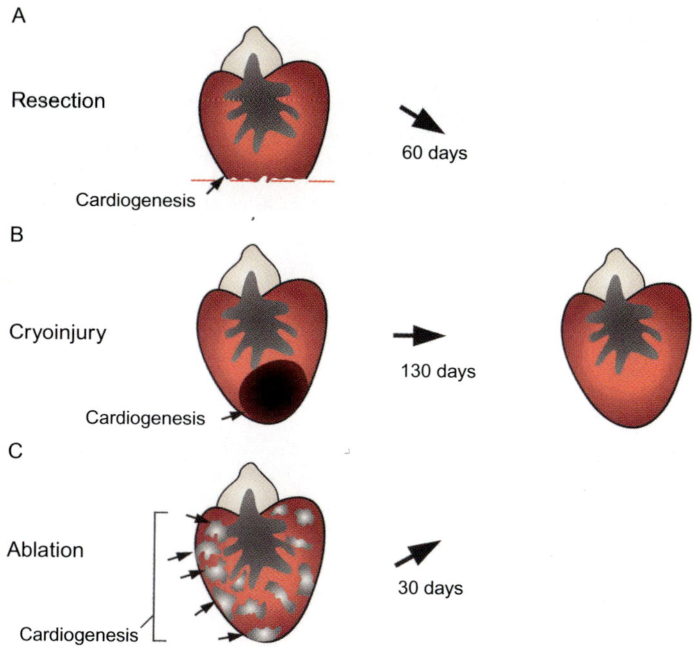

Wen-Yee Choi and Kenneth D. Poss, Figure 11.2 Injury and regeneration in zebrafish. (A) In the resection model, 20% of the ventricular apex is surgically removed. Within 2 months, the resected tissue is replaced through local cardiomyocyte proliferation. (B) In the cryoablation model, a cooled probe is applied directly to the heart to induce localized cell death affecting up to 25% of the ventricle. The lesions can remain detectable past 60 days postinjury, but the majority of collagen is eventually cleared coincident with local cardiomyocyte proliferation. (C) Using a conditional genetic approach, over 60% of the ventricular (and atrial) myocardium is destroyed throughout the heart. Regeneration is particularly rapid and robust, with full recovery within 1 month after ablation.

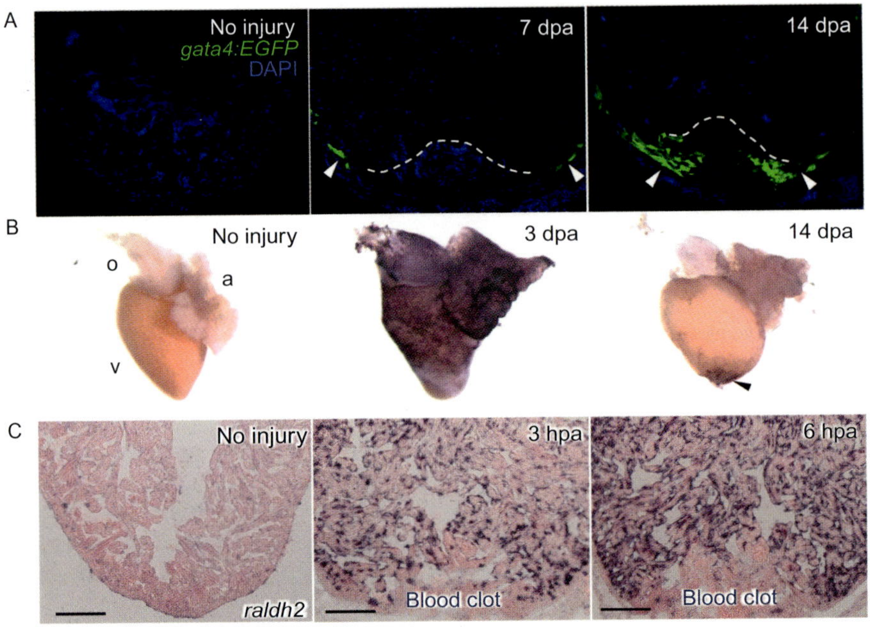

Wen-Yee Choi and Kenneth D. Poss, Figure 11.4 Three major cardiac cell types activate developmental gene expression in a chamber-wide manner after resection injury. (A) Myocardial activation, represented by the activation of a *gata4*:EGFP reporter, is first detected throughout the compact layer of the myocardium by 7 days post-amputation (dpa) and becomes localized to the regenerating myocardium by 14–30 dpa (arrowheads). Dotted lines indicate amputation planes. (B) Ostensibly, the entire epicardium activates developmental genes such as *raldh2* (violet) within several days following injury. Enhanced gene expression localizes to the wound area by 2 weeks postinjury (arrowheads). (C) Endocardial cells undergo structural remodeling following injury and upregulate *raldh2* (violet) within hours of amputation. Modified from Lepilina *et al.* (2006) and Kikuchi *et al.* (2010, 2011b).